Mechanical Vibrations

To Monika, Arnaud and Sylvain

Mechanical Vibrations

Applications to Equipment

Yvon Mori

iSTE

WILEY

First published 2017 in Great Britain and the United States by ISTE Ltd and John Wiley & Sons, Inc.

ISTE Ltd
27-37 St George's Road
London SW19 4EU
UK

www.iste.co.uk

John Wiley & Sons, Inc.
111 River Street
Hoboken, NJ 07030
USA

www.wiley.com

Library of Congress Control Number: 2016959683

British Library Cataloguing-in-Publication Data
A CIP record for this book is available from the British Library
ISBN 978-1-78630-051-5

Table of Contents

Preface

After having worked for nearly 40 years as an engineer at Thales Underwater Systems (TUS) in Sophia Antipolis, as the head of a large environmental testing laboratory, *Laboratoire d'Essais et d'Evaluation en Environnement* (LEEE), French for laboratory of environmental assessment and testing), I became interested in such diverse fields such as the mechanics of vibrations and shocks, climatic and electromagnetic problems of electromagnetic compatibility, as well as measurements. This activity naturally led me to be reached out to by COFRAC (COmité FRançais d'ACcréditation; French Certification Organization) as an expert assessor in area 38, which deals with mechanical and environmental tests in laboratories, among other domains, which has been taking place for a long time. This is what allowed me to see up close the training needs of the personnel working in these areas and that is why I thought it would be useful to put this experience to their service, by way of this book which presents itself in the form of a course with a pedagogical spirit from a confirmed engineer and therefore with a pragmatic and practical approach.

The purpose of this book is to clarify the issues related to the environment of mechanical vibrations in the life profile of materials. In particular, this is possible through their simulation in the testing laboratory, with a better knowledge of the physical phenomenon and the means to be implemented in order to simulate them, measure them and interpret the related results.

This book is aimed at developmental engineers of consulting firms and those involved in quality services, but most of all at those working in testing laboratories who are confronted with invitations to tender in testing environments, especially in terms of mechanical vibrations. Furthermore, it should also interest engineering students in the fields of development of their future occupations regarding vibrations.

We will thus define the phenomenon of vibration from the point of view of an environment specialist through some indispensable notions in order to know how to interpret the prescriptive documents published to date and perform some simple predictive calculations of behavior of the studied material.

Finally, I would like to warmly thank Henri Grzeskowiak and Christian Lalanne, my friends and companions for 30 years within the ASTE, CINEG/DGA and COFRAC for their effective support in the proofreading of this work and their pertinent and relevant corrections.

The book is divided into eight chapters:

- In Chapter 1, there is an introduction to the theory of vibrations, with a classic analytical description of the models used at one and two degrees of freedom (DOF), undamped and damped. A more practical approach of the types of vibratory phenomena referred to in the normative documents and specifications is proposed thereafter;

- Chapter 2 specifies the mathematical tools used to analyze these vibratory phenomena, and of course refers to the theory of signals. We will find the usual tools for the temporal description of signals as well as the frequency description Power Spectral Density (PSD, Fourier transform and series), but adapted to the techniques of vibrations. The proposed approach also describes specific applications to vibration signals encountered in testing laboratories. Indeed, the tests proposed by the normative documents involve sine scans, random tests in colored noise as well as composite tests such as the sine on noise, which has some peculiarities that are important to know;

- Chapter 3 offers a quality approach to address the problem of testing within testing laboratories. It is inspired by the normative documents on the subject, in particular regarding standard NF EN ISO / IEC 17025, which governs the accreditation of laboratories for testing and analysis. It allows for the initialization of an internationally recognized quality approach;

- Chapter 4 provides a relatively complete analytical description of the sine on noise vibration tests that are the most commonly used. Developments are oriented toward laboratory tests, with the usual problems that are encountered (scan speeds, level of limitation of noise testing, etc.). We will find most of what we know about the subject in this chapter;

- Chapter 5 provides a complete practical analysis of vibratory phenomena encountered in most engineering and design departments and offers simple and common conception rules that allow for a simple qualitative and quantitative approach to avoid most of the big problems that can be encountered in this type of environment. It also describes a practical approach through a particular example, which is the study of an electronic bay mounted on shock absorbers. This simplified example allows for the understanding of a practical approach (there are others) that resolves most of the technical problems and that relies on the now very common notion of “tailoring”;

- Chapter 6 provides a simplified approach to the piloting ways of testing problems, in particular, vibration exciters, in sine testing, noise and shocks, so as to educate the reader on the difficult issues typically encountered in these types of techniques;

- Chapter 7 describes succinctly the metrology and measurement aspects encountered in the use of accelerometric sensors and vibration testing means within testing laboratories;

- Chapter 8 details the operation of the two main means of vibration tests: electromagnetic generators and hydraulic vibrations. This chapter can be found in [ref. I18], but it was considered useful to reproduce it here in its entirety.

This book is developed from my professional experience and also with the help of a bibliography on the subject, which encompasses:

- specialized books in the field of shocks and vibrations;
- courses that I have followed and provided by a variety of organizations offering training on the subject, or that I have taught;
- technical reviews of manufacturers or suppliers of tests and measurements;
- articles, documentation and technical bulletins from various specialist organizations of shocks and vibrations;
- normative documents dealing with the subject;

which includes a summary that was prepared by taking all or part of some of these documents, so as to develop the themes that must be known in order to better understand the subject. Despite the amount of literature, this subject is relatively poorly perceived by development engineers and study laboratories and is generally poorly developed in its practical aspects in schools and universities.

The bibliography includes most documents that are used unanimously to date, but that are not necessarily easily accessible. This is why certain concepts or calculations are detailed and developed throughout the book so as to facilitate their comprehension.

In the appendices, information more or less developed to aid in the understanding of some concepts used throughout the document will be found.

1

Vibration Theory

This chapter is an introduction to the theory of vibrations, a term commonly used to define the study of representation models, which are used to describe the behavior of systems in a vibration regime. The scope of this theory is relatively wide and still is in constant development, mainly within the framework of dynamic analysis of structures (modal analysis or other ones), which expands beyond the framework of the course and for which there are already very good books.

Here, we will strive to develop the traditional approach of systems at 1 and 2 degree of freedom (DOF) with coupling in a free or forced regime that allows a good introduction to other, more sophisticated, techniques, and especially allows the comprehension and optimization of methods for vibration testing enforced by normative documents and carried out in testing laboratories.

1.1. Problem

1.1.1. Justification of tests in a mechanical environment

Any industrial system or equipment is subject during its lifetime to the attack of environments that may degrade its performance with the appearance of different failures. The now classic concepts of quality and reliability are therefore used to define its availability involved in all phases of its design. It seems natural and even necessary to minimize as much as possible the causes of these failures.

Therefore, the specifications of the product base are generally guaranteed by a quality assurance plan by acquiring the conviction that the specifications are kept, and this belief is maintained in time through *a priori* and *a posteriori* actions (measurements, tests, preventive maintenance, etc.).

Yet, the development of a part of this belief must be done from environmental tests. Here, the need to simulate experimentally is thus clear and thus through the appropriate tests, the constraints of operating the most representative of real conditions that the material meets during its service use. In particular in the case of equipment with strong evolution of design, and/or conditions of use, and/or technologies (new components and/or report processes, and/or materials).

For questions related to reproducibility, accessibility and cost, we will be limited here to laboratory tests, except in special cases.

Thus, we generally define several categories of tests according to the usual chronology:

- design phase: feasibility and validation of tests.
- development phase: model tests, qualification, formula, accelerated tests.
- production phase: batch testing and debugging.

- storage phase: storage testing, ageing.
- end of lifetime phase: withdrawal of service.

These tests are subject to the choice of the successful normative documents, whose development can be done through a variety of methods. Through the analysis and synthesis of data of the severity levels of the real environment, if we perfectly know the conditions of use, this results in the concept of *"tailorization"* (the latter being typically used when the stakes are high as well as the risks of an under- or overvaluation and if the material does not pass the bar of standard tests), or failing that by the adaptation of "standard" levels of severity usually found in the common normative documents edited by (non-exhaustive) organizations such as:

- at a national level:
    - the AFNOR (*Association Française de NORmalisation*, French Standardization Association), NF standards.
    - the UTE (*Union Technique de l'Electricité*, Technical Union of Electricity).
    - the CCT (*Comité de Coordination des Télécommunications*, Coordinating Committee of Telecommunications), CCTU standards.
    - the Ministry of Defense: formerly GAMEG13, AIR and SEFT standards, etc.
- at a European level:
    - the CENELEC (*Comité Européen de Normalisation ELECtronique*, European Committee of Electronic Normalization), CECC and EN standards and currently the STANAG 4370 within the NATO area. See, in particular, "Best Practices" by EDSTAR (https://edstar.eda.europa.eu/best-practice-recommendations).
- at an international level:

    institutional:
    - ISO (International Standard Organization).
    - IEC (Comité Electrotechnique International, International Electro-technical Committee).
    - UIC (Union Internationale des Chemins de Fer, International Union of Railways).

    sectorial:
    - DO 160 or EUROCAE ED 14.
    - ANSI.
- at the foreign military standards level:

    national level:
    - the DEF STAN 0035 (United Kingdom).
    - the MIL STD 810 (United States).
    - the STANAG 4370 (NATO area).
    - BV Standards (Germany), etc.

    international scale:
    - MIL STD 810 (because it is considered a reference in all regions of the world).

1.1.2. Quality of environmental tests

- RG Aero 00011 "quality of environmental tests".
- Re Aero 601-11 "quality provisions in mechanical environment testing".
- Re Aero 601-12 "quality provisions in climatic environment-specific tests".

1.1.3. Generating sets of vibrations

- NF E90-200: Generating sets of vibration – chart intended for the layout of characteristics.

- NF E90-210: Electrodynamic test facilities used for the generation of vibrations – methods of description of features.

- NF E90-220: Servohydraulic vibration tests resources – methods of description of features.

- NF E90-230: Auxiliary tables for vibration generators – methods of description of features.

1.1.4. "Shock and vibration" terminology

- ISO 2041: Vibration and shock – vocabulary.

1.1.5. Testing methods

For the non-specialized civilian field within a brand of activity, the following are the main methods of the NF or EN or CEI:

- NF/EN/CEI 60068-2-1 to 2-53 (see detailed list in Appendix I).

- the defense sector now refers to, for the NATO area, the methods of the STANAG 4370-AECTP 300 (climatic tests) and 400 (mechanical tests). For the international perimeter outside of NATO, these are the methods of MIL STD 810 complemented by those of the DEF STAN 0035 that apply;

- the civil aviation sector is represented by the DO160 methods.

In addition, for some specific areas, we can quote:

- NF EN 61373: Railway applications – rolling material – shock and vibration tests.

1.1.6. Uncertainty in measurement

- Guide for the expression of the uncertainty in measurement.

- ISO 5725-1 to 6: Exactitude (accuracy and fidelity) of the results and methods of measurement.

- FD ISO/TR 22971: Exactitude (accuracy and fidelity) of the results and methods of measurement. Practical guidelines for the use of ISO 5725-2 for the design, implementation and statistical analysis of the results of repeatability and interlaboratory reproducibility.

- NF ISO 21748: Guidelines relating to the use of repeatability, reproducibility and accuracy estimates in the evaluation of uncertainty in measurements.

ISO 21748: provides the guidelines in order to:

- evaluate the measurement uncertainties from data obtained during interlaboratory diffusivity tests conducted in accordance to ISO 5725-2.

- compare the results of an interlaboratory test with the uncertainty of a measurement obtained by applying formal principles of uncertainty propagation.

ISO 5725-3 provides additional models for the measurement of intermediate fidelity. However, although the same general approach can be applied to the use of these extended models, the assessment of uncertainty from these models is not covered in this international standard.

ISO 21748 is applicable in all areas of measurement and testing requiring the determination of an uncertainty associated with a result. It does not describe the use of repeatability data in the absence of reproducibility data.

ISO 21748 assumes that the recognized non-negligible systematic effects are corrected, either by applying a digital correction in the context of the measurement method, or by searching for and eliminating the origin of these effects.

ISO 21748 recommendations are tentative above all else. It is recognized that, even if they constitute a valid method of uncertainty evaluation, in many aspects, other appropriate methods may also be adopted. In general, it is understood that the references made in ISO 21748 toward results, methods and measurement processes, also apply to results, methods and testing processes.

1.1.7. Interlaboratory comparison and proficiency testing

- NF EN ISO/CEI 17043: Conformity assessment – general requirements regarding proficiency tests.

- NF EN DIS 13528: Statistical methods for use in proficiency testing by interlaboratory comparisons.

1.1.8. Metrology management

- NF EN ISO 10012: Systems of measurement management – requirements for processes and measuring equipment.

The following concepts are also defined:

- *environmental agent*: One of the physical, chemical, biological, etc., phenomena that comprises the environment which is likely to take on the characteristics or the behavior of a material or a living being;

- *quality*: The "quality" of a product can be defined as its ability to meet its operational objectives. The ISO 8402-94 standard defines quality as the set of the characteristics of an entity that has the ability to satisfy *stated* and *implied* needs. In practice, it comes in two forms: external quality, which corresponds to the satisfaction of the customer (listening to the customer) and internal quality, which corresponds to the improvement of the internal workings of the company, for example through the ISO 9000 standard or NF EN ISO/CEI 17025 (with the associated concepts of technical specifications of need [STB, *Spécification techniques de besoin*], quality assurance plan, specification compliance, etc.);

- *reliability*: Characteristic of a device expressed by the probability that it performs a required function under given conditions, for a given period of time (with the associated concepts of availability, life profile and mission, environmental and functional implementation conditions, schedule and construction of reliability, Mean time between failures (MTBF) calculation [repairable systems], Mean time to failure (MTTF) [non-repairable systems], etc.).

The environmental agent that is proposed to be mastered in this book is that of mechanical vibration.

We can note that the widespread of environmental tests have made standardization necessary, the influence of the latter on the design of tests and test facilities became fundamental and their development generally precedes the technique.

1.2. Different types of mechanical signals

1.2.1. Overview

Once the need for simulating vibrations tests by appropriate means in the laboratory is established, the interest in usual specifications of the tests used have to be specified, as well as the corresponding testing machines. In the case of vibrations, it is generally quite easy to reproduce on a machine (vibration generators), the same vibration as the one measured and recorded on site. It is, nevertheless, essential to be able to define equivalent vibrations in terms of damage afflicted on structures, so as to be able to define equivalent tests in terms of severity (amplitude, duration, type of vibration, etc.). This approach, commonly used in the development of the test specifications, is rigorously and widely described in bibliography books [ref. 11 (5 volumes)], which constitute references on the subject. It is therefore appropriate to know the different types of signals to be analyzed, knowing that a vibration can be described as an oscillation maintained around an average value that lasts long enough compared to the *time constant* of the excited system.

1.2.2. Mathematical prerequisites

1.2.2.1. *Unit circle*

It is a circle of a unit radius that is used to model usual trigonometric functions. It is shown in Figure1.1 (to also bring closer the model of circle C(0,1) of the complex plane):

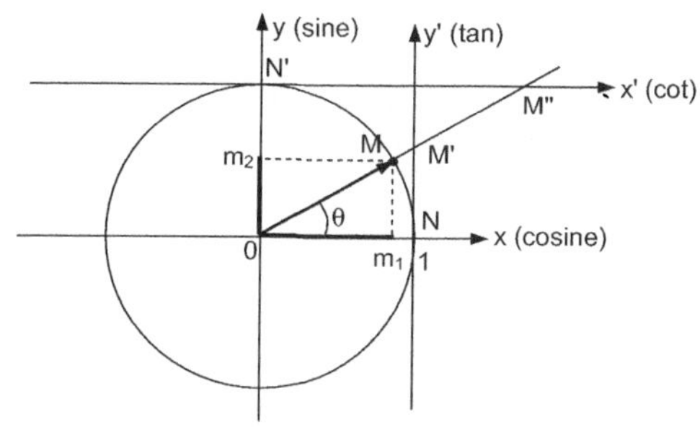

Figure 1.1. *Unit circle*

$$\cos\theta = \frac{\overline{0m_1}}{\overline{0M}} = \frac{\overline{0m_1}}{1} = \overline{0m_1}$$

$$\sin\theta = \frac{\overline{0m_2}}{\overline{0M}} = \frac{\overline{0m_2}}{1} = \overline{0m_2}$$

$$tg\theta = \frac{\sin\theta}{\cos\theta} = \frac{\overline{0m_2}}{\overline{0m_1}} = \frac{\overline{NM'}}{\overline{0N}} = \frac{\overline{NM'}}{1} = \overline{NM'}$$

$$\cot g\theta = \frac{1}{tg\theta} = \frac{\cos\theta}{\sin\theta} = \frac{\overline{0m_1}}{\overline{0m_2}} = \frac{\overline{N'M''}}{\overline{0N'}} = \frac{\overline{N'M''}}{1} = \overline{N'M''}$$

Note that the measurements on the *x* axis support the cosines, the *y* axis the sines, the *x'* axis the cotangents and the *y'* axis the tangents.

Useful trigonometric relationships:

$\cos\theta = \cos(-\theta)$ even function

$\sin\theta = \text{-}\sin(-\theta)$ uneven function

$\cos\theta = \sin\left(\frac{\pi}{2} - \theta\right) = \sin\left(\theta + \frac{\pi}{2}\right)$ the cosine is $\pi/2$ in advance compared to the sine

$-\sin\theta = \sin(\theta + \pi)$

1.2.2.2. *Vectors and Fresnel representation*

Consider a mobile M describing the circle of center 0 and radius A C(0,A), in uniform movement and angular velocity ω, modeled by the diagram shown in Figure1.2.

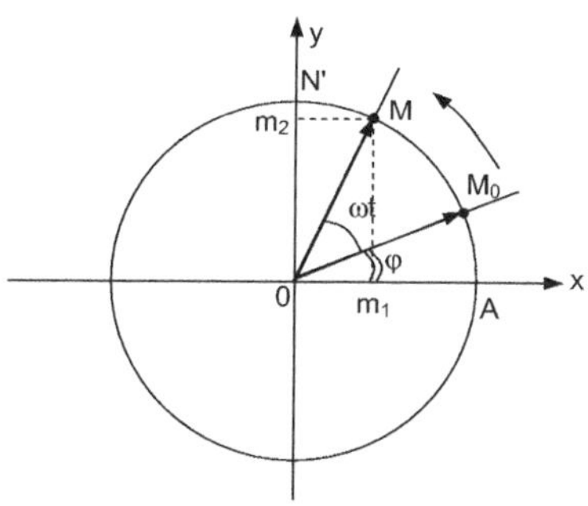

Figure 1.2. *Fresnel model*

At the moment where t = 0, the mobile is in M_0 with an initial phase angle given by $\left(\overrightarrow{0x}, \overrightarrow{0M_0}\right) = \varphi$. At the moment t, the mobile is in M with $\left(\overrightarrow{0x}, \overrightarrow{0M}\right) = \omega t + \varphi$. Therefore, we have:

$$\begin{cases} \overline{0m_1} = A.\cos(\omega t + \varphi) \\ \overline{0m_2} = A.\sin(\omega t + \varphi) \end{cases} \rightarrow \omega t + \varphi = \text{arctg}\frac{\overline{0m_2}}{\overline{0m_1}}$$

Thus with any sinusoidal function y(t) = A.sin (ωt+φ), we can match a vector $\overrightarrow{0M}$ revolving around 0, with an angular velocity ω with phase at the beginning φ. This is the model of Fresnel's rotating vector.

Consider T as the duration of a rotation and f the number of laps per second. T is thereby defined as the period of the function, f, the frequency and ω the pulse:

$$\omega = 2\pi f = \frac{2\pi}{T} \text{ in radians per second [rd / s]}$$
$$f = \frac{1}{T} \text{ in Hertz [Hz]} = [s^{-1}] \text{ and T in seconds [s]} \tag{1.1}$$

1.2.2.3. *Rectilinear sinusoidal movement*

In this case, the elongation y(t) is a sinusoidal function of time, with:

$y(t) = y_{max}.\sin(\omega t - \varphi) = y_{max}.\sin(2\pi ft - \varphi) = y_{max}.\sin\left(\frac{2\pi}{T}t - \varphi\right) = y_{max}\sin\left[\omega\left(t - \frac{\varphi}{\omega}\right)\right]$, where we can see that the concept φ/ω here represents a delay and therefore the phase is closely related to the notion of delay and thus of time (or advance, depending on the sign φ). With y_{max} = peak value = maximum compared to the reference (motion [m], speed [m/s = $m\cdot s^{-1}$] and acceleration [m/s/s = $m\cdot s^{-2}$]).

Several parameters are used to describe this signal. We commonly define three amplitude values:

- *peak value*: the maximum value from the point of balance. It has an instant character, but it is hardly noticeable when in motion;

- *peak-to-peak value:* the maximum value of either side of the point of balance. It is also instantaneous, but it can be perceived in an easier manner in a motion movement. Consequently, movements are often specified peak to peak (pk/pk);

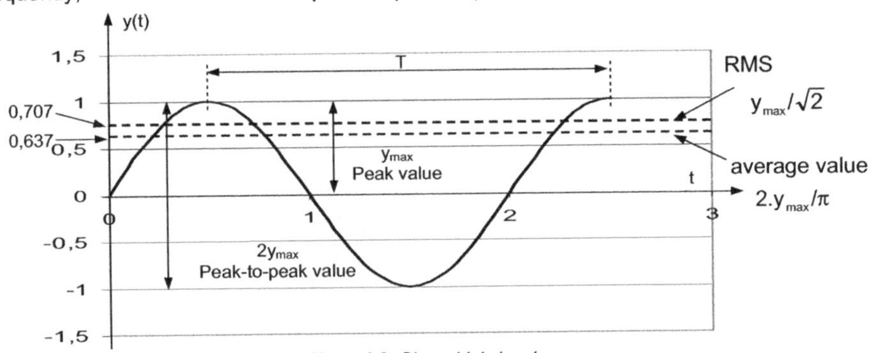

Figure1.3. *Sinusoidal signal*

- *average value:* defined by:

$$\overline{y(t)} = \bar{y} = \frac{1}{T}.\int_0^T y(t).dt \tag{1.2}$$

This value represents the continuous value that would result in the same surface between 0 and T. It is obviously zero for a sinusoidal signal due to its symmetry compared to the time axis over a period;

- *the average absolute value*: usually called "average value", defined by:

$$|\overline{y}| = y_{moy} = \frac{1}{T}.\int_0^T |y(t)|.dt \qquad (1.3)$$

which helps calculating the average value on a half-period. These expressions are independent of the phase φ. So we find:

$$y_{moy} = \frac{1}{T/2}.\int_0^{T/2} y_{max}.\sin\frac{2\pi}{T}t.dt = \frac{y_{max}}{T/2}.\frac{T}{2\pi}.\left[-\cos\frac{2\pi}{T}\right]_0^{T/2} = \frac{2y_{max}}{2\pi}[(-\cos\pi)-(-1)] = \frac{2y_{max}}{\pi},$$

therefore:

$$y_{moy} = \frac{2y_{max}}{\pi} \qquad (1.4)$$

- *effective value*: one of the most interesting concepts because it is directly connected to the average power contained in the signal. It is defined by the square root $(x)^{1/2}$ of the quadratic value $(y^2(t))$ average (1/T):

$$y_{eff} = \left\{\frac{1}{T}.\int_0^T y^2(t).dt\right\}^{1/2} \qquad (1.5)$$

In order to be practical, it represents, for example, in electricity, the value of a direct current that would dissipate the same power, in a passive resistive load, than the alternating signal y(t). We can define this concept for any function. Thereby we have:

$$y_{eff}^2 = \frac{1}{T}.\int_0^T y_{max}^2.\sin^2\frac{2\pi}{T}dt = \frac{y_{max}^2}{T}.\int_0^T \frac{1}{2}.\left(1-\cos\frac{4\pi}{T}\right)dt = \frac{y_{max}^2}{T}.\int_0^T \frac{dt}{2} = \frac{y_{max}^2}{2T}.[t]_0^T = \frac{y_{max}^2}{2}.$$ Therefore:

$$y_{eff} = \frac{y_{max}}{\sqrt{2}} \qquad (1.6)$$

We can infer the following relationships between these values:

$$y_{eff} = \frac{y_{max}}{\sqrt{2}} = \frac{y_{moy}.\pi}{2.\sqrt{2}} \qquad (1.7)$$

We can also write the following relations:

$$y_{eff} = F_f.y_{moy} = \frac{1}{F_c}.y_{max}$$

With:

$$F_c = \frac{y_{max}}{y_{eff}} = \text{peak factor} \qquad (1.8)$$

$$F_f = \frac{y_{eff}}{y_{moy}} = \text{form factor} \qquad (1.9)$$

In the sinusoidal case, we have:

$$\begin{cases} F_c = \sqrt{2} = 1{,}414 \approx 3\text{dB} \\ F_f = \dfrac{\pi}{2.\sqrt{2}} \approx 1{,}11 \approx 1\text{dB} \end{cases} \qquad (1.10)$$

In the general case of a distorted periodic signal harmonic or not, the effective value becomes the true effective value, or root means square (RMS) value. It is then necessary to take harmonics into account when there are some (harmonic frequencies of amplitude and phase that are integer *multiples* of the fundamental).

The relations therefore become:

$$F_c = \frac{y_{max}}{y_{RMS}} \quad \text{and} \quad F_f = \frac{y_{RMS}}{y_{moy}}$$

The Fourier series decomposition of the periodic signal of period $T = 1/f_0$ is then given by:

$$y(t) = a_0 + \sum_{n=1}^{\infty}(a_n.\cos 2\pi n f_0 t + b_n.\sin 2\pi n f_0 t) = a_0 + \sum_{n=1}^{\infty} c_n.\cos(2\pi f_0 n t - \varphi_n) = \sum_{n=-\infty}^{\infty} \alpha_n.e^{2\pi j n f_0 t}$$

With:

$$a_0 = \text{continuous component or average value} = \frac{1}{T}\int_0^T y(t).dt$$

$$a_n = \frac{2}{T}\int_0^T y(t).\cos 2\pi n f_0 t.dt \qquad b_n = \frac{2}{T}\int_0^T y(t).\sin 2\pi n f_0 t.dt$$

$$c_n = \sqrt{a_n + b_n} \qquad \varphi_n = \text{arctg}\frac{b_n}{a_n}$$

$$\alpha_n = \left|\alpha_n.e^{j\phi_n}\right| = \frac{1}{T}\int_0^T y(t).e^{-2\pi j n f_0 t}.dt = \frac{a_n - jb_n}{2} \qquad 2\left|\alpha_n\right| = c_n \quad \text{and} \quad \varphi_n = -\phi_n$$

- n defines the harmonic rank that varies from 1 to infinite (except for the complex expression);
- if n = 1, the harmonic is called fundamental;
- c_n is the amplitude of the nth harmonic;
- φ_n is the relative phase of the fundamental harmonic n.

a) Definitions related to the distorted sinusoidal signal

(a) The effective value or RMS:

$$y_{eff} = \sqrt{a_0^2 + \sum_{n=1}^{\infty}\frac{1}{2}c_n^2} = \sqrt{a_0^2 + \sum_{n=1}^{\infty} c_{neff}^2} = \sqrt{a_0^2 + \frac{1}{2}(a_1^2 + a_2^2 + .. + a_n^2 + ..) + \frac{1}{2}(b_1^2 + b_2^2 + .. + b_n^2 + ..)}$$

$$y_{eff} = \sqrt{a_0^2 + c_{1eff}^2 + c_{2eff}^2 + .. + c_{neff}^2 + ..} \quad c_{neff} = \frac{c_n}{\sqrt{2}}$$

(b) The peak factor: $F_c = \frac{y_{max}}{y_{RMS}}$

If the peak factor is different from $\sqrt{2}$, the signal is hence deformed by harmonics.

(c) The total distortion factor (including the continuous component):

$$FD = \frac{\sqrt{a_0^2 + \sum_{n=2}^{\infty} c_{neff}^2}}{y_{eff}}$$

NB: We note that the sum starts at n = 2 and so it does not include the fundamental power.

(d) The harmonic rate of distortion compared to the fundamental THD (or TDH), excluding the continuous component (the most common):

$$THD\% = 100.\frac{\sqrt{\sum_{n=2}^{\infty} c_{neff}^2}}{c_{1eff}}$$

And for each harmonic separately: $\tau_n\% = 100.\frac{c_{neff}}{c_{1eff}} = 100.\frac{c_2}{c_1}$.

(e) In sine vibration, the amplitude of the fundamental of the piloting signal is obtained by using a tracking filter. Its effective value a_1 (also noted as F) is compared to the total effective value a_{tot} of the unfiltered global signal (also denoted NF) measured in a wide frequency range (at least five times the excitation frequency). To stay within the usual tolerance T, of a deviation between global signal and filtered signal of ±5% (of the filtered signal), given the accuracy of the measuring equipment, the ratio a_{tot}/a_1 should not exceed 1.05, or the relation:

$$T_{\%} = \left(\frac{NF}{F} - 1\right).100$$

For that reason, we show that the value of this ratio corresponds to a total distortion of 32% according to the definition:

$$D\% = \frac{\sqrt{a_{tot}^2 - a_1^2}}{a_1} \quad \text{or} \quad \frac{D\%}{100} = \sqrt{\left(\frac{T}{100}\right)^2 + \frac{2T}{100}}$$

This happens if we neglect the share of the broad band noise compared to that relating to harmonics (hypothesis generally verified in sine scanning). Refer to section 41 of GAMEG13 and the NF INTO 60068-2-6 standard.

1.2.2.4. *Kinematic parameters*

Physically, the visual aspect that an oscillation takes and that is directly perceived, is that of peak to peak displacement. But in addition to this parameter, we also introduce the mathematical concepts of velocity and acceleration. This is, of course, for all types of signals, which we will define, with the specific case of the sinusoidal signal.

a) Displacement

Noted $d(t) = x(t) = x_{max}.\sin\omega t$.

Unit = [m] often provided in [mm] in the specifications, because this is the order of magnitude of the displacements usually encountered (it is easily measured in c/c with a sticky paper comprising a right-angled triangle of sufficient slope allowing the appreciation of the $1/10^e$ mm to a few millimeter).

b) Velocity

It is by definition the variation of the displacement during a given time. It is thus a gradient of displacement, either a rate of displacement variation in a direction during a given time and therefore the derivative of displacement in function of time, or:

$$v(t) = \dot{x}(t) = \frac{dx(t)}{dt} = \omega x_{max}.\cos\omega t = \omega x_{max}.\sin(\omega t + \pi/2)$$

Velocity is also a sinusoidal function in advance of $\pi/2$ = 90° on the displacement. Its amplitude is inferred from that of the displacement by the product with ω.

Unit = [m/s] = [$m.s^{-1}$].

c) Acceleration

It is, by definition, the velocity variation during a given time, specifically, a rate of velocity variation for a given time; therefore, the derivative of velocity in a direction in relation to time is:

$$\gamma(t)=\frac{dv(t)}{dt}=\frac{d\dot{x}(t)}{dt}=\ddot{x}(t)=\frac{d^2x(t)}{dt^2}=-\omega^2.x_{max}.\sin\omega t=-\omega^2.x(t)$$

Acceleration is also a sinusoidal function in phase opposition with displacement.

Unit = [m/s/s] = [m.s^{-2}].

NB: We often normalize acceleration to that of gravity with:

$$1g=9{,}81ms^{-2}\approx 10ms^{-2}$$

$$\text{Therefore:}\qquad \ddot{x}(t)\ [m.s^{-2}]\approx \ddot{x}(t)\ [g].10$$

For example, 1 500 m^{-2} # 150 g. This normalization allows quick access to the efforts when we know the mass by simply applying, in the case of rigid bodies, F (t) = m.γ(t). The maximum force for a sinusoidal signal is thereby given by:

$$|F_{max}|\,[daN]=m.|\ddot{x}|\,[g]\qquad \text{or}\qquad |F_{max}|\,[N]=m.|\ddot{x}|\,[m.s^{-2}]$$

In practice, vibratory measurements are often made with accelerometers that measure absolute acceleration. As a result, by simple and double integration (analogical or numerical), we reach velocity and displacement, which means:

$$\ddot{x}(t)=\gamma(t)=\gamma_{max}.\sin\omega t$$

$$\dot{x}(t)=v(t)=\int\gamma(t).dt=-\frac{\gamma_{max}}{\omega}.\cos\omega t=-\frac{\gamma_{max}}{\omega}.\sin(\omega t+\pi/2)=\frac{\gamma_{max}}{\omega}.\sin(\omega t-\pi/2)$$

$$x(t)=\int v(t).dt=-\frac{\gamma_{max}}{\omega^2}.\sin\omega t$$

The mnemonic relation to remember is as follows:

$$\left\{\left|\frac{x(t)}{v(t)}\right|=\frac{1}{\omega}\qquad \left|\frac{v(t)}{\gamma(t)}\right|=\frac{1}{\omega}\qquad \left|\frac{x(t)}{\gamma(t)}\right|=\frac{1}{\omega^2}\right. \tag{1.11}$$

We note that for a pure or sinusoidal harmonic movement, it is very easy to go from one parameter to the other. There are also charts that directly provide these parameters depending on the frequency, often in the form of a four-variable nomogram in displacement, velocity and acceleration, in the same manner that electronics is expressed in the relation between resistance capacity and inductor. An example is given hereafter in Figure 1.4.

It is noted that we pass from one parameter to another by removing 6 dB at the slope in the order displacement (slope +1), velocity (slope 0), and acceleration (slope –1).

The most used relationship is the one that expresses the peak-to-peak displacement (because it is easier to measure) with the peak acceleration g (directly linked to the maximum effort). Moreover, when speaking of g in vibrations, it is understood that all gs are a peak. Having:

$$x(t) = x_{max}.\sin\omega t \rightarrow \ddot{x}(t) = -\omega^2.x_{max}.\sin\omega t \rightarrow \frac{|x(t)|}{|\ddot{x}(t)|} = \frac{1}{\omega^2} = \frac{1}{4\pi^2 f^2} = \frac{d}{\gamma}$$

$$\gamma = d.\omega^2 = d.4\pi^2 f^2 \text{ [m.s}^{-2}\text{c / c] with d in [mc / c]}$$

$$\gamma = d.4\pi^2 f^2.10^{-3}.\frac{1}{2.9{,}81}.\text{[gc/] with d in [mmc / c]}$$

So:

$$\gamma\text{[gc/]} = d\text{[mmc / c]}.f^2.\frac{1}{497}$$

And:

$$\gamma\text{[gc /]} \approx \frac{d.\text{[mmc / c]}.f^2}{500} \qquad (1.12)$$

Therefore: $\begin{cases} 1gc/ \approx 1mmc/c \text{ at } 23Hz \approx 500^{1/2}, \text{ or :} \\ 1gc/ \approx 5mmc/c \text{ at } 10Hz \end{cases}$

which is a traditional breaking point of sine vibrations specifications. See the point M on the nomogram (see Figure 1.4).

For example, to know the displacement at f = 50 Hz with γ = 2gc/, we carry out a simple rule of three, knowing that the displacement varies as f^2. Therefore, when mounting frequency:

$$f = 50Hz \qquad \gamma = 2gc/ \qquad \rightarrow \qquad d = 2.\frac{5}{5^2} \approx 0{,}4mmc/c$$

So, by maintaining a level of constant acceleration, we note that the velocity and displacement decrease (respectively, increase) when the frequency increases (respectively, decrease), which takes place quickly since these parameters are function of $1/f^2$ (respectively, f^2).

These considerations allow for choosing the parameter measured according to the frequential domain studied. The magnitude scales of common amplitudes are the following:

$$\begin{cases} 1gc/ & 5mmc/ & 10Hz \\ 1gc/ & 12{,}5\mu mc/c & 2000Hz \\ 0{,}01gc/ & 5mmc/c & 1Hz \end{cases}$$

NB: It should be noted that in the nomogram shown in Figure 1.4, it is necessary to express the values of displacements, velocities, and acceleration in coherent measurements, either all in c/ or all in c/c.

d) Logarithmic scales

We have seen that the amplitudes of vibrations vary in broad proportions. In order to exploit this great dynamic easily, we use logarithmic scales in amplitude and frequency, with the following relations:

- in terms of amplitudes: A_{dB} = 20LogA [amplitude]
- in terms of power: A_{dB} = 10LogA [power] = 10LogA^2 [amplitude].

The graphics used are graduated in a scale Log/Log and thus promote low amplitudes by compressing strong amplitudes. The following frequential representations are thereby found (including a representation in nomogram or quadrilog): (a) with a typical template; (b) a test specification to sinusoidal vibrations.

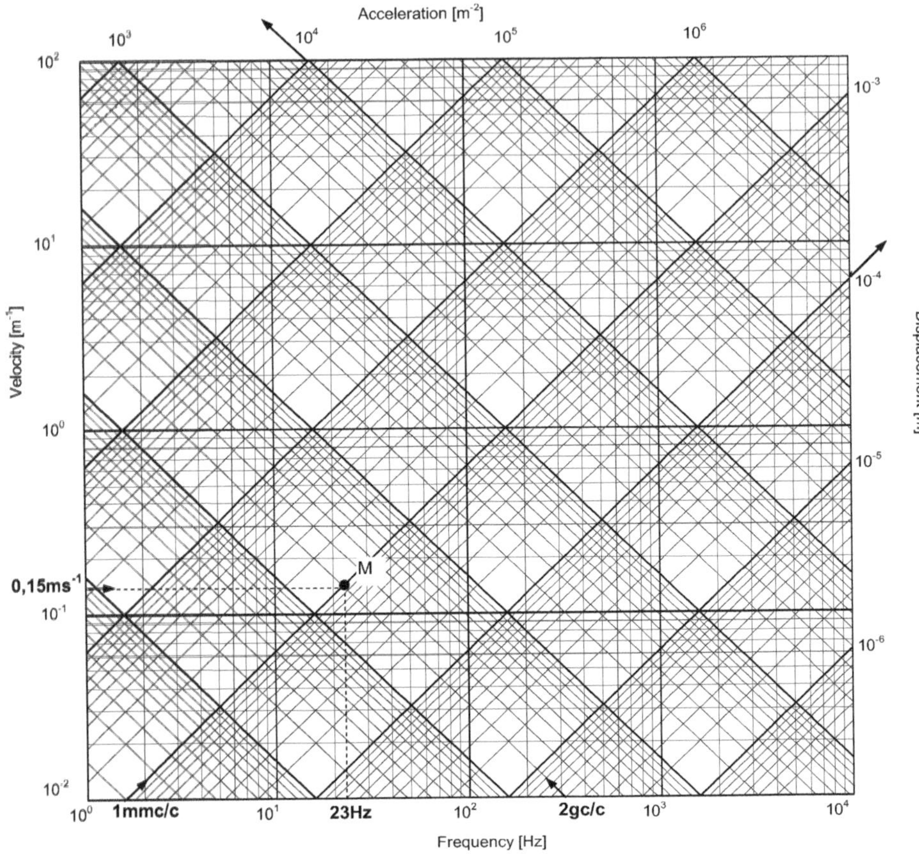

Figure 1.4. *Four-variable nomogram, displacement, velocity and acceleration*

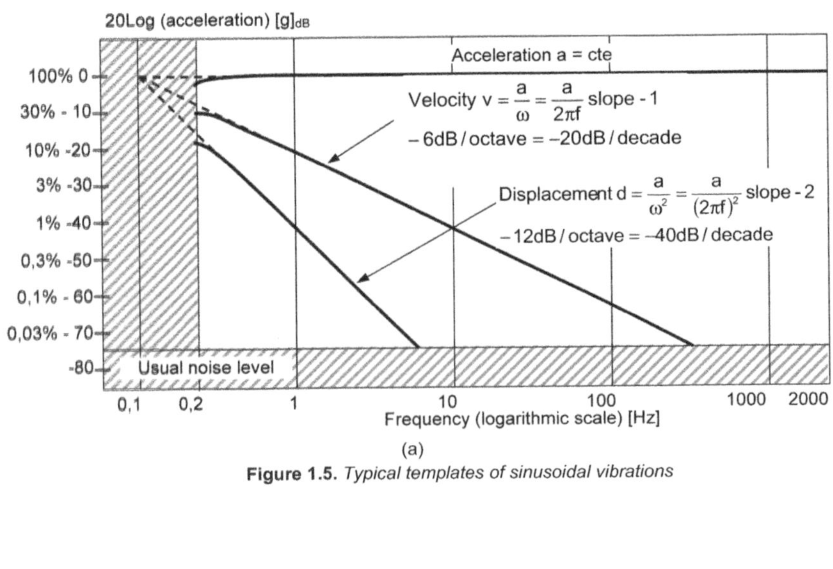

Figure 1.5. *Typical templates of sinusoidal vibrations*

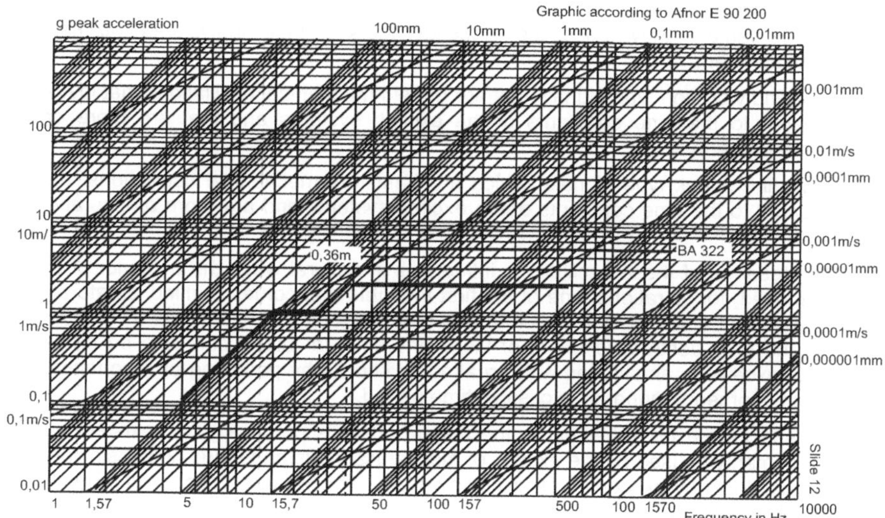

(b)

Figure 1.5. (cont.) *Typical templates of sinusoidal vibrations*

e) Temporal linear scale

The graphes linking displacements, velocities, and accelerations are the following ones (we consider here f = 0,2 Hz, which means T = 5 s).

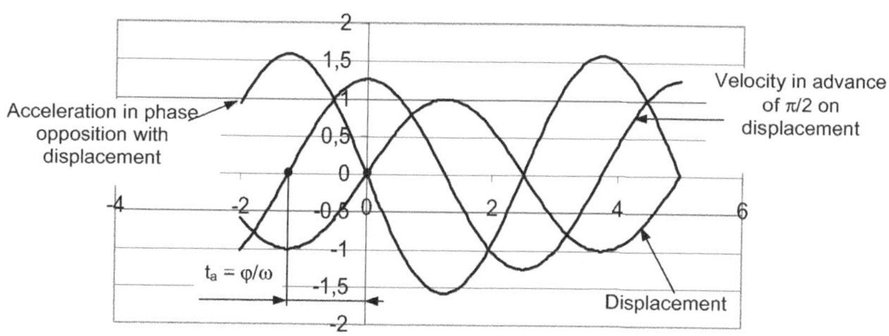

Figure 1.6. *Time characteristic of displacement, speed and acceleration*

We saw previously that the advance of time t_a equals φ/ω, so in this particular case here:

$$t_a = \frac{T}{4} = \frac{\varphi}{\omega} \rightarrow \varphi = \frac{T\omega}{4} = \frac{T2\pi f}{4} = \frac{T2\pi}{4T} = \frac{\pi}{2} \qquad \text{cqfd}$$

1.2.3. Types of mechanical signals

The mechanical signals, which appear in practice on structures during their life profile, usually result from a combination of signals acting together and giving a more or less complicated displacement function with respect to time. These are usually phenomena that are desired to be removed, except in a few special cases where their same properties are used (sorted by sieve, vibrated concrete, bell, tuning, polishing, etc.).

We usually separate them into two classes according to the way in which they affect structures, the damage they cause, the necessary tools for their analysis, or the machines used to simulate them.

Thus, we find vibration type signals that are upheld mechanical oscillations that last a long time compared to the time constants of the excited systems and that are more or less stationary phenomena.

Then, we find shock type signals, which are transitory mechanical oscillations, that are short compared to the time constants of the excited systems and are more or less aperiodic phenomena.

So, the presence of these oscillations influences in several ways the behavior of excited materials. Indeed, for vibrations, one of the first thoughts that comes to mind is naturally that of mechanical fatigue stored by a structure and that has a direct impact over its lifespan. But that is not the only destructive effect of vibrations since a failure can also be caused by vibrations of excessive amplitudes with breakages, shocks by pounding or resonance, anti-resonance phenomena during an excessive time period.

As with shocks, on the other hand, the fatigue effect is practically non-existent and the damage is primarily caused by the excess of mechanical resistance thresholds, similar to the dielectric breakdown by exceeding breakdown voltage (see volume "Mechanical Shocks and Material Applications" [ref. I18]).

Also, according to the estimate of the prevalence of such or such phenomenon, we will use the method of study deemed most appropriate.

Mechanical vibrations defined as "*oscillations more or less long around a point of reference*" with respect to the time constant of the excited system may therefore arise under several different deterministic or random aspects and thus introduce various types of signals to be analyzed. Those that we will usually find are as follows.

1.2.3.1. *Stationary periodic signals*

They are as follows:

$$y(t) = y(t+T)$$
$$T = \text{signal period} = \text{constant in [s]}$$

They are part of the deterministic signals, i.e. defined regardless of time t by an analytical relation. Generally, we show that such a function for stationary periodic signals is constituted by the sum of an infinite number of sinusoids in harmonic relations (their frequencies are of multiple *integers* of the fundamental frequency) and hence can be developed in a Fourier series with each harmonic component that will be reproduced for all values of t = T. We will then write:

$$y(t) = A_0 + A_1.\sin(\omega_0 t + \varphi_1) + A_2.\sin(\omega_2 t + \varphi_2).....+A_n.\sin(\omega_n t + \varphi_n) + ... \quad (1.13)$$

The most important particular case is that of the sinusoidal signal.

The decomposition in a Fourier series is modeled by spectra of amplitude and phase. We can see it with the classic example of the following periodic temporal signal:

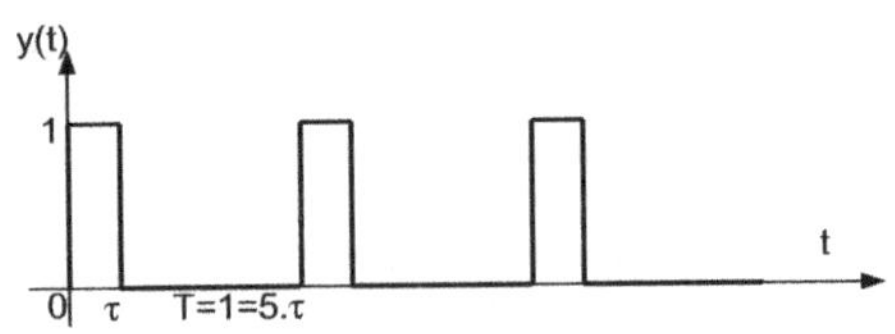

The decomposition in a Fourier series in exponential form gives:

$$\alpha_n = \frac{1}{T}.\int_0^{\tau} 1.e^{-j2\pi f_0 nt}.dt = \frac{1}{T}\left[\frac{e^{-j2\pi f_0 nt}}{-j2\pi f_0 n}\right]_0^{\tau} = \frac{1}{T}.\left[\frac{e^{-j2\pi f_0 n\tau}-1}{-j2\pi f_0 n}\right] = \frac{1}{T}.e^{-j\pi f_0 n\tau}.\left[\frac{e^{-j\pi f_0 n\tau}-e^{+j\pi f_0 n\tau}}{-2j}\frac{1}{\pi f_0 n}\right]$$

$$\alpha_n = \frac{\tau}{T}.\frac{\sin \pi f_0 n\tau}{\pi f_0 n\tau}.e^{-j\pi f_0 n\tau} = |\alpha_n|.e^{j\varphi(\alpha_n)}$$

$$|\alpha_n| = \frac{\tau}{T}.\frac{\sin \pi f_0 n\tau}{\pi f_0 n\tau} = \frac{1}{5}.\frac{\sin \pi f_0 n\tau}{\pi f_0 n\tau} \quad \text{and} \quad \varphi(\alpha_n) = -\pi f_0 n\tau \text{ linear phase (pure delay)}$$

with f_0 = fundamental and nf_0 = harmonic of rank n.

Considering the following curves:

n		n	
0	0,2000	11	0,0170
1	0,1871	12	0,0252
2	0,1514	13	0,0233
3	0,1009	14	0,0134
4	0,0468	15	0,0000
5	0,0000	16	0,0117
6	0,0312	17	0,0178
7	0,0432	18	0,0168
8	0,0378	19	0,0098
9	0,0208	20	0,0000
10	0,0000		

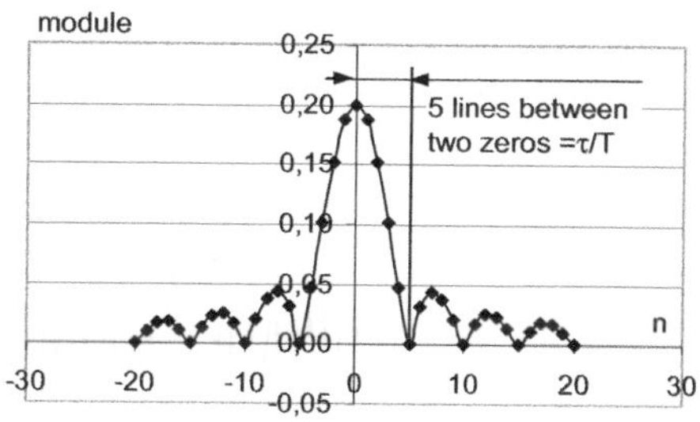

n		n	
0	0,2000	11	0,0340
1	0,3742	12	0,0505
2	0,3027	13	0,0466
3	0,2018	14	0,0267
4	0,0935	15	0,0000
5	0,0000	16	0,0234
6	0,0624	17	0,0356
7	0,0865	18	0,0336
8	0,0757	19	0,0197
9	0,0416	20	0,0000
10	0,0000		

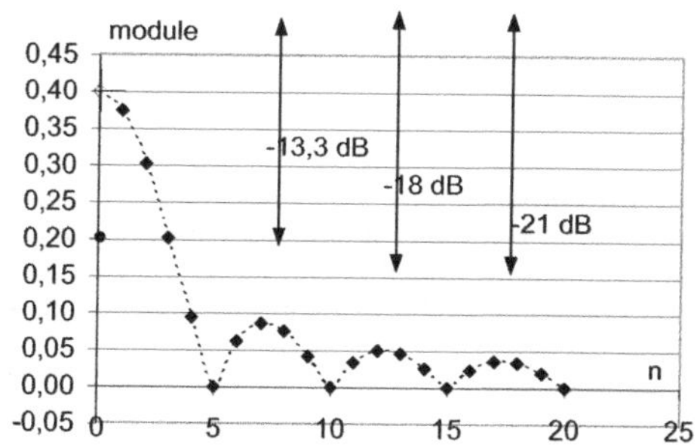

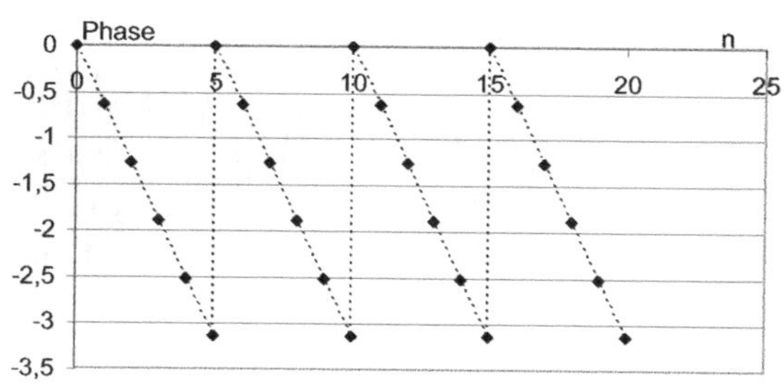

We find the cyclic ratio $\delta = \tau/T$ in n = 0, which is the average value of the signal.

"*The rate of harmonic distortion compared to fundamental*" is then given by the relative value of the total power of the different lines of the fundamental (and the power of the average), compared to the power of fundamental, therefore we have the following relation (which is used in vibrations tests and which measures the purity of the excitation signal, so it is in accordance with the standards' requirements):

$$\mathrm{THD} = \frac{\sqrt{\sum_{i=2} A_{ieff}^2}}{A_{1eff}} \tag{1.14}$$

1.2.3.2. *"Quasi sinusoidal" stationary signals*

They are generally composed of a mixture of sinusoidal signals of constant amplitude, without harmonic relation among them (incommensurable periods). The mixture can be produced by the addition or multiplication of sinusoidal signals, or any other combination.

In the following manner, for example: $y(t) = A_1.\sin\omega_1 t + A_2.\sin\omega_2 t + A_3.\sin\omega_3 t.A_4.\sin\omega_4 t + ...$

We can, for example, find this type of signals in structures moving at different velocities. An example is shown in Figure 1.7.

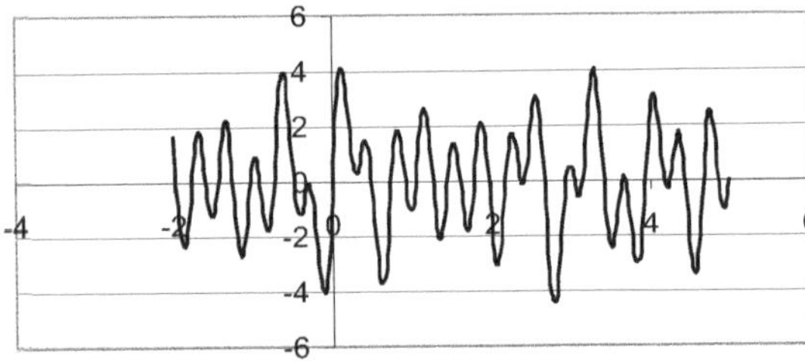

Figure 1.7. "Quasi-periodic" signals

1.2.3.3. *"Complex" signals*

The term complex here does not mean "complex bodies", but rather differentiates these types of signals between sinusoidal and periodic signals. These are sinusoidal functions of which the amplitude and the frequency vary slowly over time and can be reproduced under identical conditions. In the following manner, for example:

$$y(t) = y_1(t).\sin[\omega(t).t+\varphi]$$

$y_1(t)$ and $\omega(t)$ are times curves

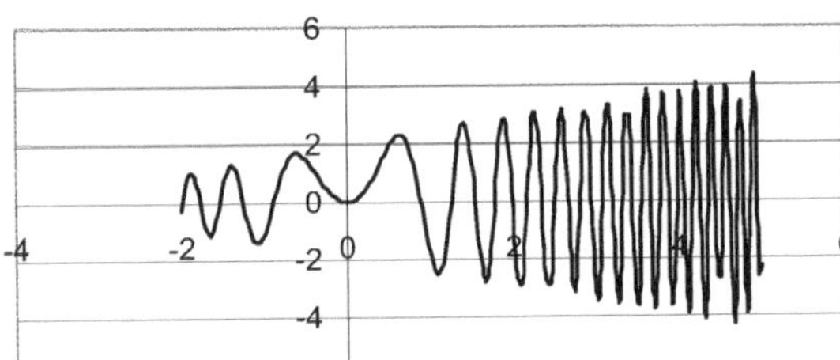

Figure.1.8. *Quasi-sinusoidal signals*

1.2.3.4. *Random stationary signals*

In previous cases, the amplitude function of time can be reproduced identically indefinitely, regardless of time t, starting from a known analytical relation. For random signals, functions used can only be determined on a probability basis. Contrary to a deterministic signal as previously defined, if we know its value at a moment t, at a moment t+ε we cannot know its value other than in probabilistic terms ∀ t and ε.

Thus, a random signal is described in the field of amplitudes by its probability density and in the field of frequencies by its power spectral density (PSD).

1.2.3.5. *Random non-stationary signals*

These signals are more difficult to analyze because their amplitude and frequency characteristics change according to the origin of time in a unforeseeable statistically manner. These signals can be studied by making them stationary by periodic repetition after registration of the studied sequence, for example. Other, more elaborate and more complex techniques exist, such as the use of wavelets, Cepstrum (see Appendix E) or the Hilbert transform.

These signals can have two main origins:

1) the source of noise or vibrations during normal operation.

2) a simulated condition of vibration testing (which is particularly interesting to us).

Signals of 2 are generally easier to control and analyze. Indeed, in the signals described in 1, some external factors can affect the analysis, such as:

- a random signal that contains periodic components;
- a stationary signal;
- a signal containing transients in an isolated manner or in bursts.

This is why we will focus especially on how stationary signals are treated.

1.2.3.6. *Transitory signals*

They are signals of the aperiodic type that occur during a limited time period in front of the time constant of the excited systems such as for example the shocks signals. Their somewhat deadened form generally tends toward zero at the end of a certain time. Henceforth, they belong to the class of signals with finished energy and no average power.

The following basic concepts are pointed out:

$$\begin{cases}\text{Energy}\\ \text{Joule [J]}\end{cases} \equiv \begin{cases}\text{Work}\\ \text{Newton} - \text{meter [N.m]}\end{cases} \qquad \begin{cases}\text{Power}\\ \text{Watts [W]}\end{cases} \equiv \begin{cases}\text{Energy debit}\\ \text{Joule per second [J.s}^{-1}\text{]}\end{cases}$$

- The stationary signals of sine type and random ones, for example, are signals with infinite energy and finite average power (there are no signals with infinite power in our universe).
- Pulse signals have finite energy and zero average power.

1.2.4. The different types of mechanical signals – a practical approach

Within the framework of the normative documents that strive to offer standard tests, but which are as close to reality as possible, the types of signals proposed are adapted to the different life patterns of materials and the situations encountered by them, primarily during their missions. There are a few differences with the types of signals defined earlier, even if they call upon the previous definitions. In practice, we find the following types of synthesis signals:

- variable frequency signal generators:
  - step by step;
  - linear sweeping;

- logarithmic sweeping;
- with programmed law of sweeping;

- geometrical generators of simple form functions:
  - half-sine impulse signals;
  - square signals;
  - rectangular signals;
  - triangular signals (known as saw tooth);
  - trapezoidal signals, etc.

with pre- and postpulses;

- of more complex forms:
  - trains of impulses (jolts);
  - geometrical combination of forms.
- random generators of noise or random pseudo-noises.
  - with fixed or variable bandwidths.
  - with constant PSD (white noise), weighted, for example, with slopes at ±X dB/octave and respecting a given template;
  - combined noise signals and sine lines (lines in harmonic relations to simulate the vibrations caused by a rotor or a propeller or other ones without harmonic relations to simulate fire cannons on aircrafts), etc.

We can see here the additional difficulty of analyzing these practical signals.

It is common to classify the deterministic and random signals, in the following manner (see Figures 1.9 and 1.10).

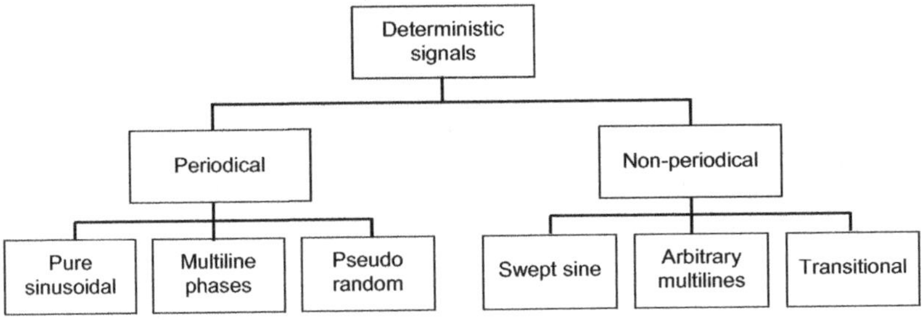

Figure 1.9. *Classification of practical deterministic signals*

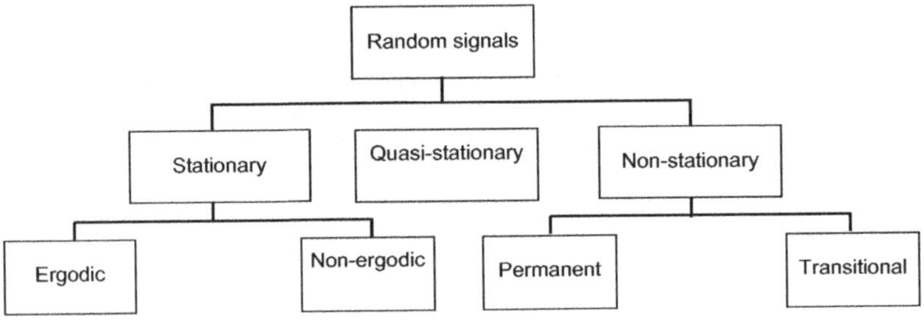

Figure 1.10. *Classification of random practical signals*

1.3. Theory of vibration – reminders

1.3.1. Fundamental law of dynamics

When a constant force in magnitude, direction and sense affects a material point, initially at rest:

- it communicates a uniformly accelerated motion;

- the acceleration and the force have the same support, the same sense and their numerical values are proportional.

The fundamental law of dynamics is expressed by the following relation:

$$\vec{F} = m.\frac{d^2\vec{x}(t)}{dt^2} = m.\ddot{\vec{x}}(t) = m.\vec{\gamma} \tag{1.15}$$

where F is the resultant of the forces applied to the material point of mass m and γ is the acceleration communicated to the material point.

For systems under rotation, the same law applies in the following form:

$$\vec{M_t} = I.\frac{d^2\vec{\theta}(t)}{dt^2} = m.\ddot{\vec{\theta}} \tag{1.16}$$

where

I = moment of inertia

M_t = moment

$\ddot{\theta}(t)$ = angular acceleration

1.3.2. Concept of DOF

A mechanical system with 1 Degrees-Of-Freedom (DOF) is a system whose position in space is defined by only one quantity q and whose movement, under the influence of forces that are applied to it, is determined by the only variation of this quantity q in function of time.

This can be naturally generalized to systems with several DOF. The number of DOF of a system will be the number of *independent* parameters that allow to fully determine the movement of the mechanical system considered.

Let us take, for example, a system that can only move along one axis, and in which only one dimension is necessary to know its position at any moment, compared to a point of reference, therefore this system has only 1 DOF.

Consider, for example, a non-deformable solid (assumed rigid in the frequency band used), spotted in a Cartesian system (0, x, y, z). The possible movements of this body can be described with three translation movements (q_1, q_2, q_3) or (x, y, z) from its center of gravity, as well as three rotations (q_4, q_5, q_6) or (θ_x, θ_y, θ_z) around orthonormal axes passing through its center of gravity. The movement of each point M of the solid can be determined from these six quantities. So, the system here has 6 DOF.

$$x(M) = x(q_1, q_2, q_3, q_4, q_5, q_6, t)$$

These different independent quantities q_i, which allow to characterize the system requirements, are called the generalized coordinates and do not necessarily have the dimension of a length. Consider the following examples:

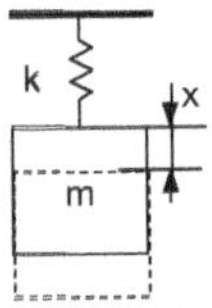

Mass and spring

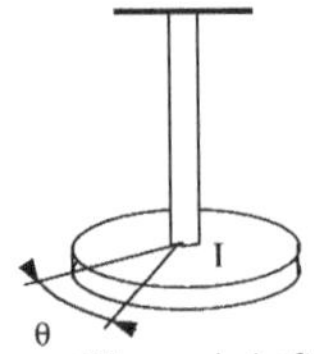

Disc and shaft

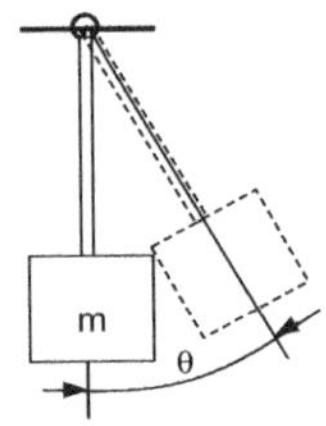

Pendulum

Systems with 1 DOF

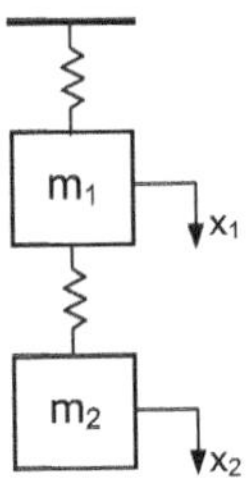

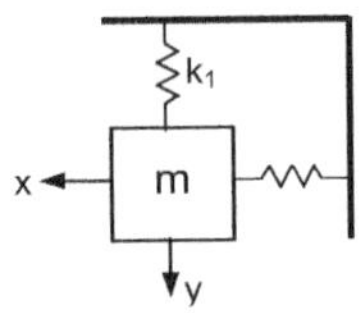

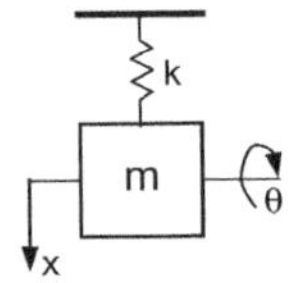

Systems with 2 DOF

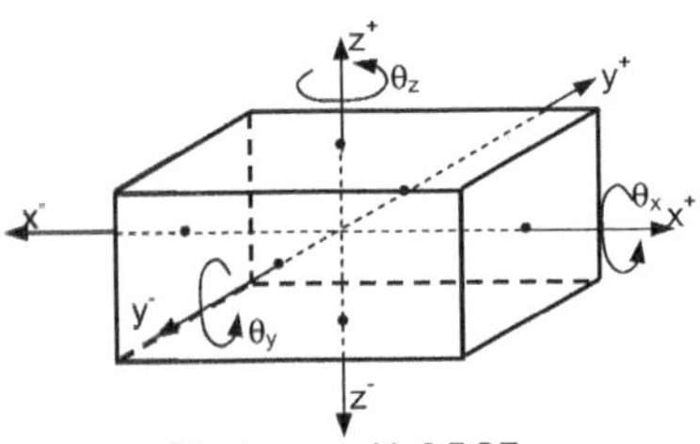

Rigid mass with 6 DOF

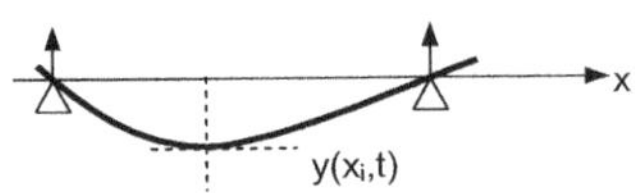

Support beam
An infinite number of DOF

The beam can indeed have an infinite number of DOF since it can theoretically be deformed in an infinite number of ways.

1.3.3. Systems comprising 1 DOF

1.3.3.1. *Components of the model with 1 DOF*

They are modeled by the following localized elements:

- the spring: stores potential energy of deformation;

- the mass: stores kinetic energy;

- the damper: dissipates energy.

In the case of continuous systems, these settings are distributed, but the hypothesis of modal recombination techniques associates a system with 1 DOF with each mode.

This is why the 1 DOF system is so important.

1.3.3.2. *Vibrating systems in translation*

a) The spring

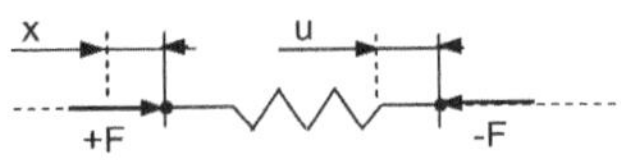

- An ideal spring does not have a mass.

- The length of the spring is proportional to the force F applied:

$$F = k.(x - u) \tag{1.17}$$

- The proportionality coefficient k is the stiffness expressed in [N/m].

b) The mass

- The mass is a body presumably rigid for the model considered (in reality, a mass is only rigid *in a field of frequency given*, which quickly becomes false even for relatively low frequencies, just a few kHz). This is the assumption used for this model of representation that is associated with restrictive assumptions. It is expressed in [kg].

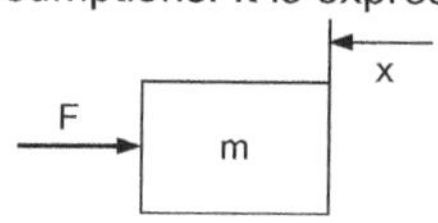

- The acceleration to which the mass is subjected to is proportional to the net force applied.

$$F(t) = m\ddot{x}(t) \tag{1.18}$$

c) Damper

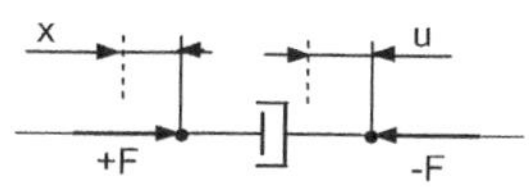

- The applied force F is proportional to the relative speed of the two ends:

$$F = c.(\dot{x} - \dot{u}) \tag{1.19}$$

- c = viscous damping coefficient. It is expressed in [$Nm^{-1}s^2$].

There are other types of dampening adapted to the physical models chosen. Dampening is related to the powers of speed or other parameters.

1.3.3.3. *Vibrating systems in rotation*

The elements of the representation model that form the systems in rotation are identical to those used in translation. We find the following parameters correspondences:

- Linear displacement:	x [m]	- Displacement in rotation:	θ [rd]
- Force:	F [N]	- Moment of force:	M_t [Nm]
- Stiffness:	k [N/m]	- Torsion constant:	C [N.m/Rd]
- Viscous damping:	c [$Nm^{-1}s^2$]	- Viscous damping:	c_r [$Nm^{-1}s^2$)]
- Mass:	m [kg]	- Moment of inertia:	I [kgm^2]

With the following corresponding relations:

$$F = k(x_1 - x_2) \qquad M_t = C(\theta_1 - \theta_2) \tag{1.20}$$

$$F = c(\dot{x}_1 - \dot{x}_2) \qquad M_t = c_r(\dot{\theta}_1 - \dot{\theta}_2)$$

$$F(t) = m\ddot{x}(t) \qquad M_t(t) = I\ddot{\theta}(t)$$

1.3.4. Free oscillations of an undamped system

1.3.4.1. *Systems in translation*

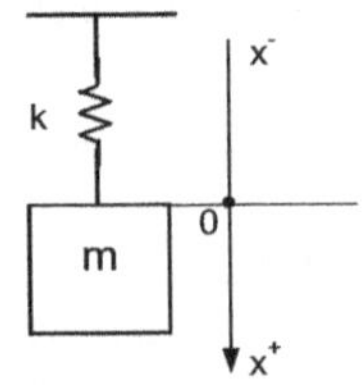

Such a system is modeled by the following diagram:

(a) At rest: The assessment of forces applied is therefore (for a deformation L_0 of the spring):

- the weight: $P = mg$
- the tension of the spring: $T = kL_0$
- the system being at rest, the net force is zero, and the system is still, therefore:

$$P + T = 0 \quad \rightarrow \quad mg = kL_0$$

(b) Let us dismiss the system from its balance position and let it oscillate freely (release technique). At a given time t, the length of the spring is L and the assessment of the applied forces gives:

- the weight: $P = mg$
- the tension of the spring: $T = kL$
- the net force applied is therefore given by:

$$F = -kL + kL_0 = k(L_0 - L) = kx$$

The origin at 0 is understood as the position of balance. F is proportional to the elongation of the spring. The equation of motion is then given by the fundamental relation of dynamic, which means:

$$F = -kx(t) = m\frac{d^2x(t)}{dt^2} \rightarrow m\frac{d^2x(t)}{dt^2} + kx(t) = 0$$

The solution of this differential equation is as follows:

$$x(t) = x_0.\sin(\omega_0 t\varphi) \qquad (1.21)$$

The amplitude x_0 and the phase at the origin φ, depend on the initial conditions (ICs) imposed on the movement. Thus, the vertical oscillations of the mass are sinusoidal, with the following parameters:

- natural pulse:

$$\omega_0 = \sqrt{\frac{k}{m}} \qquad (1.22a)$$

- natural frequency:

$$f_0 = \frac{1}{2\pi}\sqrt{\frac{k}{m}} \qquad (1.22b)$$

- natural period:

$$T = \frac{1}{f_0} = 2\pi\sqrt{\frac{m}{k}} \qquad (1.22c)$$

Thus, any sinusoidal rectilinear movement is generated by a return force proportional to the elongation x.

1.3.4.2. *Systems in rotation*

The torsion pendulum is the equivalent model of representation of the mass spring model previously defined for rotation systems.

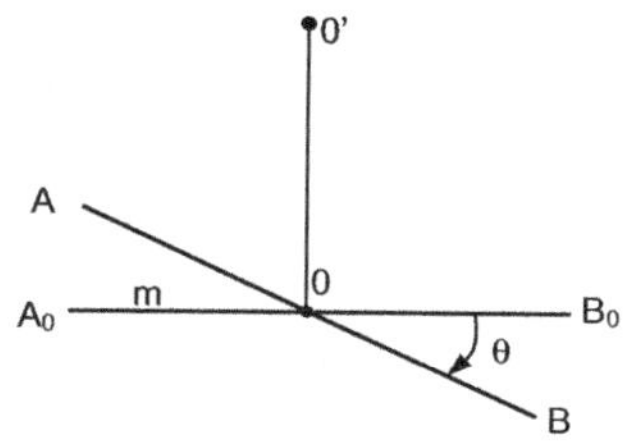

A torsion wire supports a heavy bar of mass m, movable around an axis 0'0. I is the moment of inertia of the bar relative to the axis 0'0. It is the constant of torsion of the wire.

(a) At rest, the bar is in A_0B_0. The moment of the forces applied is zero.

(b) Let us alter the system from its balance position. Recall forces appear. The return moment is therefore proportional to the angular elongation, which is the angle θ in this case, with: $M_t = -C\theta$. The equation of movement is provided by the fundamental relation of dynamics, with:

$$-C\theta(t) = I\frac{d^2\theta(t)}{dt^2} \rightarrow I\frac{d^2\theta(t)}{dt^2} + C\theta(t) = 0$$

The solution of this differential equation is as follows:

$$\theta(t) = \theta_0.\sin(\omega_0 t \ \varphi) \tag{1.23}$$

Thus, the vertical oscillations of the bar are still sinusoidal, with the following parameters:

- natural pulse: $\omega_0 = \sqrt{\frac{C}{I}}$

- natural frequency: $f_0 = \frac{1}{2\pi}\sqrt{\frac{C}{I}}$

- natural period: $T = \frac{1}{f_0} = 2\pi\sqrt{\frac{I}{C}}$

Thus, any sinusoidal circular movement is due to a return moment proportional to the angular elongation θ.

1.3.5. Free oscillations from an undamped system

In the previous section, dampening is zero and the system oscillates indefinitely, in theory. In reality, there always is a damping, no matter how small. As described in the following model of 1 DOF system (Figure 1.11). Note that on the model's drawing, the infinitely rigid frame does not physically exist in reality (Appendix B).

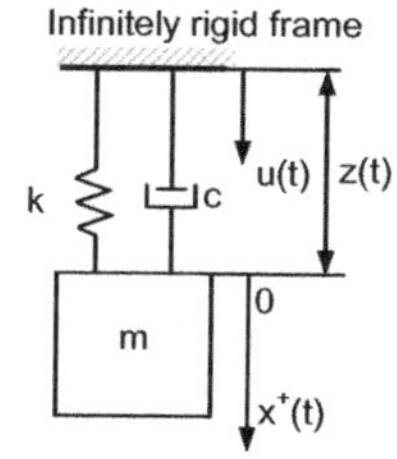

Figure 1.11. *Basic model with damping*

(a) At rest, the weight in mg is balanced by the tension of the spring.

(b) Let us dismiss the system from its balance position and release it, which amounts to imposing the non-zero IC of displacement and thus to load the system into potential deformation energy.

The assessment of the forces applied is the following (with the origin of elongations in 0, position of balance):

- return force of the spring: $-k.x(t)$
- viscous friction force: $-c.\dot{x}(t)$
- resultant of the applied forces: $F = -k.x(t) - c.\dot{x}(t)$

The system's equation of motion is given by the fundamental relation of dynamics, which means

$$F(t) = m\ddot{x}(t) = -kx(t) - c\dot{x}(t) \rightarrow m\frac{d^2x(t)}{dt^2} + c\dot{x}(t) + kx(t) = 0$$

Or: $$m\ddot{x} + c\dot{x} + x = 0$$

Relation that we generally find in the following reduced form:

$$\ddot{x} + 2\xi\omega_0.\dot{x} + \omega_0^2.x = 0 \quad (1.24)$$

With:

$\omega_0 = \left(\frac{k}{m}\right)^{1/2}$ = own pulsation of the undamped system

$\xi = \frac{c}{2(km)^{1/2}} = \frac{c}{c_c}$ = reduced damping coefficient (dimensionless parameter)

c_c = critical damping for which the differential equation has a double root, which corresponds to the end of an oscillatory movement and the passing of an aperiodic movement.

We recall:

- $[k] = [N.m^{-1}]$
- $[c] = [c_c] = [N.m^{-1}.s^2]$

a) Damped oscillatory motion

Case where $\xi < 1$ and $c^2 - 4km < 0$

The roots of the characteristic equation are imaginary combined, with:

$$\begin{cases} r_1 = -\xi\omega_0 + j\omega_0\left(1-\xi^2\right)^{1/2} \\ r_1 = -\xi\omega_0 - j\omega_0\left(1-\xi^2\right)^{1/2} \end{cases}$$

The solution x(t) is as follows:

$$x(t) = A.\exp\left\{\left[-\xi\omega_0 + j\omega_0\left(1-\xi^2\right)^{1/2}\right].t\right\} + B.\exp\left\{\left[-\xi\omega_0 - j\omega_0\left(1-\xi^2\right)^{1/2}\right].t\right\} \quad (1.25)$$

Consider then: $x(t) = C.e^{-\xi\omega_0 t}.\sin\left[\omega_0\left(1-\xi^2\right)^{1/2}.t + \Phi\right]$

C and Φ depending on the IC on x, $\dot{x}$, $\ddot{x}$ imposed on the system.

The appearance of a damped sinusoidal oscillating movement is the following:

f_0 = 10Hz, ξ = 1/6, C = 1 and $\Phi = \pi/2$

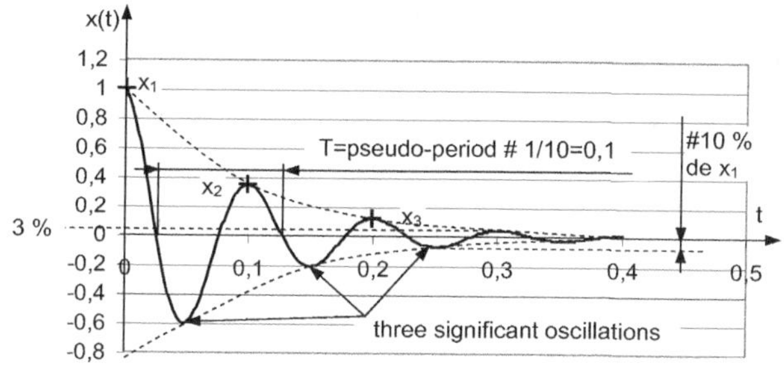

Figure.1.12. *Damped oscillations*

Then, the following concepts are also defined:

- pulsation in a damped-free mode, pseudo-pulsation or natural pulsation:

$$\begin{cases} \omega_L = \omega_n = \omega_0.\left(1-\xi^2\right)^{1/2} \\ \dfrac{\omega_n}{\omega_0} = \left(1-\xi^2\right)^{1/2} \end{cases} \tag{1.26}$$

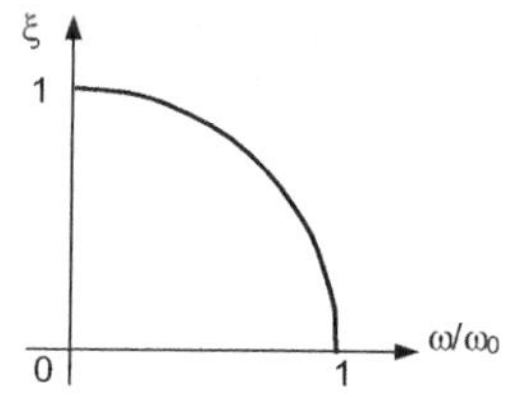

Not to be confused with the pulsation of the resonance in forced regime, which we will study later on, and is given by: $\omega_R = \omega_0.\left(1-2\xi^2\right)^{1/2}$;

- natural pseudo-frequency in a dampening-free regime:

$$f_0 = \frac{1}{2\pi}.\omega_0.\left(1-\xi^2\right)^{1/2} \tag{1.27}$$

- natural pseudo-period in a dampening-free regime:

$$T = 2\pi/\omega_0.\left(1-\xi^2\right)^{1/2} \tag{1.28}$$

- logarithmic decrement: The degree of dampening of a system can be determined precisely from the various peak values taken by x(t). Indeed, let us consider two displacements maximum (either for velocity $\dot{x}(t)=0$), distances of $nT = n2\pi/\omega_0.\left(1-\xi^2\right)^{1/2}$. Consider for the first n = 0 and the sine equal to 1 (Φ= 0 to simplify the calculations)

$$x_M(t+0.T)x(t) = x_0 = C.e^{-\xi\omega_0 t}.1$$

For the maximum, with nT, the sine is always equal to 1 and becomes:

$$x_M(t+nT) = x_n = C.e^{-\xi\omega_0(t+nT)} = C.e^{-\xi\omega_0 t-\xi\omega_0 nT} = C.e^{-\xi\omega_0 t-\xi\omega_0 n2\pi/\omega_0.\left(1-\xi^2\right)^{1/2}} = C.e^{-\xi\omega_0 t-2\pi n\xi/\left(1-\xi^2\right)^{1/2}}$$

Therefore:

$$\frac{x_n}{x_0} = e^{-2\pi n\xi/\left(1-\xi^2\right)^{1/2}}$$

Or, by definition, it is then called logarithmic decrement, the natural logarithm of two successive maximum amplitudes, which is for n = 1:

$$\Delta = -\ln\frac{x_1}{x_0} = \ln\frac{x_0}{x_1} \rightarrow e^{-\Delta} = \frac{x_0}{x_1} \rightarrow x_0 = x_1.e^{-\Delta}$$

$$\text{And}: \qquad \Delta = \frac{2\pi\xi}{\left(1-\xi^2\right)^{1/2}} \qquad (1.29)$$

And for very little damped systems where ξ<;<;1 or ξ # 0:

$$\Delta \approx 2\pi\xi \rightarrow \xi = \frac{\Delta}{2\pi} \qquad (1.30)$$

Another simpler technique allows to directly access the value of the reduced damping ξ. Indeed, we can recall here, as in the case of forced regimes (or in harmonic excitation), the equations providing the Q-factor and resonance frequency are the following:

$$Q = \frac{1}{2\xi\left(1-\xi^2\right)^{1/2}} \approx \frac{1}{2\xi} \quad \text{When } \xi \text{ is small before 1, meaning from around } Q = 3, \text{ we have } \xi = 0{,}17$$

$$\omega_R = \omega_0.\left(1-2\xi^2\right)^{1/2}$$

It is possible, for these low values of ξ, to determine the value of Q and thus of ξ starting from a release test. Indeed, *Q is practically equal to the number of free oscillations significantly dampened*, i.e. those for which the peak amplitude remains greater than approximately 5% of the initial maximum amplitude. We can see from Figure 1.12 that we have about three significant oscillations for a Q of three, and thus a dampening of 1/6 # 0,166.

We can note that for such a system, which corresponds in practice to the response of standard dampers and thus envisaged *a priori* to have an interesting damping coefficient, we can find an error of $\left(1-\xi^2\right)^{1/2} = \left(1-(1/6)^2\right)^{1/2} \approx 0{,}014$. Consider an error of 1,4% only and for real practical cases, approximations made for a *small damping* are more than sufficient. For a system deemed undamped, with, for example Q = 10, we find ξ # 1/20 = 0,05 and an error of 1,3 × 10^{-3}.

b) Critical movement

Case where $\xi = 1$ and $c = 1.2(km)^{1/2}$

So, in this case the characteristic equation $r^2 + 2\xi\omega_0 r + \omega_0^2 = 0$ admits a double root given by:

$$r_{1,2} = -\xi.\omega_0 = -1.\omega_0 = -\omega_0$$

The solution x(t) has, as an expression, in this case:

$$x(t) = (A + Bt)e^{-\omega_0 t} \qquad (1.31)$$

A and B depend on the IC and f_0 = 10 Hz. It is the borderline case where there is no overage.

c) Aperiodic movement

Case where $\xi > 1$ and $c^2 - 4km > 0$

In this case, the characteristic equation admits two distinct real roots, with:

$$\begin{cases} r_1 = -\xi\omega_0 + \omega_0.\left(\xi^2 - 1\right)^{1/2} \\ r_2 = -\xi\omega_0 - \omega_0.\left(\xi^2 - 1\right)^{1/2} \end{cases}$$

The solution x(t) has, as an expression, in this case:

$$x(t) = e^{-\xi\omega_0 t}.\left(A.e^{\omega_0\left(\xi^2-1\right)^{1/2}t} + B.e^{-\omega_0\left(\xi^2-1\right)^{1/2}t}\right) \quad (1.32)$$

A and B depend on the IC. The shape of the curves is shown in the following figure:

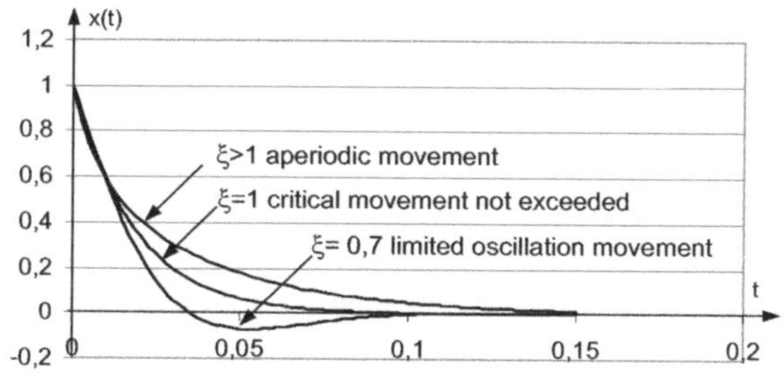

1.3.6. Forced oscillations of a damped system

1.3.6.1. *Equation of motion*

a) System in translation mass and spring

In this case, as previously done with the technique of releasing, the excitation is supposed to be applied to the mass, the frame being always infinitely rigid here as well. In the case where the latter is applied on the frame (case of equipment mounted on dampers), the development is carried out in Appendix B. The model here is shown in the following figure:

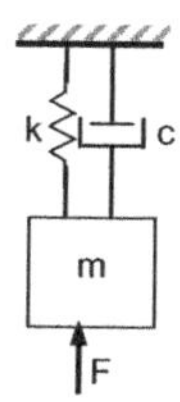

We have already seen that the equation of motion is given by the fundamental relation of dynamics, which means:

$$\sum F(t) = m.\ddot{x}(t)$$

NB: The movement of mass is different according to the point of application of the force on the mass or on the construct.

Let us suppose that we apply a sinusoidal force, which shall be referred to as:

$$F(t) = F.e^{j\omega t}$$

Knowing that only the real part of the force interests us, it thus becomes:

$$\sum F = F.e^{j\omega t} - k.x(t) - c.\dot{x}(t) = m.\ddot{x}(t)$$

Therefore :

$$\ddot{x}(t) + \frac{c}{m}.\dot{x}(t) + \frac{k}{m}.x(t) = \frac{F}{m}.e^{j\omega t}$$

b) System in rotation (torsion oscillations)

The model is as follows: the excitation being here a sinusoidal moment $M(t) = M.e^{j\omega t}$, from which we will only retain the real part.

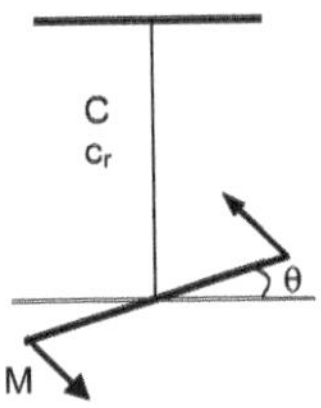

The equation of motion is obtained here by writing the fundamental relation of dynamics in rotation:

$$\sum M(t) = I.\ddot{\theta}(t)$$

So, if the system is excited by a sinusoidal moment M(t) = $M.e^{j\omega t}$, it becomes:

$$\sum M(t) = M.e^{j\omega t} - C.\theta(t) - c_r.\dot{\theta}(t) = I.\ddot{\theta}(t)$$

Of the form: $$\ddot{\theta}(t) + \frac{c_c}{I}.\dot{\theta}(t) + \frac{C}{I}.\theta(t) = \frac{M}{I}.e^{j\omega t}$$

c) RLC oscillating electric system

We can show that we obtain the same type of relation with an RLC electric circuit, which prepares for electromechanical analogies. The model is shown as follows:

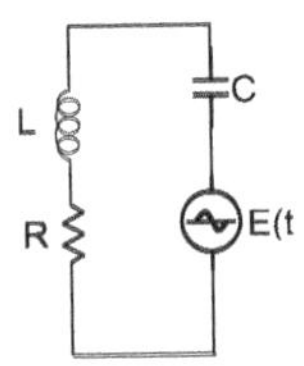

The relations of the circuit are the following ones with an excitation given here by e(t) = $E.e^{j\omega t}$:

-e(t) = Z.i(t)

- $Z = R + jL\omega + \dfrac{1}{jC\omega}$

Let us work, for example, with the loads q(t), and it becomes:

$$q(t) = C.v(t) \text{ with: } v(t) = Z_C.i(t) = \frac{1}{jC\omega} \rightarrow q(t) = \frac{1}{j\omega}.i(t)$$

So, by successive derivations, we obtain:

- $\dot{q}(t) = i(t)$
- $\ddot{q}(t) = j\omega i(t)$

From where the differential equation: $e(t) = R.i(t) + jL\omega.i(t) + \dfrac{1}{jC\omega}.i(t) = R.\dot{q}(t) + L.\ddot{q}(t) + \dfrac{q(t)}{C} = E.e^{j\omega t}$

Is of the type: $$\ddot{q}(t) + \frac{R}{L}.\dot{q}(t) + \frac{1}{LC}.q(t) = \frac{E}{L}.e^{j\omega t}$$

In all cases, we obtain a second-order differential relation of the *general form* as follows:

$$\ddot{x}(t) + 2\xi\omega_0.\dot{x}(t) + \omega_0^2.x(t) = A.e^{j\omega t} \tag{1.33}$$

where:

- ω_0 = natural frequency of the undamped system;
- ξ = reduced damping coefficient.

1.3.6.2. *Pulsation, amplitude and phase of the response*

The complete solution of the differential equation of the second order is the sum of the general solution of the equation without a second member, plus a particular solution of the equation with a second member. The solution without a second member is that of the free answer that we have previously evaluated, and which, in practice, disappears quickly. It corresponds to the transitional regime. We will only be interested in the forced permanent regime, thus the phenomenon after its phase of establishment. Let us consider for a given excitation frequency, the following type of response:

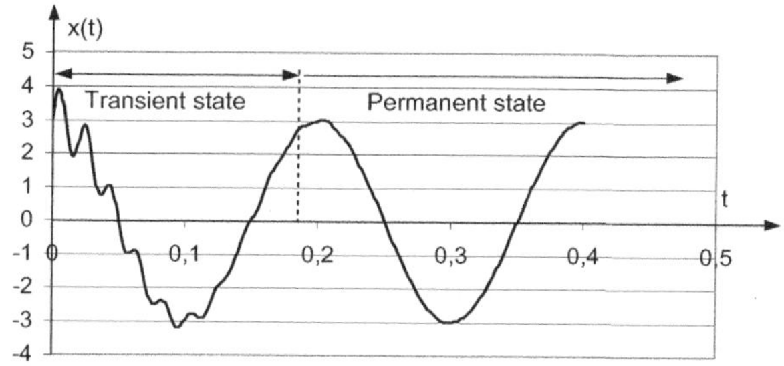

Let us look for solutions in the form of complex exponentials of the following traditional form:

$x(t) = X.e^{j(\omega t+j)} = \underline{X}.e^{j\omega t}$ with : $\underline{X} = X.e^{jj}$ complex number in amplitude and phase

Therefore :

$\dot{x}(t) = j\omega\underline{X}.e^{j\omega t}$ and $\ddot{x}(t) = -\omega^2\underline{X}.e^{j\omega t}$

With these values in the general equation, we obtain:

$$\ddot{x}(t) + 2\xi\omega_0.\dot{x}(t) + \omega_0^2.x(t) = A.e^{j\omega t} = -\omega^2\underline{X}.e^{j\omega t} + 2\xi\omega_0.j\omega.\underline{X}.e^{j\omega t} + \omega_0^2.\underline{X}.e^{j\omega t}$$

Having :
$$\underline{X} = \frac{A}{\omega_0^2 - \omega^2 + j2\xi\omega_0\omega} = X.e^{j\varphi}$$

The solution is therefore of the form $x(t) = \underline{X}.e^{j\omega t}$, which means that the response's pulsation equals to that of the excitation.

We can then study the characteristics of the response in several ways by:

- studying the amplitude, the most common phase;
- studying the imaginary part and real part;
- studying the Nyquist plot.

It should be noted that it is convenient to use standardized variables without dimensions in order to obtain networks universal curves valid in all the systems of variables (the parameter of the curves being the damping ξ). For the displacement, we will be interested in the amplitude and in the traditional phase, with:

- amplitude (module):

$$X = \frac{A}{\left\{\left(\omega_0^2 - \omega^2\right)^2 + (2\xi\omega_0\omega)^2\right\}^{1/2}} \tag{1.34}$$

- phase:

$$tg\varphi = \frac{-2\xi\omega_0\omega}{\omega_0^2 - \omega^2} \tag{1.35}$$

Thus, for a system and a given reduced damping ξ, these are functions of ω_0 and ω.

1.3.6.3. *Expressions in standardized variables*

Let us consider the static elongation that is obtained for $\omega = 0$ or $f = 0$, it becomes:

$x_{stat} = \underline{X}(\omega = 0) = \dfrac{A}{\omega_0^2}$; we can then express the module of displacement relative to the static elongation and in a standardized variable (ω/ω_0) with:

$$|X_N(\omega/\omega_0)| = \frac{X}{X_{stat}} = \frac{1}{\left\{\left[1-\left(\frac{\omega}{\omega_0}\right)^2\right]^2 + \left(2\xi\frac{\omega}{\omega_0}\right)^2\right\}^{1/2}} \quad (1.36)$$

We define a ratio without dimensions, which does not depend on the amplitude of the excitation, which is a function of the standardized variable (ω/ω_0) and is parameterized by the reduced damping ξ. It is thus a universal curve.

The research of the extremums, by canceling the derivative, gives its numerator:

$$\omega\left[\omega^2 - \omega_0^2(1-2\xi^2)\right] = 0$$

Having : $\omega = 0$ horizontal tangent and $\omega^2 = \omega_0^2(1-2\xi^2) \rightarrow \omega_r = \omega_0(1-2\xi^2)^{1/2}$

Having:

$$\begin{cases} X_N \max = Q = \dfrac{1}{2\xi(1-\xi^2)^{1/2}} \text{ amplitude resonance} \\ \text{for} \left(\dfrac{\omega}{\omega_0}\right) = (1-2\xi^2)^{1/2} \end{cases} \quad (1.37)$$

The maximum value of amplitude is called the Q-factor where we find the already mentioned fact that if the reduced damping factor ξ (and therefore damping natural c) is small before 1, it is practically equal to:

$$Q \approx \frac{1}{2\xi} \quad (1.38)$$

The universal networks of curves are the following: amplitude and phase:

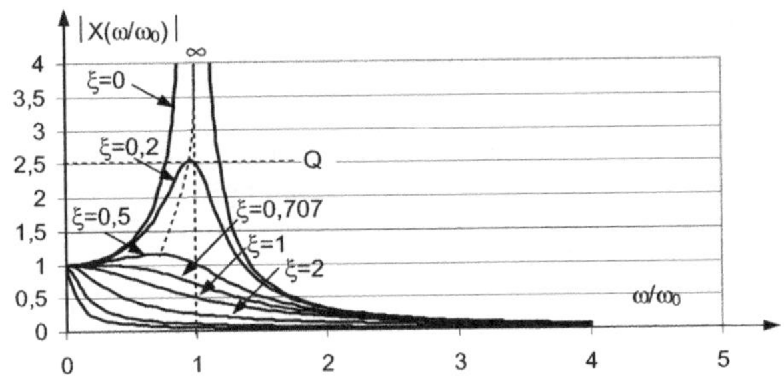

Linear coordinates

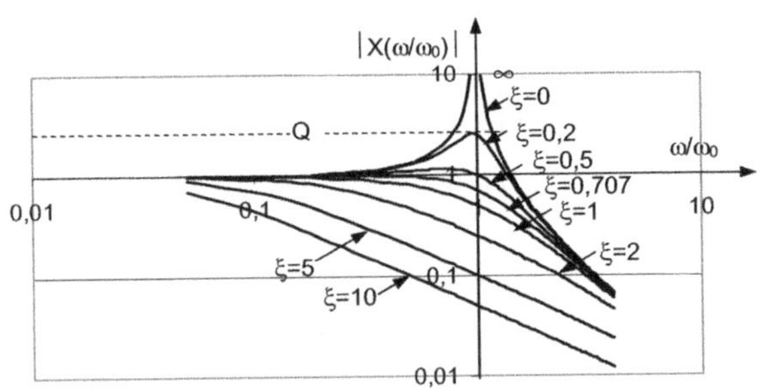

Logarithmic coordinates

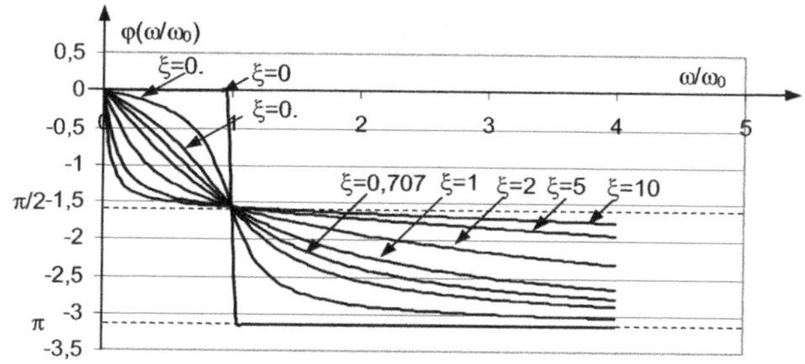

Linear coordinates

1.3.6.4. *Admittance circle of velocity*

With the same complex notations as previously, we write down V, the velocity module; therefore we have: $\dot{x} = V.e^{j(\omega t+\varphi)} = \underline{V}.e^{j\omega t}$ with: $\underline{V} = V.e^{j\varphi}$.

If we carry this value in the general equation (1.33), it becomes:

With: $\ddot{x}(t) = j\omega\underline{V}.e^{j\omega t}, \; x(t) = \dfrac{\underline{V}}{j\omega}.e^{j\omega t} = -j\dfrac{\underline{V}}{\omega}.e^{j\omega t}$

Having: $$j\omega\underline{V}.e^{j\omega t} + 2\xi\omega_0.\underline{V}.e^{j\omega t} - j\omega_0^2\frac{\underline{V}}{\omega}.e^{j\omega t} = A.e^{j\omega t} \quad (1.39)$$

And: $$\underline{V} = \frac{A.\omega}{2\xi\omega_0\omega + j(\omega^2 - \omega_0^2)} = V.e^{j\varphi}$$

This expression is used in the models of mechanical impedances for the analysis of the transfer mobility of systems, of which we will evaluate an approach in a following section. The following concepts are therefore defined:

- mechanical impedance, which is the ratio of the force over velocity with:

$$Z_m(\omega) = \frac{F(\omega)}{v(\omega)} = \frac{F.\sin\omega t}{V.\sin(\omega t + \varphi)}$$

- mechanical admittance still called mechanical mobility with:

$$Y_m(\omega) = \frac{v(\omega)}{F(\omega)}$$

The admittance velocity is thus given here by:

$\underline{Y}_v(\omega) = \dfrac{\underline{V}(\omega)}{F}$ with the excitation here, the equation $F(\omega) = m.A.e^{j\omega t}$

Therefore: $$\underline{Y}_v(\omega) = \frac{\underline{V}(\omega)}{m.A} = \frac{A.\omega}{m.A.[2\xi\omega_0\omega + j(\omega^2 - \omega_0^2)]} = \frac{\omega}{m.[2\xi\omega_0\omega + j(\omega^2 - \omega_0^2)]}$$

Let us consider the following ratio without dimensions:

$$Y_{NV}(\omega/\omega_0) = \underline{Y}_V(\omega/\omega_0).m.\omega_0 = m.\omega_0.\frac{(\omega/\omega_0)}{m.\omega_0.[2\xi(\omega/\omega_0) + j[(\omega^2/\omega_0^2 - 1)]]}$$

$$Y_{NV}(\omega/\omega_0) = \frac{(\omega/\omega_0)}{[2\xi(\omega/\omega_0) + j[(\omega^2/\omega_0^2 - 1)]]}$$

If we separate the real part and the imaginary part by reducing the same denominator, it is a complex number that is usually written in the following form:

$$Y_{NV}(\omega/\omega_0) = CO(\omega/\omega_0) + jQUAD(\omega/\omega_0)$$

With the real part in phase and the imaginary part in quadrature (+π/2 of +j):

- component in phase:

$$CO(\omega/\omega_0) = \frac{2\xi.(\omega/\omega_0)^2}{[2\xi(\omega/\omega_0)]^2 + \left[1-(\omega/\omega_0)^2\right]^2} \tag{1.40}$$

- component in quadrature:

$$QUAD(\omega/\omega_0) = \frac{\left[1-(\omega/\omega_0)^2\right](\omega/\omega_0)}{[2\xi(\omega/\omega_0)]^2 + \left[1-(\omega/\omega_0)^2\right]^2} \tag{1.41}$$

whose curves are shown in Figure 1.13.

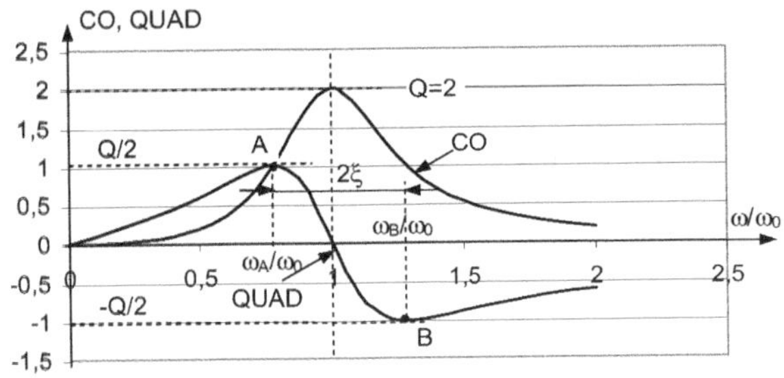

Figure 1.13. *Curves in phase and quadrature*

We can show that by eliminating (ω/ω_0) between two relations, both components are interconnected by the following relation:

$$CO^2 + QUAD^2 - Q.CO = 0 \tag{1.42}$$

which is the equation of a circle with center coordinates $\begin{cases} CO = \dfrac{Q}{2} \\ QUAD = 0 \end{cases}$, and of radius $\dfrac{QUAD}{2}$.

It is this circle that is called the admittance circle of velocity, which is a partial imaginary representation of the function based on the real part of the admittance (meaning in the complex plane) of admittance velocity $Y_{NV}(\omega/\omega_0)$.

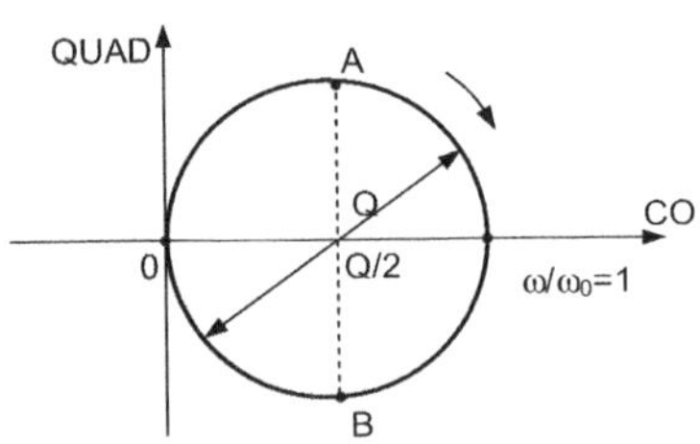

Figure 1.14. *Admittance circle of velocities*

We see that the maximum admittance velocity is therefore given here for (ω/ω_0) = 1, which means:

$$\begin{cases} Y_{NV}\max = Q \\ \text{for : } \quad \dfrac{\omega}{\omega_0} = 1 \end{cases} \tag{1.43}$$

1.3.6.5. *Bandwidth*

It is defined as usual by the frequency band comprised between the two points at half-power of the velocity response or at –3 dB of the module curve. These are the points A and B of the admittance circle. Then, consider $\Delta\omega$ or $\dfrac{\Delta\omega}{\omega_0}$ standardized and the following curve of module of $Y_{NV}(\omega/\omega_0)$, with (Q = 5 and ξ# 0,1):

$$\left|Y_{NV}(\omega/\omega_0)\right| = \frac{(\omega/\omega_0)}{\left[[2\xi(\omega/\omega_0)]^2 + [(\omega^2/\omega_0^2) - 1]^2\right]^{1/2}}$$

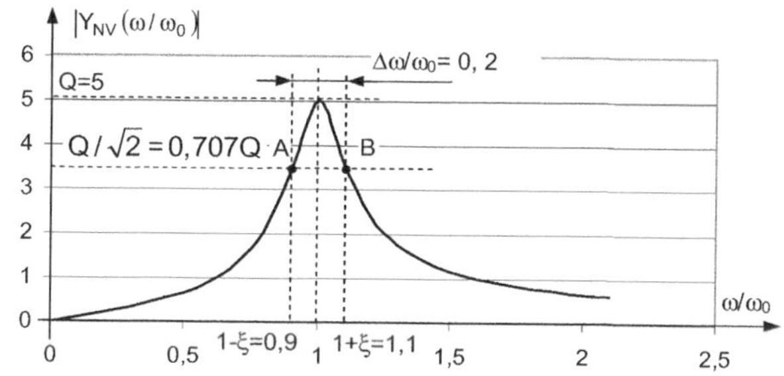

Figure 1.15. *Admittance velocity module*

We can note more easily on the curve with logarithmic coordinates that follows than the velocity curve, which is the integral of the curve of acceleration whose form is a high-pass slope of – 12 dB cut strip, as we will study it later on, present, on the other hand, in the shape of a bandpass. The high-pass slope took –6 dB around the frequency of resonance standardized to 1. So, we find on curve +6 dB before the resonance frequency and –6 dB after the resonance frequency (for displacement, an additional integration would give a low-pass slope with a slope of –12 dB band cut).

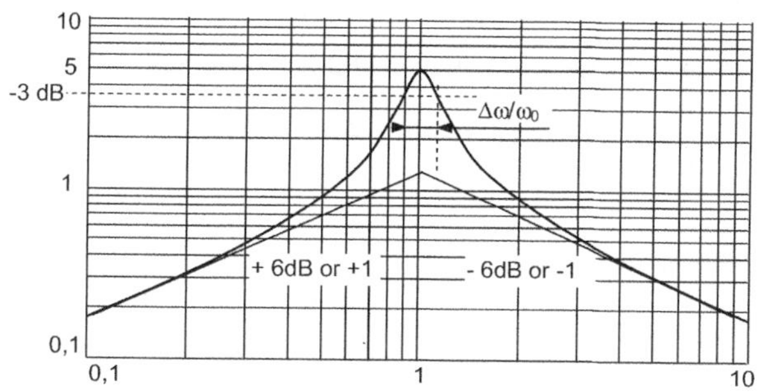

Frequencies ω_A and ω_B are then given by:

$$\left|Y_{NV}(\omega/\omega_0)\right| = \frac{Q}{\sqrt{2}}$$

$$\text{Having :} \qquad CO = \frac{Q}{2} \tag{1.44}$$

From where we can draw the values of the corresponding standardized pulsations:

$$(\omega/\omega_0)_{A,B} = 1 \pm \xi \tag{1.45}$$

Having: $$\frac{\Delta\omega}{\omega_0} = 1+\xi-(1+\xi) \quad \rightarrow \quad \frac{\Delta\omega}{\omega_0} = \frac{1}{Q} \approx 2\xi \tag{1.46}$$

The reduced damping coefficient can be calculated from points A and B, with:

$$2\xi = \frac{\omega_B - \omega_A}{\omega_0} \rightarrow \qquad \xi = \frac{1}{2}\frac{\omega_B - \omega_A}{\omega_0} = \frac{1}{2}\frac{\Delta\omega}{\omega_0} \tag{1.47}$$

We find that the result obtained in (1.43) is also valid for the amplitude

if ξ <;<; 1. Having: $X_N \max = \frac{1}{2\xi\left(1-\xi^2\right)^{1/2}} \approx \frac{1}{2\xi} = Q$. So we have for the bandwidth at –3 dB:

$$\frac{X_{N\max}}{\sqrt{2}} \approx \frac{1}{2\xi\sqrt{2}} = \frac{1}{\left\{\left[1-\omega_N^2\right]^2 + \left(2\xi\omega_N\right)^2\right\}^{1/2}} \qquad \text{with}: \frac{\omega}{\omega_0} = \omega_N$$

$$\text{Having}: \omega_N^4 - \omega_N^2\left(4\xi^2 - 2\right) + \left(1 - 8\xi^2\right) = 0 \rightarrow X^2 - X\left(4\xi^2 - 2\right) + \left(1 - 8\xi^2\right) = 0 \ \text{ with } \omega_N^2 = X$$

$$X = \frac{-4\xi^2 + 2 \pm \sqrt{16\xi^4 + 16\xi^2}}{2} \approx \frac{-4\xi^2 + 2 \pm 4\xi}{2} \approx 1 \pm 2\xi$$

$$X^{1/2} = \omega_N = (1 \pm 2\xi)^{1/2} \approx 1 \pm \frac{1}{2}2\xi = 1 \pm \xi$$

1.3.6.6. *Acceleration admittance*

With the same approach, the admittance acceleration is calculated with the following definitions (see chapter on the concept of mechanical impedances).

We express the variables of the general equation in function of the acceleration with:

$$\ddot{x}(t) = \Gamma.e^{j(\omega t + j)} = \underline{\Gamma}.e^{j\omega t} \qquad \text{with}: \underline{\Gamma} = \Gamma.e^{j\varphi}$$

$$\dot{x}(t) = \frac{\underline{\Gamma}}{j\omega}.e^{j\omega t} \qquad x(t) = -\frac{\underline{\Gamma}}{\omega^2}.e^{j\omega t}$$

Inertance is defined by: $I = \left(\frac{\text{Force}}{\text{Acceleration}}\right)^{-1} \rightarrow \underline{I} = \frac{\underline{\Gamma}}{Am} \rightarrow$ admittance in acceleration.

If we carry this value in the general equation (1.32), it becomes:

$$\underline{\Gamma}.e^{j\omega t} + \frac{\underline{\Gamma}}{j\omega}.e^{j\omega t}.2\xi\omega_0 - \omega_0^2\frac{\underline{\Gamma}}{\omega^2}.e^{j\omega t} = A.e^{j\omega t} \rightarrow \underline{\Gamma} = \frac{A}{\left(1 - \frac{\omega_0^2}{\omega^2}\right) - j2\xi\frac{\omega_0}{\omega}}$$

Admittance in acceleration is given by:

$$Y_{NA}(\omega/\omega_0) = \frac{\underline{\Gamma}}{Am} = \frac{1/m}{\left(1 - \frac{\omega_0^2}{\omega^2}\right) - j2\xi\frac{\omega_0}{\omega}} \tag{1.48}$$

It is then noted that this expression is of course connected to that obtained in displacement through the square of the pulsation (double derivation) with:

$$\text{Compliance} = C = \left(\frac{\text{Force}}{\text{Displacement}}\right)^{-1} = \frac{\text{Displacement}}{\text{Force}}$$

$$C = \text{admittance in displacement} = Y_{ND}\left(\frac{\omega}{\omega_0}\right)$$

Indeed: $$Y_{ND}\left(\frac{\omega}{\omega_0}\right)=\frac{X}{Am}=\frac{1}{Am}\frac{1}{\omega_0^2}\frac{A}{\left(1-\frac{\omega^2}{\omega_0^2}\right)+j2\xi\frac{\omega}{\omega_0}}=\frac{1}{\omega_0^2}\frac{1/m}{\left(1-\frac{\omega^2}{\omega_0^2}\right)+j2\xi\frac{\omega}{\omega_0}}$$

So: $$Y_{NA}\left(\frac{\omega}{\omega_0}\right)=\frac{\omega^2}{\omega_0^2}\frac{1/m}{\left(\frac{\omega^2}{\omega_0^2}-1\right)-j2\xi\frac{\omega}{\omega_0}}=-\omega^2.Y_{ND}\left(\frac{\omega}{\omega_0}\right)$$

The module of the ratio without dimension standardized with 1 is given by:

So: $$Y_{NA}\left(\frac{\omega}{\omega_0}\right).m=\frac{1}{\left(1-\frac{\omega_0^2}{\omega^2}\right)-j2\xi\frac{\omega_0}{\omega}}\rightarrow\left|Y_{NA}\left(\frac{\omega}{\omega_0}\right).m\right|=\frac{1}{\left[\left(1-\frac{\omega_0^2}{\omega^2}\right)^2+\left(2\xi\frac{\omega_0}{\omega}\right)^2\right]^{1/2}}$$

With: $\left|Y_{NA}\left(\frac{\omega}{\omega_0}\right).m\right|=\left|Y_{ND}\left(\frac{\omega}{\omega_0}\right).m\right|.\left(\frac{\omega_0}{\omega}\right)^2$, the sign not intervening for the module.

Consider the network of curves shown in Figure 1.16.

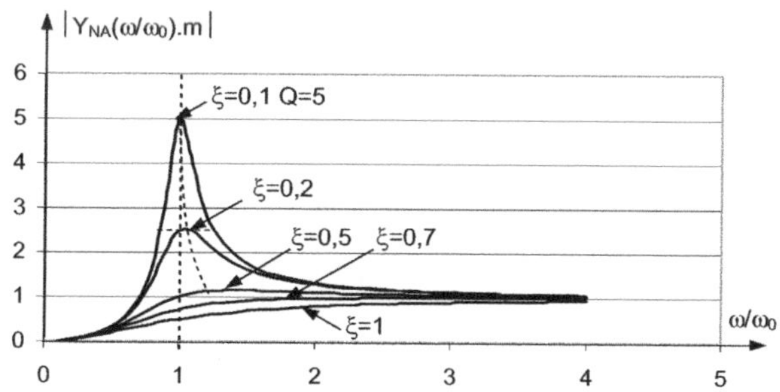

Figure 1.16. *Acceleration admittance module*

As expected, we note the high-pass form of acceleration admittance, which, compared to the band pass of the velocity admittance (from which it derives), takes +6 dB around the frequency of resonance standardized to 1.

NB: Be careful not to lose sight that the excitation is on the mass. The remarks are different if the excitement is on the frame.

We also show here that:

- the maximum value $Y_{NA}\,\max=\frac{1}{2\xi\left(1-\xi^2\right)}=Q\approx\frac{1}{2\xi}$, for here, $\frac{\omega}{\omega_0}=\frac{1}{\left(1-2\xi^2\right)^{1/2}}$;

- so we always have $Y_{NA}\max=Q$, and therefore the bandwidth is given by:

$$\frac{\Delta\omega}{\omega_0}=\frac{1}{Q}$$

1.3.6.7. *Transmissibility*

We talk here about transmissibility of the force conveyed by the system over the frame (characteristic approach of the study of insulations of led generating vibration machinery). Then, we will see, in the following, the notion of transmissibility of the frame into the system, which is characteristic of the study of insulation with respect to external attacks.

The transmitted power is given here by: $F_T = c.\dot{x} + k.x$

Or in a complex notation with:

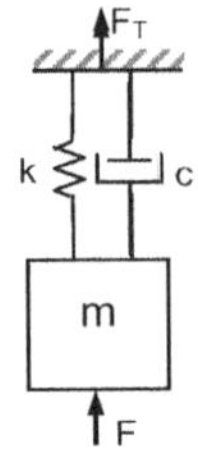

$c = 2\xi(km)^{1/2} \quad \omega_0^2 = \dfrac{k}{m} \qquad \dot{x} = j\omega\underline{X}.e^{j\omega t} \quad x = \underline{X}.e^{j\omega t}$

$\underline{F}_T.e^{j\omega t} = j2\xi(km)^{1/2}\omega\underline{X}.e^{j\omega t} + k\underline{X}.e^{j\omega t}$

$\underline{F}_T = j2\xi(\omega_0^2 m^2)^{1/2}\omega\underline{X} + m\omega_0^2\underline{X} \rightarrow \underline{F}_T = (j2\xi\omega_0.\omega + \omega_0^2).m\underline{X}$

We have seen that the excitation force is given by:

$\underline{F} = \underline{A}.m = (\omega_0^2 - \omega^2 + j2\xi\omega_0\omega).m\underline{X}$

Therefore:
$$\underline{T} = \frac{\underline{F}_T}{\underline{F}} = \frac{1 + j2\xi\dfrac{\omega}{\omega_0}}{1 - \left(\dfrac{\omega}{\omega_0}\right)^2 + j2\xi\dfrac{\omega}{\omega_0}} \qquad (1.49)$$

and the ratio of the transmitted force to the excitation force is the coefficient (or factor) of transmissibility.

We are mainly interested in its amplitude (in module), with:

$$T = \frac{\left\{1 + \left(2\xi\dfrac{\omega}{\omega_0}\right)^2\right\}^{1/2}}{\left\{\left[1 - \left(\dfrac{\omega}{\omega_0}\right)^2\right]^2 + \left(2\xi\dfrac{\omega}{\omega_0}\right)^2\right\}^{1/2}} \qquad (1.50)$$

from where the traditional network of curves is derived.

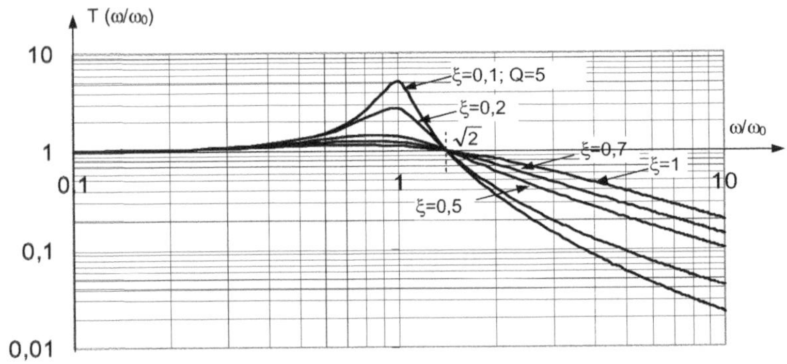

Figure 1.17. *Coefficient of transmissibility module*

1.3.7. Systems with several coupled DOF

1.3.7.1. *Free oscillations of an undamped system*

1.3.7.1.1. *Equation of motion*

The example chosen is a system with 2 DOF, knowing that by definition the number of DOF is the number of position variables of the system. The model is shown in the following figure; it does not include dampening in order to simplify the approach (it is said that we then obtain the equation of the associated conservative system).

The position variables x_1 and x_2 are said to be coupled because the motion of the mass m_1 causes a known force on mass m_2 and vice versa. Indeed, if we move mass m_1 a quantity x_1, mass m_2 moves to a quantity x_2 under the effect of the force equal to $k_{12}.(x_1 - x_2)$.

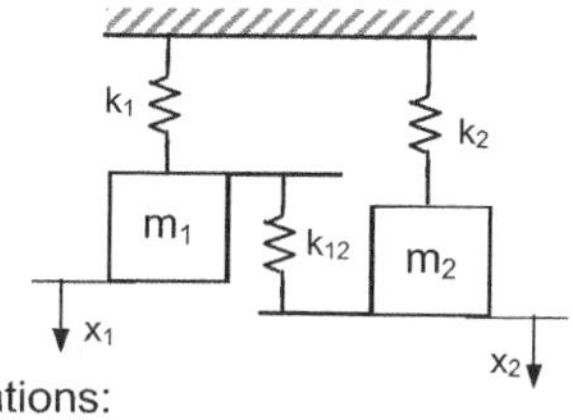

It is enough to introduce this additional force into the equation of motion of mass m we have already determined, which means:

$$-k_1x_1 - k_{12}(x_1 - x_2) = m_1\ddot{x}_1$$

Likewise: $$-k_2x_2 - k_{12}(x_2 - x_1) = m_2\ddot{x}_2$$

The movement of the system is thus defined by the two equations:

$$\begin{cases} \ddot{x}_1 + \dfrac{k_1 + k_{12}}{m_1} x_1 = \dfrac{k_{12}}{m_1} x_2 \\ \ddot{x}_2 + \dfrac{k_2 + k_{12}}{m_2} x_2 = \dfrac{k_{12}}{m_2} x_1 \end{cases} \quad (1.51)$$

1.3.7.1.2. *Natural frequencies*

The determination of natural frequencies of the complete second-order system depends on the knowledge of the natural frequencies of both systems of the first order m_1 and m_2, taken separately, are simply here when one of both is blocked; therefore, we calculate:

- natural frequency of m_1 when m_2 is blocked: $\omega_1^2 = \dfrac{k_1 + k_{12}}{m_1}$;

- natural frequency of m_2 when m_1 is blocked: $\omega_2^2 = \dfrac{k_2 + k_{12}}{m_2}$.

By substituting $\alpha_1 = \dfrac{k_{12}}{m_1}$ and $\alpha_2 = \dfrac{k_{12}}{m_2}$, the equations of motion become:

$$\begin{cases} \ddot{x}_1 + \omega_1^2.x_1 = \alpha_1.x_2 \\ \ddot{x}_2 + \omega_2^2.x_2 = \alpha_2.x_1 \end{cases} \quad (1.52)$$

As before, we will seek solutions in a complex form as follows:

$$\begin{cases} x_1 = X_1.e^{j(\omega t + \varphi_1)} = \underline{X}_1.e^{j\omega t} \text{ with: } \underline{X}_1 = X_1.e^{j\varphi_1} \\ x_2 = X_2.e^{j(\omega t + \varphi_2)} = \underline{X}_2.e^{j\omega t} \text{ with: } \underline{X}_2 = X_2.e^{j\varphi_2} \end{cases}$$

Equation (1.50) gives:

$$\left(\omega_1^2 - \omega^2\right)\underline{X}_1.e^{j\omega t} = \alpha_1.x_2 \rightarrow x_2 = \frac{\left(\omega_1^2 - \omega^2\right)}{\alpha_1}.\underline{X}_1.e^{j\omega t} = \underline{X}_2.e^{j\omega t} \rightarrow \frac{\left(\omega_1^2 - \omega^2\right)}{\alpha_1}.X_1.e^{j\varphi_1} = X_2.e^{j\varphi_2}$$

Having:

$$\begin{cases} \text{In module: } & \dfrac{\left(\omega_1^2 - \omega^2\right)}{\alpha_1}.X_1 = X_2 \\ \text{In phase: } & e^{j\varphi_1} = \pm e^{j\varphi_2} \rightarrow \varphi_1 = \varphi_2 + k\pi \end{cases} \quad (1.53)$$

This result is important because it shows that at a natural frequency, all the points of the system vibrate in phase (near π).

The natural frequencies are then obtained by solving the system in ω according to:

$$\begin{cases} (\omega_1^2 - \omega^2).\underline{X}_1 = \alpha_1.\underline{X}_2 \\ (\omega_2^2 - \omega^2).\underline{X}_2 = \alpha_2.\underline{X}_1 \end{cases} \tag{1.54a}$$

It is by replacing $\underline{X}_{1,2}$ by their expressions, that it is suddenly solved:

$$(\omega_1^2 - \omega^2)(\omega_2^2 - \omega^2) = \alpha_2\alpha_1 = \frac{k_{12}^2}{(k_1 + k_{12})(k_2 + k_{12})}\omega_1^2\omega_2^2 = K^2.\omega_1^2\omega_2^2$$

$$\text{With:} \quad K^2 = \frac{k_{12}^2}{(k_1 + k_{12})(k_2 + k_{12})} \tag{1.54b}$$

which can be interpreted like a coupling coefficient between two masses m_1 and m_2.

The expression obtained in ω does not contain anything else but parameters in function of pulsations and stiffness. The solution in ω of this equation is long and tedious. Without changing the general nature of the result, we can simplify this resolution by taking the special case where the pulsations of both systems with 1 DOF are equal, which means:

$$\omega_1 = \omega_2 = \omega_0$$

$$\text{So:} \quad (\omega_0^2 - \omega^2)^2 = K^2.\omega_0^4 \rightarrow (\omega_0^2 - \omega^2) = \pm K.\omega_0^2$$

The equation with its own pulsations is: $\omega = \omega_0.\sqrt{1 \pm K}$

The system has two own pulsations and two natural frequencies, given by:

$$\begin{cases} \omega' = \omega_0.\sqrt{1-K} \\ \omega'' = \omega_0.\sqrt{1+K} \end{cases} \tag{1.55}$$

These pulsations are therefore as follows: $\omega' < \omega_0 < \omega''$

They are thus all the more distant as the coefficient of coupling K is large.

1.3.7.1.3. *Analysis of movements with normal modes – concept of normal modes*

This analysis allows making a few important notes about the behavior of coupled modes.

- With the pulsation ω', we have: $(\omega_0^2 - \omega'^2) = +K.\omega_0^2$

And:

$$x_2 = \frac{(\omega_0^2 - \omega'^2)}{\alpha_1}.x_1 = +\frac{K.\omega_0^2}{\alpha_1}.x_1$$

The movements of the two masses are in phase.

- With the pulsation ω'', we have: $(\omega_0^2 - \omega''^2) = -K.\omega_0^2$

And:

$$x_2 = \frac{(\omega_0^2 - \omega''^2)}{\alpha_1}.x_1 = -\frac{K.\omega_0^2}{\alpha_1}.x_1$$

The movements of the two masses are in antiphase.

- In conclusion, the natural modes are particular free movements, for which all the points of the system vibrate sinusoidally at the same frequency, in phase or antiphase. In our case, these are real modes (because there is no dampening by assumption, otherwise we would have to deal with complex modes).

- The number of normal modes is also equal to the number of DOF of the system.

A simple example is given by the behavior of two identical simple pendulums, coupled in the following free mode:

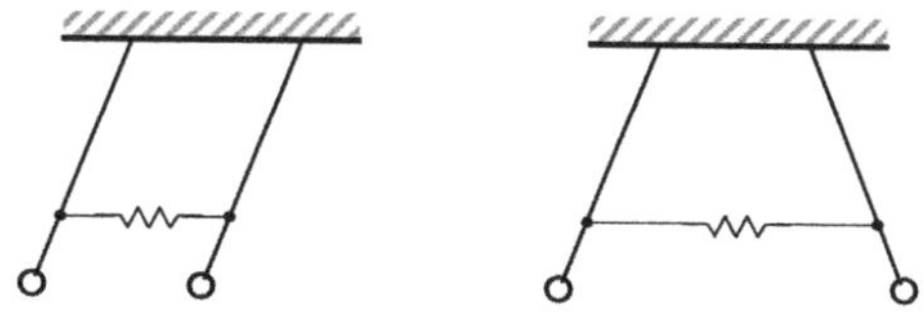

First mode in phase

Second mode in antiphase

1.3.7.1.4. *Vibrations of systems with many coupled DOF*

We saw that our simplified system ($\omega_1 = \omega_2 = \omega_0$) with two coupled DOF has two solutions of two normal modes given by ω' and ω''. We calculated these two solutions, which are as follows:

- for the first mode, we have: $\begin{cases} x_1 = \underline{X}'.e^{j\omega' t} \\ x_1 = \underline{X}''.e^{j\omega'' t} \end{cases}$

- for the second mode, we have: $\begin{cases} x_2 = \underline{X}'.e^{j\omega' t} \\ x_2 = -\underline{X}''.e^{j\omega'' t} \text{ (in antiphase)} \end{cases}$

The differential system governing the movement is linear, a general solution can thus be obtained by linear combination of these two solutions, which means that for the movements of the two modes or of the two masses m_1 and m_2, we have the simple solution:

$$\begin{cases} x_1 = \underline{X}'.e^{j\omega' t} + \underline{X}''.e^{j\omega'' t} \\ x_2 = C.\underline{X}'.e^{j\omega' t} - C.\underline{X}''.e^{j\omega'' t} \end{cases} \quad \text{with : } \begin{cases} \underline{X}' = X'.e^{j\varphi'} \\ \underline{X}'' = X''.e^{j\varphi''} \end{cases}$$

The terms $\underline{X}'$ and $\underline{X}''$ are a function of the IC that we can arbitrarily choose or are a function of a given situation of the system.

The real parts give the following system of equations:

$$\begin{cases} x_1 = X'.\cos(\omega' t + \varphi')X'' + \cos(\omega'' t + \varphi'') \\ x_2 = X'.\cos(\omega' t + \varphi') - X''.\cos(\omega'' t + \varphi'') \end{cases}$$

NB: It should be noted that in a general case, we can put these equations in a more general vectorial form (and matrix) to obtain the following forms:

$$\begin{bmatrix} x_1 \\ x_2 \end{bmatrix} = X'.\begin{bmatrix} 1 \\ 1 \end{bmatrix}.\cos(\omega' t + \varphi') + X''.\begin{bmatrix} 1 \\ -1 \end{bmatrix}.\cos(\omega'' t + \varphi'')$$

Let us suppose that we have the IC as follows:

$$x_1(0) = X_0 \qquad \dot{x}_1(0) = 0$$
$$x_2(0) = 0 \qquad \dot{x}_2(0) = 0$$

So, we then obtain the system of equations utilizing the following IC:

$$\begin{cases} X'\cos\varphi' + X''\cos\varphi'' = X_0 \\ X'\cos\varphi' - X''\cos\varphi'' = 0 \\ \omega' X'\sin\varphi' + \omega'' X''\sin\varphi'' = 0 \\ \omega' X'\sin\varphi' - \omega'' X''\sin\varphi'' = 0 \end{cases}$$

It is a system of four equations with four unknown factors. Knowing *a priori* that the parameters ω', ω", X', and X" are different from zero, the sum and the difference of the two last equations show that φ' = φ" = 0. The sum and the difference of the two first equations show that X' = X "= $X_0$2. Having:

$$\varphi' = \varphi'' = 0 \qquad\qquad X' = X'' = \frac{X_0}{2}$$

The solution will be:

$$\begin{cases} x_1 = \dfrac{X_0}{2}(\cos\omega' t + \cos\omega'' t) \\ x_2 = \dfrac{X_0}{2}(\cos\omega' t - \cos\omega'' t) \end{cases} \qquad (1.56)$$

In conclusion, the movements of a system with several coupled DOF can be considered as the superposition of movements at their natural frequencies. They are also called forms associated with the modes.

Amplitudes and phases are thus determined by the IC.

1.3.7.1.5. *Sympathetic pendulum – beats*

The simple example of the two pendulums coupled in a free mode, having neighboring frequencies, makes it possible to introduce the concept of a beat. By assumption, we also ask that at the initial moment t = 0, the two pendulums are released without an initial speed, with the following elongations:

$$\begin{cases} x_1 = X_m \\ x_2 = 0 \end{cases}$$

The equations then give to the search of the terms $\underline{X}'$ and $\underline{X}''$ by using the assumption of zero initial velocity:

$$\begin{cases} \dot{x}_1 = j\omega'\underline{X}'e^{j\omega' t} + j\omega''\underline{X}''e^{j\omega'' t} \\ \dot{x}_2 = j\omega' C\underline{X}'e^{j\omega' t} - j\omega'' C\underline{X}''e^{j\omega'' t} \end{cases} \qquad \text{So, for } t = 0, \dot{x}_1 = \dot{x}_2 = 0, \text{we have :}$$

$$\begin{cases} \omega'\underline{X}' + \omega''\underline{X}'' = \omega' X' e^{j\varphi'} + \omega'' X'' e^{j\varphi''} = 0 \\ \omega' C\underline{X}' - \omega'' C\underline{X}'' = \omega' CX' e^{j\varphi'} - \omega'' CX'' e^{j\varphi''} = 0 \end{cases}$$

The resolution of this system gives:

$$(\omega' X' + \omega' CX')e^{j\varphi'} \pm (\omega'' X'' + \omega'' CX'')e^{j\varphi''} = 0$$

$$(\omega' X' + \omega' CX')\cos\varphi' + (\omega'' X'' + \omega'' CX'')\cos\varphi'' \pm j.[(\omega' X' + \omega' CX')\sin\varphi' + (\omega'' X'' + \omega'' CX'')\sin\varphi''] = 0$$

The imaginary part must be zero and as *a priori* ω', X', C and ω", X" are different from zero, we can deduce that the only solution is: φ' = φ" = 0

At t = 0, we therefore find that:

$$\begin{cases} x_1(0) = X' + X'' = X_m \\ x_2(0) = X' - X'' = 0 \end{cases} \quad \rightarrow \quad X' = X'' = \frac{X_m}{2}$$

The equations of motion are thus written as:

$$\begin{cases} x_1(t) = \dfrac{X_m}{2}\left(e^{j\omega' t} + e^{j\omega'' t}\right) \\ x_2(t) = \dfrac{X_m}{2}\left(e^{j\omega' t} - e^{j\omega'' t}\right) \end{cases},$$

$$\left(e^{j\omega' t}+e^{j\omega'' t}\right)=e^{j\omega' t/2}.e^{j\omega'' t/2}\left(e^{j\omega' t/2}.e^{-j\omega'' t/2}+e^{-j\omega' t/2}.e^{j\omega'' t/2}\right)$$

$$\left(e^{j\omega' t}+e^{j\omega'' t}\right)=e^{j(\omega'+\omega'')t/2}.2.\cos\left(\frac{\omega'-\omega''}{2}\right)t$$

Yet:

$$\left(e^{j\omega' t}-e^{j\omega'' t}\right)=e^{j\omega' t/2}.e^{j\omega'' t/2}\left(e^{j\omega' t/2}.e^{-j\omega'' t/2}-e^{-j\omega' t/2}.e^{j\omega'' t/2}\right)$$

$$\left(e^{j\omega' t}-e^{j\omega'' t}\right)=je^{j(\omega'+\omega'')t/2}.2.\sin\left(\frac{\omega'-\omega''}{2}\right)t=e^{j\pi/2}e^{j(\omega'+\omega'')t/2}.2.\sin\left(\frac{\omega'-\omega''}{2}\right)t$$

Thus, by keeping the real part:

$$\begin{cases} x_1(t)=X_m.\cos\left(\dfrac{\omega'+\omega''}{2}.t\right).\cos\left(\dfrac{\omega'-\omega''}{2}.t\right) \\ x_2(t)=X_m.\sin\left(\dfrac{\omega'+\omega''}{2}.t\right).\sin\left(\dfrac{\omega'-\omega''}{2}.t\right) \end{cases} \quad (1.57)$$

We have, in addition, the following relations see 1.54a:

$$\omega''=\omega_0.\sqrt{1+K}\approx\omega_0.\left(1+\frac{K}{2}\right) \quad \text{K is small before 1 by assumption}$$

$$\omega'=\omega_0.\left(1-\frac{K}{2}\right)$$

Having:

$$\begin{cases} \dfrac{\omega''+\omega'}{2}\approx\omega_0 \\ \dfrac{\omega''-\omega'}{2}\approx\dfrac{K}{2}.\omega_0 << \omega_0 \end{cases} \quad (1.58)$$

The motions of the two pendulums are given by the following equations:

$$\begin{cases} x_1(t)=X_m.\cos\dfrac{K\omega_0}{2}t.\cos\omega_0 t\approx A_1.\cos\omega_0 t \\ x_1(t)=X_m.\sin\dfrac{K\omega_0}{2}t.\sin\omega_0 t\approx A_1.\sin\omega_0 t \end{cases} \quad (1.59)$$

Therefore, beats are observed and the oscillations of the two modes (or pendulums) are out of phase by $\pi/2$, taking place at a pulsation ω or frequency $f_0 = \omega_0/2\pi$. Their amplitudes are slowly modulated with pulsation $\omega_m = K\omega_0/2$ or at frequency $f_m = K\omega_0/4\pi$. When one is maximum, the other one is minimum and vice versa.

So there is a transfer of energy of one pendulum to the other.

1.3.7.2. *Free oscillations with a 2 DOF damped system*

The model is the following:

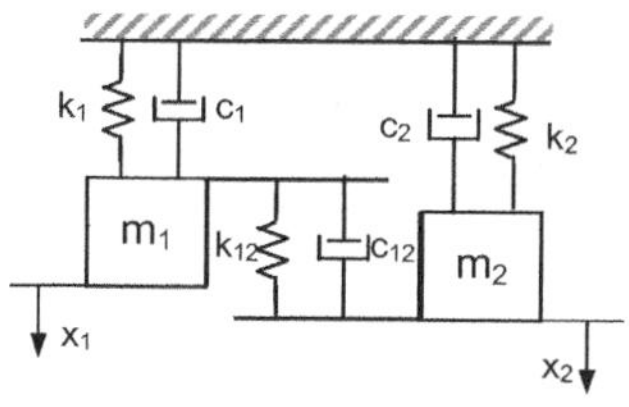

Relation (1.49) is therefore enough to add forces due to dampening (presumably viscous), which means:

$$\begin{cases} -k_1x_1-k_{12}(x_1-x_2)-c_1\dot{x}_1-c_{12}(\dot{x}_1-\dot{x}_2)=m_1\ddot{x}_1 \\ -k_2x_2-k_{12}(x_2-x_1)-c_2\dot{x}_2-c_{12}(\dot{x}_2-\dot{x}_1)=m_2\ddot{x}_2 \end{cases}$$

The movement of the system is thus defined by these two equations:

$$\begin{cases} \ddot{x}_1 + \dfrac{c_1 + c_{12}}{m_1}\dot{x}_1 + \dfrac{k_1 + k_{12}}{m_1}x_1 = \dfrac{k_{12}}{m_1}x_2 + \dfrac{c_{12}}{m_1}\dot{x}_2 \\ \ddot{x}_2 + \dfrac{c_1 + c_{12}}{m_2}\dot{x}_2 + \dfrac{k_1 + k_{12}}{m_2}x_2 = \dfrac{k_{12}}{m_2}x_1 + \dfrac{c_{12}}{m_2}\dot{x}_1 \end{cases}$$

Without affecting the generality of the results and conclusions, and to simplify the calculations, we will have:

$$\begin{cases} k_1 = k_2 = k_{12} = k \\ c_1 = c_2 = c_{12} = c \\ m_1 = m_2 = m \end{cases}$$

Obtaining:

$$\begin{cases} \ddot{x}_1 + \dfrac{2c}{m}\dot{x}_1 + \dfrac{2k}{m}x_1 = \dfrac{k}{m}x_2 + \dfrac{c}{m}\dot{x}_2 \rightarrow m\ddot{x}_1 + 2c\dot{x}_1 + 2kx_1 = kx_2 + c\dot{x}_2 \\ \ddot{x}_2 + \dfrac{2c}{m}\dot{x}_2 + \dfrac{2k}{m}x_2 = \dfrac{k}{m}x_1 + \dfrac{c}{m}\dot{x}_1 \rightarrow m\ddot{x}_2 + 2c\dot{x}_2 + 2kx_2 = kx_1 + c\dot{x}_1 \end{cases} \quad (1.60)$$

a) If we study the undamped system

We need to find:

$$\begin{cases} m\ddot{x}_1 + 2kx_1 = kx_2 \\ m\ddot{x}_2 + 2kx_2 = kx_1 \end{cases}$$

We will use here a different notation and approach by asking for the solution of the modes in the following form:

$$x_{1,2} = X_{1,2}.\cos(\omega t + \theta)$$

which becomes: $\begin{cases} m\ddot{x}_1 + 2kx_1 = kx_2 \rightarrow (2k - m\omega^2)X_1 - kX_2 = 0 \\ m\ddot{x}_2 + 2kx_2 = kx_1 \rightarrow (2k - m\omega^2)X_2 - kX_1 = 0 \end{cases}$

The frequency equation giving the specific values is therefore:

$$(2k - m\omega^2)^2 - k^2 = 0$$

We find the previous pulsations of the undamped system with here:

$$\omega_1^2 = \frac{k}{m} \qquad \text{and} \qquad \omega_2^2 = \frac{3k}{m}$$

This is another way of expressing the previous results. The same values are found knowing that we were asked to simplify: $\omega_0 = \sqrt{\dfrac{2k}{m}}$ and that thus with K = 1/2:

$$\begin{cases} \omega' \equiv \omega_1 = \omega_0.\sqrt{1-K} = \sqrt{\dfrac{2k}{m}.\dfrac{1}{2}} = \sqrt{\dfrac{k}{m}} \\ \omega'' \equiv \omega_2 = \omega_0.\sqrt{1+K} = \sqrt{\dfrac{2k}{m}.\dfrac{3}{2}} = \sqrt{\dfrac{3k}{m}} \end{cases} \quad \text{c.q.f.d}$$

The general solution of free vibrations will thus be of the following form, with the same choice of the forms:

- natural mode form 1: both modes have the same amplitude and move in phase;

- natural mode form 2: the two modes have the same amplitude and move in antiphase.

Having: $\begin{cases} x_1 = \lambda.\cos(\omega_1 t + \theta_1) + \mu.\cos(\omega_2 t + \theta_2) \\ x_2 = \lambda.\cos(\omega_1 t + \theta_1) - \mu.\cos(\omega_2 t + \theta_2) \end{cases}$

where λ, μ, θ_1, θ_2 are constants depending on the IC. With the same ones as before, which means:

$\begin{cases} x_1(0) = X_0 & ; & \dot{x}_1(0) = 0 \\ x_2(0) = 0 & ; & \dot{x}(0) = 0 \end{cases}$, the following system is obtained:

$$\begin{cases} \lambda\cos\theta_1 + \mu\cos\theta_2 = X_0 \\ \lambda\cos\theta_1 - \mu\cos\theta_2 = 0 \\ \omega_1\lambda\sin\theta_1 + \omega_2\mu\sin\theta_2 = 0 \\ \omega_1\lambda\sin\theta_1 - \omega_2\mu\cos\theta_2 = 0 \end{cases} \rightarrow \theta_1 = \theta_2 = 0 \quad \text{and} \quad \lambda = \mu = \frac{X_0}{2}$$

Consider the same values as in the preceding method and previous equation (equation 1.56). And we get the same solution: $\begin{cases} x_1(t) = \dfrac{X_0}{2}.\cos\omega_1 t + \dfrac{X_0}{2}.\cos\omega_2 t \\ x_2(t) = \dfrac{X_0}{2}.\cos\omega_1 t - \dfrac{X_0}{2}.\cos\omega_2 \end{cases}$

b) Let us study the damped-free system

So, we need to solve the following system: $\begin{cases} m\ddot{x}_1 + 2c\dot{x}_1 + 2kx_1 = kx_2 + c\dot{x}_2 \\ m\ddot{x}_2 + 2c\dot{x}_2 + 2kx_2 = kx_1 + c\dot{x}_1 \end{cases}$. This type of coupled system is generally difficult to solve directly (sometimes even impossible). The easy way consists of using "generalized" coordinates, not introduced in this volume, which gives here a change in variables by using the following linear relations ((1.61) resulting from the matrix calculation, which uses the matrix model and connects the coordinates in x to the generalized coordinates). This allows working with "independent" variables artificially removing the effect of coupling in differential relationships:

$$\begin{cases} x_1 = \delta_1 + \delta_2 \\ x_2 = \delta_1 - \delta_2 \end{cases} \tag{1.61}$$

The system is then simplified (we replace x_i by their values and perform the sum and the difference of the two modified equations) and it has suddenly been solved with the new coordinates:

$$\begin{cases} m\ddot{\delta}_1 + c\dot{\delta}_1 + k\delta_1 = 0 \\ m\ddot{\delta}_2 + 3c\dot{\delta}_2 + 3k\delta_2 = 0 \end{cases}$$

Which is a system of uncoupled differential equations from the second order, which we can then solve through their characteristic equations. So we find:

$$\begin{cases} \delta_1 = A_1.e^{-\xi_1\omega_1 t}.\sin\left[\left(\omega_1\sqrt{1-\xi^2}\right) + \theta_1\right] \\ \delta_2 = A_2.e^{-\xi_2\omega_2 t}.\sin\left[\left(\omega_2\sqrt{1-\xi^2}\right) + \theta_2\right] \end{cases} \tag{1.62}$$

By imposing the same IC as previously, it then comes for the constants:

$$\begin{cases} \theta_1 = \operatorname{arctg}\dfrac{\sqrt{1-\xi_1^2}}{\xi_1} & ; \quad \theta_2 = \operatorname{arctg}\dfrac{\sqrt{1-\xi_2^2}}{\xi_2} \\ A_1 = \dfrac{X_0}{2\sqrt{1-\xi_1^2}} & ; \quad A_2 = \dfrac{X_0}{2\sqrt{1-\xi_2^2}} \\ \text{With: } \xi_1 = \dfrac{c}{2m\omega_1} & ; \quad \xi_2 = \dfrac{3c}{2m\omega_2} \end{cases}$$

And finally the general solution in x is given by:

$$\begin{aligned} x_1(t) &= A_1..e^{-\xi_1\omega_1 t}.\sin\left[\left(\omega_1\sqrt{1-\xi_1^2}\right)t+\theta_1\right] + A_2.e^{-\xi_2\omega_2 t}.\sin\left[\left(\omega_2\sqrt{1-\xi_2^2}\right)t+\theta_2\right] \\ x_2(t) &= A_1..e^{-\xi_1\omega_1 t}.\sin\left[\left(\omega_1\sqrt{1-\xi_1^2}\right)t+\theta_1\right] - A_2.e^{-\xi_2\omega_2 t}.\sin\left[\left(\omega_2\sqrt{1-\xi_2^2}\right)t+\theta_2\right] \end{aligned} \quad (1.63)$$

1.3.7.3. *Forced oscillations of a damped system with 2 DOF*

We can see that calculations quickly become complicated if the number of modes becomes important. This is why the formalism generally used is the one that uses the matrices and vectors (of mass, stiffness, damping, force, displacement, velocity, etc.), it is the modal approach that goes beyond the framework of the volume. We will still deal with our problem here through a more educational and direct approach. So we have to resolve the following system in a forced regime.

We thus suppose, for example, that we apply a harmonic force to the displacement mass x_2 such as:

$$\underline{F_2} = \underline{F_{02}}.e^{j\omega t} \quad \rightarrow F_2 = F_{02}.\sin\omega t \qquad \text{with } F_{02} = \text{real quantity}$$

$$\underline{F_1} = 0 \quad \rightarrow F_1 = 0$$

The real harmonic force represents here the imaginary part of the general complex expression.

Similarly, for displacements of the two masses:

$$\underline{x_{1,2}} = \underline{X_{1,2}}.e^{j\omega t} \rightarrow x_{1,2} = \left(X_{1,2}^R + jX_{1,2}^I\right).\sin\omega t$$

$\underline{X_{1,2}}$ is expressed under the form of the sum of a real part and an imaginary part because there is a module and a phase in the answer.

The system to be solved then takes the following form: $\begin{cases} m\ddot{x}_1 + 2c\dot{x}_1 + 2kx_1 - kx_2 - c\dot{x}_2 = 0 \\ m\ddot{x}_2 + 2c\dot{x}_2 + 2kx_2 - kx_1 - c\dot{x}_1 = F_2 \end{cases}$

Indeed, the system can be solved directly through the use of complex expressions of the parameters. It then becomes, after development and elimination of $e^{j\omega t}$:

$$\begin{cases} -\omega^2 m\left(X_1^R + jX_1^I\right) + 2c\omega\left(X_1^R + jX_1^I\right) + 2k\left(X_1^R + jX_1^I\right) - k\left(X_2^R + jX_2^I\right) - c\omega\left(X_2^R + jX_2^I\right) = 0 \\ -\omega^2 m\left(X_2^R + jX_2^I\right) + 2c\omega\left(X_2^R + jX_2^I\right) + 2k\left(X_2^R + jX_2^I\right) - k\left(X_1^R + jX_1^I\right) - c\omega\left(X_1^R + jX_1^I\right) = F_{02} \end{cases}$$

It is thus necessary to determine the four unknown factors making it possible to build displacements x_1 and x_2. The development of the previous system into real and imaginary parts gives out the following four equations:

$$\begin{cases}(2k-m\omega^2)X_1^R - kX_2^R - 2c\omega X_1^I + c\omega X_2^I = 0 & \text{real part}\\ 2c\omega X_1^R - c\omega X_2^R + (2k-m\omega^2)X_1^I - kX_2^I = 0 & \text{imaginary part after the elimination of j}\\ -kX_1^R + (2k-m\omega^2)X_2^R + c\omega X_1^I - 2c\omega X_2^I = F_{02} & \text{real part}\\ -c\omega X_1^R + 2c\omega X_2^R - kX_1^I + (2k-m\omega^2)X_2^I = 0 & \text{imaginary part after the elimination of j}\end{cases}$$

The solution of this system allows to calculate expressions of $X_{1,2}^R$ and $X_{1,2}^I$. The answer of the system is then given by:

$$\begin{cases}x_1 = (X_1^R + jX_1^I).\sin\omega t\\ x_2 = (X_2^R + jX_2^I).\sin\omega t\end{cases}$$

Here, again, to simplify the approach, we will assume that any damping equals zero, the system is reduced to:

$$\begin{cases}(2k-m\omega^2)X_1^R - kX_2^R = 0\\ -kX_1^R + (2k-m\omega^2)X_2^R = F_{02}\end{cases} \quad \text{(a) and:} \quad \begin{cases}(2k-m\omega^2)X_1^I - kX_2^I = 0\\ -kX_1^I + (2k-m\omega^2)X_2^I = 0\end{cases} \quad \text{(b)}$$

- From system (a), we can deduct the real parts:

$$\begin{cases}X_1^R = F_{02}.\dfrac{k}{(k-m\omega^2)(3k-m\omega^2)}\\ X_2^R = F_{02}.\dfrac{(2k-m\omega^2)}{(k-m\omega^2)(3k-m\omega^2)}\end{cases} \tag{1.64}$$

- From system (b), we can deduct the imaginary parts:

$X_1^I = X_2^I = 0 \quad if \quad \begin{cases}\omega \neq \omega_1\\ \omega \neq \omega_2\end{cases}$ When $\omega = \omega_1$, or $\omega = \omega_2$, there is indetermination.

The module is given here directly by the standardized expressions (1.64), which means:

$$\begin{cases}\dfrac{x_1}{F_{02}} = \dfrac{k}{(k-m\omega^2)(3k-m\omega^2)}.\sin\omega t \rightarrow \left|\dfrac{x_1(\omega)}{F_{02}}\right| = \left|\dfrac{k}{(k-m\omega^2)(3k-m\omega^2)}\right|\\ \dfrac{x_2}{F_{02}} = \dfrac{(2k-m\omega^2)}{(k-m\omega^2)(3k-m\omega^2)}.\sin\omega t \rightarrow \left|\dfrac{x_2(\omega)}{F_{02}}\right| = \left|\dfrac{(2k-m\omega^2)}{(k-m\omega^2)(3k-m\omega^2)}\right|\end{cases} \tag{1.65}$$

EXAMPLE 1.1. k = m = 1, so: $\omega_1 = \sqrt{\dfrac{1}{1}} = 1,\ \omega_2 = \sqrt{\dfrac{3}{1}} = 1{,}732,\ \omega' = \sqrt{\dfrac{2}{1}} = 1{,}414$.

Of course, in the case of equipment, the reduced damping coefficients are generally low and at least higher than 0,16 (with Q = 3). Thus, calculations that neglect the latter provide interesting results. Indeed, taking into account a slight damping shifts the resonance frequencies and limits amplitudes with the easily calculable Q-factor. Traces can be relatively well approached.

REMARK.– Even when the system vibrates according to one of its vibration modes, its behavior is practically identical to that of a system with 1 DOF and the free movement of the system will be obtained by the superposition of its various modes. Participation of different being "weighted" modes by the corresponding natural form. This means that there is little or no transfer of energy from one mode to the other (matrices do not have cross or low-value terms); diagonal terms are predominant.

This is the basis, as already mentioned, of the modal superposition techniques.

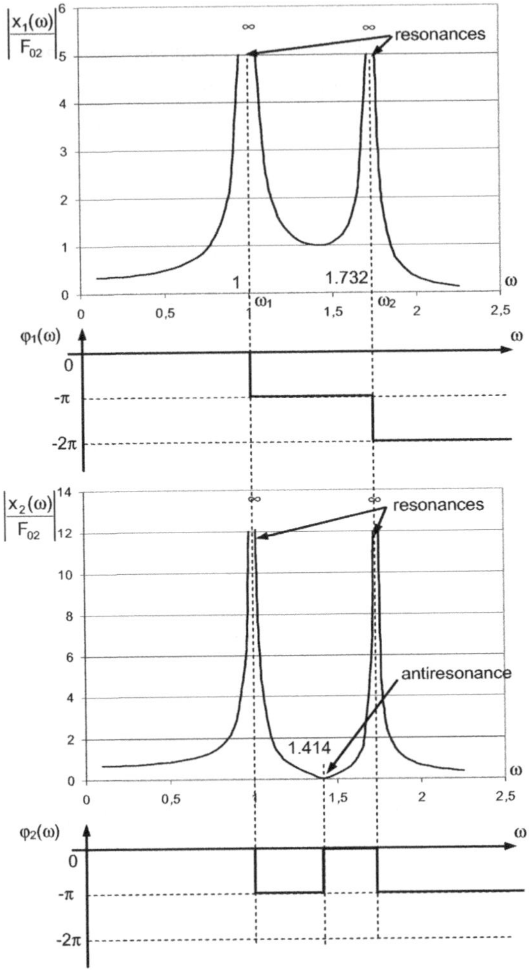

Figure 1.18. *Amplitude responses and phase of a coupled and damped system with 2 DOF*

1.4. Concept of mechanical impedance

1.4.1. Overview

The concept of mechanical impedance in the dynamic study of structures is very useful in the field of shocks and vibrations. Although the theory dates back to early 20th Century, its development is relatively recent because of the evolution of sensor technology and the measuring devices in recent years. The application of these techniques is essentially made, more and more, in the following areas:

- the determination of the necessary power for a vibration test;
- the influence of the reactions of structures under testing;
- the study of support and adaptation parts;
- the establishment of a test routine starting from vibrations analyzed on a vehicle;
- the relation between control under a constant force or under constant acceleration, etc.

1.4.2. Definitions

If a forced vibration in a stationary or harmonic regime is applied to a structure, a resulting movement of the same frequency exists at the point of application of the force as well as on the other points of the structure. The mechanical impedance in velocity and in sinusoidal excitation is therefore defined as the ratio between the applied force F(t) = F.sinωt and the speed of the resulting motion v(t) = V.sin(ωt +φ). If the velocity is measured at the point of application of the force, the impedance is said to be punctual, it represents the ability of a solid to resist the movement we want to transmit to it. If the speed is measured in another point of the structure, the impedance is known as of transfer; in this case, it represents the ability of a solid to resist the transmission of the movement at the point of transfer. In both cases, the mechanical impedance will therefore have as a value:

$$Z = \frac{F \sin \omega t}{V \sin(\omega t + \varphi)} \tag{1.66}$$

NB: We also define the concept of mechanical impedance in "displacement and acceleration", but it is especially that of velocity that is used. This concept involves the same approach as those developed in all the fields where we talk about impedance, which means that we find the concepts of adaptation and transmission associated with energy or power. If it takes a large force to set in motion a solid, this one will have an important mechanical impedance. We will study later on that this definition results from an electromechanical analogy, since, indeed, we can note that by simplifying relationships, we have the point of view of power.

In mechanics, in a general manner, power can be given by: P = F.v

So we can write with: $Z = \frac{F}{v} \quad \rightarrow \quad \begin{cases} P = \frac{F^2}{Z} \\ P = v^2 Z \end{cases}$

From where an immediate analogy between force and tension and velocity and intensity comes in, as well as the concepts of internal impedances of generators and the associated problems of adaptation.

The solution of relation (1.53) can be expressed in the following manner:

$$Z = \frac{F}{V}\cos\varphi + j\frac{F}{V}\sin\varphi \tag{1.67}$$

By noticing that in expression (1.67), the module provides the maximum amplitude (or module) of (1.66) and the phase gives out φ again.

From these data, information can either be treated by the reading of diagrams or curves, or by the use of complex conjugates or matrix calculus.

We note that it is possible, and very interesting, starting from the mechanical values of impedances between various points, to determine the energy transfer in a structure.

The opposite of a mechanical impedance is called the mobility or mechanical admittance, which is also widely used.

1.4.2.1. *Parameters associated with mechanical impedance*

The usual parameters associated with the notion of mechanical impedance are:

- mechanical impedance: $\frac{\text{Force}}{\text{velocity}} = \frac{F}{\dot{x}}$ $[Nm^{-1}s]$

- mobility: $\left(\frac{\text{Force}}{\text{velocity}}\right)^{-1} = \frac{\text{velocity}}{\text{Force}} = \frac{\dot{x}}{F}$ $[N^{-1}ms^{-1}]$

- dynamic stiffness: $\frac{\text{Force}}{\text{displacement}} = \frac{F}{x} = k$ $[Nm^{-1}]$

- compliance: $\left(\frac{\text{Force}}{\text{displacement}}\right)^{-1} = \frac{x}{F} = k'$ $[N^{-1}m]$

- apparent mass: $\frac{\text{Force}}{\text{acceleration}} = \frac{F}{\ddot{x}}$ $[Nm^{-1}s^2]$

- inertance: $\left(\frac{\text{Force}}{\text{acceleration}}\right)^{-1} = \frac{\ddot{x}}{F}$ $[N^{-1}ms^{-2}]$

In the MKSA system, the force is expressed in newton with: 1 N = 10^5 dynes.

The unit of mass of the MKSA system corresponds to a weight of 0,102 kg. It is the mass to which a force unit communicates a acceleration unit, which means:

$$F = P = mg \rightarrow m = \frac{F}{g} \rightarrow [m] = \frac{1N}{9{,}81 ms^{-2}} = 0{,}102\, Nm^{-1}s^2$$

In English units, the correspondence is:

Force	Pound-force	[Lbf]
Displacement	Inch	[In]
Velocity	Inch per second	[In/s]
Mechanical impedance	Pound per inch per second	[Lbf/in/s]
mass	Pound	[Lb]

The correspondence for the units of mechanical impedance is as follows:

$$1\frac{N}{cm/s} = 1{,}75.\frac{Lb}{IN/s}$$

1.4.2.2. *Vectorial representation*

We can use the following vectorial mode to represent a mechanical impedance.

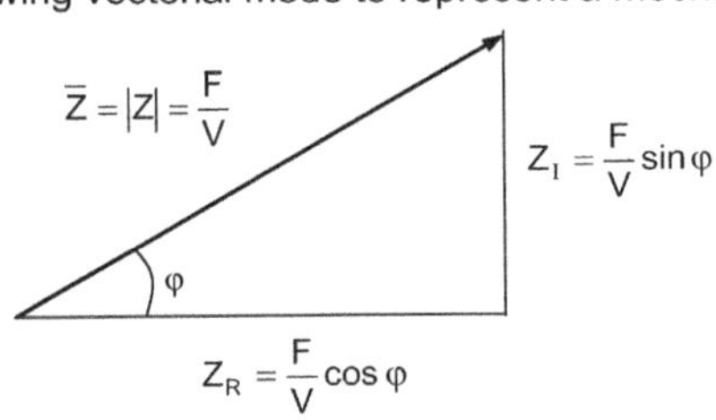

By adding $Z = \frac{F}{V}.e^{j\varphi}$, we have $Z_R = \frac{F}{V}.\cos\varphi$ for the real part and $Z_I = \frac{F}{V}.\sin\varphi$ for the imaginary part. From where the following relations derive and make it possible to calculate it:

$$\begin{cases} tg\varphi = \frac{Z_I}{Z_R} \\ |Z| = \sqrt{Z_R^2 + Z_I^2} \end{cases} \quad (1.68)$$

NB: The mechanical impedance has the following advantages:

- there is a relation between vibratory velocity and the level of noise;

- velocity is the magnitude that varies less in function of the frequency (for low frequencies, acceleration becomes very low, and for the high frequencies, it becomes displacement).

1.4.2.3. *Fourier transform and Laplace transform of non-sinusoidal functions*

The response of mechanical systems to non-sinusoidal excitations such as the impulses of shocks, for example, can also be considered with the concept of mechanical impedance and the use of the traditional tools of TF, TL and the transfer function [ref. I19] (systems being studied in their linear zones). We can write, for mobility, the following remarkable relation (same approach for the other parameters):

$$F(t) = \text{Force} \rightarrow \left\{ TF[F(t)] = F(f) = \int_{-\infty}^{\infty} F(t).e^{-2\pi jft}.dt \quad \text{and} \quad TL[F(t)] = F(p) = \int_{0}^{\infty} F(t).e^{-pt}.dt \right\}$$

$$\dot{x}(t) = \text{Velocity} \rightarrow \left\{ TF[\dot{x}(t)] = \dot{X}(f) = \int_{-\infty}^{\infty} \dot{x}(t).e^{-2\pi jft}.dt \quad \text{and} \quad TL[\dot{x}(t)] = \dot{X}(p) = \int_{0}^{\infty} \dot{x}(t).e^{-pt}.dt \right\}$$

$$M_o(f) = 1/Z(f) = \text{Mobility}$$

Then:

$$\begin{cases} \dot{X}(f) = F(f).M_o(f) = \dfrac{F(f)}{Z(f)} \\ \dot{X}(p) = F(p).M_o(p) = \dfrac{F(p)}{Z(p)} \end{cases} \qquad (1.69)$$

We can extend the notions of filtering in mechanical impedances.

1.4.2.4. *Calculations of simple impedances of usual discreet elements*

a) Case of pure mass

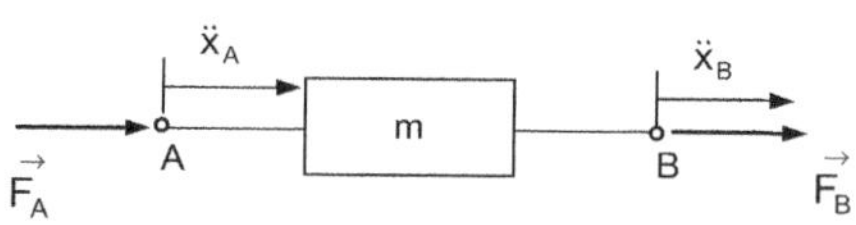

We have $F_A - F_B = m(\ddot{x}_A - \ddot{x}_B)$, yet in the case of pure mass $\ddot{x}_A = \ddot{x}_B = \ddot{x} \rightarrow F = 0$. This means that a mass is not able to transmit effort, instead such an element will correspond to a loss of force which is required to put the mass in motion. If F(t) is the force applied, it becomes:

$F(t) = m.\gamma(t)$ for a sinusoidal excitation, it becomes:

$$F(t) = \text{Re}\left\{F_m.e^{j\omega t}\right\} = \text{Re}\left\{F_m(\cos\omega t + j\sin\omega t)\right\} = F_m \cos\omega t$$

With:

$$\gamma(t) = \gamma_m.e^{j(\omega t + \varphi_1)}$$

$$v(t) = v_m.e^{j(\omega t + \varphi_2)} \qquad \text{So, like this: } \gamma(t) = \frac{dv(t)}{dt} = j\omega v_m.e^{j(\omega t + \varphi_2)} = j\omega v(t)$$

$$x(t) = x_m.e^{j(\omega t + \varphi_3)}$$

$$F(t) = jm\omega v(t) \rightarrow \qquad Z(\omega) = \frac{F(t)}{v(t)} = jm\omega \qquad (1.70)$$

It is a pure imaginary number. From where following diagrams derive:

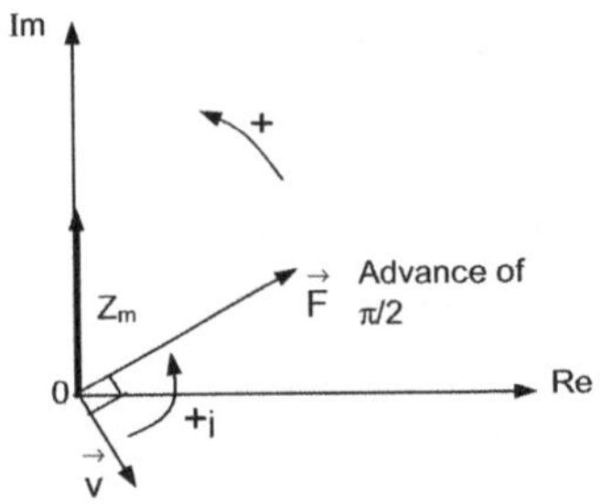

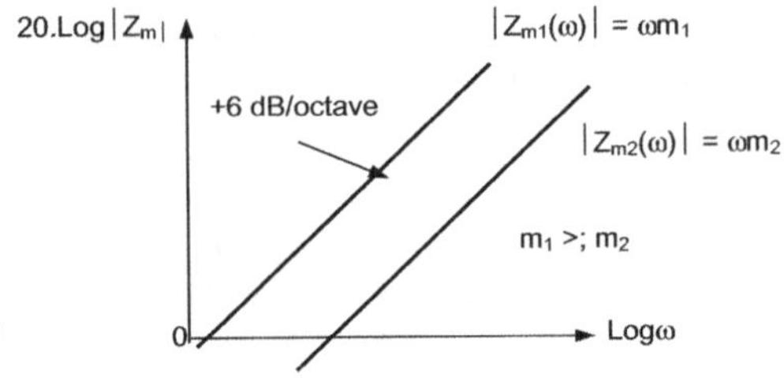

We can extend this concept to unspecified signals in isomorphous regime with: $p = j\omega$, so:

$$Z(p) = mp \tag{1.71}$$

b) Case of viscous friction

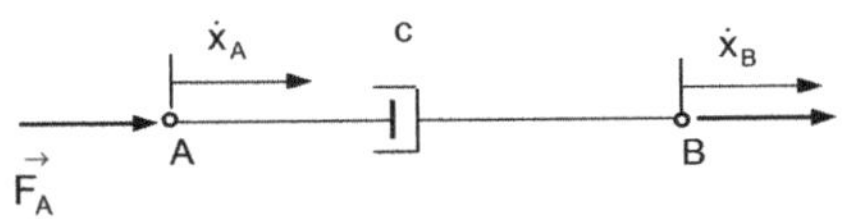

If we apply the forces F_A and F_B at the ends of the damper, these points will acquire speeds $\dot{x}_A$ and $\dot{x}_B$ such as: $F_a - F_b = c.(\dot{x}_A - \dot{x}_B)$. So if F(t) is the force transmitted by the damper and $\dot{x}$ the relative velocity of points A and B, it becomes: $F = c\dot{x}$.
Therefore:

$$Z = c \text{ [Ns/m]} \tag{1.72}$$

It is a pure real number, which is translating the proportionality between velocity and force for a viscous damping, i.e. meaning that mechanical energy is transformed into heat. From where the following diagrams derive:

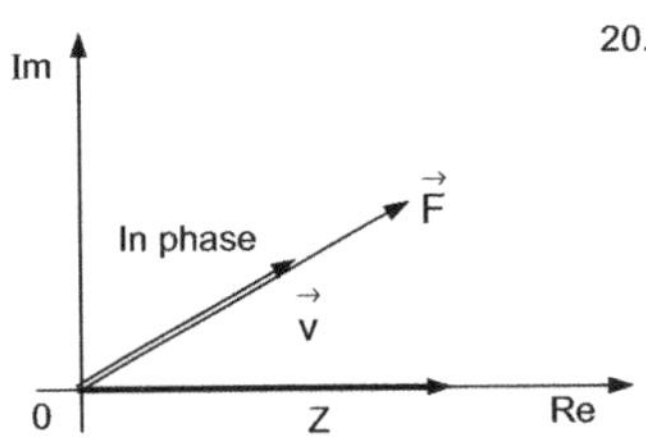

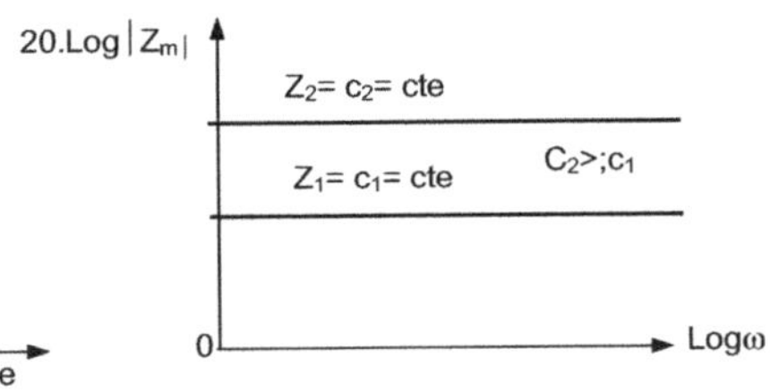

c) Case of pure rigidity

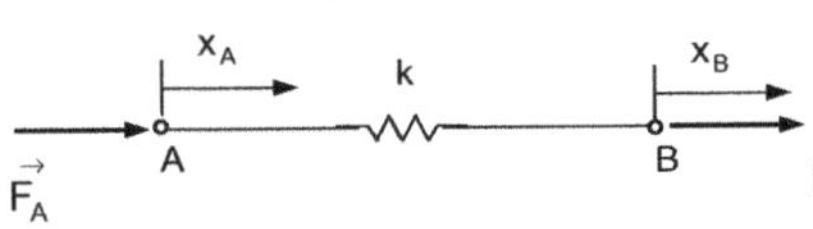

In the same way, as mentioned previously, we will have: $F_A - F_B = k.(x_A - x_B)$.

The transmitted force F will thus be connected to the relative displacement x by: F = k.x. As in the harmonic regime:

$$x(t) = \int_t^{t+dt} v(t).dt = \frac{1}{j\omega} v_m.e^{j(\omega t + \varphi_2)} = \frac{v(t)}{j\omega} = -j\frac{v}{\omega}, \quad \text{so :} \quad F(t) = -j\frac{kv}{\omega}$$

$$Z = -j\frac{k}{\omega} = \frac{k}{p} \tag{1.73}$$

From where following diagrams derive:

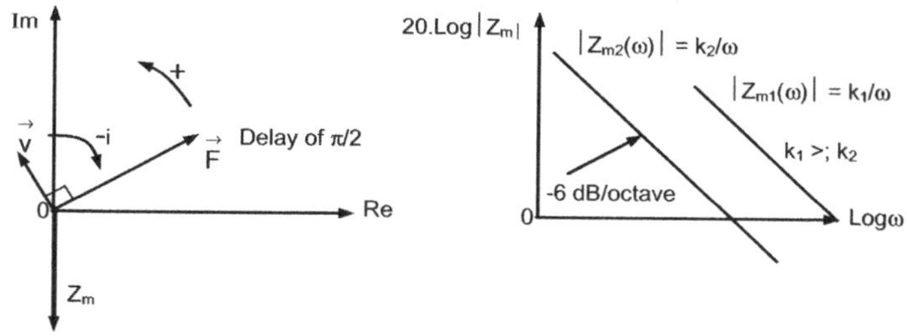

1.4.2.5. *Principles of use*

We have just defined the impedances of basic elements, which make it possible to compose a mechanical or mathematical discrete model. It is now necessary to assemble the different elements to obtain the equivalent network model of impedance. To build this network, we will provide a few simple rules of construction.

1 - It is necessary to identify the two ends of each basic element and to connect the elements having in common a given end, i.e. having same velocity.

2 - We will connect to a node, which we will identify later, all ends of the elements which have this common node (same velocity).

3 - We will link to a reference velocity zero, the nodes that are considered as motionless.

4 - Each mass will have an end connected to this reference; the other end is connected to a node of the network corresponding to a velocity equal to that of the mass.

5 - The excitation applied to the system is a force F acting on the node, which constitutes the input of the system; the other end reacts against the ground (reference) and will thus be connected to it.

6 - It is necessary to assign a coordinate (DOF) to each free node to be driven and given a positive direction. The direction and the sense of this movement will be obtained in reference to the force applied.

EXAMPLE 1.2. Equivalent circuit in term of mechanical impedance for the system with 2 coupled DOF (excitation on the mass, model of section 1.3.7.2).

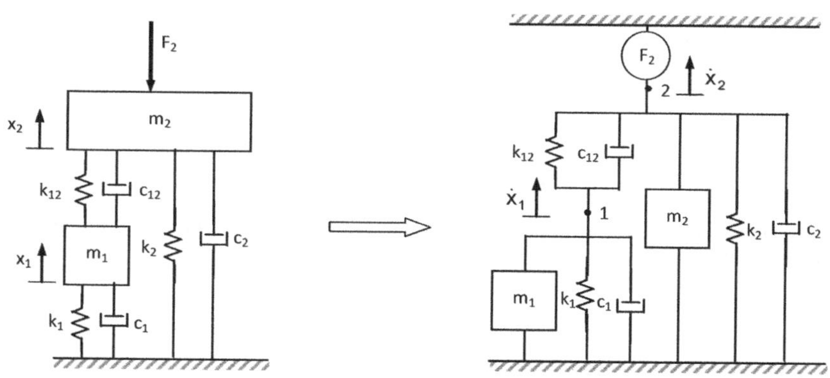

We note as many points or nodes as DOFs.

1.4.2.6. *Association rules*

Once the mechanical network of impedance is established, it can be interesting to be able to calculate the equivalent impedance (or the mobility) of such a system at the entrance of the force (entry impedance), from which we will be able, knowing F, to know the answer directly.

a) Parallel elements

The model is as follows:

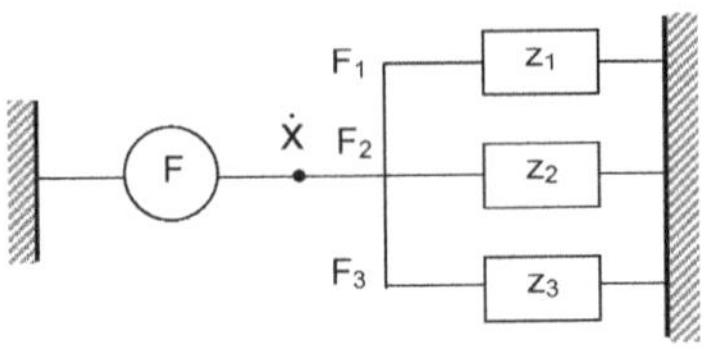

We obtain: $\begin{cases} F_1 = z_1.\dot{x} \\ F_2 = z_2.\dot{x} \\ F_3 = z_3.\dot{x} \end{cases}$ with: $F = F_1 + F_2 + F_3 = z_{eq}.\dot{x} = (z_1 + z_2 + z_3).\dot{x}$

So if n impedances are in parallel (respectively, of mobilities in series), the equivalent impedance will be given by:

$$Z_{eq} = \sum_n Z_n \tag{1.74}$$

Similarly, the equivalent mobility is given by:

$$\frac{1}{M_{eq}} = \sum_n \frac{1}{M_n} \tag{1.75}$$

b) Elements in series

The model is as follows:

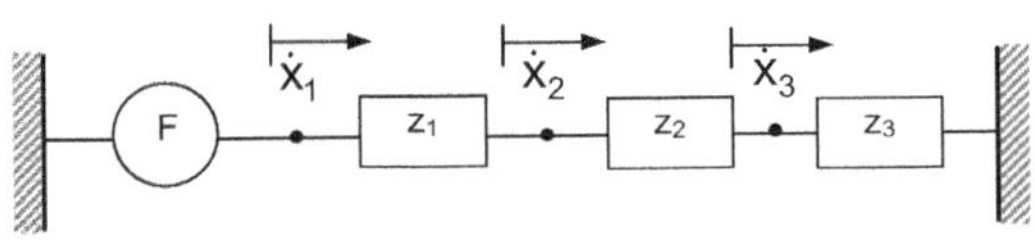

We obtain: $\begin{cases} F = z_1.(\dot{x}_1 - \dot{x}_2) \\ F = z_2.(\dot{x}_2 - \dot{x}_3) \\ F = z_3.\dot{x}_3 \end{cases}$ such...as: $\dot{x}_1 = (\dot{x}_1 - \dot{x}_2) + (\dot{x}_2 - \dot{x}_3) + \dot{x}_3$

It becomes: $$\dot{x}_1 = F.\left(\frac{1}{z_1} + \frac{1}{z_2} + \frac{1}{z_3}\right) = \frac{F}{z_{eq}}$$

So, if n impedances are in series (respectively, of mobilities in parallel), the equivalent impedance z_{eq} will be given by:

$$\frac{1}{Z_{eq}} = \sum_i \frac{1}{Z_i} \tag{1.76}$$

Similarly, the equivalent mobility is given by:

$$M_{eq} = \sum_n M_n \tag{1.77}$$

EXAMPLE 1.3. Let us reconsider the previous mechanical network model of the system with 2 coupled DOF and to simplify calculations, we will neglect the dampening; then, we obtain, through successive associations of different impedances, the following models:

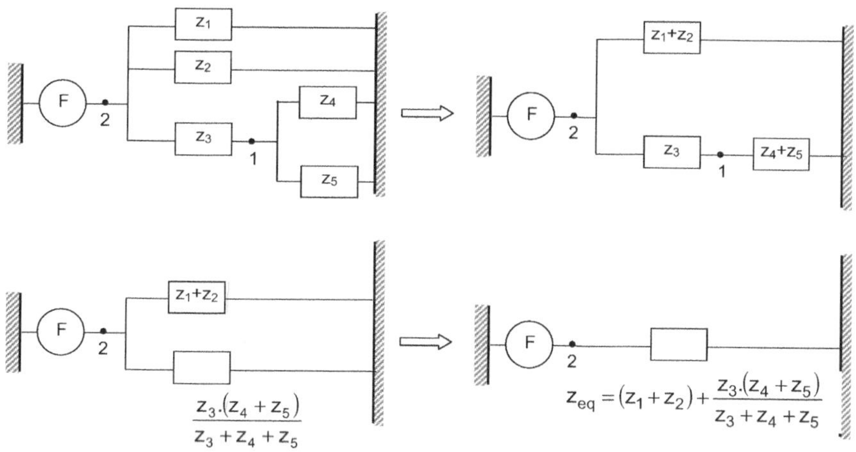

If we replace now the different impedances by their values, while assuming:

$m = m_1 = m_2$ and $k = k_1 = k_2$ it becomes:

$$z_1 = jm\omega \qquad z_2 = \frac{k}{j\omega} \qquad z_3 = \frac{k}{j\omega} \qquad z_4 = jm\omega \qquad z_5 = \frac{k}{j\omega}$$

Having: $$z_{eq} = jm\omega + \frac{k}{j\omega} + \frac{\frac{k}{j\omega}.\left(\frac{k}{j\omega} + jm\omega\right)}{\frac{2k}{j\omega} + jm\omega}$$

After reduction it becomes: $$z_{eq} = \frac{1}{j\omega}.\frac{\left(k - m\omega^2\right).\left(3k - m\omega^2\right)}{2k - m\omega^2}$$

Therefore: $$x_2 = \frac{\dot{x}_2}{j\omega} = \frac{F}{j\omega.z_{eq}} = \frac{F.\left(2k - m\omega^2\right)}{\left(k - m\omega^2\right)\left(3k - m\omega^2\right)}$$

We get the same expression obtained by the direct method in section 1.3.7.2 (equation (1.65)). We can similarly calculate the displacement x; from the network of the figure, we get the same relations:

$$z_3.\left(\dot{x}_2 - \dot{x}_1\right) = \left(z_4 + z_5\right).\dot{x}_1 \qquad \rightarrow \qquad \frac{\dot{x}_1}{\dot{x}_2} \equiv \frac{x_1}{x_2} = \frac{z_3}{z_3 + z_4 + z_5} = \frac{k}{2k - m\omega^2}$$

Having: $$x_1 = \frac{F.k}{\left(k - m\omega^2\right)\left(2k - m\omega^2\right)}$$

This expression is also identical to the one found previously.

We can see the interest offered by such method, which allows very simply and quickly to solve discrete mechanical models.

1.4.2.7. *General theorems of mechanics*

The principles of electromechanical analogy and even from the definition of mechanical impedance, the following properties can be exploited. They are applied only to the equivalent networks established to account for systems composed of linear elements.

a) Kirchhoff's laws

First law: The sum of the forces acting on a point of common velocity is zero:

$$\sum_i F_i = 0 \tag{1.78}$$

Second law: The sum of velocities at different connections to a series of items taken while following a closed loop is zero:

$$\sum_i \dot{x}_i = 0 \tag{1.79}$$

b) Association in series of mechanical impedances (respectively, of mobilities in parallel)

$$\frac{1}{Z_{eq}} = \sum_i \frac{1}{Z_i} \tag{1.80}$$

c) Association in parallel of mechanical impedances (respectively, of mobilities in series)

$$Z_{eq} = \sum_i Z_i \tag{1.81}$$

d) Superposition

If a mechanical system composed of linear elements is controlled by several sources of excitations, which can be at the same frequency and of different amplitudes, the total velocity at an unspecified point is equal to the vectorial sum of velocities resulting from each source functioning separately, with the other generators replaced by their internal impedances. Thus, the sum of responses at a point is equal to the sum of responses induced at this point by each source taken separately, with the other sources not providing energy and being replaced by their internal impedances.

EXAMPLE 1.4.

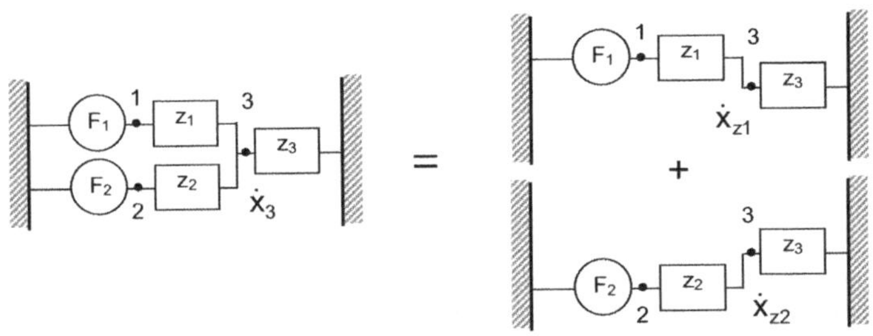

Consider in the first network: $\dot{x}_{z1} = \frac{F_1}{Z_3}$

In the second network: $\dot{x}_{z2} = \frac{F_2}{Z_3}$

From where: $\dot{x}_3 = \dot{x}_{z1} + \dot{x}_{z2} = \frac{F_1 + F_2}{Z_3}$

e) Reciprocity

If an excitation applied to point A of a network of linear elements, creates in another point B a velocity $\dot{x}_B$, the same force applied in B will create the same velocity on point A, while impedances remain unchanged.

In the transfer impedance measures, force and velocity points can be reversed without changing the result, which means:

$$z_{AB} = z_{BA} \tag{1.82}$$

EXAMPLE 1.5.

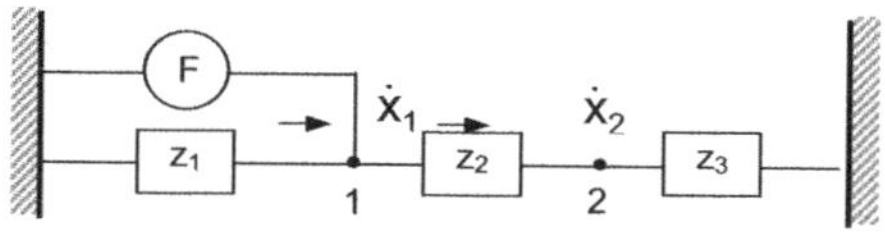

Let us apply a force F to node 1 of the network, and it becomes:

$$F = \left(z_1 + \frac{z_2 z_3}{z_2 + z_3} \right) . \dot{x}_1 \text{ and as } \dot{x}_2 = \dot{x}_1 \frac{z_2}{z_2 + z_3} \rightarrow \frac{\dot{x}_1}{\dot{x}_2} = \frac{z_2 + z_3}{z_2}$$

$$\text{Therefore}: \ \dot{x}_2 = \frac{F z_2}{z_1 z_2 + z_2 z_3 + z_1 z_3} \circ \frac{F}{z_{12}} \text{ by definition of } z_{12}$$

Consider now a force F applied to point 2, it becomes:

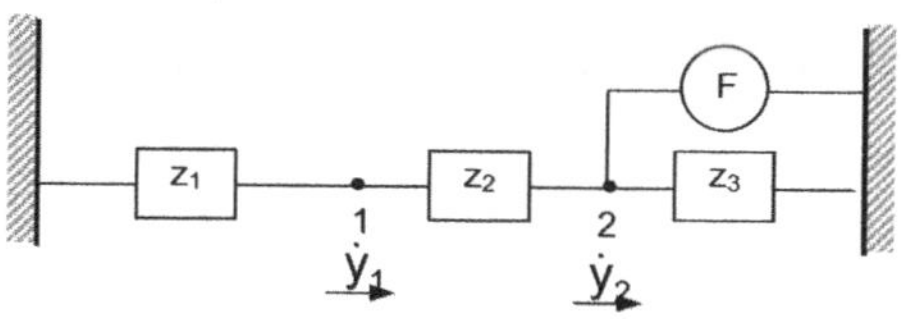

$$F = \left(z_3 + \frac{z_1 z_2}{z_1 + z_2} \right) . \dot{y}_2 \quad \text{and} \frac{\dot{y}_2}{\dot{y}_1} = \frac{z_1 + z_2}{z_2}$$

$$\text{So}: \ \dot{y}_1 = \frac{F z_2}{z_1 z_2 + z_2 z_3 + z_1 z_3} \circ \frac{F}{z_{21}} \text{ by definition of } z_{21}$$

We check that we have: $z_{12} = z_{21}$.

f) Thevenin's theorem

If we find ourselves in the presence of a network of linear elements attacked by sources of excitations containing a variety of impedances and are connected to a load external to the network, we can replace this entire system by a generator delivering a constant force F_{eq}, mounted in parallel with a single impedance z_{eq} on the same external load (z_{eq} is then the internal impedance of a single generator). To determine these two quantities F_{eq} and z_{eq}, we take the following example:

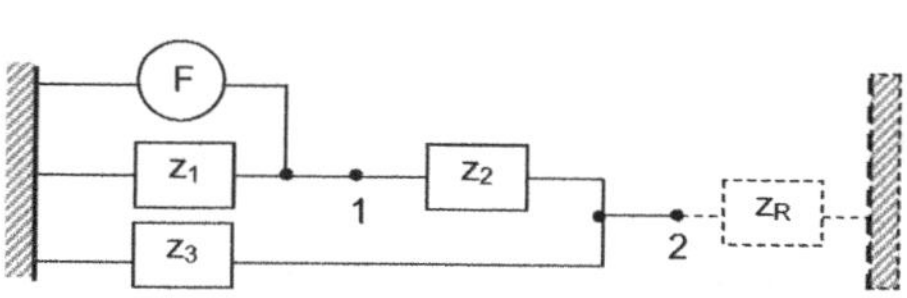

It is assumed that this network represents a given structure connected to point 2 and to a second structure of impedance Z_R.

(a) Determination of the equivalent force F_{eq}

For this, let us fix point 2 and evaluate the transmitted force to the ground by this point. Considering the diagram:

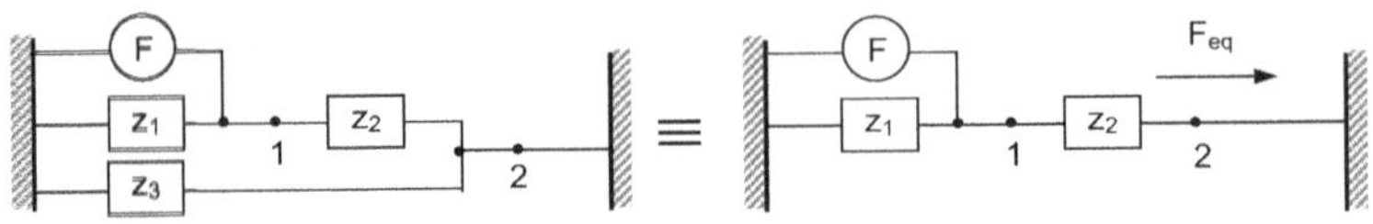

Therefore: $F_{eq} = z_2.\dot{x}_1$ with : $\dot{x}_1 = \dfrac{F}{z_1 + z_2}$ $\rightarrow$ $F_{eq} = \dfrac{z_2}{z_1 + z_2}.F$ (1.83)

(b) Determination of impedance z_{eq}

It will be the equivalent impedance seen from point 2, which is shown in the following diagram:

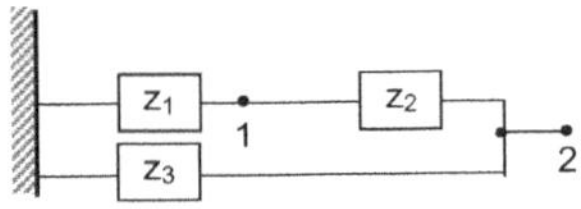

Having: $$z_{eq} = z_3 + \frac{z_1 z_2}{z_1 + z_2} \tag{1.84}$$

The equivalent system to the original system is then as follows:

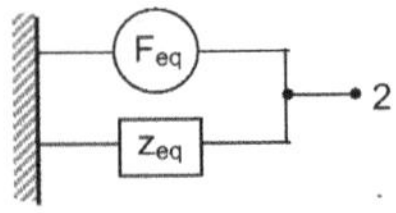

So we brought back the system to point 2 and if Z represents the load, the velocity at point 2 will be given by the relation:

$$\dot{x}_2 = \frac{F_{eq}}{z_{eq} + z_R}$$

Thus, a complex system can be represented in a simplified way. Indeed, if we consider a point of interest in a system comprising multiple generators and mechanical impedances, the system can be replaced in this point by a single generator with a mechanical impedance in parallel.

- The force of the equivalent generator is the same as the force measured in this point of interest when it is blocked (velocity is equal to 0).

- The equivalent mechanical impedance equals the impedance as seen from this point toward the free-motion system. Consider the following general model:

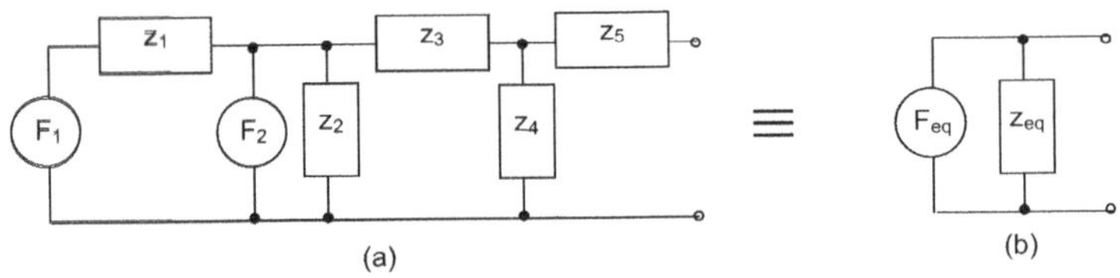

(a) = Complex system: the case of a vibrator with its adaptation parts, for example.
(b) = Equivalent system: F_{eq} is measured in a blocked movement.
Z_{eq} is measured in free movement.

g) Norton's theorem

A mechanical system composed of linear elements and comprising sources of vibrations and an output connection can be represented by a generator working at a constant velocity

$\dot{x}_{eq}$ mounted in series with an internal impedance z_{eq}. This theorem is equivalent to the Thevenin's theorem and leads to the same internal impedance z_{eq}.

Let us reconsider the same network, with:

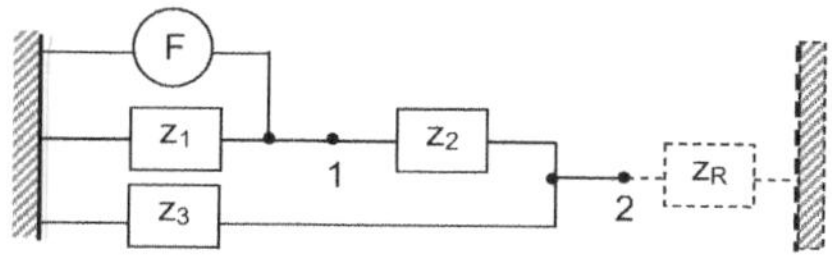

(a) Determination of the equivalent velocity $\dot{x}_{eq}$, which will be the system's velocity not connected (free) to the load, meaning the velocity at point 2 is $\dot{x}_2$. It becomes:

$$\left.\begin{array}{l} F=(z_1+z_2).\dot{x}_1-z_2\dot{x}_{eq} \\ z_2(\dot{x}_1-\dot{x}_{eq})=z_2\dot{x}_{eq} \end{array}\right\} \rightarrow \dot{x}_{eq}=\frac{F.z_2}{z_1z_2+z_2z_3+z_3z_1} \qquad (1.85)$$

The internal impedance is always z_{eq} previously calculated; the equivalent system brought back to point 2 is then given by:

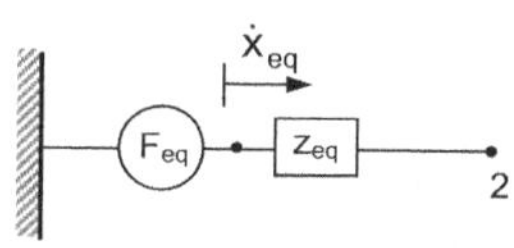

If z_R is connected in 2, we obtain: $\dot{x}_2=\frac{z_{eq}}{z_{eq}+z_R}.\dot{x}_{eq}$

Of course, both methods give out identical results.

EXAMPLE 1.6. Thus, Norton's theorem makes it possible to determine the response characteristics of a system when an additional impedance is added to it. The model is the following for a vibrator that is loaded by a specimen:

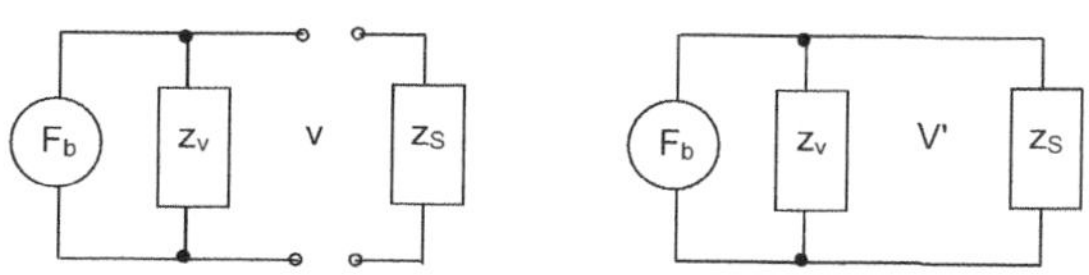

F_b = Blocked force
Z_v= Impedance of the vibrator and adaptation parts

v = free velocity
Z_S = Impedance of the test structure
v' = coupled system velocity

- Total impedance z_t is equal to $z_v + z_S$.
- Velocity v' in the connected system has as a value:

$$\left.\begin{array}{l} v'=F_b\left(\frac{1}{z_v+z_S}\right), \text{Or}: F_b=z_v.v \\ \text{Therefore}: \qquad v'=v\left(\frac{z_v}{z_v+z_S}\right) \end{array}\right\} \rightarrow \text{The equivalent circuit}:$$

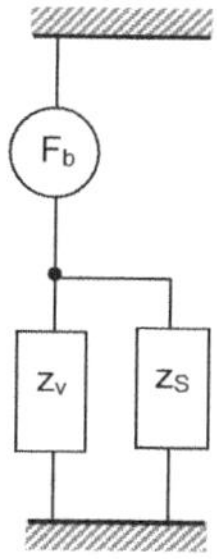

h) Transfer impedance

In the case of the adaptation pieces between the vibrator and the structure under testing, it is necessary to study the transfer impedance of these pieces. This transfer impedance z_{12} will be determined on the piece alone; the influence of the fixed structure will change the value of the transfer impedance and the new value will be given by:

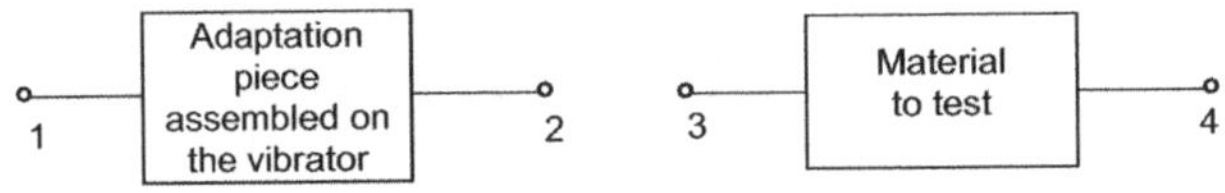

with:

- z_{12} = transfer impedance of 1 toward 2, non-complete system.
- z'_{12} = transfer impedance of 1 toward 2, complete system.
- z_{33} = specific impedance point 3.
- z_{22} = specific impedance point 2.

Then, we have the following relations:

$$\frac{F_1}{v_2} = z_{12} \qquad \frac{F_1}{v'_2} = z'_{12} \qquad v_2 = \frac{F_1}{z_{22}} \qquad v'_2 = \frac{F_1}{z_{22} + z_{33}} \rightarrow z'_{12} = z_{12} \cdot \left(1 + \frac{z_{33}}{z_{22}}\right)$$

1.4.2.8. *General modeling of structures by a network of impedances*

When it is difficult to model *a priori* a mechanical structure by discrete elements, it can be convenient to represent it by a network of mixed impedances that can be determined experimentally. Indeed, it is shown that any mechanical system having an input and output connection can be represented (or modeled) by a system in T or in Π.

(a) System in Π

The structure may be represented by three impedances z_1, z_{12}, z_2 forming a network such as the one shown in the following figure:

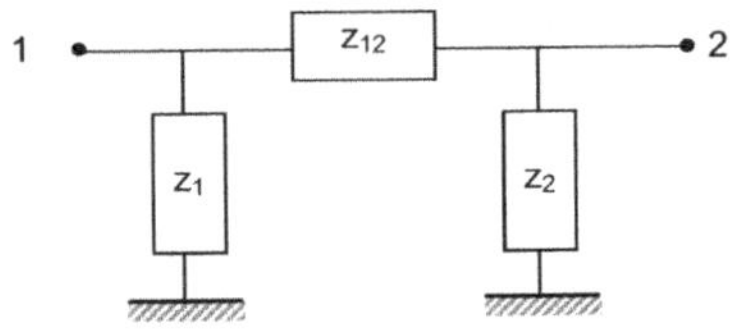

To determine the different impedances, it will be enough measurement, for example:

$$Z_{11} = \frac{F_1}{\dot{x}_1} \qquad \text{with: } \dot{x}_2 = 0$$

$$Z_{22} = \frac{F_2}{\dot{x}_2} \qquad \text{with: } \dot{x}_1 = 0$$

$$Z_{12} = \frac{F_1}{\dot{x}_2} \qquad \text{with: } \dot{x}_1 = 0$$

We have, in fact: $Z_{11} = z_1 + z_{12}$ $\qquad Z_{22} = z_2 + z_{12}$ $\qquad Z_{12} = z_{12}$

Having: $z_1 = Z_{11} - Z_{12}$ $\qquad z_2 = Z_{22} - Z_{12}$ $\qquad z_{12} = Z_{12}$

We can thus assemble several structures connected on a point (a point which can be generalized on a surface of which all the points have an identical movement, being infinitely rigid in the explored frequency band).

EXAMPLE 1.7. Let us suppose that the first structure is excited at point 1 by a force F of internal impedance z_i and that at point 2 connected to a second structure. The impedance network will be modeled by the following diagram:

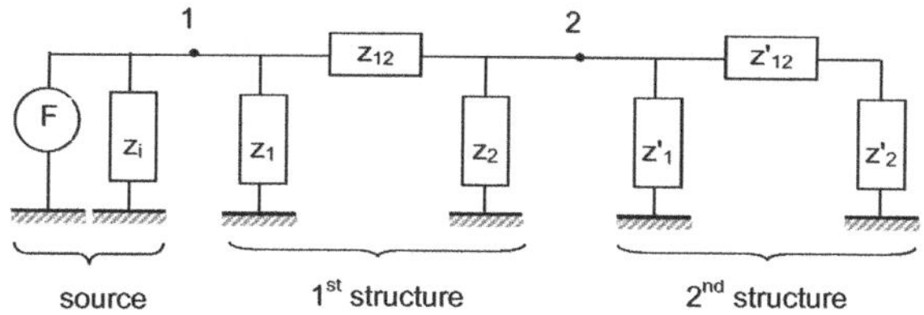

If we want to calculate the response to item 2, we can then apply the concepts studied before. Let us bring back the generator force F at point 2. We have to establish a system equivalent to the network. Therefore:

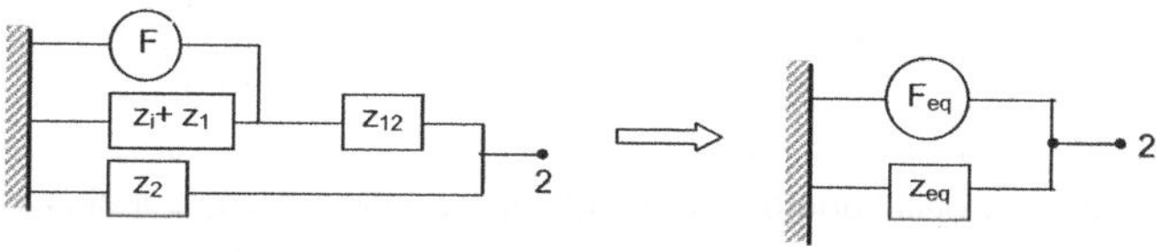

Similarly, the entry impedance presented by the second structure will be: $z_R = z'_1 + \dfrac{z'_{12}.z'_2}{z'_{12}+z'_2}$.

We will thus have the following model:

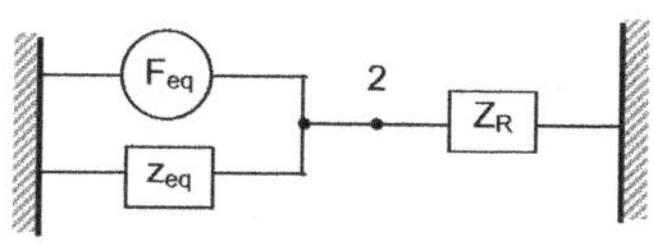

With: $\begin{cases} F_{eq} = \dfrac{F.z_{12}}{z_i + z_1 + z_{12}} \\ z_{eq} = z_2 + \dfrac{z_{12}.(z_i + z_1)}{z_{12} + z_i + z_1} \end{cases} \quad \rightarrow \quad \dot{x}_2 = \dfrac{F_{eq}}{z_{eq} + z_R}$

So, in case where there are several connection points, it will be possible to generalize what we have just explained, by inserting, for example, between each pair of points, a system in PI. Such an approach will allow to experimentally model the structure, but it is obvious that its resolution will require calculation techniques of appropriate networks (using a software named SPICE, for example).

b) System in T

In the same manner as previously explained, we will be able to represent any mechanical system presenting an entry and an exit by an equivalent network in T. It is particularly desirable here to work in terms of mobility and the equivalent network will be then in conformity with the following diagram:

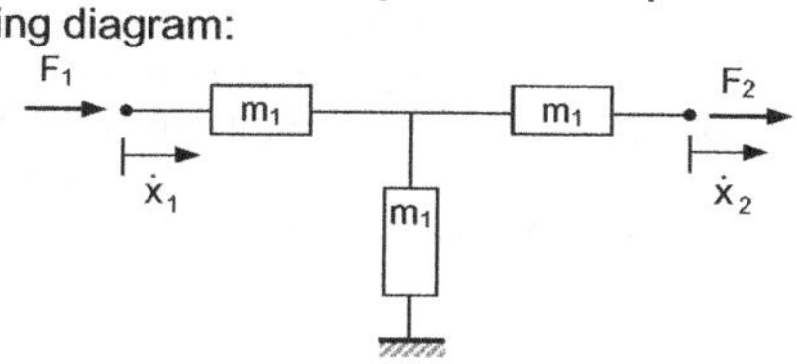

To determine these different amounts, we can measure, for example:

$$M_{11} = \frac{\dot{x}_1}{F_1} \qquad \text{for } F_2 = 0$$

$$M_{22} = \frac{\dot{x}_2}{F_2} \qquad \text{for } F_1 = 0$$

$$M_{21} = \frac{\dot{x}_2}{F_1} \qquad \text{for } F_2 = 0$$

Indeed, we have the following relationships in the T:

$$M_{11} = m_1 + m_{12} \qquad M_{22} = m_2 + m_{12} \qquad M_{12} = M_{21} = m_{12}$$

Therefore: $m_1 = M_{11} - M_{12}$ $\quad m_2 = M_{22} - M_{12}$ $\quad m_{12} = M_{12}$

Thus, in the same manner as previously explained, we will be able to consider the connection of several structures and we will use here Norton's theorem, preferably, to define a generator velocity to calculate, for example, the response in a point of several coupled structures, starting from the particular characteristics of each structure.

In short, the concept of mechanical impedance offers, due to its convenience of employment and its theoretical and experimental manipulation, many benefits that make it an invaluable tool in the study of the structures and allows, in particular, in a nonrestrictive way:

- the quick study of the discrete models while allowing calculating the various answers and the transfers between different points;

- the study of adaptation conditions of the structures amongst them, which is interesting when we have a large-size total structure, on which a direct experimentation would be difficult, even impossible;

- the modeling of structures by networks of approximate impedances (T shaped, Pi), which we will be able to determine directly by the experiment.

1.4.3. Resonance and antiresonance

If we consider an elementary system including a mass and a stiffness (spring), according to the point of application of the force we will obtain a resonance or an antiresonance.

1.4.3.1. *Antiresonance*

The model is the following:

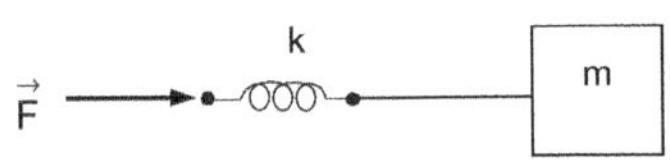

If the force F is applied at the end of the spring K, for low frequencies, the spring will completely transmit the force to the mass m (infinitely slow displacement). The impedance diagram will follow a line of constant mass equal to the mass of the system. For the frequency corresponding to $\omega_0 = \sqrt{\frac{k}{m}} \rightarrow f_0 = \frac{1}{2\pi}\sqrt{\frac{k}{m}}$, the vibration level becomes very low to the point of application of force; this is a very important value of F, the antiresonance frequency. The apparent mass is therefore very large. Beyond this antiresonance frequency, the mass m can be regarded as a fixed point, the system behaves as if the spring was maintained and the diagram follows a line of constant stiffness (this approach is used in acoustic discretion of ships to determine the behavior of the very rigid cables capable of transmitting noises that can be found in a distant sound field).

So we have the following diagram:

Equivalent circuit Corresponding diagram

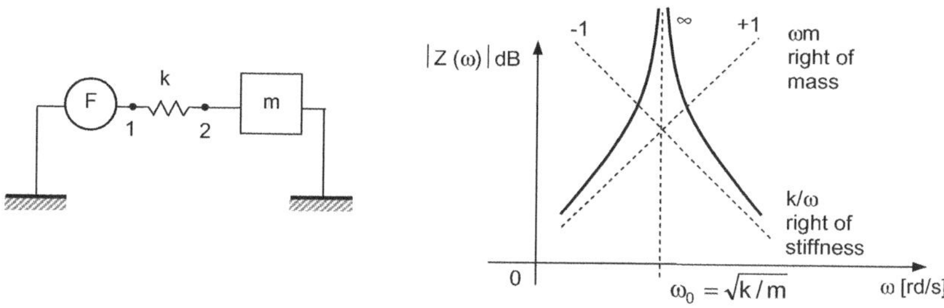

And we have:

$$\frac{1}{Z_2} = \frac{1}{Z_k} + \frac{1}{Z_m} = \frac{j\omega}{k} + \frac{1}{j\omega m} \tag{1.86}$$

1.4.3.2. *Resonance*

The model is the following:

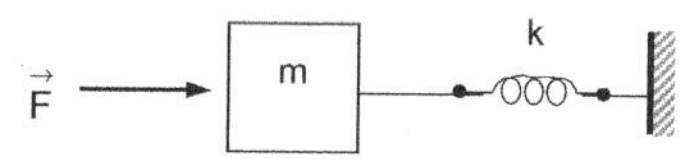

If we consider the same circuit with a fixed point at the end of the spring (this fixed point may be a mass beyond its resonance frequency), the force is then applied to the mass m, which transmits it to the spring k (m has very large and infinite stiffness in the explored frequency band).

At low frequencies, the system's impedance is about equal to that of the spring K and it follows a line of constant stiffness. At high frequencies, the impedance is roughly equal to that of the mass and follows a line of constant mass.

At resonance frequency, the impedance is controlled by the damping that always exists physically, the mechanical impedance is very low and the level of vibration is very important for a weak force (it is the very destructive effect of resonance in equipment). And we have the diagram:

Equivalent circuit Corresponding diagram

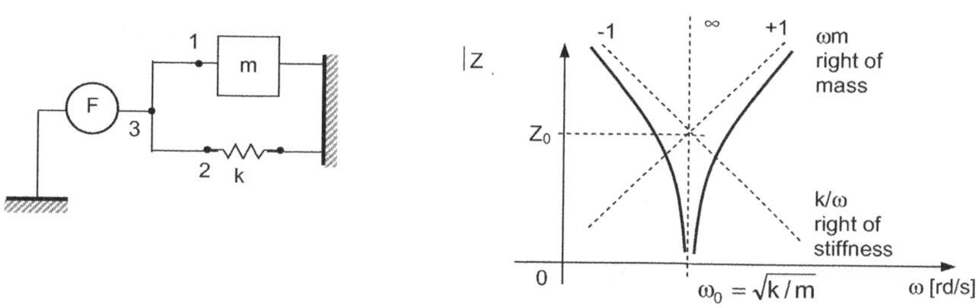

And we have:

$$Z_3 = Z_2 + Z_1 = jm\omega + \frac{k}{j\omega} \tag{1.87}$$

Thus, in a system with 1 DOF, the stiffness line and the line of mass intersect in a point Z_0, for a frequency, which has as a value:

$$f_0 = \frac{1}{2\pi}\sqrt{\frac{k}{m}} \tag{1.88}$$

and which corresponds to the resonant frequency.

In practice, there is always a damping c that can be represented in parallel with k. The value of c is read directly for the resonant frequency. This value is expressed in F/v [$N.s.m^{-1}$]. If we represent an equivalent circuit with damping, it becomes:

Equivalent circuit

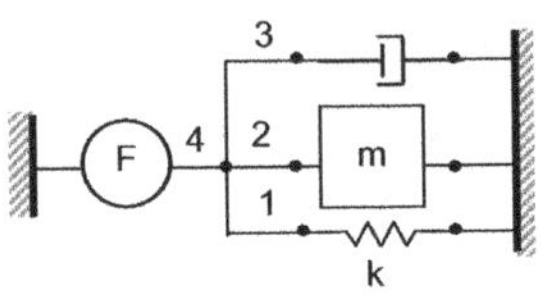

Corresponding diagram

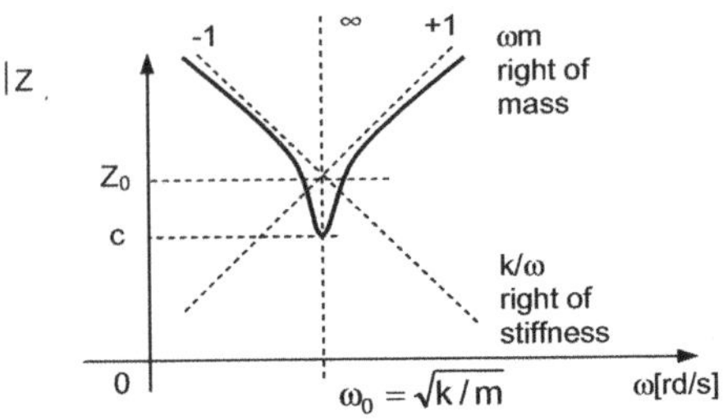

Having:

$$Z_4 = j\omega m + \frac{k}{j\omega} + c = \frac{jZ_0\omega_0}{\omega}\left(\frac{\omega^2}{\omega_0^2} - 1\right) + c \tag{1.89}$$

$$\begin{cases} \omega = \text{excitation pulse} \\ \omega_0 = \text{resonance pulse} \\ Z_0 = \sqrt{km} = \text{value of the impedance to the resonance } \omega = \omega_0 \end{cases}$$

We see that for $\omega = \omega_0$, the term between brackets is null and that there remains only c, the term of damping. So at the frequency or resonance pulse, the impedance Z = c. The curve of phase is shown in the following figure:

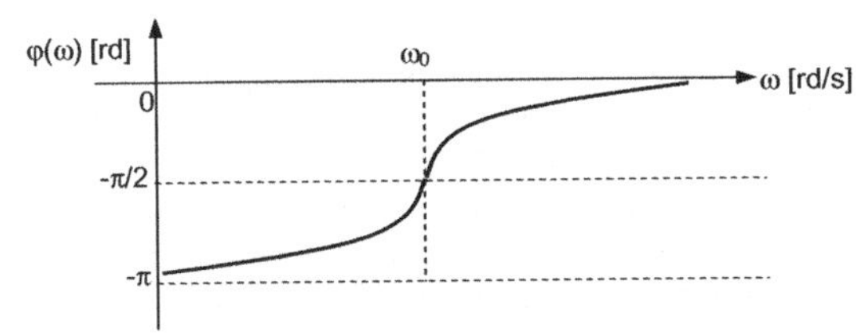

For frequencies lower than the resonance frequency, dephasing corresponds to stiffness (–90°), it goes to zero at the resonance and corresponds to the mass (+90°) beyond this frequency. The rate of change in the phase angle in the vicinity of the resonance $\frac{d\varphi(\omega)}{d\omega}$ also depends on amortization c. It is much faster than "c" is small.

NB: Damping measure of a system with 1 DOF. We have already defined the parameters to measure this damping. So we find:

- the reduced damping which is given by: $\xi = \frac{c}{c_c}$.

With: $\begin{cases} c = \text{depreciation value to measure} \\ c_C = 2m\sqrt{\frac{k}{m}} = 2\sqrt{km} = \text{critical damping} \end{cases}$

that we can connect to the maximum amplitude resonance equal to the Q-factor coefficient,

with: $$\xi \approx \frac{1}{2Q} \rightarrow 2\xi = \frac{1}{Q} \quad \text{and} \quad \frac{c}{Z_0} = 2\xi \tag{1.90}$$

1.4.3.3. *Impedance of a system with damping and 2 DOF*

The simplified model is shown in the following figure:

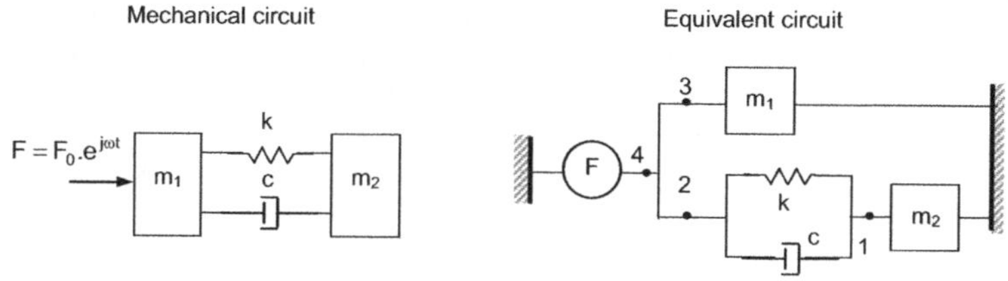

It is a system with two free masses. In this system, the impedance will have a value:

$$z_4 = z_3 + z_2 = z_3 + \frac{1}{\dfrac{1}{z_1} + \dfrac{1}{z_k + z_c}} = z_3 + \frac{z_1.(z_k + z_c)}{z_1 + (z_k + z_c)} = \frac{z_3.z_1 + z_3.(z_k + z_c) + z_1.(z_k + z_c)}{z_1 + (z_k + z_c)}$$

$$z_4 = \frac{z_3.(z_1 + z_k + z_c)z_1.(z_k + z_c)}{z_1 + z_k + z_c}$$

With the values of the different impedances, it becomes:

$$z_4(\omega) = \frac{j\omega.(k - m_1 k - \omega^2 m_2) - (\omega^2 c + m_1 \omega^2 c)}{j(\omega c) + (k - \omega m_1)}$$

$$\rightarrow |z_4(\omega)| = \frac{\left\{\left[\omega.(k - m_1 k - \omega^2 m_2)\right]^2 + \left[(\omega^2 c + m_1 \omega^2 c)\right]^2\right\}^{1/2}}{\left\{(\omega c)^2 + (k - \omega m_1)^2\right\}^{1/2}}$$

In the following figure, consider the type of curve with numerical application:

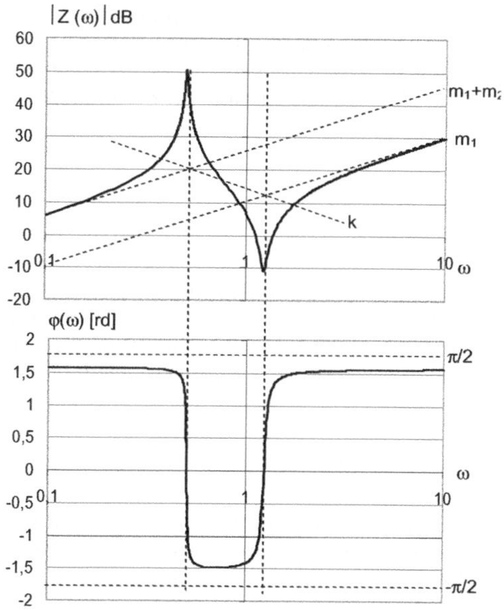

$$\begin{cases} m_1 = 2 \\ m_2 = 1 \\ k = 20 \\ c = 0{,}15 \end{cases}$$

It is noted that such a system presents initially a frequency of antiresonance and then a frequency of resonance.

At low frequencies, the system has an impedance practically equal to that of the total of the masses m_1 and m_2, and at high frequencies the impedance of the mass m_1.

The phase angle goes from 90° to –90° at antiresonance and returns to +90° at resonance, passing by zero for these two frequencies. More complex systems can be determined according to this method, but the following figure is particularly interesting because it matches the impedance of a free bar to the first inflection mode that can be modeled as a continuous system by a succession of discrete systems at 1 DOF, allowing to model all the other modes. For each one of the modes, we get a succession of resonance and antiresonance frequencies in accordance with the following figure:

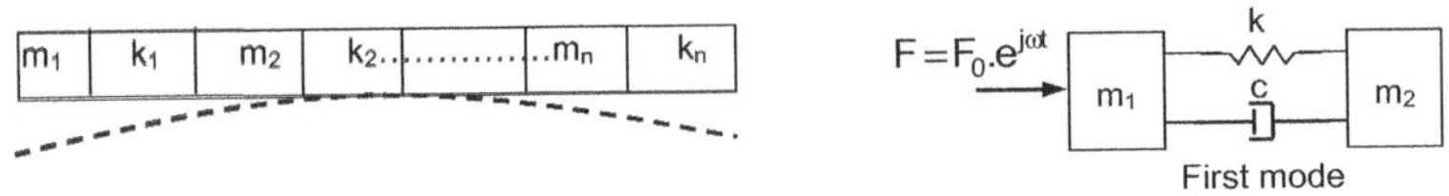

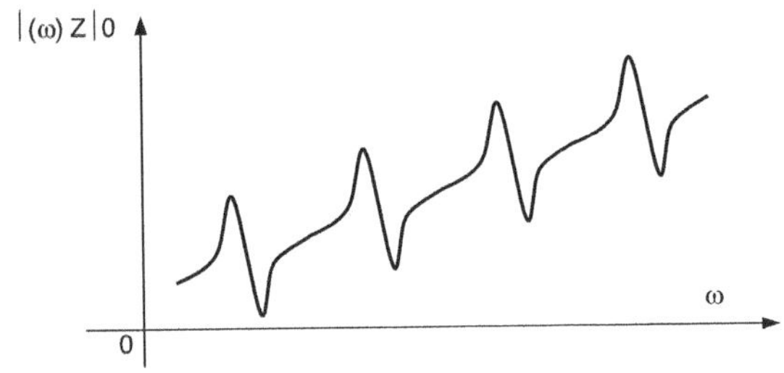

After a resonant frequency, there is always an anti-resonant frequency.

a) Sample application

With mechanical structures becoming increasingly important and more and more complex, it is often difficult to measure their properties when they are completely assembled. By measuring the mechanical impedance of the various parties, it will be possible to determine the characteristics of the complete system. In fact, Norton's theorem will allow us to determine a system's response characteristics when additional impedance is added. Applications are very important for very large materials. For example:

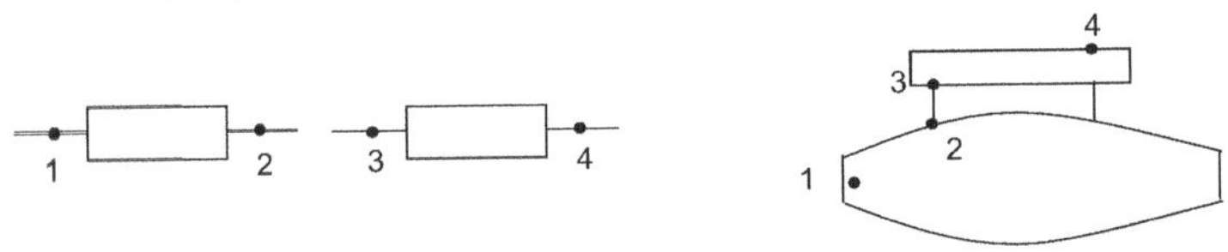

A force F_1 is applied to item 1, it will produce a vibratory speed v_2 in point 2. Transfer impedance Z_{12} will be equal to F/v_2. If we set now the auxiliary on the vehicle (that is, it establishes a connection between points 2 and 3 of the scheme), Norton's theorem allows to know velocity v'_2 in point 2 as a result of force F_1 as a function of this additional impedance. Having:

$$v'_2 = v_2 \cdot \left(\frac{z_{22}}{z_{22} + z_{33}} \right)$$

z_{22} and z_{33} corresponding, respectively, to the impedances of points 2 and 3 measured without a connection between the two systems. Transfer impedance z'_{12} of the connected system will then be equal to:

$$z'_{12} = \frac{F_1}{v'_2} = \frac{F_1}{v_2 \left(\frac{z_{22}}{z_{22} + z_{33}} \right)} = \frac{F_1 \cdot (z_{22} + z_{33})}{v_2 \cdot (z_{22})} = Z_{12} \cdot \left(1 + \frac{z_{33}}{z_{22}} \right) \tag{1.91}$$

And we find an expression similar to that found for the vibrating pot loaded by a system to be tested. Impedances of the connected systems can therefore be determined from the impedances of isolated systems and other combinations z_{14}, $z_{34,...}$, will be obtained in the same way.

NB: It is important and often essential in the determination of standards of vibration tests to consider the mechanical impedance of the source of the exciter. Indeed, when a frequency spectrum is measured on a specimen that will be studied, the most important points of the spectrum generally correspond to resonant frequencies, therefore, to very low impedances. When the equipment that must be mounted on the specimen is tested on vibrators whose impedance is practically infinite (due to the control system that usually does not control forces but rather other constant parameters), it can result in levels of very important constraints that have nothing to do with those who would exist when the systems are connected.

In fact, when studying the responses of the structure on vibrators, tests are generally carried out at a constant level of excitation and by measuring the response in one or more points of the structure. The curve therefore presents the following form:

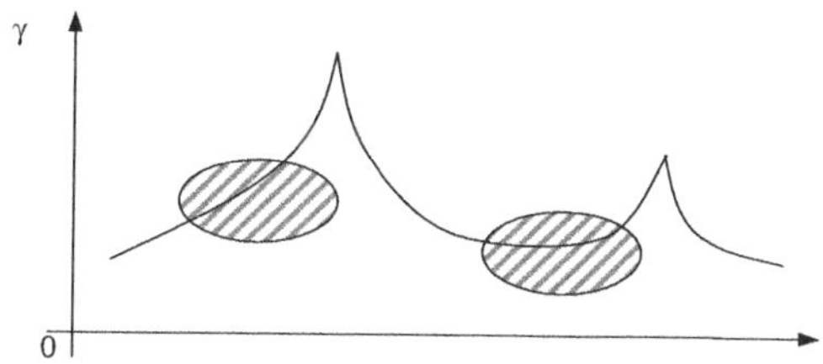

However, very interesting information is missing in this representation. Indeed, to maintain a constant acceleration before the resonant frequency (shaded areas), the vibrator delivers a large force (due to the reaction of the structure, which then presents a very high apparent mass) and because of the control system level of the pilotage system. Henceforth, common systems that have a dynamic range of the order of 80 dB allow ratios from 1 to 10,000. However, if the test was performed with a control at a constant force (piloting starting from a force sensor), the vibratory level of the vibrator would be more constant and would diminish considerably before each resonant frequency because of the reactions of the structures (constant excitation force) and the curve obtained would have the following usual form:

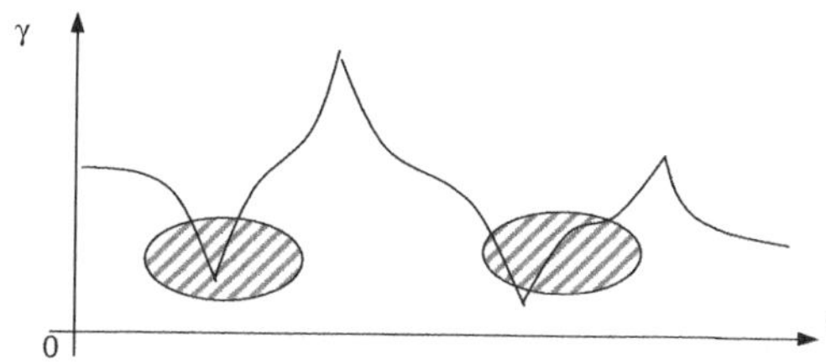

The results obtained starting from this curve are already much more interesting in the studies of the structures, since it is rather easy to know (or to measure) the forces applied and therefore to determine the answers. Measurements of impedance by taking into account the source impedance of the discharger will be then more representative of reality. Indeed, if the total impedance, including parts of the adaptation, has a finite value, the level of acceleration to which the structure is subjected will be affected. The influence of this finite impedance will appear in the following form:

$$\gamma_s = \frac{1}{\left(1+\frac{Z_s}{Z_v}\right)} \cdot \gamma_1 \tag{1.92}$$

With:

$\frac{Z_S}{Z_V}$ = as the ratio of the impedances of the fibrous structure to that of the vibrator.

γ_1 = acceleration of the vibrator in non-charged free mobile system.

γ_s = acceleration on the structure at the point of connection with the vibrator charged.

b) Study of test frame

The mechanical impedance or the apparent mass of the adaptation parts can be measured directly on the vibrator by using the following assembly. This arrangement allows a very accurate measurement of apparent masses and mechanical impedances of adapter pieces. This single method is valid because it helps determining resonances and antiresonances.

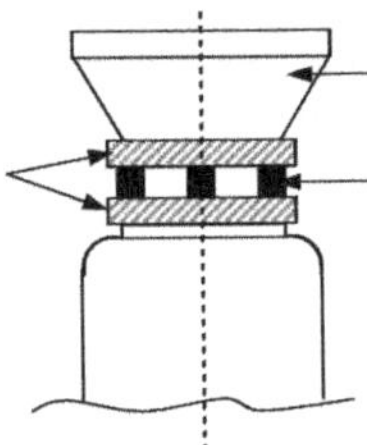

A second test with the connected structure to be tested will determine from previous formulas the mechanical mass and the reactions of this structure on the fixing points. From the signal forces and acceleration measured by the impedance heads, it will be possible to carry out mass cancellations. In case the apparent mass is not constant, it will be possible to make a cancellation from a prerecorded signal.

1.4.4. Problems with environmental laboratory simulation

This simulation is generally done from a frequency spectrum recorded in DSP on the specimen that will be studied or from an imposed normative document. The test program will be based on these data. With the examples:

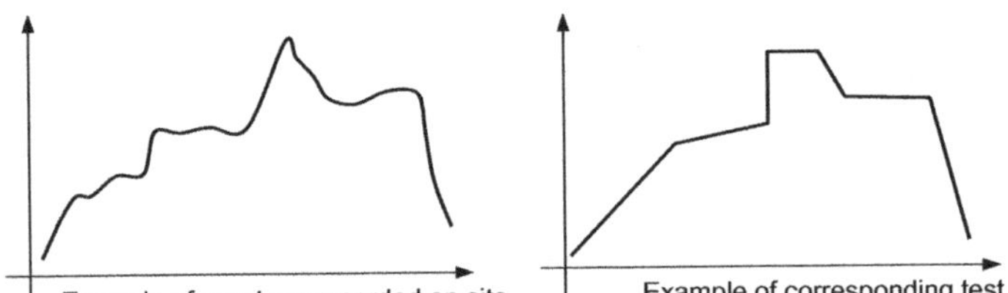

It is necessary to admit in this case that the impedance of the support is very large compared to the impedance of the equipment and that the inertia of the latter will not influence the movement. The level of vibration of the equipment will be different if the fixing impedance or the medium impedance is changed, as studied previously. Force control techniques can then be used but they are quite tricky to implement. The assembly is that of the preceding figure, using force sensors instead of impedance heads, which makes it possible to use a control method of the vibrations by limitation of the force infected on the specimen at the level of conjunction with the frame.

Another solution is to use the "notching" technique that allows limiting on the material in the test or on the vibrating table, the acceleration to a preset level with measurements into operational, when they are available, which then takes account of the real power injected on the equipment tested by the carrier, which is of course limited (while an electrodynamic vibration generator, for example, has an *a priori* infinite impedance and is limited only by its own power). This is the case of test materials that, by their resonances, are likely to influence the received excitement of the carrier. Particularly, when the mass of the material in relation to that of the carrier is important.

Moreover, as the possible failure is at a low level of assembly (tests of feasibility of qualification, locally on subsets), a correction of the failure does not challenge its high level of assembly (i.e. the whole of the structure, the carrier plus subsets). That makes it possible to usually carry out these tests at the end of the development, contrary to the static and fatigue tests, in general dimensioning on a high level of assembly, and which are carried out at the beginning of development.

If a defect revealed under vibration test questioned (which should be the exception that proves the rule) the high level of assembly, it would probably be a design or manufacturing defect.

From the fact that failure is normally expected at a low level assembly, the entry conditions of the excitement in the test structure are little critical. The difference between the real conditions of operational use and those of the test has relatively little importance, which explains the usual shapes of the trial frames: squares, plates, wafers, not very representative in terms of mechanical impedance compared to the real conditions and yet, of course, the tests are very useful.

A random vibration signal is analyzed in PSD (g^2/Hz) and a transitory shock spectrum (SRS). The obtained spectra are usually very important points that, as already stated, correspond to the resonant frequencies of the structure with or without equipment. The level of vibration of the test will be determined from these points of the spectrum, which correspond to the frequencies where the impedance is low and that can be tested properly.

We can, in the same way, note that at antiresonance frequencies of the equipment, it behaves like an absorbing circuit, and the vibration is very weak on the level of fixing. These frequencies are dangerous for fragile equipment if a significant level is applied with a very high value impedance due to the control system (using then notching). Thus, a rigorous test should involve not only the vibrational spectrum at the point of attachment of the equipment, but also the impedance source (or apparent mass) that allows determining the transmitted force to the equipment. The most important levels generally correspond to local vibrations involving low masses.

Thus, if we have the records of the excitation forces injected with the interface on the material as well as the knowledge of the boundary conditions, records which will be supplemented by a measurement of apparent mass are also measured on the vehicle at the fixing points. The combination of these settings will allow obtaining the spectrum of force, which we will simulate by approximation (noticing that even for a spectrum of strength, it is possible to get to the slopes of 6 dB or 12 dB per octave by single and double integration).

The spectrum of force will be obtained by the following equation:

$$F = m_a \cdot \gamma \tag{1.93}$$

F = force transmitted to the equipment.

γ = acceleration measured without equipment.

m_a = apparent mass-side source at the attachment point.

NB: For random vibrations, if the PSD S(f) is applied to the point (a), the DSP to point (b) shall be given by:

$$S_b(f) = \frac{S_a(f)}{|Z_{ab}(f)|^2} \tag{1.94}$$

And the effective value of the velocity at point (b) will be given by:

$$v_{beff} = \left(\int_0^\infty \frac{S_a(f)}{|Z_{ab}(f)|^2} . df \right)^{1/2} \tag{1.95}$$

1.4.5. Conclusion

In summary, it appears that the techniques of mechanical impedances can be useful for very important systems with complex geometries and their study allows to better adjust the testing standards involving the source impedance as well as the controlled force vibration tests that allow a better simulation of the vibratory environment. They can also be used for the study of adaptation parts and testing frames that often cause design problems. In addition, some tests with mechanical impedance simulation can be more practical and easier to implement.

Mechanical impedances techniques therefore constitute a complement to the vibration and shock tests since they make it possible to study the dynamic behavior of the structure, and also to determine the test routine, which will be very much in conformity with reality and very often less dangerous for the equipment. The study of adaption pieces on great power vibrators on materials of large masses justifies the use of these methods. The implementation of these techniques is relatively complex and delicate.

It should be noted that from measurements of vibrations (easily measurable accelerations), it is very difficult to directly trace the forces except in particular cases, which involve a consequent work based on modeling (see mechanical general annex of the GAMEG13 in Chapter 11). On the other hand, if we can measure the accelerations at the points of entry of the equipment and we inject these measurements in the same entry points of the equipment represented by its model, we take into account the interactions between this equipment and its carrier support. Therefore, usually the tests of vibrations are carried out in acceleration on generators of electrodynamic or hydraulic vibrations.

Of course, for the materials whose mass does not exceed a few tens, even hundreds of kilograms, the placement of test frames follows the few following usual rules, which generally makes it possible to work in controlled usual parameters, meaning: displacement, velocity, acceleration (see Appendix F for more precision), with:

- the mass of the test frame must be at least five times that of the material to test;

- its first mode of vibration must be beyond the maximum frequency explored by taking into account the mass reported by the specimen under testing, or at least not interfering with one of the first three modes of the equipment under testing.

When it is difficult to hold these constraints, it is advisable to pay detailed attention at the level of the cockpit by adapting available strategies (with the strongest, average, Notching, etc.).

1.5. Electromechanical analogies

1.5.1. Overview

We already evoked this method of study of the mechanical systems. It is indeed well known that many mechanical problems lead to the same systems of differential equations as those governing the electrical supply networks. However, among all oscillating systems, the electronic systems are those which were studied and thus the best known ones. It seems that the study of some mechanical systems is easier to undertake with the use of electromechanical analogies known, especially since there are synthesis and analysis software of high-performance electrical networks (such as SPICE, for example).

We will not elaborate on its history that goes back to 1860 with Maxwell and then with Darrieus in 1929, but it should be noted, nevertheless, that this method can be very useful. We will specifically deal with the so-called analogy "force intensity" due to its more practical application. We will therefore elaborate on this type of method at a greater length.

The study of equivalent networks (in the sense of differential equations that govern them, as we will see later) shows that we can indeed perform the following analogies:

$$\begin{array}{llll} -\mathrm{F}\,(\text{force}) & \leftrightarrow & \mathrm{i}\,(\text{intensity}) & \mathrm{i}=\mathrm{F} \\ -\dot{x}\,(\text{velocity}) & \leftrightarrow & \mathrm{u}\,(\text{voltage}) & \mathrm{u}=\dot{x} \\ -\mathrm{m}\,(\text{mass}) & \leftrightarrow & \mathrm{C}\,(\text{capacity}) & \mathrm{C}=\mathrm{m} \\ -\mathrm{c}\,(\text{dampening}) & \leftrightarrow & \mathrm{G}\,(\text{conductance}) & \mathrm{R}=1/\mathrm{c} \\ -\mathrm{k}^{-1}\,(\text{compliance}) & \leftrightarrow & \mathrm{L}\,(\text{self}) & \mathrm{L}=1/\mathrm{k}=\mathrm{k}' \end{array} \quad (1.96)$$

Moreover, due to the laws of association, mechanical mobilities in this analogy are the equivalent of the electric impedances. It will be enough to simply transpose the mechanical network in an electrical supply network, from which it will be very convenient to solve the mathematical model given.

EXAMPLE 1.8. The following mechanical system, and its similar electrical network:

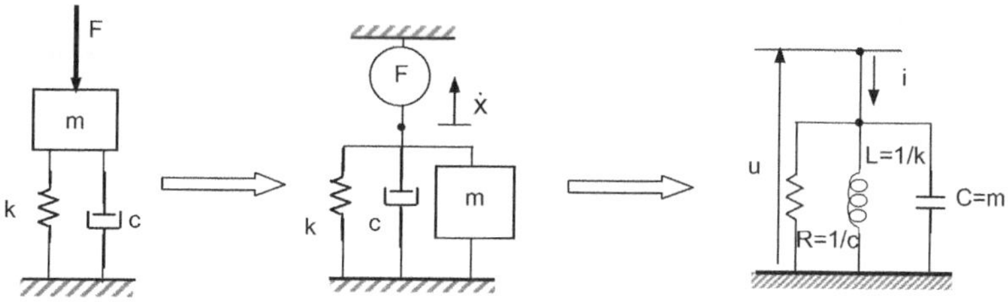

It thus becomes:

$$u = Z_e.i \qquad \text{with}: \frac{1}{Z_e} = jm\omega + \frac{k}{j\omega} + c$$

Expression is identical to the mechanical impedance of the system.

The following analogy also exists:

$$\begin{array}{llll} -\mathrm{F}\,(\text{force}) & \leftrightarrow & \mathrm{u}\,(\text{voltage}) & \mathrm{u}=\mathrm{F} \\ -\dot{x}\,(\text{velocity}) & \leftrightarrow & \mathrm{i}\,(\text{intensity}) & \mathrm{i}=\dot{x} \\ -\mathrm{m}\,(\text{mass}) & \leftrightarrow & \mathrm{L}\,(\text{self}) & \mathrm{L}=\mathrm{m} \\ -\mathrm{c}\,(\text{dampening}) & \leftrightarrow & \mathrm{R}\,(\text{resistance}) & \mathrm{R}=\mathrm{c} \\ -\mathrm{k}^{-1}\,(\text{compliance}) & \leftrightarrow & \mathrm{C}\,(\text{capacity}) & \mathrm{C}=1/\mathrm{k}=\mathrm{k}' \end{array} \quad (1.97)$$

The advantage of the strength/intensity analogy comes from the fact that the speed frame zero has an equivalent electric zero potential, ground reference. In addition, the mass m will thus have a capacity with a terminal on the ground (voltage zero), the other terminal having an equal potential of the velocity $\dot{x}$.

Such an analogy is at the base of analogical simulators that allow very simply, by using primary electric circuits, the study of discrete dynamic mechanical models. This historical approach tends to be replaced by tools for circuit simulations as already suggested (SPICE, etc.).

1.5.2. Maxwell's analogy of the "force tension" type

It is the first electromechanical analogy carried out, which equates the condenser charged to a tense spring and introduces the force tension analogy. In this analogy, we match the equations related to a mechanical node to the existing relationships in a corresponding electric branch. Indeed, consider the following mechanical node where a mass, a spring and a damper come together at the same time.

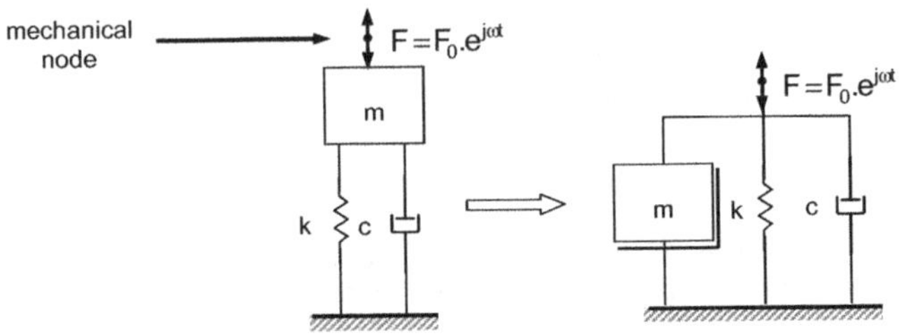

NB: The representation of the mass in the previous model is the one used in the United States; it requires having a connection with the reference of the ground mass. It will be used more subsequently.

Correlatively with a mechanical node, let us consider the following electric branch, involving, in series, a resistance, a capacity and an inductance. This branch is subject to a dpp U.

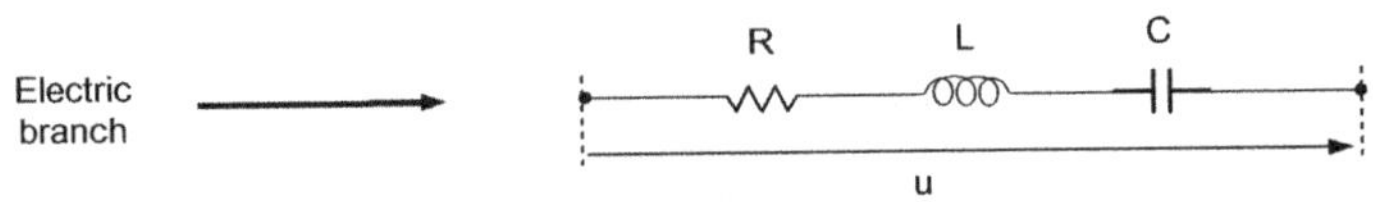

Let us write the equality of forces acting at the point of application of F in the mechanical system; by the D'Alembert's principle, it becomes:

$$m\frac{dv}{dt} + cv + k\int v.dt = F \qquad (1)$$

Yet in the electric branch we have the following equation :

$$L\frac{di}{dt} + Ri + \frac{1}{C}\int idt = u \qquad (2)$$

So, if we consider the two equations (1) and (2), we can admit that the electrical grid is the transformed by duality of the corresponding mechanical network. The correspondence between mechanical and electrical elements will be the following:

Inductance	mass
Damper	resistance
Stiffness	capacity
Force	tension
speed	intensity
Displacement	quantity of electricity or load

From this simple example, we see a rule of construction of networks that spreads very quickly:

"Mechanical elements arranged in parallel match their electrical counterparts in series and vice versa".

1.5.2.1. *Advantages and disadvantages of Maxwell's analogy*

a) Disadvantages

- We are already seeing one of the disadvantages of this analogy in the rule of construction of the similar network, since the electrical system tends to move away from the physical mechanical system from a structural point of view.

- Similarly, the equivalent power of a frame (velocity zero) is an element in which the intensity is null. It is thus an element that can be removed in the electric diagram.

- The mechanical equivalent of the electric ground (voltage zero) is an element on which no force must act; it is difficult to obtain its physical significance.

b) Advantages

- An electrical circuit has many DOF (sufficient number of variables to characterize the state of the system) that has independent meshes.

- The corresponding mechanical circuit must thus have the same number of DOF.

- Consequently, a mechanical system with 1 DOF will have to correspond an electrical supply network made up of only one mesh.

1.5.3. Darrieus' Analogy "force intensity'

a) Mechanical mass element

The mass known as ideal is such as the acceleration of this assumed infinitely rigid body (in the explored frequency band) is proportional to the force applied. Having the model:

$$F = m\frac{dv}{dt} \qquad (1)$$

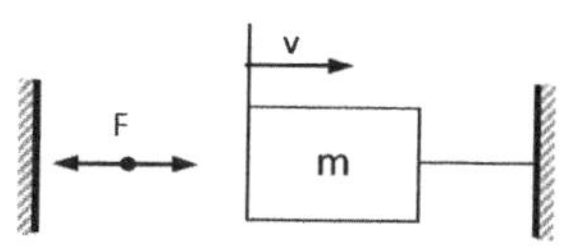

The kinetic energy of this mass at every moment is given by: $W_{cin} = \frac{1}{2}m.v^2$

The electric element of comparison is the capacitor that we assume to be ideal (perfect dielectric, inductance zero, zero and infinite escape series resistance), for which we have the following equation to balance:

$$u - \frac{q}{C} = 0 \qquad \text{with}: q = \int i dt$$

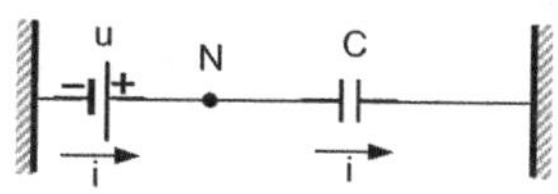

And if one expresses the equation of continuity to the node N, we have: $i = C.\frac{du}{dt}$ (2)

The stored energy is given by: $W_e = \frac{1}{2}Cu^2$

So there is complete identity between the mechanical element mass and the electrical capacitor element.

b) Mechanical damper element

We will suppose that we deal with a damping of the viscous type, which is used the most. In this case, the damper element is such that the relative speed between its extreme points is proportional to the applied force. Having the model:

$$F = c.v$$

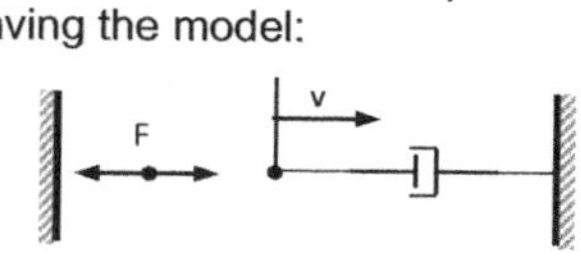

The corresponding electric element in the intensity analogy force is then the resistance element. We have in fact in the following model:

$$i = \frac{u}{R}$$

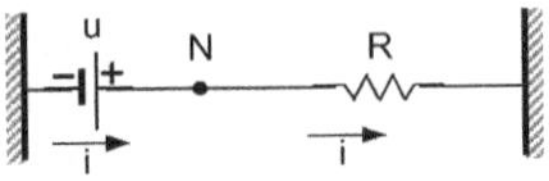

In reality, the ideal dampening element does not exist, because as the study of the rheological modules shows, it is more complex and is made up of a connection in parallel and in series of stiffness or viscous damper elements, for example:

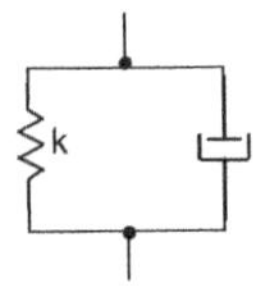

Model of
Kelvin - Voigt

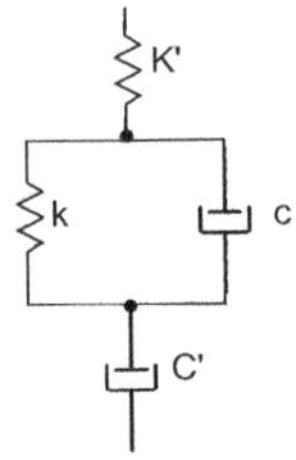

Burghers' Model

These types of damping are easily transposable in the equivalent electric diagram, as we will be able to study later on. However, nonlinear depreciation such as the Saint–Venant model or the elastoplastic model, asks the use of active electrical components (diodes) to get the desired response curves. Or, for example, the following models:

Model of
Saint–Venant

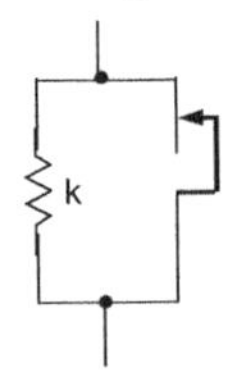

elastoplastic
model

c) Element of mechanical stiffness

The ideal mechanical spring is such that the relative movement of its extreme points is proportional to the applied force. Having the model:

$$F = k.x = k.\int v.dt$$

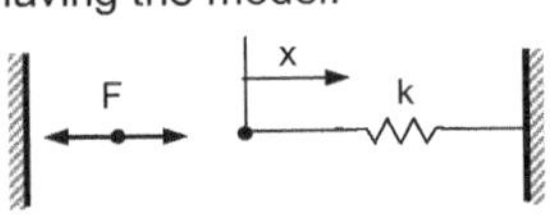

If we express the voltage applied to the terminals of an inductor based on intensity, we have:

$$u = L.\frac{di}{dt} \rightarrow i = \frac{1}{L}.\int u.dt$$

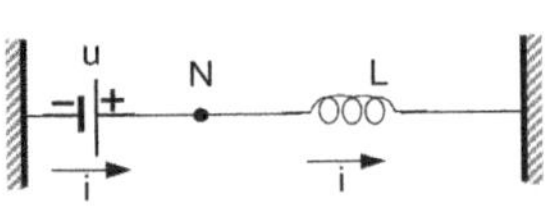

We can deduct from it the correspondence between compliance $k' = \frac{1}{k} = k^{-1}$ and inductance L.

The nonlinear mechanical stiffness elements are replaced in the similar electric network by active elements (operational amplifiers, for example).

1.5.3.1. *Construction of Darrieus' analogy*

Darrieus attempted to translate a similarity of laws. Thus, from the structure mass spring damping, studied previously, he wrote the equation of the nodes in the electrical system with the following rule of construction:

"A mechanical node matches with an electrical node".

Let us consider a mechanical node where a mass, a damper and a spring, come together at the same time. We find the model described earlier namely:

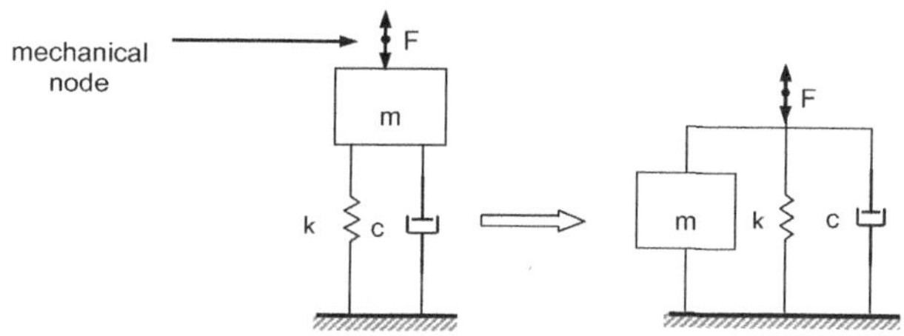

Let us consider the following electric node now, where an inductance, a resistance and a capacity are in parallel.

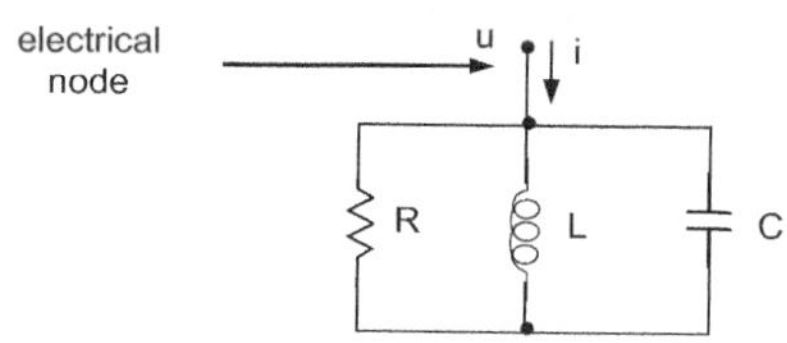

In the case of the mechanical node, we have seen that we have the following relations:

$$m.\frac{dv}{dt} + c.v + k.\int v.dt = F \qquad (1)$$

In the case of the electrical node, we have:

$$C.\frac{du}{dt} + \frac{1}{R}.u + \frac{1}{L}.\int u.dt = i \qquad (2)$$

The correspondence between mechanical and electrical elements is deduced from two equalities (1) and (2), with:

mass	capacity
damping	resistance
stiffness	inductance

force	intensity
velocity	voltage
displacement	impulse

Therefore:

- Any machine elements placed in parallel have corresponding electric homologous elements placed in parallel too.

- Any mechanical elements placed in series have matching electrical counterpart items placed also in series.

Important rule

We find that the fact that to carry out the decomposition of the structure in simple electric elements, the rules applied in the study of mechanical impedance of a system are valid. In particular, a very significant one that requires "mechanical mass" components to have a reference, namely "the earth or ground here", which means that in all electric diagrams the "capacitive" elements, homologous with the "mechanical mass" elements must have one of the two terminals connected to an electric ground.

1.5.4. Units and electrical mechanical correspondence

We directly connect the units of the mechanical system to the units of the electric system. Thus, in the MKSA system, it results in:

- a force expressed in N, an intensity expressed in A corresponds.
- at a velocity expressed in m/s, a voltage expressed in V corresponds.

Two calculation parameters are introduced naturally, which makes it possible to match the mechanical and electric operating ranges in terms of frequency and dimensioning of the various elements, which means:

$$\lambda = \frac{\text{mechanical frequency}}{\text{electrical frequency}}$$

$$\rho = \frac{\text{mechanical mobility}}{\text{electrical impedance}}$$

For the correspondence masses springs damping, it becomes:

- mass and capacity:
  - mechanical mobility: $M_m = \frac{1}{jm\omega_m} = Z_m^{-1}$
  - electric impedance: $Z_e = \frac{1}{jC\omega_e}$

Obtaining: $$m = \frac{1}{\rho.\lambda}.C \qquad (1.98)$$

- damping and resistance:
  - mechanical mobility: $M_m = \frac{1}{c} = Z_m^{-1}$
  - electric impedance: $Z_e = R$

Obtaining: $$c = \frac{1}{\rho}.\frac{1}{R} \qquad (1.99)$$

- stiffness and inductance:
  - mechanical mobility: $M_m = j\frac{\omega_m}{k} = Z_m^{-1}$
  - electric impedance: $Z_e = j\omega_e L$

Obtaining: $$k = \frac{\lambda}{\rho}.\frac{1}{L} \qquad (1.100)$$

Thus, the choice of the corresponding ranges of variation of electrical components will provide a range of variation of mechanical components. The choice of λ and ρ will allow to precisely define this range and get realistic values of the components (see Chapter 8 for applications).

Let us point out the anomaly due to the choice of the first analogy due to Maxwell who resulted in defining the mechanical impedance as the F/v ratio. Indeed, in the analogy that we adopted "force intensity", we define the electric impedance by Z_e = u/i, as usually with Z_e in ohms. However, the equivalent in the mechanical diagram would then be the impedance defined by v/F, which in fact by "the use" is called "mechanical mobility", that is to say the opposite of the usual mechanical impedance, which remains defined by $_m$ = F/v.

1.5.5. Establishment of electric diagrams

a) Systems in traction compression

The models are as follows:

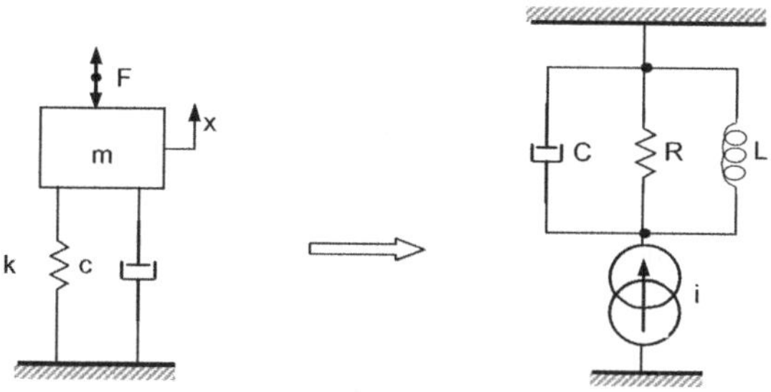

With:

mass	m	- capacity	c
damping	c	- conductance	$\frac{1}{R}$
compliance	$\frac{1}{k}$	- inductance	L

force	F	- intensity	i
velocity	v	- voltage	u
displacement	X	- impulse	$\int vdt$

b) Systems in rotation

The models are as follows:

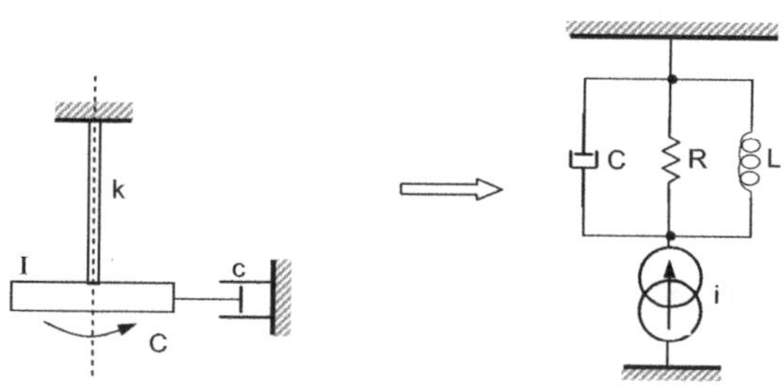

inertia	I -	capacity	c
damping	c -	conductance	$\frac{1}{R}$
compliance	$\frac{1}{K}$ -	inductance	L

match	c -	intensity	i
angular velocity	Ω -	voltage	u
angular displacement	θ -	momentum	$\int vdt$

c) Mechanical and transformer lever analogy

Having the model of lever according to the following figure, we have, from the point of view of mechanical powers:

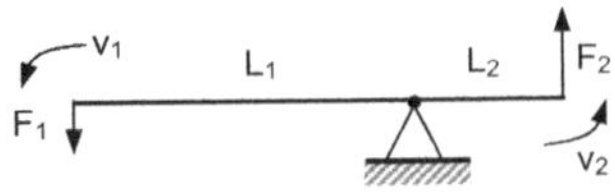

$$F_1 v_1 = F_2 v_2 \qquad \text{Hence:} \quad \frac{F_1}{F_2} = \frac{v_2}{v_1} = \frac{L_2}{L_1}$$

In electricity, the transformer provides us with a similar expression:

$$\frac{i_1}{i_2} = \frac{u_2}{u_1} = \frac{n_2}{n_1}$$

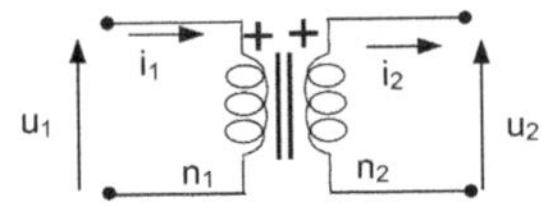

This result is also valid for pulleys, gears, etc. In all cases, we will characterize the transformer by its current transformation ratio.

d) Elimination of the transformer

(a) Resistance

The models are as follows:

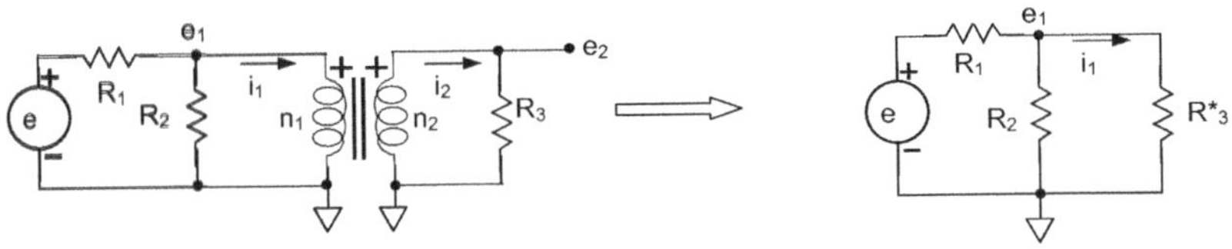

The electric elements extracted from the secondary (2) to the primary (1) are:

Electrical elements	Mechanical elements
Resistance: $R_3^* = R_3 \cdot \left(\frac{n_1}{n_2}\right)^2$	Dampening: $c^* = \frac{c}{\left(\frac{n_1}{n_2}\right)^2}$

(b) Inductance

The models are as follows:

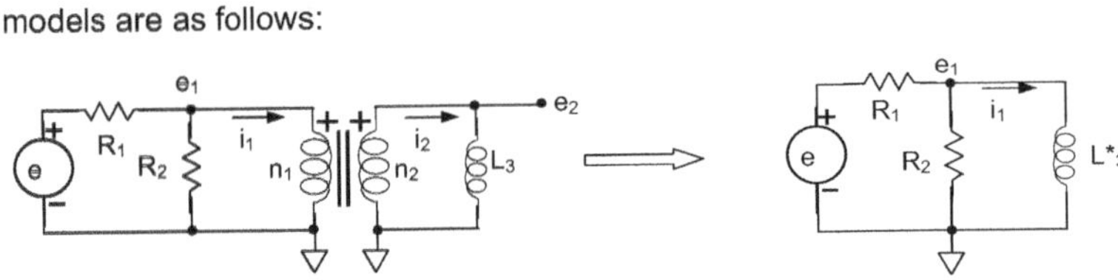

The electric elements extracted from the secondary (2) to the primary (1) are:

Electrical elements	Mechanical elements
Inductance: $L_3^* = L_3 \cdot \left(\frac{n_1}{n_2}\right)^2$	Stiffness: $k^* = \frac{k}{\left(\frac{n_1}{n_2}\right)^2}$

(c) Capacity

The models are as follows:

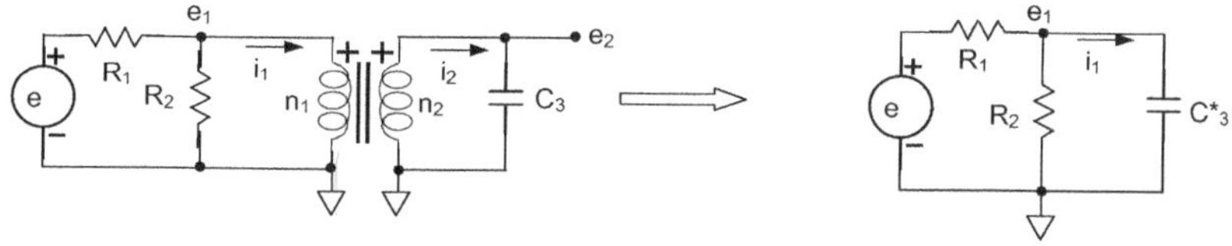

The electric elements extracted from the secondary (2) to the primary (1) are:

	Electrical elements	Mechanical elements
Capacity :	$C_3^* = \dfrac{C_3}{\left(\dfrac{n_1}{n_2}\right)^2}$	Mass : $m^* = \dfrac{m}{\left(\dfrac{n_1}{n_2}\right)^2}$

e) Application: example of the electrodynamic generator of vibrations

The simplified models (neglected mechanical damping) are then the following:

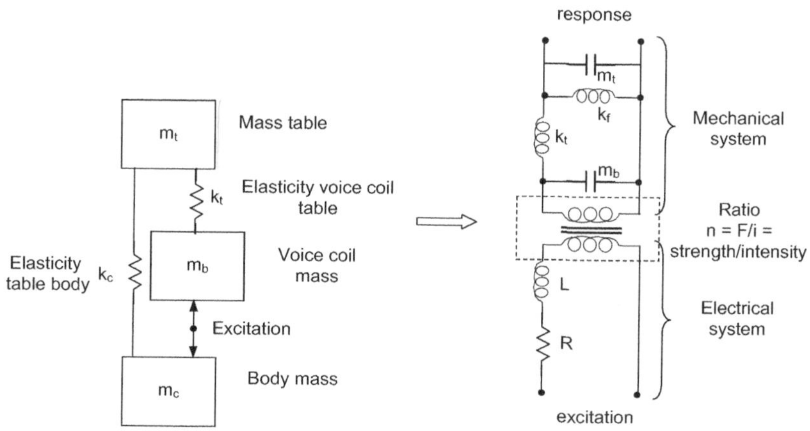

A comprehensive study is provided in Chapter 8.

The importance of these electrical analogies can be summarized as follows:

- ease of realization of the model;
- ease of measurement performance;
- ease of study of the responses.

It is obvious, on the other hand, that this method quickly becomes very complicated as soon as the mechanical system comprises many disparate geometries. In addition, with the progress of calculators, there are currently finite elements coupled with calculation programs along with modal analysis software and drawing software that also meet the need for more complex structures. Despite everything, in the case of simple systems, these methods are very effective.

It should be noted that the same remark could be made for analog and logic calculators, which were largely used before the appearance of the large current computer codes.

1.6. Analog and logic computer simulation

1.6.1. Overview

The preceding analogies can be interesting in certain cases of study of systems, because of the possible association of analogy methods and those of simulation. Indeed, electronic analog computers may be used for solving many mathematical and physical problems as well as for the simulation of various processes or systems. The computing power of these methods can be extremely useful in vibration problems.

1.6.2. Structure of an analog and logic calculator

We perform arithmetic unit standards able to form the differential equations to study which one interconnects accordingly. They are usually made up of:

- Linear elements: attenuators, operators made with AO, amplifier inverters, summoners, etc.

- Nonlinear elements: black boxes performing functions such as multipliers, dividers, squaring and square root extraction, sine, cosine, logarithm, exponential, etc.

- Reading and control units.

- Operating modes.
  - manual;
  - repetitive;
  - hybrid (calculation programs);
  - bodies and elements of sequential logic, etc.

NB: It should be noted that in the types of processed differential equations, a single independent variable is treated in the calculation of "pure" analogs. It is in fact the time that is used for the dynamic study of systems with synthesis of the process giving place to oscillatory phenomena in vibration, for example.

1.6.3. Resolution of various systems

a) Linear systems with constant coefficients

Simple mechanical systems such as: $F(t) = ay''+by'+cy$

And for physical systems involving more than 1 DOF, there will be several equations.

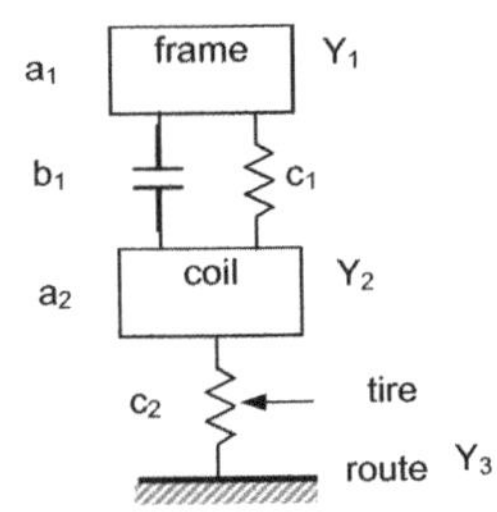

- Example of the simulation of a suspension of car.

System:

$$\begin{cases} a_1 y_1'' + b_1.(y_1' - y_2') + c_1.(y_1 - y_2) = 0 \\ a_2 y_2'' + (y_2' - y_1') + c_1.(y_2 - y_1) + c_2.(y_2 - y_3) = 0 \end{cases}$$

From where the shape of the analog calculator derives:

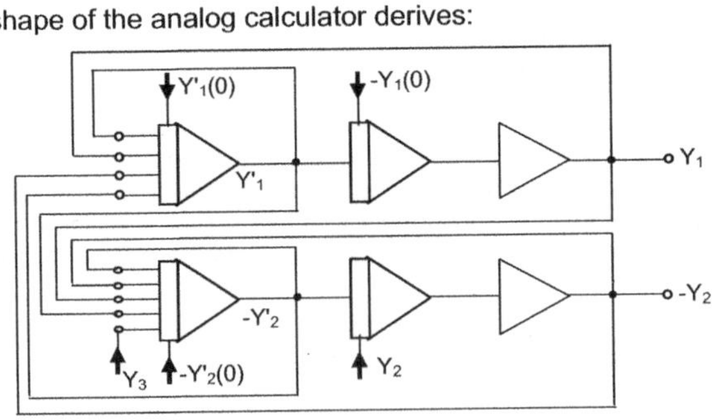

It should be noted that we can utilize the IC.

- Example of solution of a linear differential equation of the second order with constant coefficients.

$$F(t) = ay'' + by' + cy$$

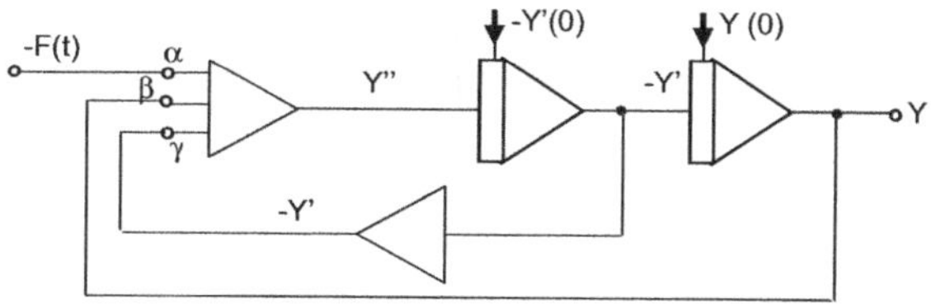

$$\alpha = \frac{1}{a} \quad \beta = \frac{c}{a} \quad \gamma = \frac{b}{a}$$

We converted the traditional writing by: $y'' = \frac{F(t)}{a} - \frac{b}{a}.y' - \frac{c}{a}.y$

b) Nonlinear systems – example of the Van Der Pol's equation

$$ay'' + b.(y^2 - 1).y' + c.y = 0$$

It intervenes in certain problems including nonlinear devices of dampers (threshold phenomena, non-viscous frictions, saturation, hysteresis, etc.), whose simulations are done starting from diodes, relays, etc.

c) Systems with variable coefficients – example of Ricatti equation

$$y'' + f(t).y' + a.g(t).y = 0$$

This equation describes the oscillations of a simple pendulum suspended with a vibrating spring (phenomena with variable environments, randomized, etc.), which are very difficult to integrate.

d) Simulation of transfer functions

e) Systems governed by partial derivative equations

They are solved by iterative methods. Under repetitive operation, we manually recycle the calculation in order to evaluate, for example, the influence of a parameter on the solution of the problem.

Under iterative mode, a lot of times we reproduce the calculation cycle, but by changing each time or "passes", a condition of the problem and this automatically.

The main advantage is to implement buckled procedures in which the variation of parameters for which the structure of the calculation is controlled on the result obtained in progress or at the end (it should be noted that in this case, the use of memory is essential). Essential opportunities include:

- parametric sweeping;
- problems of boundary conditions;
- parameter optimization (method of the gradient);
- stage process, systems with distributed constants;
- research of mathematical models, etc.

1.6.4. Sequential programming of analog computers

There are many natural processes that bring into play continuous functions and switching functions. In this case, an analog representation is not enough; it must be complemented by decision structures of combinatorial and sequential nature. It is the "widespread" operation of calculators.

- For example, the Nyquist layout of a system defined by its transfer function.
- Determination of signal correlation functions.
- Analysis of continuous and discrete systems.
- Problems of regulation and subjugation.
- Optimization problems (continuous periodic disturbance method), etc.

1.7. Conclusion

We will retain of all these methods and their utility is especially found in the following fields, without being exhaustive:

- vibration of structures;
- calculation of suspensions of structures (vehicles, cabinets, etc.);
- calculation of braking systems;
- study of movement of movable parts, frequencies;
- vibration efforts;
- study of the vibrations on vehicles (air, sea, ground);

The expected educational effect and the purpose of this introduction to all these modeling methods was to understand the mechanical structures and the general concepts that need to be known. They remain effective for the areas for which the studied mechanical systems are relatively simple. There is, of course, as already mentioned, means of heavier calculations (finite element calculations and modal analysis) that allow a finer and more comprehensive analysis of large systems, but which are beyond the scope of this book.

2

Signal Analysis

In this chapter, the main mathematical tools for the description of vibrating signals are discussed. Our approach is a traditional one and essentially presents a harmonic or sinusoidal regime and the random-type signals. Temporal and spectral analysis techniques will be also addressed.

An introduction to typical vibration signals in laboratory testing will also be performed. Indeed, the latter have some specific features that are essential to be known. A few examples of practical applications will be given from the description of analog and numerical analyzers. Due to pedagogical reasons, emphasis will be placed on the analog approach. Numerical developments will be limited.

Moreover, the main constraints of calculation, in terms of signal processing, are common to both approaches.

2.1. Overview

There are different analysis methods for random signals or not, meaning deterministic ones or not. They are based on temporal or spectral approaches.

2.1.1. Temporal or spectral form

In the case of deterministic signals, a signal can be represented in three dimensions. Thus, the representation of a sine wave on an amplitude frequency plane will give out a line of amplitude A_i at frequency f_i. This represents the spectrum of a sinusoidal wave. The projection on an amplitude–time plane will reproduce the original temporary shape of the signal.

Therefore, in the event that the signal is represented by its frequency spectrum, it will be integrated during a certain time interval and weighted by the appropriate Fourier function, Fourier series (FS) or Fourier transform (FT). The result is a function of amplitude, frequency and phase frequency. Of course, with digital technology, the equivalent will be carried out with techniques of fast FTs.

2.1.2. Correlation functions

If instead of representing this waveform in real time, a pseudo-time τ equal to the difference between two sample points is used, we have formed a correlation function $R_{11}(\tau)$. They are also the only functions that are likely to exist in case of stationary random signals.

These techniques have been used for many years to extract a signal from a noise, or to detect the presence of periodicity in a random signal.

If a signal is correlated with itself (autocorrelation), random components are quickly reduced to zero in the field of correlation and the periodic components will be immediately detected as a continuous function in the field of time delay τ. This correlation function or autocorrelation function, in a general case, is defined by:

$$R_{11}(\tau) = \lim_{T\to\infty} \frac{1}{2T} \cdot \int_{-T}^{T} f_1(t).f_1(t+\tau).dt \tag{2.1}$$

Similarly, the correlation of two signals or intercorrelation, cross-correlation, or mutual correlation, will highlight the mutual properties that these signals can have. The cross-correlation function is defined by:

$$R_{12}(\tau) = \lim_{T\to\infty} \frac{1}{2T} \cdot \int_{-T}^{T} f_1(t).f_2(t+\tau).dt \tag{2.2}$$

T being duration of the window of observation of the signal [s] as well as the integration time.

An important application of cross-correlation is to determine, for example, between several sources, the cause of vibrations measured at a given point. Cross-correlation is then performed between this point and each possible source. It is clear here that the choice of these sources will provide the most interesting information. In this case, an average value is still obtained by integration in function of time. The result is a function represented in the pseudo-field of time τ.

2.1.3. Probabilities

If the time amplitude function is represented as a percentage of time during which the function remains contained between certain amplitude limits, the result will be a function of probability. This function is independent of frequency, and it is then represented in the field of amplitude, as a time percentage (histogram concept).

NB: The previously defined stationary complex signals can be decomposed into a sum of sine functions and will be analyzed in the same manner.

Probability functions are extremely useful to predict behavior and responses from many mechanical structures. Therefore, in random vibration tests, probability features allow obtaining the estimated levels of constraints of structures compared to those encountered in real operating conditions.

The probability density function can be defined by:

$$P(y) = \lim_{\substack{T\to\infty \\ \Delta y\to 0}} \frac{1}{T.\Delta y} \cdot \sum_{i=1}^{\infty} t_i(y, y+\Delta y) \tag{2.3}$$

- where t_i is the amount of time during which the signal is understood between y and y + dy.

Although probability and correlation functions are extremely interesting studies, by far the most used technique and the better understood one is that of spectral representation of a signal.

2.1.4. Spectral analysis

The main factors that have determined this method's generalization are first the long history of Fourier analysis, the rather old development of wave analysis of different types, the improvement of the signal to noise ratio by the use of appropriate frequency filters and the relationship between the spectrum of a signal and general specifications tests in an environment, as well as the quantitative and qualitative values that they allow to obtain.

This is quite evident in the case of random signals. So, when it was decided that the best description of an arbitrary signal is a spectral representation, the first question was to determine which parameter in function of the frequency must be recorded, for example:

Peak value	Average value	Average value detected
True effective value	Higher order values	

However, we note that if the comparison is made between a correlation function or a probability function and a spectral density function, the correlation coefficient for $\tau = 0$, or the standard deviation σ, has a correlation with the effective value of the signal. From where following typical curves arise:

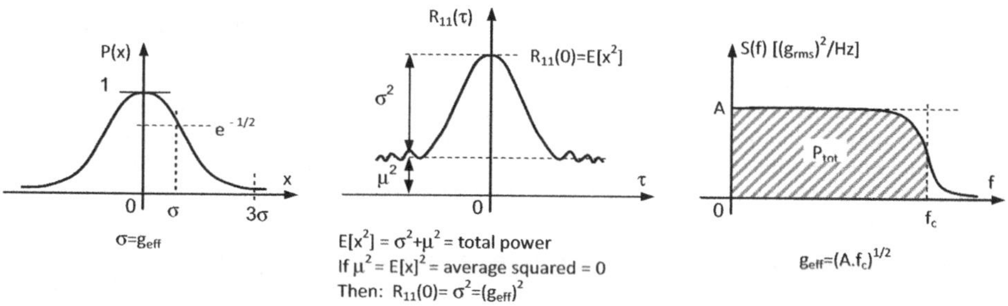

Figure 2.1. *Relations of functions with effective g*

Additionally, in general, most of the devices are scaled to give out the true effective value of a signal (or Root Means Square (RMS) value) and this value was generally accepted by everyone, long vibration spectra were recorded in their true effective value. However, in most cases, it is necessary to know the relationship of spectral distribution of the spectrum of a signal with the effective value of this signal. This, in an arbitrary section of this spectrum between f_1 and f_2, for example, is shown in the following figure.

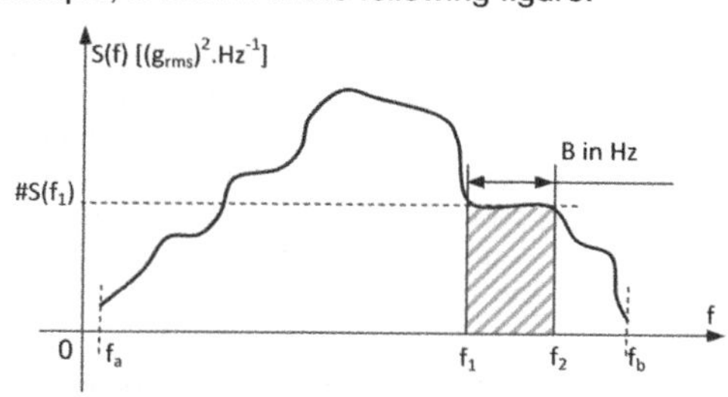

Yet, for a spectrum represented in effective value, the determination of the corresponding acceleration at a given frequency band requires a complex calculation. Since the RMS values are added only in a quadratic sum it is thus difficult to simply manipulate these values in the study of the functions of transfer between two points (there is no direct relationship between input and output of a linear system here, like with the Power Spectral Density (PSD)). On the other hand, powers are algebraically added. We will learn that for the given specification templates in the form of lines, it is possible to find simple equations. For these reasons, it seems that the RMS value is not the best representation of random signals.

Finally, we note that in a random signal analysis, the registered level is also proportional to the bandwidth of the analysis filter, which can be a drawback. Thus, to get comparable results regardless of the bandwidth of the analysis filter used and especially for values that are easy to add up or subtract, the most interesting parameter setting seems to be the *DSP power spectral density*.

As we will study, it is expressed in $[g_{RMS}^2.Hz^{-1}]$ or $[g_{RMS}^2/Hz]$, which is the square of the effective acceleration, divided by the bandwidth of the analysis filter (density!).

We can also note that with the definition of PSD, we lose any sense of phase. This is why we also introduce the notion of "QUADrature" spectrum and "Coincidence" spectrum.

So let us look specifically at this notion of PSD.

2.1.5. Notion of shock spectrum

We also use more and more the notion of shock response spectrum [ref. I1 and I18] Shock response Spectrum (SRS) in the study of mechanical signals. Not to be confused with the previous spectral concepts, which are defined mathematical tools, from causal integral biunivocal transformations. Indeed, a shock spectrum is also a spectral representation, but it is defined from a mechanical model of representation (prone to restrictive assumptions of modeling), which is univocal, in the sense that we lose the phase in this calculation and that for a given SRS there is *a priori* an infinity of possible temporary signals, whereas for a given temporal signal, there is only one [ref. I18].

These definitions allow accessing, from extreme values of relative displacements made by the model adopted, notions of extreme damage primarily used to describe the destructive potential of mechanical shocks (extreme response spectrum). They have been developed and extended to the sinusoidal and random signals from mechanical vibrations. Thus, with the condition of having means of adapted calculations and taking the precautions of use, it is possible to determine *"trustworthy equivalent tests"*, which wrap several shock situations and even of vibrations of a life profile given. It is this aspect of *"synthesis of real environment of a life profile"*, which is currently the most attractive, since it allows saving up several tests by combining them into one considered equivalent, with the following possibilities:

- Replacing several shock tests of amplitudes and periods by the temporary signal whose SRS encompasses the involved SRS shocks.

- With the notion of "spectrum of robustness and spectrum of associated damage by fatigue" applied to vibratory signals and shocks, it is even possible to replace impact tests by random tests of vibrations, for example. This will allow to save a considerable amount of time for the tests (for relatively small amplitudes from shocks). This very rich approach is beyond the scope of the course and is very well developed in [ref. 11]; we will provide a summary for further information. We talk about *"extreme response spectrum or robustness spectrum"* for signals of shocks and vibrations, and *"spectrum of damage by fatigue"* for vibration signals exclusively.

These notions are defined by the response in maximum relative displacement, weighted, taken by all basic systems to 1 DOF, decoupled, linear, whose mass elements, spring and damping have frequencies of resonances extending from 0 to infinity, theoretically (but a few thousand hertz in practice), and which form the mechanical model chosen, when it is excited by the signal we want to study.

Thus, for each f_i of each of the base systems, we carry the ordinate value of $(2\pi f_i).z_{max}$, where z_{max} is the relative displacement by the mass compared to the base of the model, which is the excitement, and it is represented in Figure 2.2.

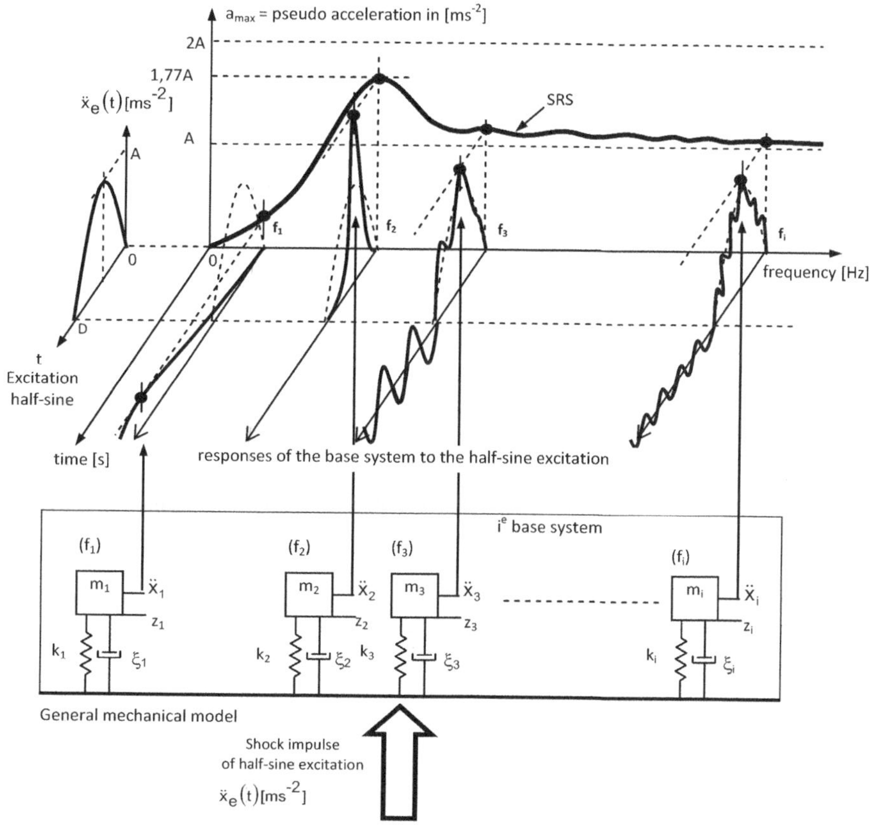

Figure 2.2. *Model of spectrum for a half-sinusoidal excitation spectrum*

2.2. Spectral density of power

2.2.1. Overview

The amplitude of a single frequency signal coming from a random signal filtered through a narrowband filter is aleatory. We cannot define such a signal by its peak value, but it may be defined by its real effective value or RMS, if it does not vary in time, which is the case for a stationary signal.

A very important remark: since we will know how to transform a signal by its repetition on a magnetic loop or any other process of repetition, we will be able to, always after recording an unspecified random signal, identify it with a stationary random signal.

But we know that the effective value of a sinusoidal signal is easily defined. Let us imagine then from a theoretical point of view, a band-pass filter with infinitely stiff faces and of unitary bandwidth (the unit being obviously Hz), with zero attenuation in the band and infinite on the outside (moreover, we show that its distribution is therefore Gaussian).

For the signal v(t) from the filter, we will have: $v_{RMS}^2(t) = \frac{1}{T}\int_0^T v^2(t)dt$

This model for an infinitely narrow filter is shown in Figure 2.3:

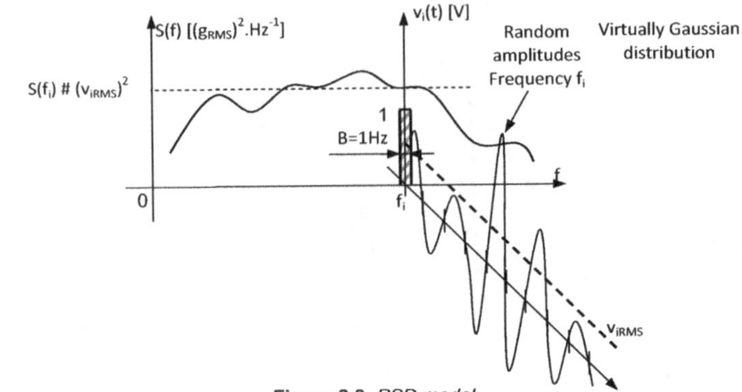

Figure 2.3. *PSD model*

By analogy, the RMS/Hz of the random signal, i.e. its contribution value, will be defined at a frequency of 1 Hz. The corresponding power in this band of 1 Hz is then given by the square of the RMS value. Thus, if the studied signal is an acceleration, we will define its DSP in terms of $g^2_{RMS}.Hz^{-1}$. It is usually abbreviated, although rather improperly, in $g^2_{eff}.Hz^{-1}$ or g^2/Hz (with g being the acceleration of gravity in Paris equaling 9,81 ms^{-2}).

Let us now analyze a random signal with the filter defined previously by moving its central frequency in the range of the frequency spectrum of the signal. We obtain its curve of power spectral density or PSD, S(f). The following typical curve is obtained.

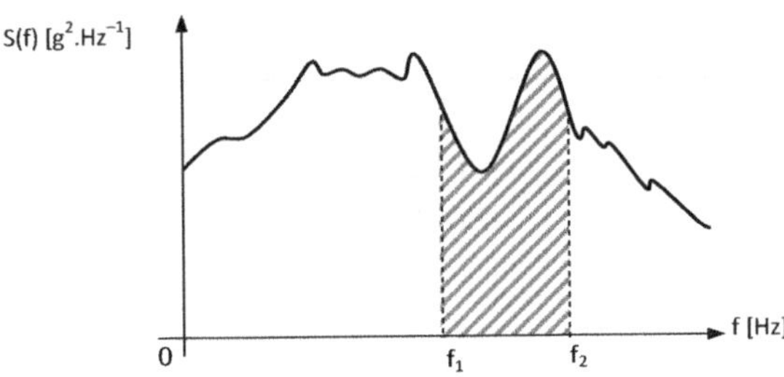

The unit of S(f) represents very well a power per unit of frequency, meaning, a density of power. The power between f_1 and f_2 (or the shaded area under the curve) is then given by the following integral:

$$g^2_{RMS} = \int_{f_1}^{f_2} S(f).df \qquad \text{and} \qquad g_{RMS} = \sqrt{\int_{f_1}^{f_2} S(f).df}$$

The true effective value is therefore given by the square root of PSD.

Thus, the mathematical definition of DSP is given by the mean square of the signal based on time x(t) (here, a random process), which is contained in an infinitely narrow frequency band B centered at a frequency f and integrated over an infinitely long time. We get:

$$S(f) = \lim_{\substack{T \to \infty \\ B \to 0}} \frac{1}{T} \int_0^T \frac{x_B^2(t)}{B}.dt \tag{2.4}$$

In practice, of course, B and T are finite and we can then define the power spectral density estimated by the following equation:

$$\tilde{S}(f) = \frac{1}{T}\int_0^T \frac{x_B^2(t)}{B}.dt = \frac{\overline{x_B^2}}{B} \qquad (2.5)$$

$$\begin{cases} \tilde{S}(f) = \text{ estimated PSD } [g_{RMS}^2 / Hz] \\ T = \text{ integration time [s]} \\ B = \text{ noise bandwidth of the analysis filter [Hz]} \\ \overline{x_B^2} = \text{average of the square of the amplitudes in the analysis bandwith } [g_{RMS}^2] \end{cases}$$

NB: We note that the PSD of a signal is independent of the phase compared to an arbitrary reference. Thus, two signals of the same nature, but of different phase, will have the same PSD (this is also true for the autocorrelation function).

For the description of a random process, the PSD function is very important:

- It represents a continuous spectrum that can be represented by a density function.

- The integration of a spectrum can be obtained by the sum of the powers (throughout an integral or by arithmetic).

- There is a relation between the PSD and the autocorrelation function.

- For stationary processes that can often be reduced, the PSD function brings as much information as the autocorrelation function, of which is the FT.

NB: It should be noted that the notion of PSD is applicable only to random signals and that the PSD notion applied to sinusoidal signals has no practical meaning, it only takes a mathematical and analytical significance with the use of Dirac distributions, as well as to determine the real effective value (RMS) at a frequency determined from the PSD, which has no meaning. Indeed, at a given moment there is obviously an instantaneous value but not an effective value. The latter requires the observation of a given frequency band in order to be measured.

2.2.2. PSD Study

2.2.2.1. *Introduction*

Initially, this study will be an analog approach that seems more educational and allow to study a variety of parameters used in the construction of the PSD of a random signal, although currently devices are, in most cases, numerical.

The signal is analyzed by a narrow band filter sweeping the considered frequency range. A power function $\left(v_{eff}^2(t)\right)$ is formed from the output signal of the filter. Then, a standardized average of the product is formed by integration by dividing the effective bandwidth of the filter. For example, in the portion of frequency of B Hz, where we have the practically constant PSD S(f) = 0,25 [g^2Hz], we can write in this case of a particular figure that the effective acceleration in the B band is:

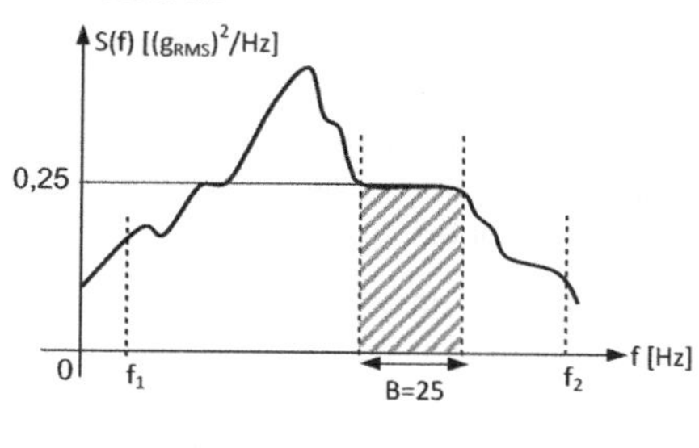

$$\begin{cases} S(f) \approx 0{,}25 g_{RMS}^2.Hz^{-1} \\ B = 25Hz \end{cases} \rightarrow g_{RMS}^2 = 0{,}25.25 = 6{,}25 \rightarrow g_{RMS} = \sqrt{6{,}25} \approx 2{,}5$$

If we want to calculate the total effective acceleration between f_1 and f_2, we note that we must take the square root of the sum of the mean square of acceleration between f_1 and f_2, meaning:

$$\underset{f1,f2}{g_{RMS}} = \left(\int_{f_1}^{f_2} S(f).df \right)^{1/2}$$

In practice, PSD curves are rather given in the specifications, by envelopes in an asymptotic form, with slopes in dB/octave and parts of constant amplitude. For example, the following general pace (STANAG 4370 AECTP400, method 401 vibrations) is shown.

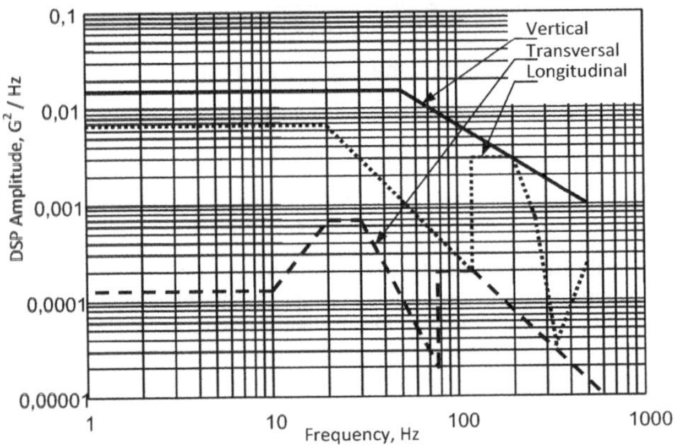

It is possible to find an analytical equation to calculate the effective value based on the slope of the asymptotes in dB/octave. Of course, there are calculators that do this work, but from an educational point of view, it is good to remember the calculation. So, the general model is the following:

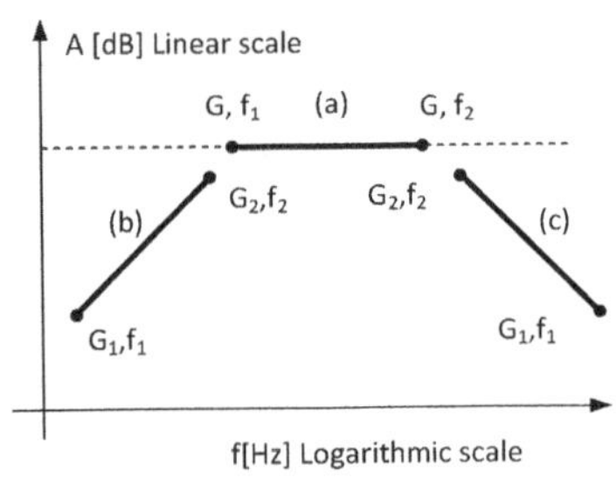

The general form of the function that renders the asymptotes in the form of straight lines of slopes ±n (n = 1, 2, 3...) in logarithmic coordinates is given by: $G = K.f^{\pm n}$. So n sets the value of the slope in logarithmic coordinates. So, we have:

$$G_1 = K.f_1^n \text{, therefore, for an octave: } G_2 = K.(2.f_1)^n = K.2^n.f_1^n$$

Meaning in dB power:

$$G_{1dB} = 10LogK + n.10Logf_1 \quad \text{and} \quad G_{2dB} = 10LogK + n.10Logf_1 + n.10Log2$$

$$G_{2dB} - G_{1dB} = n.10Log2$$

If we want to formulate the dB/octave slope, it is necessary that: m [dB/octave] = n.10Log2

Therefore: $$n = \frac{m}{10\text{Log}2} \approx \frac{m}{3}$$

It thus becomes:

a) $G_1 = G_2 = G = \text{cte}$: $$g_{eff}^2 = G.\int_{f_1}^{f_2} df = G.(f_2 - f_1) \tag{2.6}$$

b) Positive slope n: $$G = K.f^n \rightarrow \begin{cases} G_2 = K.f_2^n \rightarrow K = \dfrac{G_2}{f_2^n} \\ G_1 = K.f_1^n \rightarrow K = \dfrac{G_1}{f_1^n} \end{cases} \rightarrow \frac{G_1}{G_2} = \frac{f_1^n}{f_2^n}$$

$$\text{Therefore}: g_{eff}^2 = \int_{f_1}^{f_2} K.f^n.df = \frac{G_2}{f_2^n}\int_{f_1}^{f_2} f^n.df = \frac{G_2}{f_2^n}\frac{1}{n+1}\left[f^{n+1}\right]_{f_1}^{f_2} = \frac{G_2}{f_2^n}\frac{1}{n+1}\left(f_2^{n+1} - f_1^{n+1}\right)$$

$$= \frac{G_2}{n+1} f_2 \left[1 - \left(\frac{f_1}{f_2}\right)^{n+1}\right] \tag{2.7a}$$

$$\text{Or}: g_{eff}^2 = \frac{G_2}{n+1} f_2 \left(1 - \frac{G_1}{G_2}\frac{f_1}{f_2}\right) \quad \rightarrow \quad g_{eff}^2 = \frac{1}{n+1}(G_2 f_2 - G_1 f_1)$$

c) negative slope n: $$G = K.f^n \rightarrow \begin{cases} G_2 = K.f_2^{-n} \rightarrow K = G_2 f_2^n \\ G_1 = K.f_1^{-n} \rightarrow K = G_1 f_1^n \end{cases} \rightarrow \frac{G_1}{G_2} = \frac{f_2^n}{f_1^n}$$

$$\text{Therefore}: g_{eff}^2 = \int_{f_1}^{f_2} K.f^{-n}.df = G_1 f_1^n \int_{f_1}^{f_2} f^{-n}.df = G_1 f_1^n \frac{1}{-n+1}\left[f^{-n+1}\right]_{f_1}^{f_2} = \frac{G_1 f_1^n}{-n+1}\left(f_2^{-n+1} - f_1^{-n+1}\right)$$

$$= \frac{G_1}{-(n-1)} f_1 \left[\left(\frac{f_2}{f_1}\right)^{-(n-1)} - 1\right] = \frac{G_1 f_1}{(n-1)}\left[1 - \left(\frac{f_2}{f_1}\right)^{-(n-1)}\right] = \frac{G_1 f_1}{(n-1)}\left[1 - \left(\frac{f_1}{f_2}\right)^n \left(\frac{f_2}{f_1}\right)\right]$$

$$= \frac{G_1 f_1}{(n-1)}\left(1 - \frac{G_2 f_2}{G_1 f_1}\right)$$

$$\text{Or}: \quad \text{with } n \neq 1 \qquad g_{eff}^2 = \frac{1}{n-1}(G_1 f_1 - G_2 f_2) \tag{2.7b}$$

So excepting for n = 1, where we have:

$$g_{eff}^2 = G_1 f_1 \int_{f_1}^{f_2} f^{-1}.df \rightarrow \qquad g_{eff}^2 = G_1 f_1.\text{Log}\frac{f_2}{f_1} \tag{2.8}$$

Expressions (2.7a) and (2.7b) are very similar. We can collect all the expressions in a single one noticing that (2.7b) was calculated with n positive and a - sign to indicate an asymptotic decrease. It is thus enough in this expression to formulate the slope with –n dB/octave in the equation and it becomes:

$$g_{eff}^2 = \frac{1}{-n-1}(G_1 f_1 - G_2 f_2) = \frac{1}{n+1}(G_2 f_2 - G_1 f_1)$$

Having:
$$\begin{cases} g_{eff}^2 = \dfrac{1}{n+1}(G_2 f_2 - G_1 f_1) & \forall n \neq 1,\ 0 \text{ included} \\ g_{eff}^2 = G_1 f_1 . \text{Log}\dfrac{f_2}{f_1} & n = 1, \text{ or } m = 3 \quad \text{dB/octave} \end{cases} \qquad (2.9)$$

Example 2.1. Calculate the RMS acceleration (or effective acceleration) of the DSP along the vertical axis of the previous curve.

So, we have 0,015 [g^2/Hz] from 1 to 50 Hz constant and a decrease from 50 to 500 Hz to reach 0,001. Therefore:

- from 1 to 50 Hz: $g_{eff}^2 = 0{,}015.(50 - 1) = 0{,}735$

- from 50 to 500 Hz we have: $\begin{cases} 11{,}76\text{dB} \quad \text{power} \\ \dfrac{1}{\text{Log}2}\text{Log}\dfrac{500}{50} = 3{,}32 \quad \text{octave} \end{cases}$, so $m = \dfrac{11{,}76}{3{,}32} \approx 3{,}54$ dB/octave.

And: $n = m/3 \approx 1{,}18$

$$\text{Henceforth:} \quad g_{eff}^2 = \frac{1}{-1{,}18+1}(0{,}001.500 - 0{,}015.50) = -\frac{0{,}25}{-0{,}18} = 1{,}39$$

So, arithmetically adding the powers: 0,735 + 1,39 = 2,12 g_{eff}^2

$$\text{And:} \qquad g_{eff} = \sqrt{2{,}12} \approx 1{,}457$$

The verification with a calculation software (VIBKIT98) gives out 1,468 g_{eff}..

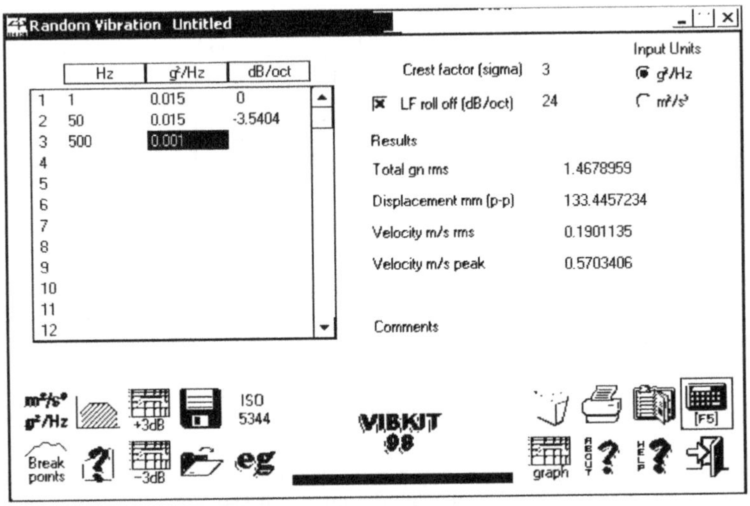

Figure 2.4. *Calculation with software Vibkit 98*

2.2.2.2. *Considerations of use*

2.2.2.2.1. *Study of filtering random signals*

Different types of analog filters were made for the measurement of the DSP (L, R filters, crystal filters, etc.). However, in this technique, the output voltages of the filters are high squared. It is therefore necessary that these filters present significant selectivity. Different studies have shown that the mitigation of these filters should be at least equal to or greater than 35 dB per octave of bandwidth.

2.2.2.2.2. *Form factor*

The form factor of the filter can be defined as the ratio of the bandwidth at –60 dB), with the bandwidth at –3 dB. Standard bandwidths used are the following: 1,5; 2; 5; 10; 20; 50; 100; 200 Hz.

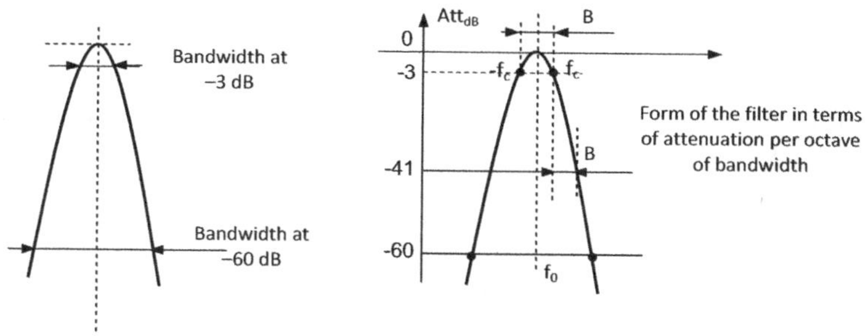

Figure 2.5. *Form factor model*

For example, for the traditional filters generally used in DSP analysis, for a length of frequency strip at –3 dB at cut-off, we have an attenuation of –41 dB, which means –38 dB (38 = 41-3) per octave of bandwidth. That is well above the 35 dB/octave strip length, previously recommended value.

For a filter of 2 Hz of bandwidth, at 2 Hz of the cut-off frequency f_c at –3 dB or f_c +2 Hz, we will have an attenuation of –41 dB.

By definition, we will then have the form factor:

$$F_f = \frac{\text{bandwidth at - 60dB}}{\text{bandwidth at - 3dB}} \tag{2.10}$$

In general, in the DSP analysis, we take as a classic form factor the following value:

$$F_f = 4/1 = 4$$

So, for a filter of a 2 Hz bandwidth at –3 dB, we have 8 Hz (4 x 2) with –60 dB.

2.2.2.2.3. *Effect of filtering random signals*

The passage of a random signal in a narrow strip filter produces very interesting modifications, which will be used in capabilities of DSP.

(a) The effective value of the signal has even larger fluctuations as the strip is narrow.

(b) From a critical bandwidth, positive values will not contain more negative peaks and *vice versa*, as was the case in the random unfiltered signal. The signal then appears as a sine wave at the center frequency of the filter, modulated in amplitude, with slight variations of phase.

(c) When the bandwidth of the analysis filter decreases, the likelihood (or probability density) of instantaneous amplitudes becomes perfectly Gaussian, whatever was in its distribution in the random signal broadband. This last feature is very important and will be used to form the DSP. In addition, the distribution of peak values follows one of Rayleigh's laws.

(d) The ratio between the true effective value (RMS value) and the average value of a Gaussian distribution signal is a factor of 1,25, we just have to apply this correction to the average value measured for the RMS value.

Therefore, the use of a logarithmic recorder will make it possible to form the DSP function through the terms: $\text{Log}\left(g^2/\text{Hz}\right) = 2\text{Log}(g) - \text{Log}(\text{Hz})$

NB: Remember the following important point: to the extent of the DSP, only the density function is to be considered, and that the power is determined for a bandwidth (that of the filter of analysis) and especially, as we already saw above, that the determination of the value of effective acceleration "at a given frequency", at a moment t, starting from the DSP, has no significance, because at one moment there is an instantaneous value, but not an effective value, of course.

2.2.2.2.4. *Time of establishment of a signal in a filter*

It depends on the form factor of the filter. For example, let us look at the response of the filter at a unit level. The following type of curve is then obtained.

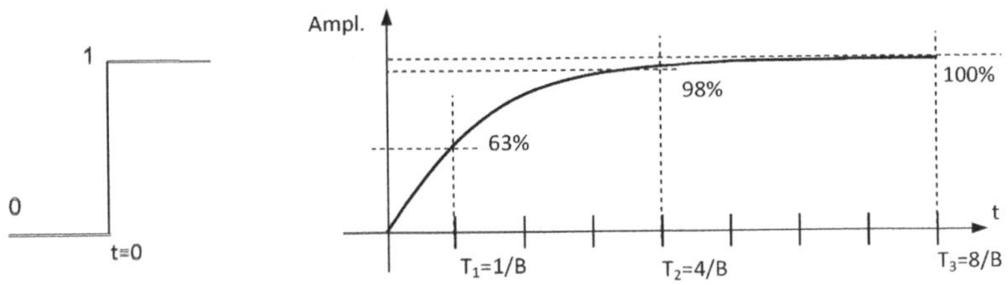

These relationships are based on the time of establishment of the signal through a crystal filter having several floors, with a form factor of 4. We can therefore empirically determine that:

- T_1 = 1/B = needed time to reach 63% of the final level of the signal.
- T_2 = 4/B = needed time to reach 98% of the final level of the signal.
- T_3 = 8/B = needed time to reach 100% of the final level of the signal.

So, after selecting the design of the filter of analysis, it is necessary to form the function of the signal's power. For this, we need for the latter to be high squared, generally difficult operation for stable multiplying circuits having traditional dynamics of 80 dB. From here necessary approximations derive.

2.2.2.2.5. *Degrees of freedom*

Like the characteristics of a random signal, the estimation of a power spectrum is governed by statistical properties. It is therefore necessary, in the first place, to consider the statistical degree of freedom (or the degree of trust) "n" associated with the data of a sample. This factor "n" corresponds to the number of samples needed to define a signal of arbitrary shape. In practice, each of the random signals on a given time path observation corresponds to a sample. The latter becomes deterministic as soon as it is recorded.

If this signal is a deterministic continuous function, one sample is enough to define it. On the other hand, for random signals, this number of samples is the bandwidth of the signal function. It is necessary to compose a great number of trajectories in order to approach better the statistics of the random signal involved.

Therefore, for a B Hz bandwidth signal, integrated during T seconds, the total number of independent samples needed to define it correctly will be 2BT (this is a consequence of the sampling theorem). We then define the statistical degree of freedom n by:

$$n = 2BT \qquad (2.11)$$

2.2.2.2.6. *Error rate in the analysis of random signals*

We defined the DSP by the following equation: $S(f) = \lim_{\substack{T\to\infty \\ B\to 0}} \frac{1}{T}\int_0^T \frac{x_B^2}{B} dt$

and the DSP is estimated by: $\tilde{S}(f) = \frac{1}{T}\int_0^T \frac{x_B^2}{B} dt = \frac{\overline{x_B^2}}{B}$

The possible error rate generated by this estimate is a function of the number of necessary and available information n to form the sample of the signal (sample in the statistical sense here) and so the duration of the recording, the bandwidth of the filter and integration time. This error can be estimated in terms of standard deviation of the measured DSP function since the latter is usually obtained by an average, that is to say σ.S(f). However, it is shown that this standard deviation is given by: $\sigma = \frac{1}{\sqrt{BT}}$.

Therefore, $\sigma.S(f) = \frac{S(f)}{\sqrt{BT}}$. The error rate is therefore given by $\varepsilon = \frac{S(f)/\sqrt{BT}}{S(f)} = \frac{1}{\sqrt{BT}}$.

Therefore,
$$\varepsilon = \frac{1}{\sqrt{BT}} \to \varepsilon^2 = \sigma^2 = \frac{1}{BT} \tag{2.12}$$

It means that for a measurement made, the error between real DSP and measured or estimated DSP will be between $\tilde{S}(f)$ to $\pm\sigma$, either with a factor of confidence of one standard deviation of 67%, or roughly two-thirds of the time. Thus, the true value of DSP will be between the measured $\tilde{S}(f)$ and this same value $\tilde{S}(f) \pm \varepsilon$.

It should be noted that this relation is only valid for $\varepsilon < 0{,}2$. If the error rate is above 20%, which is in fact rarely the case, it will be determined by other distribution law than the preceding assumption of Gaussian distribution (it is necessary that the sample size or the number of averages is ≥ 30). This law will be that of the χ^2 that we call Khi-square, Khi- two or even Pearson's law.

Its formulation is complex and irrelevant here. It depends on a parameter $\nu = n - 1$, where n is the sample size. In practice, we use a table that provides the corresponding values of χ^2 of the parameter ν degree of freedom (thus of n) and cumulative probabilities $P(\chi^2 \leq x^2)$. We use this table to find the value of x^2 under which we have a given probability to find a result χ^2.

This law is particularly used to estimate the dispersion ε of a population (assumed normal so $\varepsilon \equiv \sigma$) from an observed value ε' when the sample size is less than n = 30 or ν = 29, or to examine if the observed value ε' is consistent with a hypothesis being considered as law (normal, for example) and its value ε (or σ if normal).

It is shown that:
$$\begin{cases} \text{average} = \nu \\ \text{dispersion} = \sqrt{2\nu} \end{cases} \tag{2.13}$$

We can also use the following curves (example of curves of distribution of χ^2):

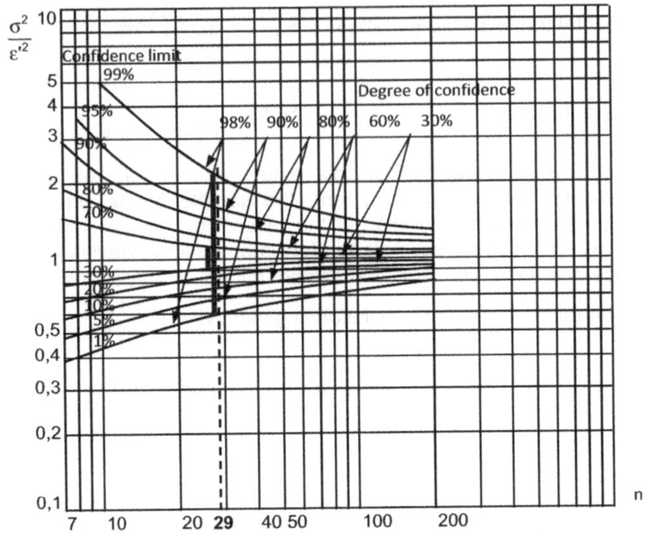

Example 2.2: Find the range of the true mean square with n = 30 degrees of freedom, meaning ν = 29. The true mean square σ^2 with 98% of confidence lies between 2,1 and 0,59 times the observed mean square ε'^2. With 30% confidence, the true mean square is located between 0,91 and 1,17 times the observed mean square ε'^2. Similarly, for other degrees of confidence.

With:

$$\frac{\sigma^2}{\varepsilon'^2} = \frac{n}{\chi^2} \tag{2.14}$$

σ^2 = true mean square

ε'^2 = observed mean square

χ^2 = chi-square

n = degrees of freedom

In general, the accepted error rate will be below 20% and it will not be necessary to use this curve in χ^2 and formulas using the normal distribution can be used.

Another expression of the dispersion of the measure can be obtained from the number of available information to form the estimate. This number we called degree of freedom n (not to be confused with ν) may be deduced from the relation:

$$n = 2.B.T$$

Which comes down to saying that we have two information seconds per hertz of bandwidth (Shannon). The error rate will be more so low that "n" is greater and the relationship between error rate and degree of freedom used is usually the following for $\varepsilon < 0,2$:

$$\begin{cases} \varepsilon = \dfrac{1}{\sqrt{n/2}} & \rightarrow \quad n = \dfrac{2}{\varepsilon^2} \\ \varepsilon^2 = \dfrac{2}{n} & \rightarrow \quad \varepsilon^2 = \dfrac{1}{BT} \end{cases} \tag{2.15}$$

Example 2.3: Thus, for example, to get an error rate lower than ± 20%, with a bandwidth filter of 2 Hz, it will be necessary to use a constant of integration of:

$n = \dfrac{2}{\varepsilon^2} = \dfrac{2}{0,2^2} = 50$, such as n = 2BT $\rightarrow$ $T = \dfrac{n}{2B} = \dfrac{50}{2,2} = 12,5[s]$

T is the observation time, the value of necessary RC being of T/2, it becomes: RC = 6,25 [s], having an integration time of 6,25 [s].

2.2.2.2.7. *Resolution*

By definition, resolution corresponds to the ratio of the bandwidth of the analysis filter with the frequency of analysis, which means:

$$r = \frac{\Delta f}{f} \tag{2.16}$$

Thus, with 20 Hz, a filter of 2 Hz of bandwidth has a resolution of 10%. However, the necessary resolution will be greater than the frequencies to separate, which will be nearby or the signal to be analyzed will have important Q-tension coefficients.

It is allowed that for a correct analysis, the factor Q_f of the filter is at least four times superior to factor Q_m maximum of the signal to be analyzed:

$$Q_f \text{ filter} \geq 4.Q_m \text{ max resonance} \tag{2.17}$$

For example, if we consider a signal with a factor Q_m of 10 (which is generally low for a structure resonance), the filter must present a factor Q_f equivalent of 40, which means a resolution r = 1/40 or 2%. We will thus need a filter of 2 Hz of bandwidth to analyze a signal with 80 Hz. Therefore:

$$r = \frac{1}{Q_f} = \frac{1}{4.Q_m} \tag{2.18}$$

2.2.2.2.8. *Scanning rate in PSD analysis*

With a filter of constant bandwidth, if a sample data of Δt seconds of duration should be analyzed with a B analysis bandwidth filter, it is necessary to form the PSD, the entire signal passes through the filter in order to be integrated. So, by definition, the scanning rate will be given by:

$$v_b = \frac{B}{\Delta t} \tag{2.19}$$

Two cases arise:

- In the analyses without time constant RC, i.e. with a real integration, the scanning speed shall not exceed a bandwidth for a time of 4/B seconds, which corresponds to the time of establishment at 98% of the signal in a filter. It means:

$$v_b = \frac{B}{4/B} = \frac{B^2}{4} \tag{2.20}$$

with $[B^2] = [\text{Hz.Hz}] = [s^{-1}.\ \text{Hz}]$ = velocity in Hz/s.

- In the analyses with a time-constant RC, i.e. the signal is averaged by an RC network, the velocity to be used will be given by: $v_b = \dfrac{B}{4/B + 4RC}$. So we have:

$$v_b = \frac{B^2}{4(1+BRC)} \tag{2.21}$$

with:

$$\begin{cases} \dfrac{4}{B} = \text{time of establishment in the relevant filter.} \\ 4RC = \text{time necessary to reach 98\% of the final value of the signal with RC.} \end{cases}$$

Henceforth, the time required to form the estimate of the PSD of a signal will be therefore given by:

$$\Delta t = \frac{f_2 - f_1}{v_b} \tag{2.22}$$

$\begin{cases} f_2 = \text{higher frequency to analyze} \\ f_1 = \text{lower frequency to analyze} \\ v_b = \text{scanning speed in Hz/s} \end{cases}$

NB: In general, 4/B is weak before 4RC and the formula can be simplified with:

$$v_b \leq \frac{B}{4RC} \tag{2.23}$$

So, in real integration, if we take t = T = observation time of the signal, it becomes v_b = B/T; however, as we have already determined RC = T/2 in integration by an RC network, we have:

$$v_b = \frac{B}{4RC} = \frac{B}{4T/2} = \frac{B}{2T}$$

NB: In an integration by an RC network, the scanning speed will generally be half of that used in real integration.

2.2.2.2.9. *Integration of the signals in the PSD analysis*

It is important to distinguish real integration from integration by RC network. So, two methods can be used to provide the PSD of a signal, which means in the average search of the square:

- a true integration of the signal during a time given T that makes it possible to obtain a pont of DSP after each period of integration;

- a filter by RC network that allows obtaining a continuous spectrum of DSP. The signal is then averaged over a time corresponding to the time constant RC of the circuit.

In real integration (electronic integrator), the integration is performed during a time of T seconds. The value obtained is divided by the integration time and gives us a point of the diagram. The equation corresponding to real integration is thus given for f(t) by: $f(t) = \frac{1}{T}\int_0^T f(t).dt$.

In the integration by an RC network, we obtain a continuous diagram, contrary to the one obtained with true integration, which will consist of a series of points. Additionally, it is shown that when a signal is applied to an RC integrator network, for an equivalent or superior time to 4RC, the output is a weighted function of the time of all the previous amplitudes with an equivalent integration time of T = 2RC. This allows us to input in general, as a value of RC, the time T/2, which means that n = 2BT = 4BRC.

$$n = 4BRC \quad \text{and} \quad \varepsilon^2 = \frac{1}{2BRC} \tag{2.24}$$

NB: A more important time constant would not improve the statistical quality of the measurement.

Moreover, the analysis of periodic or complex signals does not require constant integration time. In this case, it is preferable that the detection for the time constant signal not to be too important. Thus, in the following filter analyzers, if the maximum signal used for detection is 100 kHz at filter exit, this makes it possible to use a very weak time constant, generally of the order of 0,1 ms. In this case the only limit is that of the establishment of the signal in the filter. That is to say, a time is equivalent to 4/B (for 98% of the end value of the signal), B being the bandwidth of the filter of analysis.

2.2.2.2.10. *Detection of the analyzed signal*

The type of detection used can give out slightly different results for signals, whether stationary or not, either periodic or complex ones. At the exit of the analysis filter, the signal will always be sinusoidal and devices can be calibrated taking into consideration the relationship between effective values, peaks and averages of these signals.

2.2.2.2.11. *Relations between RMS values of random signals and sinusoidal signals*

At the narrowband filter exit, the distribution of instantaneous values is always Gaussian, no matter what we had in the distribution originally. However, for a random signal with Gaussian distribution, the relations between average values and effective values are as follows:

$$\begin{aligned} \bar{E}_{gaussian} &= \sqrt{\frac{2}{\pi}}.\sigma = 0,798.E_{eff.Gaussian} \\ E_{effGaussian} &= 1,25.\bar{E}_{Gaussian} \end{aligned} \qquad (2.25)$$

Yet in the spectrum analyzers with the following filters, the signal is detected at average value (by a diode bridge). However, the devices can be calibrated taking into account the following relations:

$$\frac{E_{eff}\,(\text{random gaussian})}{E_{eff}\,(\text{sinusoidal})} = \frac{1,25.\bar{E}\,(\text{random gaussian})}{1,11.\bar{E}\,(\text{sinusoidal})}, \text{ therefore:}$$

$$E_{eff}(\text{random gaussian}) = 1,13.E_{eff}(\text{sinusoidal}) \qquad (2.26)$$

This correction factor can be applied during calibration of the device.

Example 2.4: Analysis of a signal recorded on a magnetic loop.

- Length of the loop: T_r = 10 s
- Range of analysis: 10 to 2 000 Hz
- Allowed error rate: 14,1% (which corresponds to n = 100 degrees of freedom, meaning BT = 50)
- Q-factor: Q_m = 5

(a) Resolution: We must first choose the width of the analysis filter, which will provide sufficient resolution.

Having, $Q_a = 4.Q_m = 4,5 = 20$. The necessary resolution will be of:

$r = \frac{1}{Q_a}.100 = \frac{1}{20}.100 = 5\%$ or 0,05. So we can now determine the commutation frequencies corresponding to the chosen resolution, that is to say, 0,05, if we have, for example, the following classic analysis filters:

1,5; 2; 5; 10; 20; 50; 100; 200 Hz.

(b) Commutation frequencies: $r = \frac{\Delta f}{f} = \frac{B}{f} \rightarrow f = \frac{B}{r}$, therefore:

- For the filter B = 1,5 Hz: $f_t = \frac{1,5}{0,05} = 30\,\text{Hz}$ (so for 10 <; f <; 30 Hz, r >; 5% in this case

  - B = 5 Hz: f_t = 100 Hz
  - B = 10 Hz: f_t = 200 Hz
  - B = 20 Hz: f_t = 400 Hz

- B = 50 Hz: f_t = 1,000 Hz
- B = 100 Hz: f_t = 2,000 Hz (out of range of analysis)
- B = 200 Hz: f_t = 4,000 Hz (out of range of analysis)

(c) Adjustment of the time constants RC: We use the following equation as ε = 14,1% is imposed, meaning:

$$\varepsilon^2 = \frac{2}{n} \rightarrow n = \frac{2}{\varepsilon^2} = \frac{2}{0{,}141^2} \rightarrow n = 102 \rightarrow n \approx 100, \text{ thus :}$$

$$n = 4BRC \qquad \rightarrow \qquad RC = \frac{n}{4B} \approx \frac{100}{4B} \approx \frac{25}{B}$$

- For B = 1,5 Hz, as r < 5%, we can take n smaller, which means n = $2BT_r$ = 2,1, 5,10 = 30, so: RC = 30/4.1,5 = 30/6 = 5 s.

For other filters, we will take n = 100 and it becomes:

- for B = 5 Hz: RC = 25/5 = 5 s
- for B = 10 Hz: RC = 2,5 s
- for B = 20 Hz: RC = 1,25 s
- for B = 50 Hz: RC = 0,5 s

(d) Determination of scanning speeds: we have in this case: $v_b = \frac{B}{4RC}$, therefore:

- for B = 1,5 Hz: $\rightarrow \frac{1{,}5}{4{,}5} = 0{,}075 \text{Hz/s}$ from 10 to 100Hz
- for B = 5 Hz: $\rightarrow \frac{5}{4{,}5} = 0{,}25 \text{Hz/s}$ from 100 to 200Hz
- for B = 10 Hz: $\rightarrow \frac{10}{4{,}5} = 1 \text{Hz/s}$ from 200 to 400Hz
- for B = 20 Hz: $\rightarrow \frac{20}{4{,}5} = 4 \text{Hz/s}$ from 400 to 1000Hz
- for B = 50 Hz: $\rightarrow \frac{50}{4{,}5} = 25 \text{Hz/s}$ from 1000 to 2 000Hz

The total analysis time will be of 1,430 s, or t = 24 mn, which is a relatively long time. We will see that with digital cameras, measurements will be faster. Nevertheless, in order to reduce this measurement time, analog devices are equipped with a controller that optimizes the switching at predetermined frequencies: analysis of filters, RC time constants, scanning speeds and mitigation networks based on the bandwidth of analysis with a certain given degree of confidence.

2.2.2.2.12. *Characterization of the laws of probability with central moments*

(a) The central moment of order 2 is the variance and it is expressed as σ_x^2, with:

$$\sigma_x^2 = E\left[(x - \bar{x})^2\right] = E\left[x^2\right] - (E[x])^2 = \overline{x^2(t)} - \overline{x(t)}^2 \qquad (2.27)$$

- The variance gives the power of the variation compared to that of the average, that is the dispersion of the power around the average with the square of the process.

- σ is called standard deviation and it gives out the average dispersion around the mean of the process and having the fineness of the probability distribution around the average.

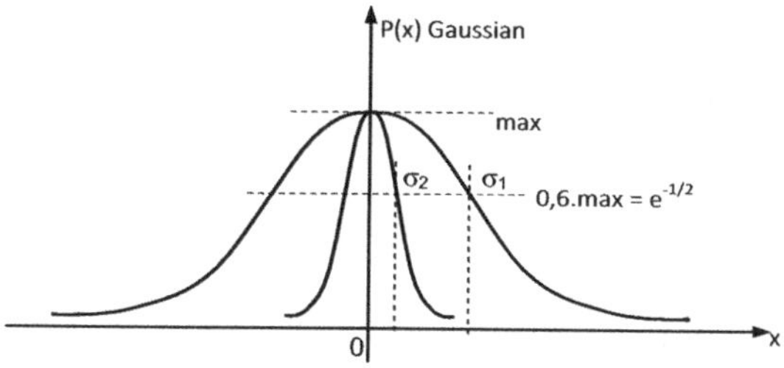

(b) The central moment of order 3 denoted as μ_3 is sometimes standardized compared to σ_x^3; it is then given by μ'_3:

$$\mu'_3 = \frac{E\left[\left(x-\overline{x}\right)^3\right]}{\sigma_x^3} = \frac{\overline{\left[x(t)-\overline{x(t)}\right]^3}}{\sigma_x^3} \tag{2.28}$$

This moment is characteristic of the symmetry of the law of probability density p(X) compared to the average $\overline{x(t)}$ (or 0 if centered); for this reason, we name it sometimes μ'_3 "asymmetry".

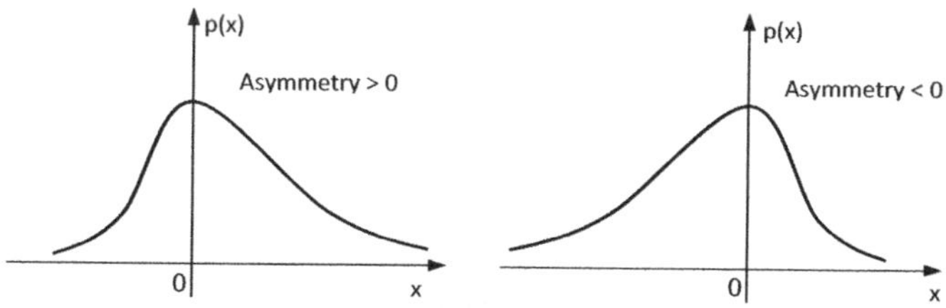

- $\mu_3 = 0$ is a Gaussian process.
- $\mu_3 > 0$ the curve of probability density has a peak to the left.
- $\mu_3 < 0$ the curve of probability density has a peak to the right.

(c) The central moment of order 4 denoted as μ_4, standardized compared to σ_x^4, is characteristic of the flatness of the curve of density of probability. It is given by μ'_4:

$$\mu'_4 = \frac{E\left[\left(x-\overline{x}\right)^4\right]}{\sigma_x^4} = \frac{\overline{\left[x(t)-\overline{x(t)}\right]^4}}{\sigma_x^4} \tag{2.29}$$

- $\mu'_4 = 0$ is a Gaussian process.
- $\mu'_4 < 3$ characterizes a tendency to the clipping of the signal or the existence of a sinusoidal component ($\mu'_4 = 1{,}5$ for a pure sine, 1 for a square, 1,75 for a triangle).
- $\mu'_4 > 3$ characterizes the presence of peaks of high values (more than in the Gaussian case).

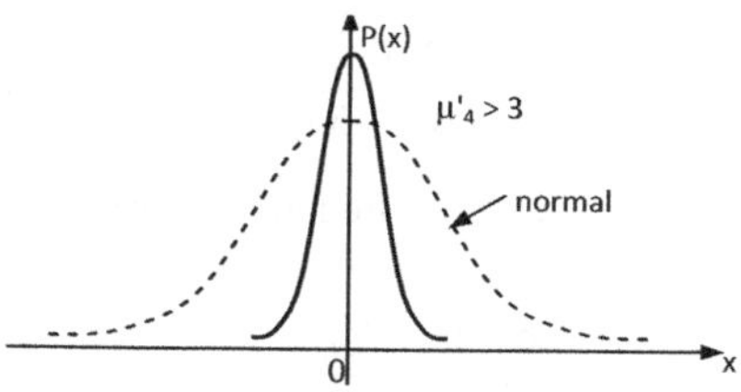

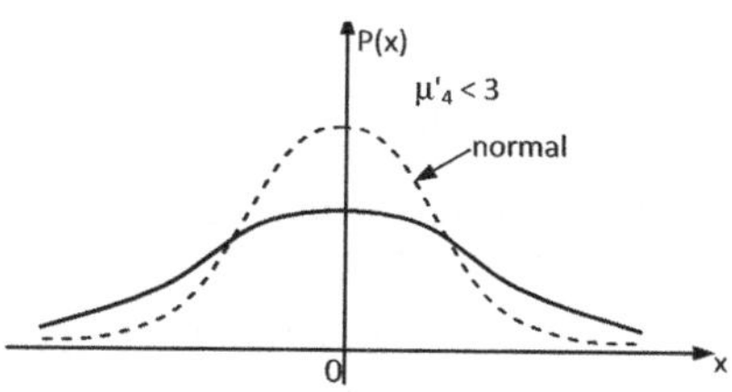

NB: The parameter μ'_4 is still called *Kurtosis*. Of course, there are tests that make it possible to characterize the laws as well as the stationarity and ergodicity.

2.2.2.2.13. *Calculation of displacement in random mode*

The relation between displacement and acceleration is $\gamma = \omega^2.d$, in which $d = \frac{\gamma}{4\pi^2 f^2} = \frac{9{,}81.g}{4\pi^2 f^2}$; thus, if g is an effective value, d will also be an effective value. In a small frequency band Δf (of DSP = g_0), we will thus have effective values:

$$d_{eff}^2 = \frac{(9{,}81)^2}{16\pi^4}.\frac{g^2}{f^4} = \frac{(9{,}81)^2}{16\pi^4}\frac{g_0.\Delta f}{f^4}, \text{ because } g^2 = g_0.\Delta f$$

Thereby, for the frequency band ranging between f_1 and f_2, it will be necessary to perform the sum of the unit surfaces represented by the previous equation, which means:

$$d_{eff}^2 = \frac{(9{,}81)^2}{16\pi^4}\sum_{f_1}^{f_2}\frac{g_0.\Delta f}{f^4} = \frac{(9{,}81)^2}{16\pi^4}\int_{f_1}^{f_2}\frac{g_0.df}{f^4} \quad (2.30)$$

In this particular case of white noise, $g_0.(\Delta f_i) = cte = G_0$, so:

$$d_{eff}^2 = \frac{(9{,}81)^2}{16\pi^4}G_0\int_{f_1}^{f_2}\frac{.df}{f^4} = -\frac{1}{3}\frac{(9{,}81)^2}{16\pi^4}G_0\left[\frac{1}{f^3}\right]_{f_1}^{f_2} = \frac{(9{,}81)^2.G_0}{3{,}16.\pi^4}\left[\frac{1}{f_1^3} - \frac{1}{f_2^3}\right]$$

As in general $f_3 >> f_1$:

$$d_{eff}^2 = \frac{(9{,}81)^2}{16{,}3.\pi^4}.\frac{G_0}{f_1^3} \to d_{eff} = \frac{9{,}81}{4\sqrt{3}\pi^2}\sqrt{\frac{G_0}{f_1^3}} \quad (2.31)$$

For $\sigma = 3$ and one displacement peak to peak, we have: $d_{c/c} = d_{eff}.3{,}2$, therefore:

$$d_{c/c} = \frac{\sqrt{3}}{2}.\frac{9{,}81}{\pi^2}.\sqrt{\frac{G_0}{f_1^3}} \quad (2.32)$$

But the piloting consoles have filters of finished slopes because they are not perfectly rectangular. It is therefore necessary to take into account this extra movement and correct formula (2.32). For example, we can take for the slopes of filters BF (only giving a significant displacement) a usual value of –24dB/octave, with m being the octave number, the ratio between f and f_1 is then written as:

$$\frac{f}{f_1} = 2^m \to 24.m = 10\text{Log}\frac{G_0}{g_0} \quad \text{and} \quad \text{Log}\frac{f}{f_1} = m\text{Log}2 \to \text{Log}\frac{G_0}{g_0} = 24.m$$

The elimination of m in the two relations gives out:

$$\text{Log}\frac{G_0}{g_0} = \frac{24}{10.\text{Log}2}.\text{Log}\frac{f}{f_1} \approx 8.\text{Log}\frac{f}{f_1} \to \frac{G_0}{g_0} \approx \left(\frac{f}{f_1}\right)^8 \to g_0 \approx G_0.\left(\frac{f}{f_1}\right)^8$$

Either by bringing in (2.30):

$$d_{eff}^2 = \frac{(9{,}81)^2}{16\pi^4}\frac{G_0}{f_1^8}\int_{0_1}^{f_1} f^4.df = \frac{(9{,}81)^2}{16\pi^4}\frac{G_0}{f_1^8}\left[\frac{f^5}{5}\right]_0^{f_1} = \frac{(9{,}81)^2}{16\pi^4}\frac{G_0}{5.f_1^3} \quad (2.33)$$

The total expression of the displacement of a white noise frequency lower than f_1 becomes as follows by adding (2.32) and (2.33):

$$d_{eff}^2 = \frac{(9{,}81)^2}{16\pi^4}\left(\frac{1}{3}+\frac{1}{5}\right).\frac{G_0}{f_1^3} \to d_{eff} = \frac{9{,}81}{\pi^2.\sqrt{2{,}3{,}5}}\sqrt{\frac{G_0}{f_1^3}}$$

For a sigma of 3 and a peak-to-peak displacement, let us multiply by 6, which means:

$$d_{c/c} \approx 109\sqrt{\frac{G_0}{f_1^3}} \qquad [\text{cm}] \tag{2.34}$$

This relation can quickly determine the value of peak-to-peak displacement in function of G_0 and f_1 ($\sigma = 3$ and filter –24dB/octave). Of course, the current piloting consoles provide this information directly, but the approach was made from an educational point of view; we can see in Example 2.1 with $f_1 = 1$ Hz and $G_0 = 0{,}015$ [g^2/Hz], which we find starting from (2.34) a displacement of 133,5 mm c/c for a calculation with VIBKIT 98 of 133,44 mm c/c.

2.2.2.3. *Spectrum of coincidence and quadrature*

2.2.2.3.1. *Overview*

We have just defined PSD in section 2.2 and its importance in the analysis of random vibration signals. However, it appears that it is not sufficient for the proper study of a system's behavior with respect to this type of vibration.

It seems indeed obvious from a mechanical point of view that the evolution of the phase between critical points of a structure has a great importance. Its existence will inform us about the constraints likely to occur within a material or a mechanical structure. Constraints may be at the origin of various destruction or exaggerated toils and limit their life spans. So, in many cases, information on the mutual properties of two signals are needed, for example, for the transfer function between two points in a system, the measurement of output–input relations or mutual properties of two signals.

Now we know that the PSD function is independent of the phase of the signal relative to an arbitrary reference. It is advisable to even better adjust our means of study for random signals.

2.2.2.3.2. *Definitions*

There are several methods that are used to determine the relationship that links the measurement results between two different points of the same structure. For example, the cross-correlation function is defined by:

$$R_{12}(\tau) = \lim_{T\to\infty} \frac{1}{2T}\int_{T}^{T} f_1(t).f_2(t+\tau)dt \tag{2.35}$$

But this method seems *a priori* unrealistic directly in the field of mechanical vibrations, both in terms of the brevity of the time delay that they introduce to that on their dependence on the frequency usually observed and that severely limits the use of the cross-correlation function (or cross-correlation).

Therefore, to study the influence of the frequency and to free themselves a little more of the considerations of time delay, we can apply the FT to the cross-correlation function. The resulting function is called interspectral density function or mutual spectrum. It is given by:

$$S_{12}(\omega) = \frac{1}{2\pi}\int_{-\infty}^{\infty} R_{12}(\tau).e^{-j\omega\tau}d\tau$$

$$S_{12}(f) = \int_{-\infty}^{\infty} R_{12}(\tau).e^{-j2\pi f\tau}d\tau \tag{2.36}$$

So, a function of mutual spectrum density is similar to a function of spectral concentration of power, except for an important exception. The power spectral density function is real and always positive, while the mutual spectrum function has a quadrature component and can be either positive or negative. It is indeed a complex function, consisting of a real part and an imaginary part. And therefore, we write:

$$\begin{aligned} S_{12}(\omega) &= C_{12}(\omega) - jQ_{12}(\omega) \\ S_{12}(f) &= C_{12}(f) - jQ_{12}(f) \end{aligned} \tag{2.37}$$

Part $C_{12}(f)$ defines a function of density in a "co-spectrum" phase or "spectrum in coincidence ", while the imaginary part $Q_{12}(f)$ corresponds to a density function in quadrature "quad-spectrum" or "spectrum in quadrature ". The module and argument are then classically given by:

$$\begin{aligned} |S_{12}(f)| &= \sqrt{C_{12}^2(f) + Q_{12}^2(f)} \\ \Phi_{12}(f) &= \text{arctg}\frac{Q_{12}(f)}{C_{12}(f)} \end{aligned} \tag{2.38}$$

The module represents the module of mutual spectrum between the two points. The phase represents the dephasing between the two points.

Just as the power spectrum, which is formed by multiplying the function by itself (squaring), the mutual spectrum is formed by multiplying one of the functions by the other one. This operation provides the spectrum of coincidence that is the real term of the conjugate pair, the other term of the conjugate pair, which is in quadrature with the previous one, and is called quadrature spectrum and it is obtained in the same way as the coincidence spectrum, but one of the functions of entry is dephased $\pi/2 = 90°$.

Thus, the mutual spectrum density function is represented by two combined terms, one indicating the real energy, common to both random signals of entry, and the other energy in quadrature of these two signals, which is important information from the point of view of the generated internal stresses.

The analysis of complex transfer functions is achieved with the same methods as those that are used for the analysis of mutual spectrum. The complex transfer function is a convenient way for the study of the behavior of a system subject to random vibrations. The complex transfer function is then given by the ratio of the mutual spectrum density input–output of a circuit or between two points of a mechanical structure, the density of power applied to the input spectrum. So, in order to form it, it is necessary to get three spectrum functions, either:

- power at the entrance spectrum;
- coincidence spectrum of input output;
- quadrature spectrum input output.

We then establish the necessary ratios to form the functions that we wish to visualize: one corresponding to the real part of the complex transfer function and the other to the imaginary part in quadrature of this same function. The presentation in terms of CO (coincidence) and QUAD (quadrature) is often preferred, and the recording of one parameter based on the other will provide a Nyquist diagram that gives out information on the stability of the system.

2.2.2.4. *Conclusion*

In conclusion, PSD information is not enough and it will be necessary to form the interspectral PSD along with obtaining the curves in CO and QUAD to have all the useful information on the system being studied.

It should be noted that, currently, PSD measurements for mechanical signals that are at relatively weak frequencies (about 100 kHz) are measured with numerical analyzers after sampling of the signal. These techniques are beyond the scope of the course, but all the comments made previously remain valid, in particular estimates of the DSP and associated errors, there are also constraints related to digitized signals. Only the analyzers of HF spectra and THF (as for measurements of electromagnetic compatibility (EMC)) still use analogical techniques based on systems with heterodyne filters and receiver filters.

2.3. FS–Fourier Integral

2.3.1. Overview

In this case, we will again perform an analog approach for educational purposes knowing that here transitional modern analyzers are also working digitally with "fast Fourier transform" techniques.

These analytical methods are generally used to study transient signals. These signals can also be periodic (the vibration of a system deadened with a degree of freedom, for example) or complex (the damping of a complex vibration). They generally come from mechanical shocks more or less dampened and may have various forms. Simple forms as the one we saw, with demi-sine, square waves, triangles, etc. In extreme cases, the transient signal will consist of a single pulse (as in the impact tests).

But we show that if the impulse is applied to a linear mechanical system, of which one knows the response h(t) to an impulse unit (Dirac $\delta(t)$ theoretically), that is to say, its response impulse, response x(t) of the system to an unspecified excitation f(t) is given by the following operation of convolution [ref. I19]):

$$x(t) = \int_{-\infty}^{\infty} f(\tau).h(t-\tau)d\tau = f(t) * h(t) \tag{2.39}$$

where:

- x(t) = time response of the system;
- f(t) = excitation function;
- h(t-τ) = the system response to the impetus shifted unit of $-\tau$;
- τ = shift and variable of integration (a time dimension).

But the calculation of integral (2.39) often involves the convolution of two very complicated functions. Its resolution is not easy. Thus, to assess the response of a system to an impulse excitation, it is often preferable to use a different mode of representation and work in the space of frequencies, which is possible by applying the FT. This operation has the following advantages:

- a representation of the excitation impulse according to the frequency highlights the important frequencies and the areas where dangerous resonances can occur;
- calculations do not involve a convolution operation but rather simple multiplications. Indeed, the FT transforms the convolution product in a simple product according to the following relationship:

$$FT[f(t) * h(t)] = F(f).H(f) \tag{2.40}$$

with H(f) = FT[h(t)] = transfer function of the system.

The FT of an f(t) function is defined mathematically by:

$$F(f) = \int_{-\infty}^{\infty} f(t)e^{-j2\pi ft}dt \tag{2.41}$$

with the condition $\int_{-\infty}^{\infty} f(t) e^{-j2\pi ft} dt < \infty$ because the integral converges and that f(t) is finite. So when f(t) is an aperiodic impulse, this condition is always met.

After choosing this method of analysis, several different principles can be used: analogical or numerical. We will introduce the analogical approach, which, as already mentioned, is still used only for HF and THF.

However, an important result on the response of the filters with narrow bands to short pulses will allow us to deduce a method. Indeed, it is shown that the peak amplitude and the integrated filter response squared are both related to the bandwidth of the filter of analysis, with:

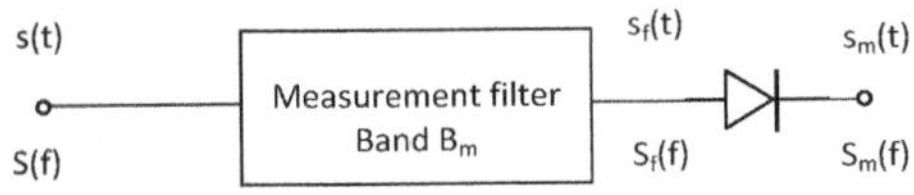

- $s(t)$ = observed signal at the entrance of the measuring device (phase $\varphi(f)$);
- $s_f(t)$ = input signal after filtering;
- $s_m(t)$ = signal measured after filtering and detection;
- H (f) = transfer function of the filter (phase $\psi(f)$);
- $S(f_m)$ = average value of the spectrum S(f) in the narrow band of analysis # cte:

$$s_f(t) = 2.\int_0^{\infty} |S(f)|.|H(f)|.\cos[2\pi ft - \Psi(f) - \varphi(f)].df = 2.\int_{f_m - \frac{B_m}{2}}^{f_m - \frac{B_m}{2}} |S(f)|.1.\cos[2\pi ft - \varphi(f)].df$$

For a rectangular input signal, for example, we find:

$$s_f(t) = 2.|S(f_m)|.\int_{f_m - \frac{B_m}{2}}^{f_m + \frac{B_m}{2}} \cos[2\pi f(t - t_0) - \varphi_0(f_m)].df$$

We have: $$s_f(t) = 2.|S(f_m)|.B_m \frac{\sin[\pi(t - t_0).b_m]}{\pi(t - t_0).B_m}.\cos[2\pi f_m.(t - t_0) - \varphi_0(f_m)]$$

Thus on a detection peak, the measured signal is equal to:

$$s_{max} = 2.B_m.|S(f_m)| \tag{2.42}$$

The output of the filter then takes the following form:

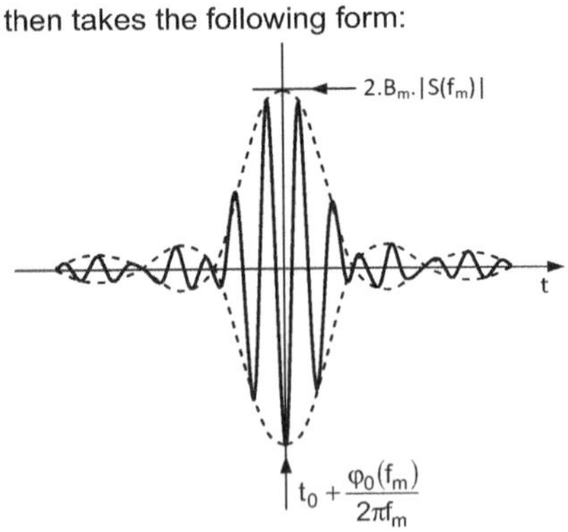

Therefore, if we trace the spectrum of the pulse with a set of adjacent constant percentage bandwidth filters, the response raised, squared and integrated, as well as the peak amplitude response increases linearly with frequency.

This as long as the condition:

$$\tau << \frac{1}{B} \tag{2.43}$$

- τ = width of the pulse;
- B = bandwidth of the filter of analysis.

It is clear that what is measured at the output of the filter must be connected directly to the part of the spectrum of the pulse that is located in the B-band by a calibration operation. So, we need to find the measurement of the Fourier spectrum of the pulse with f_0 and around f_0 central frequency of the analysis filter.

Therefore, we can have a good approximation of the Fourier spectrum with f_0 either by measuring the peak of the filter response and dividing by 2B or by measuring the output squared, integrating it, dividing it by 2B and extracting the square root of the result to get the effective value.

This method, which is therefore a direct method, requires the use of a battery of adjacent filters for a measurement practically in real time. There is another possibility which consists of transforming the isolated impulse in a periodic signal by recording a magnetic loop whose continuous scrolling will achieve the desired periodicity. We will obtain here the FS of the signal whose envelope restores the FT except for a coefficient of amplitude.

2.3.2. Considerations of use

We have studied that, according to their nature, transitory signals can be analyzed either in FS or in FT. In the analysis in FT, the increase in the period of repetition of the transient signal is limited, indeed Fourier coefficient's number is important, but in case of integration of the signal by an RC network, the output level will be lower than the repetition period of the signal ratio, which will be important. We can also confront, for example, against the problems arising from the limitations of the peak factors imposed by the various devices of the measuring equipment. The dynamic range of the measurement system analysis limits the measurement. It is therefore already a condition on the return period of the studied signal. The latter will be a function of the nature and the duration of the signal.

It is also necessary to note that these transient signals diminish over time and as the Fourier integral is deducted from the FS as the period of the series to infinity, the amplitude tends to zero. Thus it becomes necessary to assess the different elements of the integral frequency, relatively comparing the ones to the others and not in the absolute. The formation of the FT by a system of analysis is based on the fact that by increasing the repetition of the pulse relative to its time period, a growing number of Fourier coefficient are produced. Therefore which becomes infinite if this ratio tends to infinity and the Fourier integral is then formed by the envelope of these coefficients, which gives a good approximation of the FT.

For example, we know that for a periodic rectangular impulse of period T, the points of intersection with the x-axis are spaced 1/T. The number of lines between two successive zeros is then given by the ratio T/τ, so, to obtain more than one component or line between two successive zeros, the ratio must be larger than 2. In general we will choose a ratio of 5 that allows a good compromise for a correct measurement of the envelope of the FT.

2.3.3. Analysis of the results

When we have obtained the recording of the spectrum of the discrete components or lines of the periodized impulse, we must calibrate the axis value in units, corresponding to the theoretical values of the Fourier spectrum. Since what was recorded is the effective value of harmonics of periodized impulse and the Fourier spectrum of the isolated impulse is expressed in terms of spectral concentration, how can we deduct these values one from the other?

To resolve this problem, consider a periodic square wavelength τ and a period T. The coefficients of the FS can be calculated according to the following formula:

$$C_n = \frac{2}{T}\int_{-T/2}^{T/2} f(t).e^{-2\pi jnf_0 t}dt \tag{2.44}$$

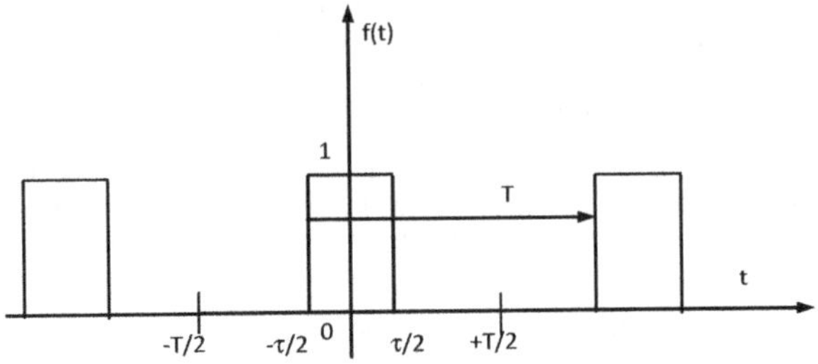

where f_0 is the fundamental frequency and nf_0 is the harmonics of that frequency. We also note that in this example, the signal intervenes only between –T/2 and +T/2 from $-\tau/2$ to $+\tau/2$ where it is different from zero. It thus becomes:

$$C_n = \frac{2}{T}\int_{-\tau/2}^{\tau/2} f(t).e^{-2\pi jnf_0 t}dt$$

So, for a specific frequency $f = nf_0$, the FT of an isolated pulse duration τ is given by:

$$F(f) = \int_{-\infty}^{\infty} f(t).e^{-2\pi jft}dt = \int_{-\tau/2}^{\tau/2} f(t).e^{-2\pi jft}dt$$

While comparing these two expressions, it becomes:

$$C_n = \frac{2}{T}F(nf_0) \equiv \frac{2}{T}F(f)$$

So C_n represents the peak value of a sinusoidal signal (the line with nf_0) and the measured values are in general effective values, therefore the relation between the theoretical value F(f) and the effective value C_{nRMS} of the line is given by:

$$F(f) = \frac{T}{\sqrt{2}}C_{nRMS} \tag{2.45}$$

with $f = nf_0$.

The value of the Fourier spectrum for $f = nf_0$ can be calculated simply by multiplying the effective value by the repetition period and dividing the result by $\sqrt{2}$.

2.3.4. Calculation example

We will take the classic example of a pulse of rectangular shape as described in the above figure.

The calculation of the FS coefficient provides:

$$C_n = \frac{2}{T}\int_{-\tau/2}^{\tau/2} 1.e^{-2\pi jnf_0 t}dt = \frac{2}{T}\left[\frac{e^{-2\pi j(nf_0)t}}{-2\pi j(nf_0))}\right]_{-\tau/2}^{\tau/2} = \frac{2}{T\pi(nf_0)}\left[\frac{-e^{-2\pi j(nf_0)\tau/2} + e^{+2\pi j(nf_0)\tau/2}}{2j}\right]$$

$$C_n = \frac{2}{T\pi(nf_0)}\sin\pi(nf_0)\tau = \frac{2\tau}{T}\frac{\sin\pi(nf_0)\tau}{\pi(nf_0)\tau} \rightarrow C_{nRMS} = \frac{2\tau}{\sqrt{2}T}\frac{\sin\pi(nf_0)\tau}{\pi(nf_0)\tau} = \frac{\tau\sqrt{2}}{T}\frac{\sin\pi(nf_0)\tau}{\pi(nf_0)\tau}$$

So with $(nf_0) \equiv f$: $$F(f) = \tau\frac{\sin\pi(nf_0)\tau}{\pi(nf_0)\tau} = \tau\frac{\sin\pi n\tau/T}{\pi nf_0\tau/T}$$

We find a relationship in sinx/x. There is also the relationship giving C_n in the following form:

$$C_n = 2A\delta\frac{\sin\pi n\delta}{\pi n\delta}$$

with: $\delta = \frac{\tau}{T}$ cyclical ratio < 1, and A amplitude of the pulse of duration τ.

- So, for τ = 2 ms and T = 40 ms, we have δ = 1/20, we get the following standardized curve to 1 (i.e. the amplitude divided by $2A\delta$) and standard compared to nf_0/f_0= n:

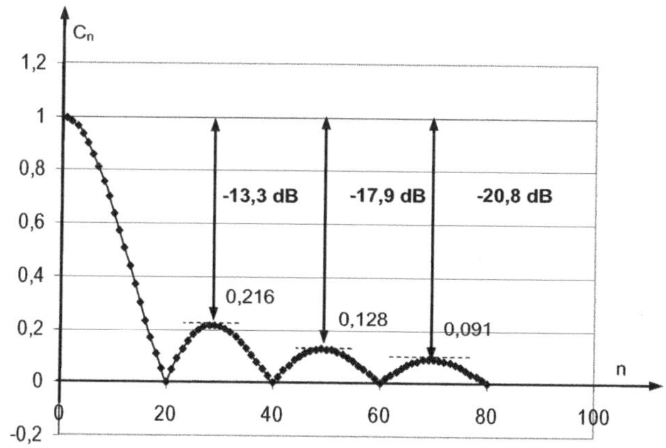

We see here the amplitudes in relative values of different harmonics with conventional attenuation of the first three lobes. Furthermore, we have 20 lines between 0 and n = 20. The casing of these lines gives the FT as per the previously closely calculated coefficient, which does not change anything if we work in both cases with an amplitude normalized to 1.

- Another example for τ = 4 ms and T = 20 ms, meaning δ = 1/5.

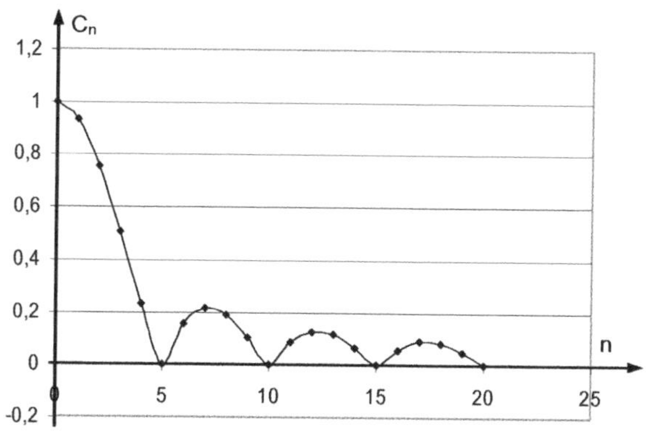

The previous curves are calculated curves, and the spectrum analyzer will provide a curve that has the following general form, the casing in dotted lines being the Fourier spectrum.

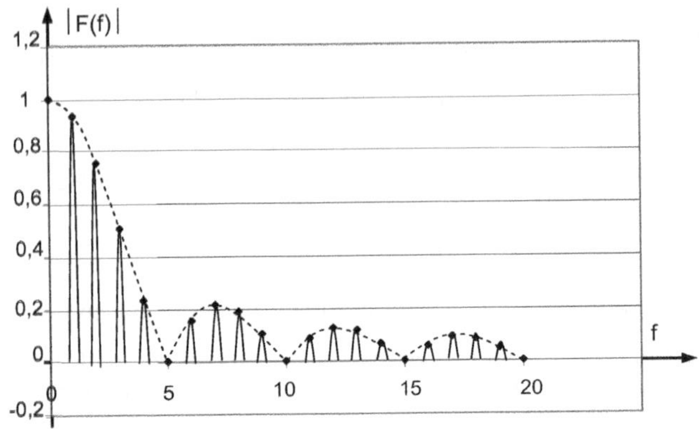

Of course, the most current acquisition fronts use digitized signals with a slightly different treatment.

3

Test Preparation

In this chapter, the different phases and methods and possible procedures that allow optimizing the preparation and course of a test are analyzed.

The proposed approach is primarily based on the standard EN NF ISO/CEI 17025 and LAB Ref 02, so as to prepare the laboratory accreditation requirements COFRAC. The hard points will be discussed and possible solutions will be proposed.

Regarding domain 38 of structural and climate tests, it is recommended to read the GTA38 that defines some important technical recommendations, repeated, in part, here. The classification of test areas mentioned here is the one used by COFRAC. All this is available on COFRAC's website www.cofrac.fr.

3.1. Test demand analysis and associated test specifications

3.1.1. Review of testing demand

After receiving a testing request, the first action to take is the test application review. This analysis must be the subject of a specific procedure (an internal document at the laboratory that is defined by the quality manual), which is intended to ensure that all the requirements, including method of testing to use, are well defined, documented and understood, and that the laboratory has the ability and the resources in terms of test facilities and qualified staff to meet these requirements. It is also advisable to take into account the financial, legal and planning aspects.

This analysis must, in particular, verify if the limits of the proposed method are such that its application can be confusing when applied in the context of the requested test. If this is not the case, it is the responsibility of the laboratory to warn the customer and possibly offer an alternative.

The laboratory can, as a guideline, refer to the following documents:

- NFX50-141-1 and 2: Supplier–customer relationships. Quality of demonstrations. Guidelines for requesting and organizing tests.

- FDX07-019: Metrology. Supplier–customer relationships in metrology.

The laboratory should seek cooperation with clients, or their representatives when it comes to clarifying the customer's request if such need arises and shall help the latter in all necessary technical approaches. Customers attach value to maintaining good communication, the advice and recommendations of technical nature, as well as the opinions and interpretations based on the results of the tests. The laboratory must keep its customers informed of its ability and willingness to offer opinions and interpretations. In case there are compliance statements, the laboratory must analyze along with the clients aspects of "customer risk" and "provider risk", which is declined when the former says that equipment is in conformity when it might not be and for the latter saying that equipment is in non-conformity when it might be.

3.1.2. Test specifications analysis

In general, testing specifications rely on internal documents with the customer or national or international standards. In a general manner:

- the laboratory must implement appropriate methods and procedures for all suggested tests. These methods must meet the needs of the customer and correspond, preferably, to the methods published in regional, international or national standards. It must ensure that it uses any latest edition, except if this is not appropriate or possible. If it is necessary, clarifications must be brought to the standard in order to ensure a consistent application;

- if the proposed method by the customer is deemed inappropriate or outdated, the laboratory must let the client know;

- we admit that it is not necessary to supplement or rewrite in the form of internal procedures the international, regional, or national standards, or other recognized specifications containing sufficient and concise information on how to carry out the tests, if these standards are written in a usable manner as performed by the laboratory operations personnel. On the other hand, it may be necessary to provide additional documentation for optional methodology stages or complementary details, especially as the laboratory must *confirm* that it can properly apply standardized methods before implementing them for testing. In case of changing the standardized methodology, confirmation must be repeated. This means that the laboratory must check with its personnel and test facilities that it is capable of ensuring the technical operation of the method with the required quality. As there are always unique characteristics of the labs in relation to the materials used, it is at least advisable to establish "*a* checklist *of technical hard points*" with the technical recommendations that go with it;

- when the client does not specify a method, the use of default methods can be drawn from international approaches such as the CEI-60068-2-XX. If the use of non-standard methodologies is needed, then they should be validated (validation is the confirmation through examination and provision of objective evidence that particular requirements for a specific intended use are fulfilled). The client must be informed of the methodology used;

- uncertainties or maximum permissible measurement error (MPE) of the specifications must be carefully analyzed and compared with the possibilities of the laboratory means (see Appendix H). Any variation must be analyzed and explained to the client, who remains in agreement of undertaking the tests.

Example of a *nonrestrictive* checklist

- Use of specific piloting sensors in VLF because of the levels of important noises in this frequency band or possible presence of strong levels, and for large duration crash tests (see ISO 60068-2-27, LF cut Figure 4).

- Specific piloting strategy to implement for this methodology.

- Use of a specific test frame with implementation instructions (attachment points and tightening torques, if it is necessary).

- Mandatory implementation of environmental witnessing, which is the measure of the pilot by an independent registration system in case an incident occurs during testing (such is the case of costly or sensitive materials, for example). We can usefully refer to RE Aero 601-11 and 12 and to the GTA38 of COFRAC (section 11.1)).

- Environmental conditions and any period of necessary stabilization.

- Specific verifications to complete before starting the tests (for material testing and measurement of devices and piloting deemed critical).

- Checking measuring equipment and tests (so that they are checked and/or calibrated) and the proper functioning of the equipment, and if necessary, calibrations and intermediate settings of critical equipment.

- Specific adjustments of piloting parameters.

- Etc.

3.1.3. Commercial proposal or contract

After the application analysis and agreeing on the test requested, it is advisable to establish a commercial proposal that generally consists of two parts.

a) Financial part

This contains the details of the cost of the services offered to the customer based on the type of requested test and associated measures.

b) Technical part

The latter is often ignored or insufficiently developed in the commercial proposal, it is essential to perform it in order to avoid ongoing conflict situations during the test and at the end of the test. It is indeed essential to analyze all technical hard points and resolve all differences between the trial application and commercial proposal before undertaking unspecified work. This is performed, of course, with the agreement of the customer and the technical assistance of the laboratory. Without being exhaustive, it contains the following topics:

- reception, preparation, packaging and shipping of the specimen under testing;
- test assembly;
- control strategies;
- points of selected measurements, records, treatments;
- test limits (in particular the estimate of uncertainties related to the MPE of the specification and test limits);
- opinions and interpretations.

Deviations from test methods must take place if the gap has been documented, justified at a technical level, authorized and accepted by the customer. Of course, relevant records of reviews and discussions with the customer must be preserved.

3.1.4. Order review

Once the commercial proposal is sent to the client, the latter responds with an order, whose review by the laboratory allows to check if the client has understood all the topics contained in the proposal and in particular all technical aspects that must be accepted. If there is a difference between the order and the commercial proposal, testing should not be started before the problem is resolved.

Afterward, authorization to prepare and conduct testing may be provided to qualified personnel whose competences are recognized and maintained. The laboratory must keep up-to-date records of authorities, skills and professional qualifications of its staff.

3.2. Test initiation

It essentially consists of checking the aptitude of the means used to carry out testing along with the required quality. Therefore:

- the equipment and corresponding software used for testing must provide the required accuracy and must conform to the relevant specifications for testing (frequency ranges, piloting types, etc.). In some particular cases, it is perhaps necessary to control and/or calibrate some critical devices before testing (concept of intermediary audits). It is thus necessary to check the labels of all devices and testing facilities to ensure compliance with the dates of calibration or verification;

- the laboratory must ensure that environmental conditions are not likely to invalidate the results of the tests and those likely to affect the results must be registered (for example electromagnetic interference of electrodynamic vibrators, etc.);

- the laboratory should have instructions for the use and operation of all relevant devices for the handling of objects under testing. These instructions must be kept up-to-date and must be easily accessible to personnel who may consult them during the preparation phase;

- testing frames must be chosen so as not to disturb the levels of severity of selected specifications;

- the laboratory must ensure the configuration condition of the material under testing;

- relevant annotations must be written down in a lab notebook (or any other support considered to be sufficient) in order to ensure the traceability of the test. It is desirable that this support is legally admissible (for example a registry with numbered pages, or any other internal support validated, not likely to be changed);

- the laboratory must implement the control of the environment, if necessary (measuring rate of transverse, harmonics levels, vibration homogeneity, etc.);

- the laboratory must ensure the confidentiality of the test and the availability of the services rendered to the customer :

  - access to the laboratory,
  - telephone availability, access to internet connection, photocopy, fax, etc.,
  - ease of restoration, transportation (taxis, buses, planes), toilets, etc.,
  - customer zoning and confidentiality protection, etc.

3.3. Test fixtures

3.3.1. Overview

Usually, the direct vibrating source (excluding acoustic vibrations) that excites the specimen to be tested is that of its support in a given mechanical structure (example of the subsets mounted into bays or electronic cabinets, whether deadened or not, or to various supports mounted in vehicles whose mission operation causes vibrations, such as land vehicles, planes, ships, etc.). The test assembly (frameworks, frames, test support, etc.) will have to be able to reproduce a vibration of comparable nature than that induced on the support level, when it is known, or equivalent in the sense of certain criteria, or contractual according to the specifications or the goal of the proposed test.

It is generally characterized by the following parameters:

- its first resonant frequency (or its first mode);

- its mass;

- in a lesser degree, its form.

a) Resonant frequency

The purpose of the test assembly will be to embed perfectly the specimen on the mobile table of the discharger of vibrations used, generally, the mobile reel for an electrodynamic vibrating pot or the piston of the hydraulic actuating cylinder for hydraulic vibrating pots. This, in an identical manner, will exist between the specimen and the vibrating source in normal operation from the point of view of the nature of the number and the provision of the points of fixing. On the other hand, as it will have to completely transmit the levels of vibrations of the discharger to the specimen in a frequency band given in a controlled way, which means that without distortions, amplifications or weakening, it will have to have transmissibility equal to 1 in the explored frequency band and with the level of all fixing points selected. It is said that:

"The frame of tests must be infinitely rigid in the explored frequency band".

The traditional model of transmissibility is shown in Figure 3.1.

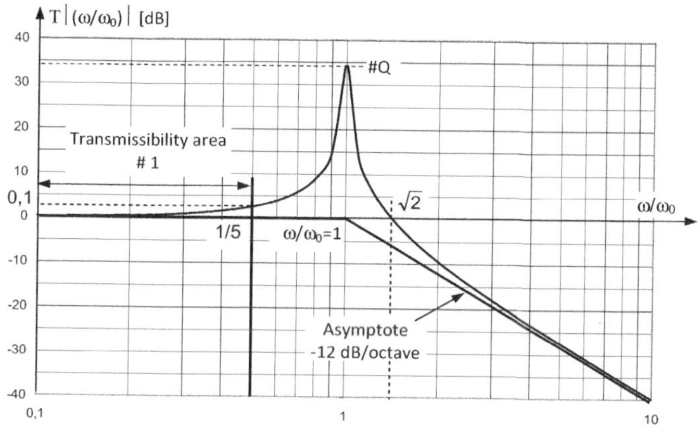

Figure 3.1. *Transmissibility model*

We note in the figure that the area, where transmissibility remains equal to 1 or at least remains below 0,1 (or 10%) in amplitude, is understood between 0 (or very low frequency (VLF)) and the maximum frequency such as $f_{max}= f_0/5$. It is this criterion that is generally accepted. It should be noted that in ∀Q, all curves are well below 0 (in dB), or 1 (linear), for $\omega / \omega_0 = \sqrt{2}$.

The rigidity of a mechanical structure is all the more important than elements are thick and massive.

b) Its mass

The parameter that is most often specified is acceleration (velocity and displacement are connected to it). The maximum acceleration that the exciter may issue to the specimen depends on the following bodies:

- the mobile element of the exciter (coil or cylinder) m_m;
- the coupling part between the mobile element and the assembly of tests when there is m_c;
- test assembly m_e;
- specimen to test m_s.

So: $m_{tot} = m_m + m_c + m_e + m_s$. The maximum force is then given by:

$$F = m_{tot}.\ddot{x} = (m_m + m_c + m_e + m_s)$$

For maximum acceleration, it is necessary for m_{tot} to be as low as possible. But the sheer mass on which we can act is the mass of the test assembly. It is necessary that:

"The mass of the test assembly must be as low as possible".

So we see here that there is a contradiction between the two requirements of rigidity and lightness of the test assembly.

c) Form

The form of the testing assembly is generally imposed by that of the specimen tested, its dimensions and its fixing points. Furthermore, it is almost always requested to excite the three rectangular dimensions of the specimen, which sometimes leads to take into account directly this requirement in the editing of tests when the test facility does not have a transverse table. Test fixtures are generally presented in the form of round or rectangular plates and brackets. The constraint here, in order to avoid couplings and to reduce to the maximum the transverse rates, consists of carrying out these most massive possible fixtures and the smallest possible dimensions. The center of gravity of the assembly must be located in the vibration axis in order to avoid couplings and cantilever, and height must be as low as possible.

3.3.2. Principles of construction

a) Compromise stiffness mass

To obtain a frequency of raised resonance, which must at least be lower than the maximum frequency of the useful bandwidth, that we assume being usually of 2,000 Hz (in traditional vibrations and shocks), the studied assembly must have great stiffness in all directions (along the axis of vibration and in transverse to minimize the rate of transverse). This stiffness will be larger if the forms become thicker and heavier. Therefore, we will have, as previously evoked, a trade-off and we know that the compromises generally bring a bit of satisfaction. Therefore, testing frames will always be imperfect tools to which it becomes necessary to adapt ourselves by correcting imperfections whenever possible. An approach on possible tolerances of test frames is given in Appendix F.

We will therefore work on geometric parameters and the nature of the materials, that is:

- Young modulus, E;
- density, ρ;
- Poisson's ratio, ν;
- damping capacity, ψ.

b) Piloting possibilities

In the case where after having chosen the best compromise in (a) we cannot get a resonant frequency greater than 2,000 Hz, or the maximum frequency explored, we can admit this situation insofar as the amortization of the material used allows the dynamics of the enslavement of the management system in order to maintain the required level at the entrance of the attachments of the specimen. This while trying to have a Q-factor coefficient at the weakest possible resonance. It is necessary to use materials with a strong damping coefficient.

c) Damping study

If we admit that for the first mode (the lowest), the latter behaves like a system with 1 DOF, we can then point out the following results:

- temporal response in the form of a damped sine wave oscillation;

- the number of significant oscillations (amplitude greater than 10% of the max value) is equal to the Q-factor coefficient;

- logarithmic decrement: by definition $\Delta = \frac{1}{n}.\mathrm{Log}\frac{x_m}{x_{m+n}} = \frac{2\pi\xi}{\left(1-\xi^2\right)^{1/2}}$ where m is the mth amplitude and n the nth amplitude after m. In case of a low damping (case of metals), $\Delta \approx 2\pi\xi$;

- damping capacity that is given by the ratio of energy absorbed by the next cycle to the following maximum energy:

$$\Psi = \frac{x_m^2 - x_{m+1}^2}{x_m^2} = 1 - \frac{x_{m+1}^2}{x_m^2} \tag{3.1}$$

We obtain:

$$\Psi = 1 - e^{-2\Delta} \tag{3.2}$$

In case of a low damping (case of metals) $\Psi \approx 2\Delta$.

- The amplification coefficient to the resonance that is also given by:

$$Q = \frac{x_{max} = \text{maximum amplitude taken by the mass m in a forced regime}}{x_0 = \text{Amplitude of the excitation in forced regime}} = \frac{4\pi^2 + \Delta^2}{4\pi\Delta}$$

Having if $\Delta << 1$ (metallic materials):

$$Q \approx \frac{\pi}{\Delta}$$

Therefore:

$$Q \approx \frac{2\pi}{\Psi} \tag{3.3}$$

So, the amplification coefficient Q at resonance is lower as damping is greater.

d) Example of a simple assembly: the plate or the circular plate

The model is the following (circular plate embedded on a circle of radius R) knowing that a precise calculation is very complex and that we provide only partial results here, of course, but they are usable in practice with a few simplifying assumptions:

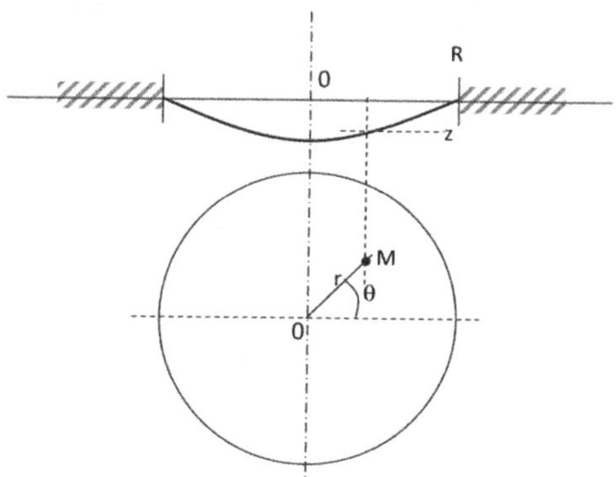

With:

- R = plate embedding radius of the circle;
- h = thickness of the plate;
- z = arrow in any point M (r,θ) of the plate;
- r and θ = polar coordinates of point M;
- E = Young's modulus of the material;
- ν = Poisson's ratio of the material;
- ρ = material density ($\pi R^2 h\rho = mg$);
- m = mass of the plate.

We therefore show (it is important to see it once) that if we impose a sinewave excitation of pulsation ω and by neglecting the effects of the side forces of connections, the movement of the plate is defined by the following equation (assumption of thin plates):

$$\Delta.\left(\frac{\partial^2 z}{\partial r^2} + \frac{1}{r}\frac{\partial z}{\partial r} + \frac{1}{r^2}\frac{\partial^2 z}{\partial \theta^2}\right) = \frac{12(1-\nu)^2 \rho}{gEh^2}\frac{\partial^2 z}{\partial t^2} = \left(\frac{\partial^2}{\partial r^2} + \frac{1}{r}\frac{\partial}{\partial r} + \frac{1}{r^2}\frac{\partial^2}{\partial \theta^2}\right)\left(\frac{\partial^2 z}{\partial r^2} + \frac{1}{r}\frac{\partial z}{\partial r} + \frac{1}{r^2}\frac{\partial^2 z}{\partial \theta^2}\right)$$

Δ being the Laplacian in polar coordinates here.

For the research of the first resonant frequency, we put in a sinusoidal regime: $\frac{\partial^2 z}{\partial t^2} = \omega^2 z$,

and we look for a solution of the following form:

$$z_n(r,\theta) = (A_{1n}\sin n\theta + A_{2n}\cos n\theta).\left[J_n(a,r) - \frac{J_n(a,R)}{I_n(a,R)}.I_n(a,r)\right]$$

$$= A_n.\cos(n\theta - \varphi).[J_n(a,r) + \lambda.I_n(a,r)]$$

J_n are Bessel functions of the first order and I_n of the modified Bessel functions.

With $a = \sqrt[4]{\frac{\rho h \omega^2}{gD}}$ and $D = \frac{Eh^3}{12(1-\nu^2)}$ = flexural rigidity

We finally found an expression of the form:

$$\omega = \frac{K}{R^2}\sqrt{\frac{Dg}{\rho h}} \tag{3.4}$$

K is a coefficient that can be calculated with the tables of Bessel functions or measured experimentally.

So, we can, from this expression, determine the optimal choice of material based on the parameters that characterize it (E, ρ, ν). It becomes with: $h = \frac{mg}{\pi R^2 \rho}$.

$$\omega = K.\frac{mg}{\pi R^4}.\sqrt{\frac{g}{12}}.\sqrt{\frac{E}{\rho^3(1-\nu^2)}} \tag{3.5}$$

Thus, for a given mass, the best material is one that will provide the value of the term under the second highest radical.

	E in 10^{10} Pa	μ	ρ in 10^4 N/m ^{3}	$10^2\sqrt{\frac{E}{\rho^3}}$	$10^2\sqrt{\frac{E}{\rho^3(1-\nu^2)}}$
Beryllium	**28,9**	**0,05**	**1,85**	**22,49**	**22,52**
LA141-A(Mg-Li)	4,22		1,35	13,10	
LA91 (Mg-Li)	4,64		1,45	12,34	
LAZ933 (Mg-Li)	4,5		1,45	12,15	
Ceramics in BeO	31,5		2,9	11,36	
Magnesium	**4,60**	**0,34**	**1,74**	**9,34**	**9,94**
Pyrex glass	6,5		2,23	7,66	
Ceramics in AL_2O_3	34,5		3,9	7,63	
AS22 UNK	**8,44**	**0,33**	**2,63**	**6,81**	**7,22**
AS20 U	**8,1**	**0,33**	**2,55**	**6,99**	**7,40**
Aluminum	**7,05**	**0,34**	**2,65**	**6,15**	**6,54**
Duralumin AU4G	**7,5**	**0,33**	**2,8**	**5,85**	**6,19**
Titanium	11,6	0,3	4,41	3,68	3,86
Stainless steel	20	0,38	7,8	2,05	2,22
Inconel	20	0,38	8,16	1,92	2,07
Molybdenum	34,3	0,31	10	1,85	1,95
Nickel	20,6	0,31	8,8	1,74	1,83
Copper	11	0,36	8,8	1,27	1,36
Zinc	8,24		7,19	1,49	
Zirconium	8,34	0,33	6,51	1,74	1,84
Casting	7	0	7,4	1,31	1,31

The previous table provides the values of this term for different materials.

In this table, we see in bold the materials that could be used. Knowing that Beryllium is very expensive and that magnesium has risks when it comes to processing it and using it, we can say that, excepting for exceptional cases, aluminum is the material more commonly used to design test frames (and sometimes magnesium). For very important structures and in particular for impact tests, fixtures with profiled steel are also used.

Another simpler and approximate relation can also be used [ref. I20].

$$f = \frac{\beta}{4\pi^2}\sqrt{\frac{Eh^2}{\rho . a^2}} \qquad (1)$$

a = diameter [m];
h = thickness [cm];
ρ = density [kg/m^3];
E = elasticity module or Young's modulus [MPa];
g # 9,81 [ms^{-2}].

Circular plate embedded around: β # 744
Free circular plate: β # 383
Circular plate fixed at the center: β # 274
Circular plate with support all around: β # 371

Or according to [ref. I21 and I2]. For circular or square plates depending on the conditions of support and here a constant A_k:

$$\omega_K = A_K\sqrt{\frac{Eh^2}{\rho D^4\left(1-\nu^2\right)}} \qquad (2)$$

where

E = Young modulus [N/m^2 or Pa];
h = Thickness [m];
D = diameter or side [m];
ρ = Density [kg/m^3];
ν = Poisson's ratio.

Conditions at the borders	Description	Mode 1	Mode 2	Mode 3	Mode 4	Mode 5
Circular, embedded on all sides	○	11,84	24,61	40,41	46,14	103,12
Circular, free–free	○	6,09	10,53	14,19	23,8	40,88
Circular simply supported	○	5,9				
Circular held in the center	⊕	4,35	24,26	70,39	138,85	
Square, embedded on all sides	□	10,4	21,21	31,29	38,04	38,22
Square free–free	□	4,07	5,94	6,91	10,39	17,8
Square, simply supported on all sides	□	5,7	14,26	22,82	28,52	37,08

Values of A_k and λ

NB: All of these relations derive from the original relation (3.5) by implication of the constants.

We note the complexity of the calculations here, especially when it comes to obtaining parameters K, β or A_k. Fortunately, there are tables providing the frequencies of the modes of the elements of beams and plates that are used to properly model the simpler configurations of mechanical fixtures as we will discuss further ahead.

Example 3.1. Having the circular plate of the following test fixed on the voice coil (of 0,6 m diameter) of an electrodynamic vibration generator:

With: a = 0,6 m
h = 10 cm
E = 75,000 MPa
ρ = 2,798 kg/m^3

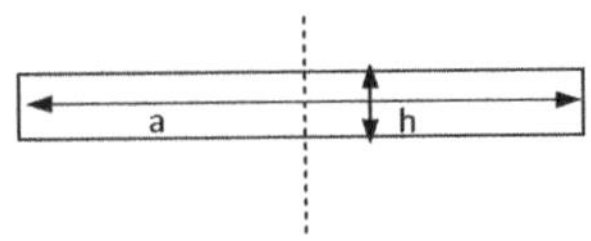

The matter is AU4G commonly used for frames and its calculation is made for the embedded mode in A. That is to say, all calculations are made with (1) for example (see above):

f_1 # 2,710 Hz so beyond the 2,000 Hz commonly used. We see that the frequencies of the first mode are relatively low with respect to the 2,000 Hz even for h = 10 cm, fortunately, in reality the plate is fixed on 17 inserts on a diameter of 30 cm and behaves practically as a rigid body that only adds a mass to the voice coil and degrades the maximum possible acceleration.

3.3.3. Concepts of strenght of materials

a) Reminders on the flexion of a free embedded beam

This example will be used to explain the problem from a simplified approach.

The model is as follows, with:

T = shearing force;

M_f = bending moment;

σ = constraint.

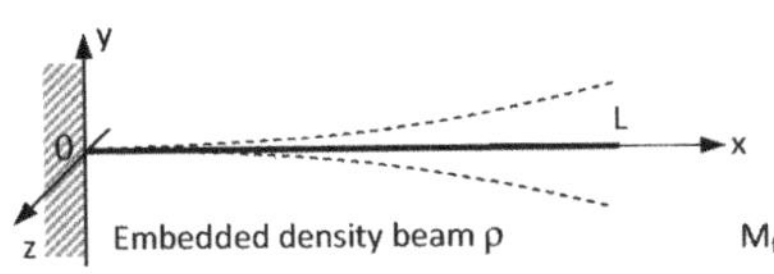

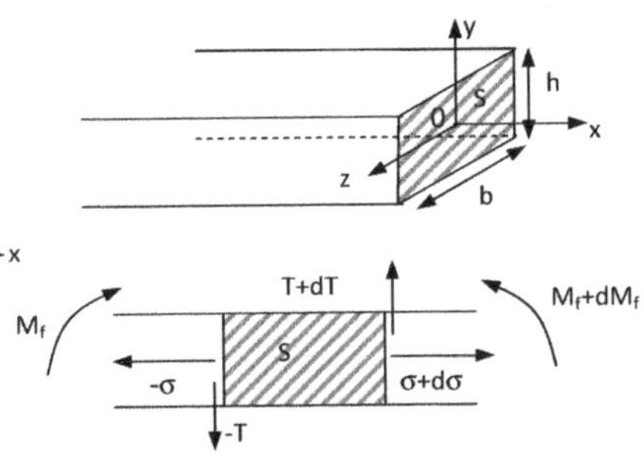

Given the inertia of the beam along the axis 0z by: $I_z = \iint_{\text{right section}} y^2 dy = \frac{bh^3}{12}$

The mechanical equations provide: $\sum_i \vec{F_i} = m.\vec{\gamma} = m.\frac{d^2\vec{y}}{dt^2}$ et $\sum_i \vec{M_{fi}} = J.\ddot{\theta}\,\vec{z} \approx 0$

Having along the vertical axis for the shearing force: $T + dT - T = \rho S dx.\frac{\partial^2 y}{\partial t^2}$

Having: $$\frac{\partial T}{\partial x} = \rho S.\frac{\partial^2 y}{\partial t^2} \qquad (3.6)$$

For the bending moment neglecting rotation inertia (valid in BF):

$$M + dM - M + (T + dt).dx = 0$$

We then show infinitely close to the second order that: $\frac{\partial M_f}{\partial x} = -T$ (3.7)

We can therefore deduce the following relation: $\frac{\partial^2 M_f}{\partial x^2} + \rho S.\frac{\partial^2 y}{\partial t^2} = 0$

It shows that the bending the moment is given by: $M_f = EI.\frac{\partial^2 y}{\partial x^2} = EI.\frac{1}{R}$ (R = curvature radius). It therefore has the general differential equation governing the beam's bending:

$$\frac{\partial^4 y(x,t)}{\partial x^4} + \frac{\rho S}{EI}.\frac{\partial^2 y(x,t)}{\partial t^2} = 0 \tag{3.8}$$

We therefore look for solutions of the following form: y(x,t) = P(x).Q(t). By substituting in (3.8), it becomes:

$$Q(t).P^{(4)}(x) + \frac{\rho S}{EI}.Q^{(2)}(t).P(x) = 0 \rightarrow \frac{P^{(4)}(x)}{P(x)} = -\frac{\rho S}{EI}.\frac{Q^{(2)}(t)}{Q(t)} \rightarrow \frac{Q^{(2)}(t)}{Q(t)} = -\frac{EI}{\rho S}.\frac{P^{(4)}(x)}{P(x)}$$

In harmonic regime $\frac{Q^{(2)}(t)}{Q(t)} = -\omega_n^2 = cte \rightarrow \frac{EI}{\rho S}.\frac{P^{(4)}(x)}{P(x)} = \omega_n^2 = cte$

So, for the expression $\frac{P^{(4)}(x)}{P(x)} = cte$, we substitute as: $\frac{P^{(4)}(x)}{P(x)} = k_n^4$ since we have four derivations, which becomes: $\omega_n^2 = k_n^4\frac{EI}{\rho S}$.

We show that while the index n is introduced as a result of the boundary conditions of displacement, as well as the continuity relations, it is necessary here (in the embedded-free beam) to have the following relation involving displacement: $\cos(kL).ch(kL) + 1 = 0$. Its resolution can be achieved through an iterative process. The possible values of $K_nL = A_n$ corresponds to the different modes of the beam. We admit being able to write:

$$\omega_n^2 = k_n^4.\frac{EI}{\rho S} = A_n^4 L^4.\frac{EI}{\rho S} \rightarrow \omega_n = A_n^2.\sqrt{\frac{EI}{\rho S L^4}} \tag{3.9}$$

We show that in the case of the embedded-free beam, we have the following values:

$A_1^2 = 3{,}52$ $A_2^2 = 22{,}4$ $A_3^2 = 61{,}7$ $A_4^2 = 121$ $n \geq 5 \rightarrow A_n^2 \approx \left[(2n-1)\frac{\pi}{2}\right]^2$

In conclusion, we note that for geometric dimension data, natural pulsations of a beam (or of a system in general, such as the disk, for example) are related to ratio $\frac{E}{\rho} = \frac{\text{Young's module}}{\text{Density}} \equiv \frac{\text{Stiffness}}{\text{Mass}}$.

We find the fact that the pulse is even higher as stiffness becomes higher (rigidity) and as the mass is lower (lightness).

We provide here, for example, and in different units, the following table:

Type	Steel	AU4G	Mg	Be	Ti	Zn
10^{-10}.E [N/m^2]	20,7	6,9	4,1	30	10,7	9,2
ρ [kg/m^3]	7,840	2,798	1,800	1,800	4,510	6,600
10^{-7}.E/ρ [N.m.kg^{-1}]	2,65	2,49	2,3	16,6	2,38	1,4

Note the light dispersions that exist on the values of the parameters that are necessary to research precise parameters with the product supplier.

We note that the fact that these materials have a close characteristic E/ρ except for beryllium, which has a ratio approximately six times higher. Also, the mass of the part and its manufacturing cost are two main parameters that determine the choice of material. We find here the following remarks as well:

- steel is cheap and works well but it is heavy;
- beryllium is only used for small structures. It is expensive and its machining is difficult;
- aluminum alloys (AU4G, AU2GN, etc.) remain the most used;
- magnesium is lightweight but expensive and requires machining precautions. It is often used in foundry instead of Zamack;
- there are aluminum and magnesium alloys (Maxium, AZ91).

Furthermore, equation (3.9) shows that if the height h is doubled, inertia I is multiplied by eight and the fundamental resonant frequency of the plate will be multiplied by # 2.8. This example shows the benefit of employment ridges as stiffening in the CI charts, of the transformation of square sections into rectangular sections and of course materials with a strong coefficient E to increase resonant frequencies.

b) Reminder of a plate's bending

We already saw the example of the circular board. The study of thin plates leads to the same result with the simplified approach given simply for comparison purposes.

The model is for example:

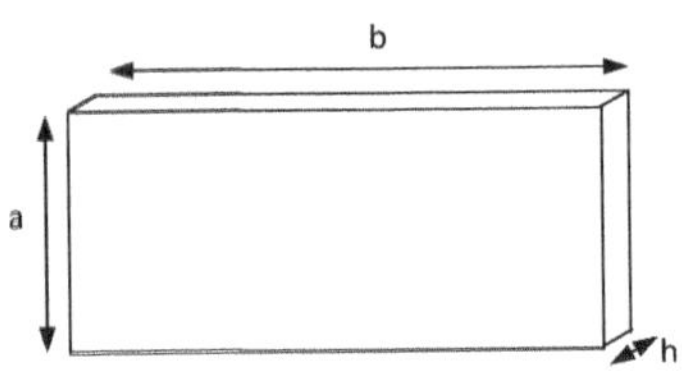

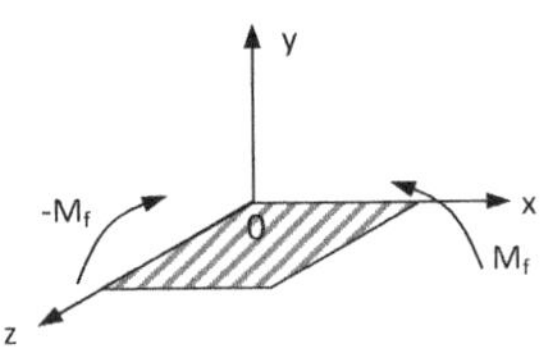

Assuming no point loads, meaning: $\varepsilon_z = 0$ and $\sigma_y = 0$.

In this case, we have the following equations:

$$\begin{cases} \varepsilon_x = \dfrac{\sigma_x}{E} - \dfrac{v}{E}(\sigma_y + \sigma_z) \\ \varepsilon_z = \dfrac{\sigma_z}{E} - \dfrac{v}{E}(\sigma_x + \sigma_y) \end{cases} \rightarrow \varepsilon_x = \sigma_x . \frac{1 - v^2}{E}$$

We admit that for the plates compared to the beam, *roughly speaking* we need to change E into $\dfrac{E}{1 - v^2}$. Moreover, using the Rayleigh–Ritz method (low damping with maximum potential energy = maximum kinetic energy), it is shown that we obtain, for different plate modes, the following equation:

$$\omega_n^2 = k^2\pi^4 \frac{Eh^2}{(1-\nu^2)12\rho b^4},$$ compared with the expression of beams and (3.5) with:

$$\omega_n^2 = A_n^2 \frac{EI}{\rho SL^4} = A_n^2 \frac{Eh^2}{12\rho L^4}, \qquad \text{with}: S = bh \text{ and } I = \frac{bh^3}{12}.$$

Here, the similarity of the expressions arise. We find, for example, for ν # 0,3 for metals and E/ρ # 23,5.10^6.N.m/kg with dimensions in centimeter, the following fundamental:

$$f_0 = 23{,}5.10^4.k.\frac{h}{b^2}$$

which shows, of course, that the frequency is in a maximum for the minimum of b or a.

ATTENTION.– There are tables providing the modes of the beams and the plates for various mounting configurations (embedded, free, supported, etc.) with units and parameters more or less different. It is necessary to accept it and analyze very well the form in which relations are given. From the point of view of units, it is recommended to use international standard units in any calculation except in special cases, explicitly.

For example, here is a table providing the modes of the beams of uniform section and uniform loading modes (knowing that any calculation that comes out of these assumptions is a complex one and is outside the scope of the course) [ref. I22].

We can note the rapid growth of frequencies of the modes that shows very well that it is especially necessary to be interested in the lowest mode.

Embedded Free	A = 3,52	0,774 A = 22,4	0,5 0,868 A = 61,7	0,356 0,644 0,906 A = 121.0
Articulated Articulated	A = 9,87	0,5 A = 39,5	0,333 0,666 A = 88,9	0,25 0,5 0,75 A = 158
Embedded Embedded	A = 22,4	0,5 A = 61,7	0,259 0,641 A = 121	0,278 0,5 0,722 A = 200
Free Free	0,224 0,776 A = 22,4	0,132 0,5 0,868 A = 61,7	0,094 0,356 0,644 0,906 A = 121	0,073 0,277 0,5 0,723 0,927 A = 200
Embedded Articulated	A = 15,4	0,560 A = 50	0,384 0,692 A = 104	0,294 0,529 0,765 A = 178
Articulated Free	0,736 A = 15,4	0,446 0,853 A = 50	0,308 0,616 0,898 A = 104	0,235 0,471 0,707 0,922 A = 178

We should also note the difficulty of experimentally obtaining perfect conditions to limits, such as the embedding, for example, which cannot be achieved by the beam's machining. This has as a consequence that sometimes experimental measurements provide approximate results.

An equivalent array is given for thin plates as previously seen, with:

$$\frac{\omega_n}{\sqrt{Dg/\rho.h.a^4}} = K_n\,;$$

$D = Eh^3/12(1-\nu^2)$;

ρ = density [kg/m^3];

h = thickness of the plate [m];

a = length of the plate (or greater length) [m].

The differences between the modes are relatively less important here. General remarks are the same as those made for the beams. For example, CI charts are neither built-in nor in simple support, we generally use slides with elastic or locked maintenance as thermal drain. In this case, only the experimentation provides the exact modes of the chart.

Nevertheless, the few templates provided here allow approaching the cases usually encountered in practice and give a good order of magnitude.

See the table below:

modes	1st mode	2nd mode	3rd mode	4th mode	5th mode	6th mode
K_n	3,494	8,547	21,44	27,46	31,17	
Modal deformation	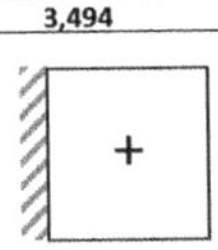	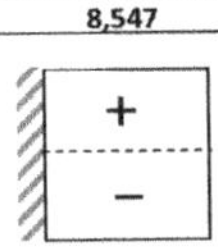	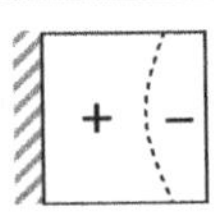	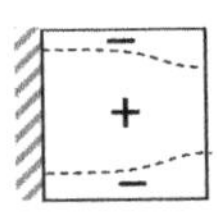	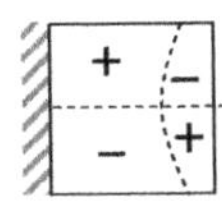	
K_n	35,99	73,41	108,27	131,46	132,25	165,15
Modal deformation	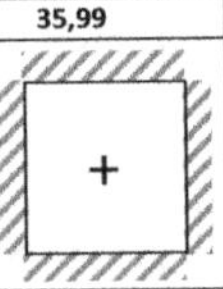	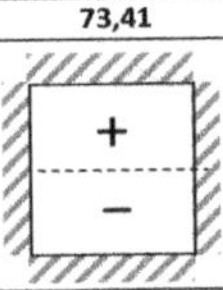	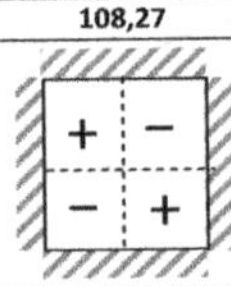	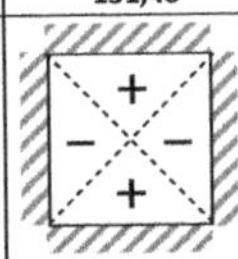	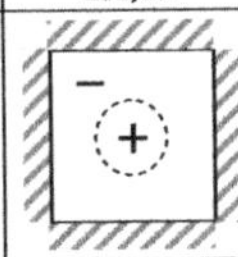	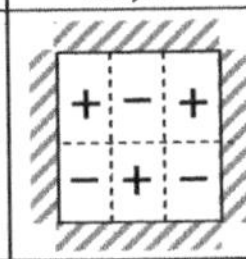
K_n	6,958	24,08	26,80	48,05	63,14	
Modal deformation	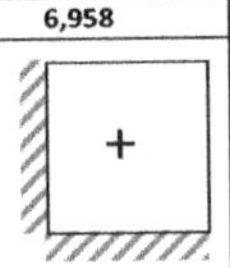	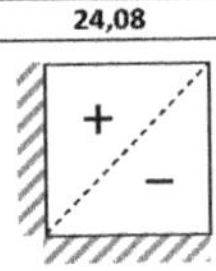	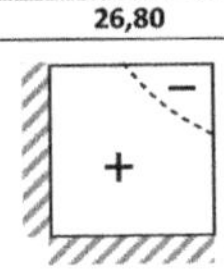	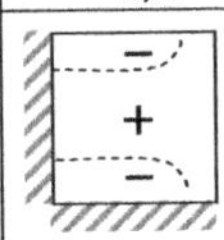	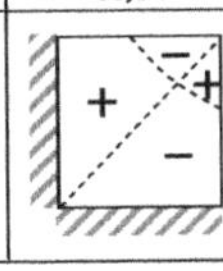	

c) Propagation phenomena

We must not forget that like any impedance medium given, the materials are prone to propagation phenomena and thus to regimes of standing waves. Compression waves appear for relatively small lengths in usual materials.

Of course, for the calculation of the structures, these are essentially bending vibrations that must be taken into account and not that of compression. The simplified model of a beam in compression is then the following:

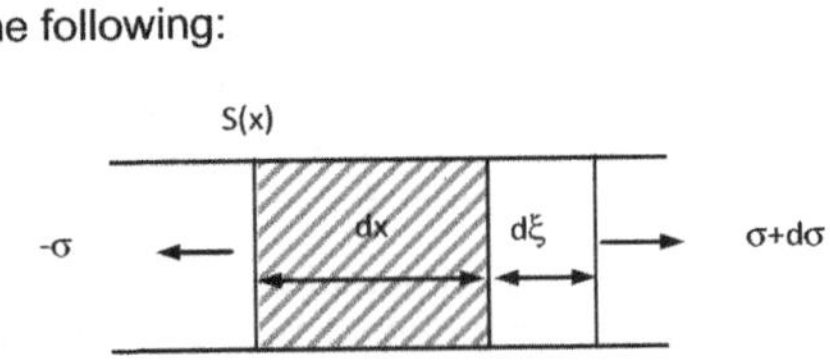

RdM relations give to balance:

$$S(x).d\sigma = (\rho S(x)dx).\frac{\partial^2\xi}{\partial t^2} \rightarrow \frac{\partial\sigma}{\partial x} = \rho.\frac{\partial^2\xi}{\partial t^2}$$

Or: $\sigma = E.\varepsilon = E.\frac{\Delta L}{L} = E.\frac{\partial\xi}{\partial x}$

From where the propagation equation derives: $\frac{\partial^2\xi}{\partial x^2} = \frac{\rho}{E}.\frac{\partial^2\xi}{\partial t^2} \equiv \frac{1}{v^2}.\frac{\partial^2\xi}{\partial t^2}$

Therefore:

$$v = \sqrt{\frac{E}{\rho}} \qquad (3.10)$$

Digital application: We have, for example, for AU4G:

$v = \sqrt{\frac{7,5.10^{10}}{2,798}} = 4,990\ [\text{m/s}]$. As the relation with the wavelength is given by:

$\lambda = \frac{v}{f} = v.T$, and for a free embedded beam the first maximum is given for $L = \lambda / 4$, it becomes: $v = f / \lambda = f / 4L$ and so for a maximum frequency of regular vibration of 2,000 Hz, we would not need massive assemblies whose largest dimension goes over $L = \frac{v}{4f} = \frac{4,991}{8,000} \approx 62,5$ cm. So for higher dimensions we are in propagation mode. In fact, the structures used as a frame for testing usually consist of assemblies screwed or welded and so, these are other phenomena that are put into play. Therefore, we see here the difficulty that exists to enslave large structures at high frequencies. We will see that in this case, it will be necessary to adapt piloting strategies (such as the use of pilotage in notching or an average).

For complex frames, it is necessary to characterize them by a master key of modal analysis to know the lowest models, at least the first mode.

3.3.4. Manufacturing influence

a) One-piece machining

In this case, we have a good fixation, because there are no bolts that act like springs in high frequency by the elasticity of the material. Embedded models are respected here and results awaited by correct calculations. On the other hand, costs are important when parts are complicated. It is necessary to hold this approach with adaptation parts that generally exist between test assembly and the variable component of the exciter.

Fatigue zones may also appear in the angles that must be cut as well as to the places where material has been sheared. In general, it is expected thicker and therefore heavier.

b) Molding

We get a contiguous, homogeneous piece with little machining (additional thickness to ensure through machining, the flatness and the correction of flat surfaces). The cost here is even more important here because it is necessary to make a mold and a model. As before, it is necessary to reserve this technique for immediate parts, for example the transverse vibration tables and the coupling parts.

c) Assembly

Systems including parts assembled with bolts are convenient, but have drawbacks in HF. Indeed, bindings are springs that lower resonant frequencies. It is necessary to take precautions on the implementation of fixings (deep counterbores to minimize the lengths of the screws and thus to have a spring stiffer effect to avoid putting on line the fixings points with layouts in quincunx, for example, and embed as well as possible with sets the weakest possible plates or stiffeners). For low and medium frequencies and with usual precautions, these fixtures are interesting because they are less expensive and can be changed. They function up until a few hundred hertz.

It is preferable to use welded parts with welding in continuous cords. Modern welding positions allow doing very good work from this point of view. The downside is that the editing is not removable or editable.

Digital application: We can note, as already mentioned, that the fixing with a bolt is virtually considered as a condition of free beam, while welding is practically considered an embedding. We can see this effect on an aluminum square bar with S = b.b = 100 mm^2 and L = 100 mm. Then, we have:

Of:

$$f_0 = \frac{1}{2\pi} A_0 \sqrt{\frac{EI}{\rho S L^4}} \qquad I = \frac{b^4}{12} \rightarrow \qquad f_0 = A_0 . \frac{1}{2\pi} . \sqrt{\frac{E}{\rho}} \sqrt{\frac{b^2}{12L^4}} = A_0 . \frac{4991}{2\pi} . \sqrt{\frac{10^{-4}}{12.0,25^4}} = A_0 .36,7$$

So, we practically double the useful range of frequency with a welded assembly.

d) Welding

This is usually the best solution for the quality--price ratio. As we have seen, it is necessary to do welding in cords. After welding and before machining, it is advisable to make some tempering (at roughly 250°C) to remove thermal stresses and thus avoid warping. Also provide an allowance for possible eventual machining avoiding sharp angles (correction of welding).

NB: It is also possible to mitigate the coefficients to the resonance with damping products such as some varnish or polymers.

3.3.5. Conclusion

In conclusion, we recall the following constraints:

- having a rigid and light structure;
- welded joints preference;
- having a well-machined bearing surface and/or rectified and limited to the bare minimum;
- tightened bolts to the torque with a torque wrench.

"Always expect a characterization by a brief (modal) harmonic analysis of the editing of tests proposed".

For example, for the characterization of a thrust nose of an electrodynamic vibrator.

The assembly of tests can be the following: it includes an accelerometric measurement on the edge of the thrust nose, with the maximum overvoltage and a piloting to the center of the thrust nose.

This simple handling allows properly measuring the first resonance mode of the thrust nose; we expect beyond 2,000 Hz for a given inert load, for example.

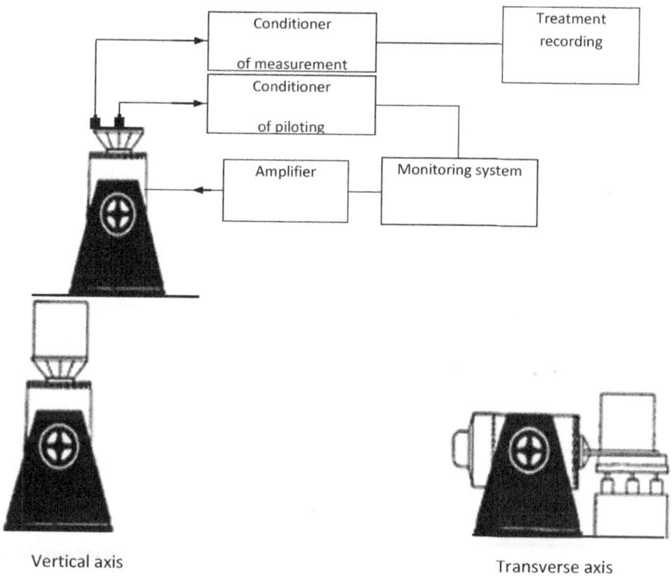

In general, for common equipment (boxes, etc.) mounted on dampers, frames made of plates of 40 mm thick AG4 screwed are enough, as well as circular plate brackets for bandwidths of a few hundred hertz. Some examples of frames are given in Figure 3.2.

Figure 3.2. *Electronic cabinet. Welded/screwed frame and electronic box. Screwed frame*

3.4. Test execution

This is the actual operation of the trial that will be specified for each type of critical test points in Chapter 5. It is, however, essential to sensitize the lab on the recordings of the various parameters of the test, whether it is for vibration controlling or good operation audits of the equipment under testing functional verification test (FVT), or of the witness of the environment, if necessary. This is to prepare the necessary elements for the preparation of testing reports and possibly the declaration of conformity or opinions and interpretations set out by the laboratory, if necessary.

3.5. Test reporting

It is in fact the most important element issued by the testing laboratory. Indeed, it is the product in the form of a document that is sold to the customer and sanctions the test. Therefore:

- the results of each test performed by the laboratory must be reported in an accurate, clear, non-ambiguous, objective way and in accordance with the specific instructions in the testing methods;

- they must be accompanied by all the information requested by the customer, necessary to the interpretation of the test results, as well as all the information required by the testing method;

- when opinions and interpretations are given, the laboratory must submit in writing the basis on which opinions and interpretations issued. Opinions and interpretations must be clearly indicated as such in a test report. They must be particularly differentiated from advice and expertise;

- they must be formulated by a designated official with thorough knowledge of the proper technical discipline and a realistic assessment of the limits of his/her own knowledge in the context of opinion and interpretations that will be carried out in a report on results.

Term	Definitions resulting from standard NF X 50-110 adapted to the case of laboratories.
Judgment	Opinion resulting from an analysis or an evaluation, in response to the question raised and having no force of decision, made by the laboratory on the basis of the results achieved and the current state of knowledge.
Interpretation	Action carried out by the laboratory, in response to the question asked to explain or give a meaning to data on the basis of results achieved and professional judgment.

For example, a possible component of the opinions and interpretations in the test report may be the statement of compliance/non-compliance of the results to the requirements of the specification.

- The mention of uncertainty in test results report is a function both of the requirements expressed by the client and those contained in the used repository.

Table 3.1 deals with the different possible scenarios.

- Whether or not they need to show the uncertainties on the reports on the test results, testing laboratories need to evaluate the uncertainties associated with these results.

Customer requirements (traceable to the level of the review of the request) / Requirements of standard(1)	The laboratory's client requests that the incertitude(3) was mentioned in the report.	The laboratory's client requests that the incertitude(3) is not mentioned in the report OR The laboratory's client does not specify anything.
The referential(1) explicitly mentions that the specification limits(2) were fixed without taking into account incertitude(3).	The incertitude(3) must be specified in the report.	
The referential(1) explicitly mentions that the specification limits(2) were fixed taking into consideration incertitude(3). OR The referential(1) does not mention anything regarding incertitudes(3).	The incertitude(3) must be specified in the report.	The incertitude(3) cannot be specified in the report.

The numbers in brackets correspond to terms under the standard.

Table 3.1. *Various cases of uncertainty declaration*

a) Use of uncertainty for purposes of conformity declaration

The manner of taking into account the measurement uncertainty in the processes of conformity declaration is not unique according to the products (word taken in the broad sense and including among others the notion of service), measurement fields, standards, occupations or even different countries, rules and practices. By declaring compliant or non-compliant, a product generates two risks:

- "customer" risk, which is the risk of wrongly declaring that a product complies;
- "provider" risk, which is the risk of wrongly declaring that the product is not in conformity.

Furthermore, the laboratory does not always know if the product being asked to check the specifications must be examined from the perspective of the "provider" or the "customer". Indeed, when it is certified, it is normally involved in a situation of independent "third party", whose works will be used by other entities, which will in fact take the risks.

The fundamental data on which the decision is based are:

- the limit(s) of the specification that is imposed on the features of the product;
- the result(s) of measurements, tests or analysis of the characteristic;
- measurement uncertainty affecting the result of measurements, tests or analysis;
- the choice of the level of risk accepted by the client of the laboratory;
- decision rule (i.e. how to combine previous data to declare or non-compliance).

The compliance declaration, when the laboratory carries it out, usually relies on a rule of the following type:

- case 1: decision = function (measurement results, associated uncertainty, specification limits or EMT). In this case, the report should include for example: "To declare or not the compliance with the specification, it was explicitly taken into account the uncertainty associated with the result". The laboratory and its client may agree to a statement explaining the rule applied;

- case 2: decision = measurements comparison to the specification limits. In this case, the report should include for example: "To declare or not the compliance with the specification, it was not explicitly taken into account the uncertainty associated with the result".

The summary is presented in Table 3.2.

Customer requirements (traceable to the level of the review of the request) / Requirements of standard[(1)]	The laboratory's client requests to take into consideration the incertitude (3) to establish conformity.	The laboratory's client requests not taking into consideration the incertitude (3) to establish conformity **OR EITHER** The laboratory's client does not specify anything.
The referential (1) mentions explicitly that the specification limits (2) were fixed without taking into consideration the incertitude (3).	Conformity declaration (4) **Case 1**	
The referential (1) mentions explicitly that the specifications (2) were fixed taking into consideration of the incertitude (3).	Conformity declaration (4) **Case 2**	
The referential (1) does not mention anything related to taking into consideration the incertitude (3).	Conformity declaration (4) **Case 1**	Conformity declaration (4) **Case 2**

The numbers in brackets correspond to terms under the standard.

Table 3.2. *Different cases of declaration of conformity*

b) Electronic transmission of the results

In case of electronic transmission of test results by telephone, telefax, fax or other electronic or electromagnetic means, the following minimum requirements must be met:

- confidentiality preservation of the data in the reports on the results;
- preservation of data integrity in the reports on the results;
- authentication of the transmitter of the reports on the results.

For example, the following methods can be used:

- e-mail with an attachment;
- extranet;
- digital media (CD, DVD, USB drive, etc.);
- simple electronic signature and electronic signature supposedly reliable.

Aside from the use of a supposedly reliable electronic signature, the use of electronic transmission means must have undergone a contract between the laboratory and its customers, still called "evidence convention" in particular defining the guarantees provided by the selected system. In case of electronic filing, the laboratory must be able to faithfully reproduce the report transmitted electronically throughout the duration of the definitive archiving life.

4

Testing

In this chapter, we will detail two main vibration testing methodologies, namely swept sine testing and noise tests on trained spectrum. The goal is to provide the main analytical description of information in order to understand the means of operation of testing and piloting. We will find the essential information concerning the processing of the signal on swept sinusoidal signals and random signals, as well as the prospects for implementation of tests.

Some specific tests will be simply referred to for information.

4.1. Sine vibration tests

There are essentially two types of sine vibration tests. Fixed frequency tests and swept frequency tests.

4.1.1. Fixed frequency testing

These are usually tests performed under frequencies considered critical that have been detected during a test in swept frequency or known *a priori* as being frequencies related to the operation of the system (frequency of rotating machines, shooting guns, helicopter frequency, etc.). The objective of the trial is an endurance test or to keep the material excited on its predominant modes for a given time.

This test is usually very severe and it is advisable to take precautions on the levels used in endurance at fixed frequencies. It is possible to excessively fatigue the material during this test even if there is no visible damage (breaks, etc.). The traditional case is the fixed frequency test suspensions of a box or an electronic bay.

One of the problems encountered during this test derives from the shift of the resonant frequency being tested. Indeed, if the vibration energy is dissipated in heat in the damper, the latter shifts in value and, as a result, the frequency too. In general, it decreases and there are test severities that require following this drift during the test in order to remain at maximum resonance. We must be very attentive with this phenomenon, as when it is too high stored fatigue becomes abnormally large and the test becomes unnecessarily destructive. Moreover, we have studied that the power of the carrier is limited (in contrast to that of the vibration generator in piloting) when a resonant frequency occurs, the level injected by the carrier is often reduced and therefore also the excitement level of the material. To excite it on a constant level can be excessive, from where the use of Notching techniques derives.

4.1.2. Swept sine tests

4.1.2.1. *Definitions*

a) Sinusoid and sweeping

By definition, a sinusoid is a signal whose amplitude and frequency remain fixed. We therefore call a signal defined by a relation of the following form sine swept:

$$y(t) = y_m . \sin[E(t) + \varphi] \qquad (4.1a)$$

- In general, the phase is taken as zero for ease of calculation.

- E(t) is a time characteristic function of the sweeping mode used. For fixed sine, it is summarized with the following expression: $E(t) = \omega_0 t = 2\pi f_0 t$ (f_0 = fixed frequency).

- y(t) is, in general, an acceleration but can also be a velocity or a displacement.

- In the case of the sinusoidal signal, the pulsation of the sinusoid can be defined as the derivative of E(t), meaning:

$$\frac{dE(t)}{dt} = \omega = 2\pi f \qquad (4.1b)$$

We will see that the most used modes are the sweepings:

- Linear sweepings or the frequency of the form: $f = bt + a$
- Logarithmic sweepings (or exponential) if: $f = f_1 . e^{t/T_1}$
- Hyperbolic sweepings (or parabolic, or log-log) if: $\frac{1}{f_1} - \frac{1}{f} = at$

These sweepings can be carried out at an increasing or decreasing frequency. This has an effect on the response of a resonant system at 1 DOF which does not give the same response in both cases.

NB: The first two laws are those used in the specifications and testing laboratories (although there are others).

b) Octaves and number of octaves in a frequency interval

- *Octave*: We have already seen what we call an octave, the interval between two frequencies whose ratio is two. So by definition the n number of octaves between two frequencies f_1 and f_2 is given by ($f_2 > f_1$):

$$2^n = \frac{f_2}{f_1} \rightarrow n = \frac{\text{Log}(f_2 / f_1)}{\text{Log}2} \qquad (4.2)$$

The two logarithms being expressed in the same base (10 or e in general), with:

Log(x) = decimal logarithm of x and log(x) = ln(x) = Natural logarithm of x.

- *Decade*: We call the interval between two frequencies whose ratio is 10 the decade. The number of decades n_d between two frequencies f_1 and f_2 is given by:

$$10^{n_d} = \frac{f_2}{f_1} \rightarrow n_d = \frac{\text{Log}(f_2 / f_1)}{\text{Log}10} = \text{Log}\frac{f_2}{f_1} \qquad (4.3)$$

The relation between decade and octave is thus given by:

$$\text{Log}\frac{f_2}{f_1} = n_d = n.\text{Log}2 \rightarrow \frac{n}{n_d} = \frac{1}{\text{Log}2} = 3.32 \qquad (4.4)$$

NB: A 6 dB/octave slope is approximately a 20 dB/decade slope (6 × 3.32 # 20).

4.1.2.2. *Vibrations in swept sine*

In a real environment, this type of vibration is relatively rare. We can primarily find it on the structures installed in the vicinity of revolving machines, during launching or shutdown, or even regime change. They have been studied, in particular to evaluate their effects during the crossing of resonant frequencies of a material and analyzing the effects to determine the deemed critical frequencies.

That is why material testing has been and still is very often performed by applying excitation of the swept sine type. The objectives are essentially as follows:

- The identification of the material: The test is performed by submitting the material to a swept sine with generally quite low and constant amplitude (in order not to strain or damage the material) of 0,5–1 g and 1 or 2 mm c/c maximum according to frequency bands. The variation in frequency with time is rather slow (next to the octave per minute in general) in order to correctly identify the different resonant frequencies and amplification factors, as well as to study the answers at various points in the material (transfer functions);

- The application of a test defined in a normative document: Flat rate test to demonstrate that the equipment has a certain standard robustness, independence or difficulty connecting with the vibrations that the equipment will face during its life profile;

- Implementation of a specification that at least covers the vibrations of its future real environment when it is clearly identified.

The swept sine can have a constant level throughout the explored frequency band, or it can be composed of several stages of constant levels of various intervals of frequency, as shown in the following figure:

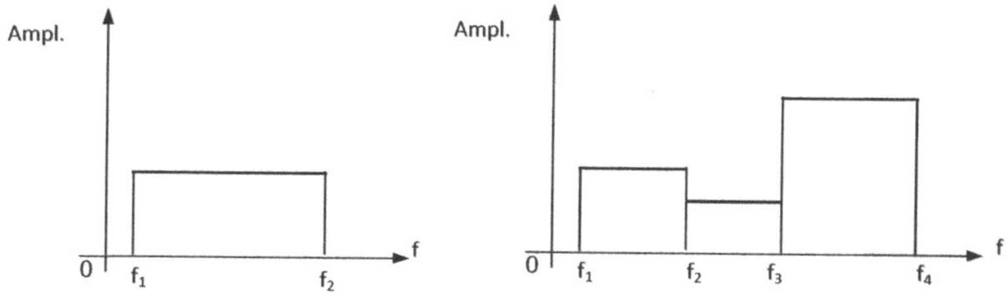

- Sweeping is frequently logarithmic. The specification sometimes specifies the direction of the sweep frequency at increasing or decreasing frequencies, generally both.

- It specifies either the scanning speed (in number of octaves per minute), or the duration of sweeping of the lowest frequency to the highest frequency, or in each frequency band at a constant level.

- The level is set either in peak value or peak-to-peak value (peak-to-peak for displacements in mm and peak for accelerations in g).

- As for the fixed frequency test, the size characterizing the amplitude can thus be either a displacement (generally in low frequency because of the weak accelerations to be controlled) or acceleration (generally at higher frequencies due to the low scales of displacement). More rarely in velocity at intermediate frequencies.

The sweeping frequency is generally chosen so that to be low enough to allow the response of the material under testing to achieve a very high percentage of the level obtained in stationary regime maintained under pure sinusoidal solicitation, or 95–98% of the value of the Q-factor coefficient.

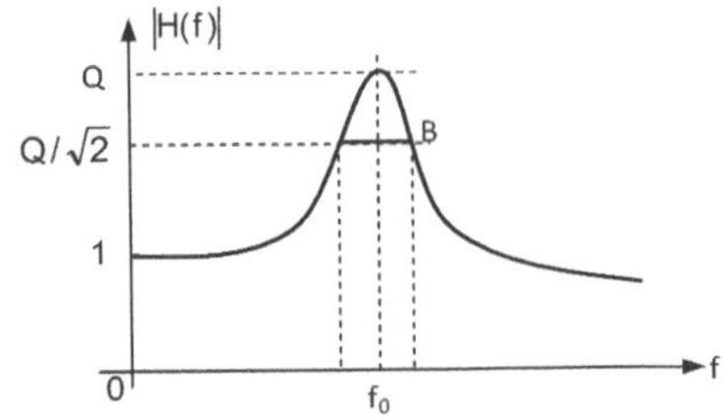

If this is not the case and the sweeping velocity is too fast, we can estimate that each resonance is excited one after the other in a transitional manner, when the frequency sweeps between points to half power of each peak of resonance of the material's transfer function. In this sense, the swept sine is an environment where effects, in the second case at extreme speeds, can be compared to those of a shock (apart from the fact that in a state of shock, all modes are excited at the same time).

We also know that the severity of an impact is most often characterized by its shock spectrum. To simulate the effects of a shock by a swept sine, some authors suggest searching for the characteristics of a swept sine of which the spectrum of extreme responses (the curve provides, as we have studied, the maximum response of a linear system with 1 DOF depending on its own frequency, for a given depreciation) includes the specified shock spectrum. The result tends to be very sensitive to the damper choice used for the calculation of the spectra. Shocks can also be simulated by fast sweeping, although this technique is not widely used.

Swept sine vibration tests are however ill-suited to simulating random vibrations, whose amplitude and phase vary in a random manner and with which frequencies are all simultaneously excited, which allows the phenomena of coupling between modes to be highlighted.

4.1.2.3. *Origin and properties of the main types of sweeping*

The objective is to analyze the different types of sweepings to measure the damage caused on the material in terms of duration, numbers of cycles amplitudes, etc. The position of the problem is the following.

We know that for a simple linear mechanical system with only 1 DOF, we have (see Appendix B):

- relative damping is given by:

$$\xi = \frac{c}{c_c} = \frac{c}{2\sqrt{km}}$$

- the Q-factor is given by:

$$Q = \frac{1}{2\xi\left(1-\xi^2\right)^{1/2}} \approx \frac{1}{2\xi} = \frac{\sqrt{km}}{c} = 2\pi f_0 \frac{m}{c} = \omega_0 \frac{m}{c}$$

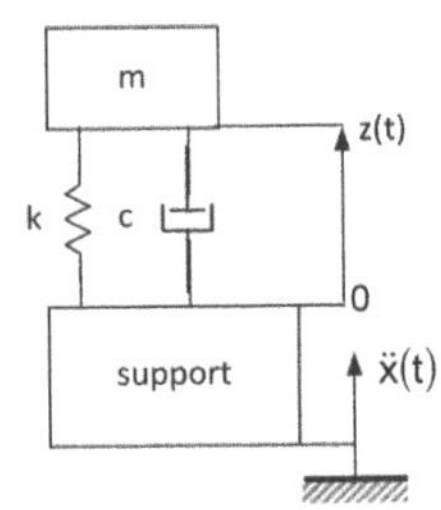

- the natural frequency of the undamped system:

$$\omega_0 = \sqrt{\frac{k}{m}} \rightarrow f_0 = \frac{1}{2\pi}\sqrt{\frac{k}{m}}$$

- resonant frequency:

$$\omega_R = \omega_0.\left(1-2\xi^2\right)^{1/2} \rightarrow f_R = f_0.\left(1-2\xi^2\right)^{1/2}$$

- the natural frequency of the free non-excited system:

$$\omega_n = \omega_0.\left(1-\xi^2\right)^{1/2} \rightarrow f_n = f_0.\left(1-\xi^2\right)^{1/2}$$

- the width of the bandwidth at half power: $B = \Delta f = \frac{f_0}{Q} \rightarrow \Delta\omega = \frac{\omega_0}{Q}$.

NB: Given a linear system with 1 DOF excited by an acceleration $\ddot{x}(t)$ and whose response is characterized by the relative displacement z(t) between the mass and the support, the normalized transfer function is therefore defined by:

$$|H(f)| = \frac{\omega_0^2 . z_{max}}{\ddot{x}_{max}} = \frac{1}{\left\{\left[1-\left(\frac{f}{f_0}\right)^2\right]^2 + \left(2\xi\frac{f}{f_0}\right)^2\right\}^{1/2}} \qquad (4.5)$$

And it passes through a maximum for:

$$|H_m(f)| = \frac{1}{2\xi\sqrt{1-\xi^2}} = \frac{Q}{\sqrt{1-\xi^2}} \approx Q \qquad \text{si } \xi << 1 \qquad (4.6)$$

Mechanical systems generally have relatively low damping (at less than 1/6, which corresponds to a Q of three, which is that of standard dampers) and so we can perform the approximation proposed in formula H_m = Q (which is, on the other hand, the exact result of the maximum when we are interested in the acceleration velocity transfer function relative to the location of acceleration relative displacement as suggested here). It should be noted that the points at half-power are defined, because of electromechanical analogy, starting from the relative acceleration velocity transfer function. That is why Q intervenes exactly in the relation.

Table 4.1 recapitulates the response r(t) generalized by the linear mechanical system to 1 DOF dampened from general equations, valid for a generalized excitation e(t) defined by a force, a velocity, an acceleration, or a displacement.

System	Excitement e(t)		Response r(t)
Fixed base	Force on the mass	F(t)/k	Relative displacement of the mass z(t)
		F(t)	Reaction force on the base $F_T(t)$
Mobile base	Displacement of the base x(t)		Absolute displacement of the mass y(t)
	Velocity of the base $\dot{x}(t)$		Absolute velocity of the mass $\dot{y}(t)$
	Base acceleration	$\ddot{x}(t)$	Absolute acceleration of the mass $\ddot{y}(t)$
		$-\frac{\ddot{x}(t)}{\omega_0^2}$	Relative displacement z(t)
		$m.\ddot{x}(t)$	Reaction force on the base $F_T(t)$

Table 4.1. *Responses in displacement, velocity and acceleration*

NB: The transfer function more widely used is that which corresponds to the "absolute acceleration of the base and the absolute acceleration of the mass" by using tests of accelerometers that measure absolute acceleration.

4.1.2.3.1. *Sweeping velocity calculation*

$\dot{f}$ is the sweeping velocity around the resonant frequency f_0: $\dot{f} = \frac{df}{dt}$.

Time Δt spent in the bandwidth Δf is approximately given by:

$$\Delta t = \frac{\Delta f}{\dot{f}}$$

And the number of cycles carried out by: $\Delta N = f_0.\Delta t = f_0.\frac{\Delta f}{\dot{f}}$. (4.7)

When such a system has departed from its position of balance and then released (or when the excitement to which it is subject to is abruptly stopped), the displacement or response of the mass can then be written in the following form:

$$z(t) = z_m.e^{-t/T}.\cos\left(2\pi f_0\sqrt{1-\xi^2}.t+\varphi\right) \qquad (4.8)$$

where T is a time constant equal to:

$$T = \frac{2m}{c} = \frac{Q}{\pi f_0} \qquad (4.9)$$

We will assume moving forward that Q is independent of the natural frequency f_0. This is in particular the case for viscoelastic materials. Another reasoning can take into account a change in Q with f_0 according to several laws, but it leads to the same laws of sweeping.

In a trial in swept sine, with the frequency depending on the time, the response of a mechanical system is never completely permanent. It is that much closer to the response of the system under permanent solicitation when the sweeping velocity is low or slow.

If we want to get the most out of this response, near to the natural frequency of the system (or practically in the frequency of resonance, if Q is large), then it is important that time Δt spent in Δf is long compared to the constant T, a condition that can be explained as follows:

$$\Delta t = \beta.T \qquad (\beta >> 1) \qquad (4.10)$$

As $|\dot{f}| = \frac{\Delta f}{\Delta t}$. If it is supposed that ΔF is sufficiently small (either ξ small or Q large), we can then confuse the slope with the curve of the sweeping rate and the slope of the cord relative to the interval Δf, meaning $|\dot{f}| \approx \frac{\Delta f}{\Delta t}$. We will see that this approximation is acceptable in practice. We can then write:

$$\begin{cases} \Delta t = \beta T = \beta\dfrac{Q}{\pi f_0} \\ \Delta f = \dfrac{f_0}{Q} \end{cases} \rightarrow |\dot{f}| = \Delta f.\frac{\pi f_0}{\beta Q} = \frac{\pi f_0^2}{\beta Q^2} \qquad (4.11)$$

The natural frequency f_0 can be unspecified in the band (f_1, f_2) studied and, whatever its value is in this band, the answer must be close to Q times the entry with f_0. We can then generalize the expression (4.11) by writing it in function to F, which means:

$$|\dot{f}| = \pm\frac{\pi.f^2}{\beta.Q^2}$$

NB: It is noted that the scanning speed varies according to $1/Q^2$. In addition, the derivative $\dot{f}$ is positive for the increasing scanning frequency and negative for the decreasing sweeping frequency. There are three classic cases to consider.

4.1.2.3.2. *Case 1: time Δt spent in Δconstant f regardless of f_0 (logarithmic)*

Type of sweeping where time Δt spent in each interval Δf is constant regardless of f_0. For that, since $\Delta t = \beta T = \frac{\beta Q}{\pi f}$, it is necessary that $\beta = \gamma.f$ is therefore proportional to f, γ having a time dimension, which becomes:

$$\Delta t = \frac{\gamma.Q}{\pi} \rightarrow \qquad \dot{f} = \pm\frac{\pi f^2}{\beta Q^2} = \pm\frac{\pi.f}{\gamma.Q^2} \qquad (4.12)$$

- For increasing scanning frequency between f_1 and f_2, we have:

$$\frac{df}{dt} = +\frac{\pi.f}{\gamma.Q^2} \rightarrow \frac{df}{f} = dt\frac{\pi}{\gamma.Q^2} \rightarrow \log f = t.\frac{\pi}{\gamma.Q^2}$$

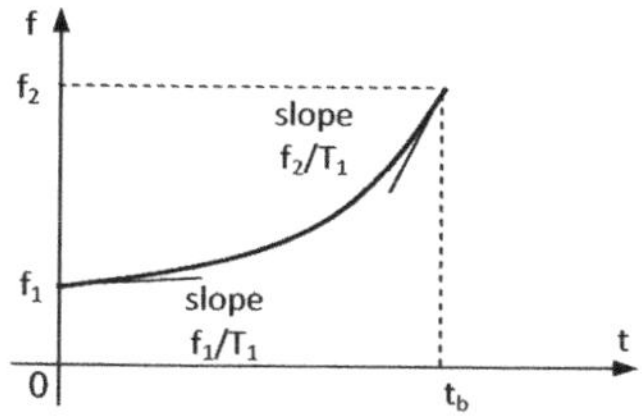

Therefore:

$$\begin{cases} f = f_1.e^{+\frac{\pi}{\gamma.Q^2}.t} = f_1.e^{\frac{t}{T_1}} \\ T_1 = \dfrac{t_b}{\log f_2 / f_1} \end{cases} \qquad (4.13)$$

The constant T_1 is such that for $t = t_b$, $f = f_2$. T is therefore the time used to sweep the interval between two frequencies whose ratio is equal to "e".

- For the sweeping at a decreasing frequency between f_1 and f_2, it becomes in this case:

$$f = f_2.e^{-\frac{\pi}{\gamma.Q^2}.t} = f_2.e^{-\frac{t}{T_1}} \qquad (4.14)$$

Constant T has the same definition as previously mentioned.

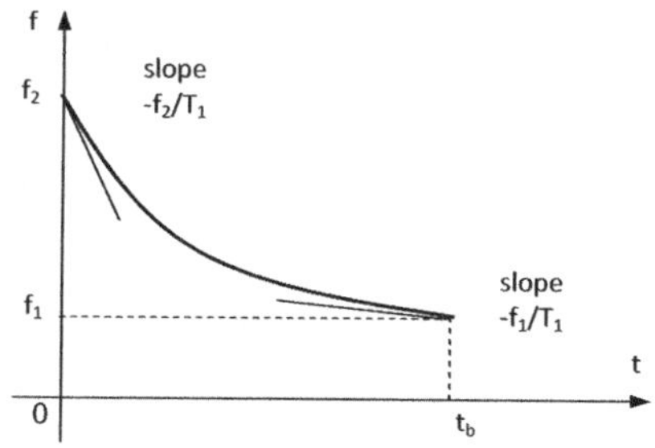

The expressions of the corresponding time functions are the following ones:

- increasing frequencies:

with: $$\frac{dE(t)}{dt} = \omega = 2\pi f \rightarrow E(t) = 2\pi.\int_0^t f_1.e^{t/T_1}.dt = 2\pi T_1 f_1\left(e^{t/T_1} - 1\right) = 2\pi T_1\left(f_1.e^{t/T_1} - f_1\right)$$

having: $$E(t) = 2\pi T_1(f - f_1) \qquad (4.15)$$

- decreasing frequencies:

$$E(t) = 2\pi.\int_0^t f_2.e^{-t/T_1}.dt = -2\pi T_1 f_2\left(e^{-t/T_1} - 1\right) = -2\pi T_1(f - f_2) \qquad (4.16)$$

NB: We previously assumed that $f_2 > f_1$ and the relations depend on the direction of the sweeping. We can free ourselves of this by supposing that f_1 is the departure frequency and f_2 is that of arrival without prejudice to relative values. Simply, the relations that follow are obtained for increasing frequencies (the decreasing frequencies are deduced easily). Then, we obtain the following relations:

a) Sweeping time

- The sweeping time to go from f_1 to any f is given by:

$$t = \frac{\gamma.Q^2}{\pi}.\log\frac{f}{f_1} = T_1.\log\frac{f}{f_1} \quad (4.17)$$

- Necessary time to go from f_1 to f_2 is given by:

$$t_b = \frac{\gamma.Q^2}{\pi}.\log\frac{f_2}{f_1} = Q.\Delta t.\log\frac{f_2}{f_1} \quad (4.18)$$

- Time Δt spent in an interval Δf around each frequency is given by:

$$\Delta t = \frac{t_b}{Q.\log(f_2/f_1)} \quad (4.19)$$

- Time spent between two unspecified frequencies in the swept area of (4.17) and (4.12). Having f_A and f_B ($f_B > f_A$) the two terminals of a frequency interval located in (f_1, f_2). Time $t_B - t_A$ passed between f_A and f_B is given by:

$$t_B - t_A = t_b.\frac{\log(f_B/f_A)}{\log(f_2/f_1)} \quad (4.20)$$

- Time Δ^*t spent between half power points of a linear system with 1 DOF. Let us calculate this time with the assumption that Q is either large or ξ very small. However, the half-power points (points of the ordinate function $Q/\sqrt{2}$) are often considered as representative of the bandwidth of the system, and have, respectively, X-coordinates as:

$$f_{c\,min} = f_0\left(1-\frac{1}{2Q}\right) \quad \text{and} \quad f_{c\,max} = f_0\left(1+\frac{1}{2Q}\right)$$

With the preceding equations, it becomes:

$$\Delta^*t = t_b.\frac{\log\{(1+1/2Q)/(1-1/2Q)\}}{\log(f_2/f_1)} \quad (4.21)$$

NB: If the curve $\frac{\Delta^*t}{\Delta t}(\xi) = \frac{1}{2\xi}.\log\left(\frac{1+\xi}{1-\xi}\right)$ is plotted, we find that for $\xi < 0{,}2$, we practically have 1 at almost 1%. When Q # 3, which is the limit, as we already saw, of a common damper system, we can practically always use the general equation.

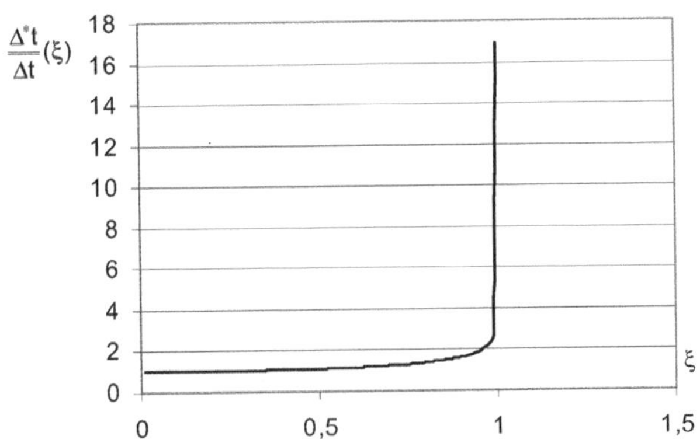

b) Numbers of cycles

The evaluation of the number of cycles is important because it allows us to estimate the damage caused by a swept sine mode test. So, we find:

- the number of cycles carried out during a time t is given by:

$N = \int_0^t f(t)dt = \int_0^t f_1.e^{\frac{\pi}{\gamma Q^2}t}dt = f_1.\frac{\gamma Q^2}{\pi}.\left(e^{\frac{\pi}{\gamma Q^2}t} - 1\right)$. So, in function of (4.13):

$$N = \frac{\gamma Q^2}{\pi}(f - f_1) \tag{4.22}$$

- the number of cycles between f_1 and f_2 is then given by, with (4.18):

$$\begin{cases} N = \frac{\gamma Q^2}{\pi}(f_2 - f_1) = T_1(f_2 - f_1) \\ N_b = t_b.\frac{f_2 - f_1}{\log(f_2 / f_1)} \end{cases} \tag{4.23}$$

The average frequency in t_b is given by:

$$f_m = \bar{f} = \frac{N_b}{t_b} = \frac{f_2 - f_1}{\log(f_2 / f_1)} \tag{4.24}$$

- the number of cycles ΔN made in the band at half-power during the time Δt is:

$$\begin{aligned} \Delta N &= \frac{\gamma Q^2}{\pi}\left[f_0\left(1 + \frac{1}{2Q}\right) - f_0\left(1 - \frac{1}{2Q}\right)\right] \\ \Delta N &= f_0.\frac{\gamma Q}{\pi} = f_0\frac{T_1}{Q} = f_0.\Delta t \end{aligned} \tag{4.25}$$

Thus, ΔN varies as f_0, from where following is derived:

$$t_b = \frac{Q\Delta N}{f_0}.\log(f_2 / f_1) \tag{4.26}$$

We can also find another expression for ΔN, with (4.23) and (4.26):

$\Delta N = f_0.\frac{N_b}{Q(f_2 - f_1)}$. Time Δt passed during a period Δf is given by:

$\Delta t = \frac{\gamma Q}{\pi} = \frac{t_b}{Q.\log(f_2 / f_1)}$. From where another expression of ΔN derives, with: $\Delta N = f_0.\Delta t$.
Meaning:

$$\Delta N = f_0.\frac{t_b}{Q.\log(f_2 / f_1)} \tag{4.27}$$

- similarly, the number of cycles N_1 to go from f_1 to f_0 is (4.22), (4.25) and (4.27)

$$N_1 = \frac{\gamma Q^2}{\pi}(f_0 - f_1) = T_1(f_0 - f_1) = Q\Delta t(f_0 - f_1) = \frac{Q\Delta N}{f_0}(f_0 - f_1)$$

So:

$$N_1 = t_b.\frac{(f_0 - f_1)}{\log(f_2 / f_1)} \tag{4.28}$$

This number of cycles is then carried out in the following time:

$$t_1 = \frac{\gamma Q^2}{\pi}\log(f_0 / f_1) = T_1 \log\frac{f_0}{f_1} = \frac{Q\Delta N}{f_0}\log(f_0 / f_1) = Q\Delta t.\log(f_0 / f_1)$$

where:
$$t_1 = t_b.\frac{\log(f_0 / f_1)}{\log(f_2 / f_1)} \tag{4.29}$$

c) Sweeping velocity

Following the sweeping direction, we have: $$\dot{f} = \frac{df}{dt} = \begin{cases} \text{de f} = \dfrac{f_1}{T_1}.e^{t/T_1} \\ \text{de f} = \dfrac{f_2}{T_1}.e^{-t/T_1} \end{cases} = \pm\frac{f}{T_1} \tag{4.30}$$

The speed is therefore linear, from where we obtain the following figures:

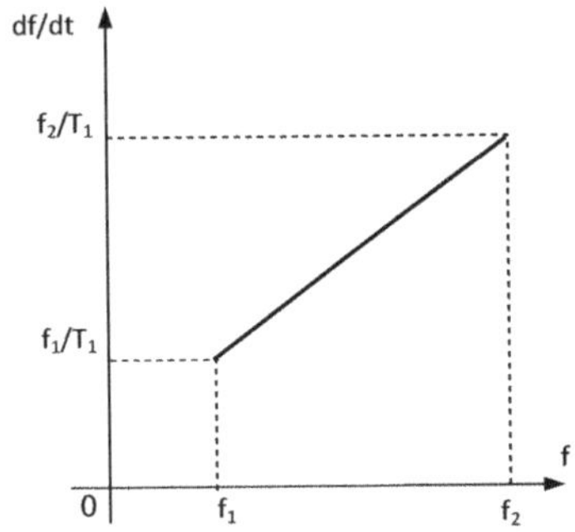

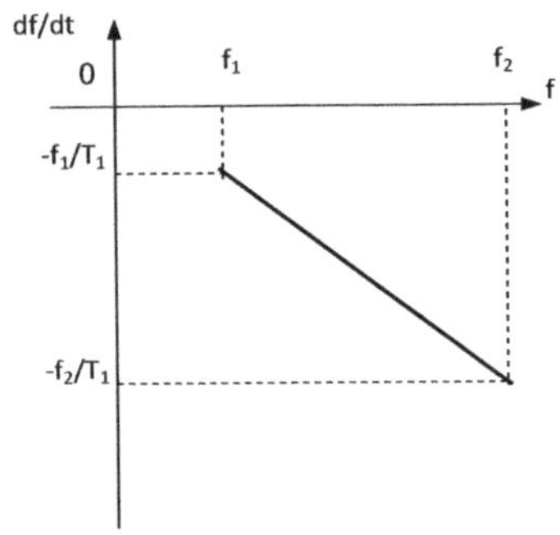

Yet, often the sweeping velocity in sine sweep tests is expressed in number of octaves per minute. We have studied that the number of octaves between two frequencies f_1 and f_2 is given by:

$$n = \frac{\text{Log}(f_2 / f_1)}{\text{Log}2} = \frac{\log(f_2 / f_1)}{\log 2}$$

So, the number of octaves R_2 per second is given by (t_b being expressed in seconds):

$$R_s = \frac{n}{t_b} = \frac{\log(f_2 / f_1)}{t_b.\log 2} \tag{4.31}$$

and the number of octaves R_m per minute:

$$R_m = \frac{60.n}{t_b} = 60.R_s \tag{4.32}$$

We also find from expression (4.13), for $t = t_b$ and $f = f_2$:

$$t_b = T_1.\log(f_2 / f_1) \tag{4.33}$$

From (4.33) and (4.31), it becomes:

$$\log(f_2 / f_1) = t_b / T_1 = R_s.t_b.\log 2 \rightarrow \qquad R_s = \frac{1}{T_1.\log 2} \tag{4.34}$$

And with (4.13) $f = f_1.e^{t/T_1} \rightarrow f = f_1.e^{tR_s\log 2}$ and equation: $a^x = e^{x\log a}$.

Another expression derives from the laws of sweeping according to that which was adopted for R:

$$\begin{cases} f = f_1.2^{R_s.t} \\ f = f_1.2^{R_m t/60} \end{cases} \tag{4.35}$$

These equations allow for an assessment on sweeping time and therefore testing time.

For example, a curve network giving t_b according to ratio f_2/f_1 was traced starting from the relation (4.31) and (4.32), which means:

$$R_s = \frac{n}{t_b} = \frac{\log(f_2/f_1)}{t_b.\log 2} \rightarrow t_b = \frac{1}{R_s.\log 2}.\log(f_2/f_1) = \frac{60.\log(f_2/f_1)}{R_m \log 2}$$

This is for values of R_m = 0,1 m; 0,2; 0,3; 0,5; 1; 2 octaves per minute.

From where the following graph derives:

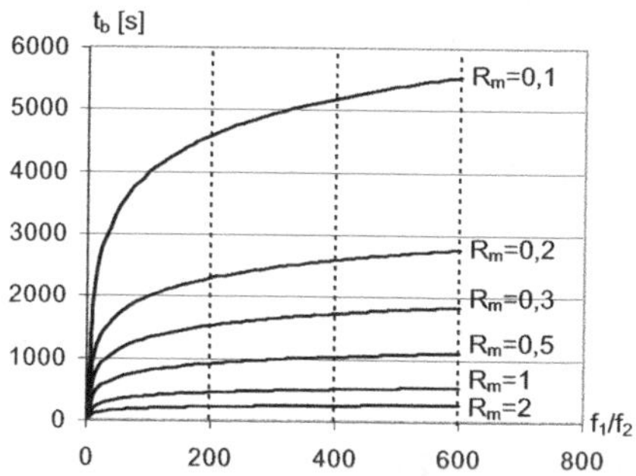

where we can see that the sweeping duration quickly becomes important for low scanning rates. We can understand that here the usual values of 1 oct/mn proposed as standard, give out an average sweeping time of about 600 s, or 10 mn maximum.

Additionally, we deduce, according to (4.30) and (4.34), the following relation:

$$\dot{f} = \frac{df}{dt} = \frac{f}{T_1} = f.R_s.\log 2 \qquad (4.36)$$

- Case where the scanning speed R_d is expressed as decade per minute:

$$R_d = \frac{n_d}{t_b/60} = \frac{60.\log(f_2/f_1)}{t_b \log 10} \quad \rightarrow \quad \text{with: } \log_a x = \frac{\log x}{\log a} = \frac{\text{Log}x}{\text{Log}a}$$

$$R_d = \frac{60.\text{Log}(f_2/f_1)}{t_b} \qquad (4.37)$$

It is important to remember here that Logx = decimal logarithm of x and logx = Natural logarithm of x.

Or:

$$R_d = R_m.\frac{\log 2}{\log 10} \approx \frac{R_m}{3,322} \approx 0,3xR_m \qquad (4.38)$$

It is also important to remember the following times:

- time spent between any two frequencies (4.20):

$$t_B - t_A = t_b.\frac{\log(f_B/f_A)}{\log(f_2/f_1)}$$

- time spent between points at half-power (4.21):

$$\Delta^* t = t_b.\frac{\log\{(1+1/2Q)/(1-1/2Q)\}}{\log(f_2/f_1)}$$

NB: It should be noted that when Q is large this expression tends toward the value given by (4.19), we can still express it in the following form:

$$\frac{t_b}{\log(f_2/f_1)} = \frac{1}{R_s \log 2} = \frac{60}{\log 2.R_m} \rightarrow \Delta t^* = \frac{60}{\log 2}\frac{\log\{(1+1/2Q)/(1-1/2Q)\}}{R_m} \qquad (4.39)$$

Time is expressed in seconds.

- The number of cycles in this interval is then given by $\Delta N^* = f_0.\Delta t^*$:

Therefore: $$\Delta N^* = \frac{60}{\log 2}\frac{f_0}{R_m}\log\left(\frac{1+1/2Q}{1-1/2Q}\right) \quad (4.40)$$

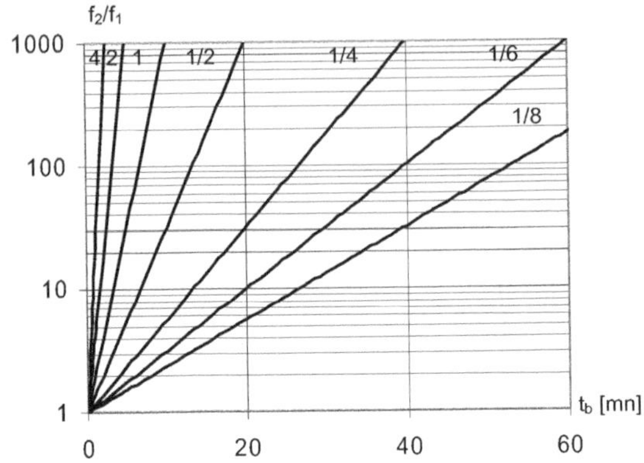

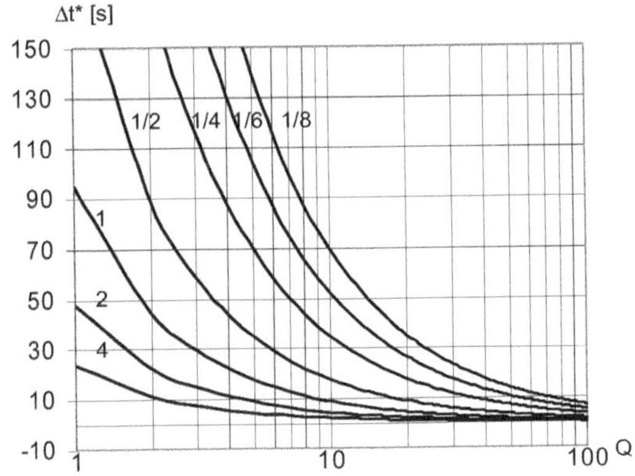

The figure above provides the evolution of Δt^* in function of Q for a variety of values of R_m in octaves per minute, meaning: 4, 2, 1, 1/2, 1/4, 1/6 and 1/8.

The figure on the right provides the sweeping time required to go from f_1 to f_2 for sweeping velocities R_m in oct/min equal to 4, 2, 1, 1/4, 1/6, 1/8.

NB: These graphs should allow us to estimate the time of a trial and check if the velocity is compatible with the Q-factor coefficients found.

The following parameters are defined:

- number of cycles per octaves: f_A and f_B are two frequencies separated by an octave:

$$N_2 = \frac{\gamma.Q^2}{\pi} f_A = \frac{f_A.t_b}{\log 2} = Q.\Delta N \frac{f_A}{f_0} \quad (4.41)$$

- time to sweep an octave:

$$t_2 = \frac{\gamma.Q^2}{\pi}.\log 2 = Q.\Delta t.\log 2 = \frac{Q.\Delta N}{f_0}.\log 2 \quad (4.42)$$

- time to sweep 1/nth of octave:

$$t_n = \frac{\gamma.Q^2}{\pi}.\log 2^{1/n} = \frac{\gamma.Q^2}{\pi.n}.\log 2 = \frac{Q.\Delta N}{f_0.n}.\log 2 \qquad (4.43)$$

4.1.2.3.3. *Case 3: constant sweeping speed (linear)*

If we want to perform a sweeping at a constant speed, it is necessary to have:

$\dot{f} = \frac{df}{dt}$ = constant, however, as in (4.12) we have: $\dot{f} = \pm\frac{\pi}{\beta.Q^2}.f^2$, it is necessary that: $\beta = \delta.f^2$.

However, if we ask: $\beta = \delta.f^2$, where δ is a constant with the dimension of a time squared, it becomes: $\left|\dot{f}\right| = \frac{\Delta f}{\Delta t} = \pm\frac{\pi}{\delta.Q^2} \rightarrow \Delta t = \pm\frac{\Delta f.\delta.Q^2}{\pi}$

If Δf represents the band at half-power, $\Delta f = \frac{f_0}{Q}$ is in increased scanning:

$$\Delta t = \frac{\delta.f_0.Q}{\pi} \qquad (4.44)$$

NB: Thus the time spent in the band at half-power Δf varies as the natural frequency f_0. For a given Q, f_0 is larger and Δt becomes more important.

We also find:

$$\dot{f} = \frac{\Delta f}{\Delta t} = \pm\frac{\pi}{Q^2}\frac{f^2}{\delta.f^2} = \mp\frac{\pi}{Q^2.\delta} = \text{constant for a given Q} \qquad (4.45)$$

- In case of increased frequency sweeping:

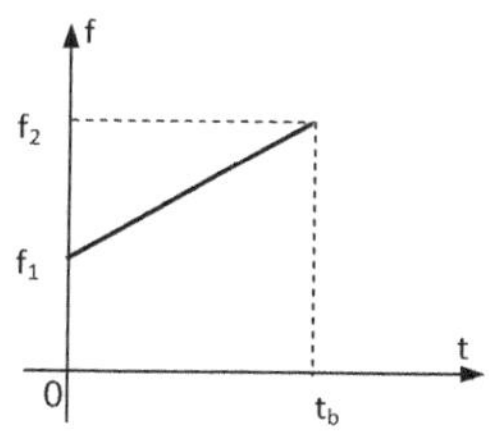

$$f = \frac{\pi}{Q^2.\delta}.t + f_1 = bt + f_1 \qquad (4.46)$$

with $b = \frac{\pi}{Q^2.\delta}$, the constant b being such as f = f_2 when t = t, from where:

$$b = \frac{f_2 - f_1}{t_b} \qquad (4.47)$$

Sweeping is known as “linear”.

- In case of decreased frequency sweeping:

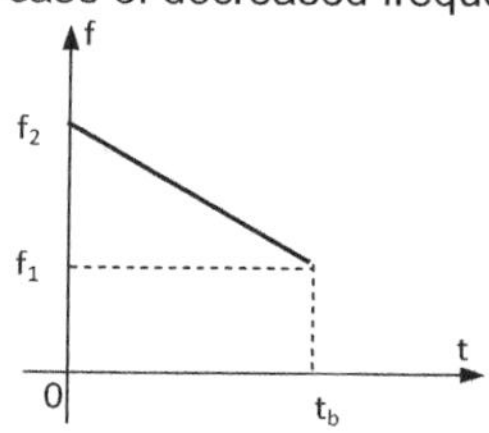

$$f = -\frac{\pi}{Q^2.\delta}.t + f_2 = -bt + f_2 \qquad (4.48)$$

Where $b = \frac{\pi}{Q^2.\delta} = \frac{f_2 - f_1}{t_b}$

Expressions of the corresponding time functions are the following ones:

- increasing frequencies, calculation of function E(t):

$$\text{with: } \frac{dE(t)}{dt} = \omega = 2\pi f \rightarrow E(t) = 2\pi.\int_0^t (bt + f_1).dt = 2\pi.t.\left(\frac{b.t}{2} + f_1\right) \qquad (4.49)$$

- decreasing frequencies, calculation of function E(t) (reminder: which characterizes the type of scan (4.0))

$$\text{with: } \frac{dE(t)}{dt} = \omega = 2\pi f \rightarrow E(t) = 2\pi.\int_0^t (-bt + f_2).dt = 2\pi.t.\left(-\frac{b.t}{2} + f_2\right) \quad (4.50)$$

a) Sweeping velocity

Following the direction of sweeping, it is equal to:

$$\dot{f} = \frac{df}{dt} = \pm b = \pm\frac{f_2 - f_1}{t_b} \quad (4.51)$$

From which the following graphs derive:

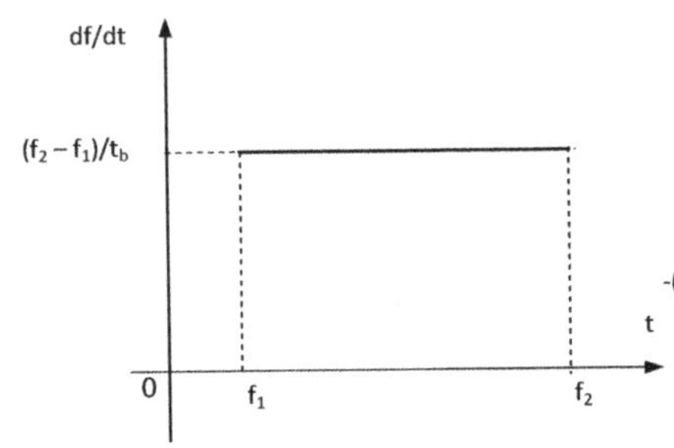

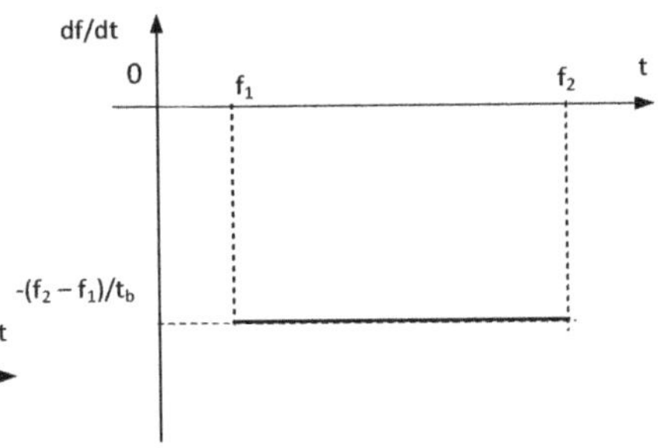

4.1.2.3.4. *Case 3: sweeping with ΔN number of cycles in Δf*

The consideration guiding this type of scanning, ensuring a number of cycles are identical in all Δf intervals (at half power) regardless of f_0, is the equivalence of damage in these intervals. With this assumption, we have the following relationship for the sweeping number:

$$\Delta N = f.\Delta t = f.\frac{\beta.Q}{\pi f} = \frac{\beta.Q}{\pi} \quad (4.52)$$

and for this relationship to be constant, parameter β itself must be constant, which means (4.12):

$$\dot{f} = \pm\frac{\pi f^2}{\beta Q^2} \quad (4.53)$$

Sweeping velocity varies as the square of instantaneous frequency, from where the following differential equation derives:

$$\frac{df}{dt} = \pm\frac{\pi f^2}{\beta Q^2} \rightarrow \frac{df}{f^2} = \pm\frac{\pi}{\beta Q^2}.dt \quad (4.54)$$

- For the increasing sweeping frequency between f_1 and f_2, we have:

A t = 0, it is assumed that $f = f_1$ at the start of the sweeping frequency. Thus, the integration of (4.54) gives out:

$$\frac{1}{f_1} - \frac{1}{f} = \frac{\pi}{\beta.Q^2}.t = a.t$$
$$a = \frac{\pi}{\beta.Q^2} \quad (4.55)$$

or:

$$f = \frac{\beta.Q^2 f_1}{\beta.Q^2 - \pi.f_1 t} = \frac{f_1}{1 - a.f_1.t} \quad (4.56)$$

We can also write the following form for $t = t_b$, $f = f_2$:

$$a = \frac{f_2 - f_1}{f_1.f_2.t_b} \tag{4.57}$$

From which the following graph derives:

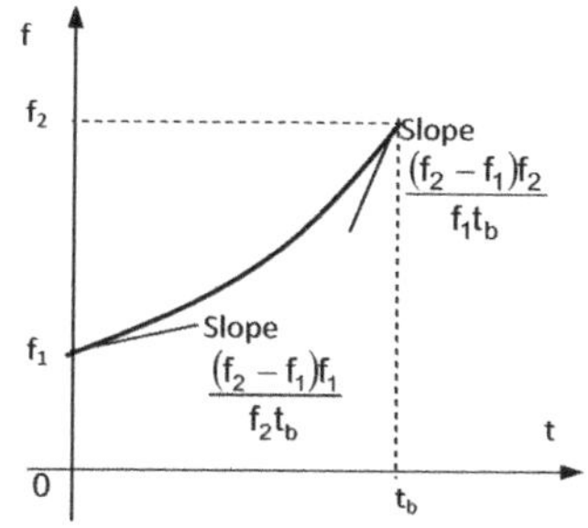

In this case, rarely used in practice, sweeping is said to be "hyperbolic", the name "parabolic sweeping" is also used, undoubtedly because of the shape of relationship (4.53) as well as "log – log" sweeping.

NB: Despite the form of denominator f in equation (4.56), frequency f cannot be negative. It would be necessary that $1 - a.f_1.t < 0$, meaning that: $t > \frac{1}{af_1} = \frac{t_b f_2}{f_2 - f_1}$, which is never the case, since we must always have $0 \le t \le t_b$. These two conditions would require that we have $\frac{t_b f_2}{f_2 - f_1} < t_b \rightarrow f_2 < f_2 - f_1$, or $f_1 < 0$, which is impossible since by assumption f_1 is always positive.

- For the decreasing sweeping frequency between f_1 and f_2, we have:

The equation becomes:

$$\frac{df}{f^2} = -\frac{\pi}{\beta.Q^2}.dt \tag{4.58}$$

$$\frac{1}{f_2} - \frac{1}{f} = -\frac{\pi}{\beta.Q^2}.t = -a.t$$

Therefore:

$$f = \frac{f_2}{1 + af_2 t} \tag{4.59}$$

$$a = \frac{f_2 - f_1}{f_1 f_2 t_b}$$

As $t = t_b$, then $f = f_1$. From which the following graph derives:

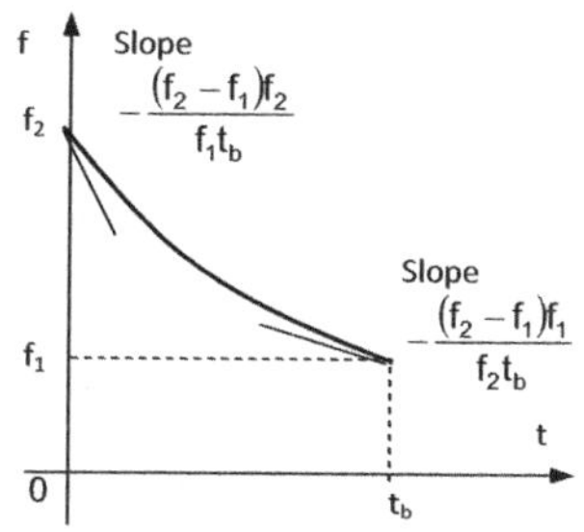

Expressions of the corresponding time functions are found below:

Function E(t) under the sine can be calculated from definition (4.1), which means:

$$\frac{dE(t)}{dt} = 2\pi f \rightarrow E(t) = \int_0^t 2\pi f(t).dt \tag{4.60}$$

- increasing frequency:

$$E(t) = 2\pi \int_0^t \frac{f_1}{1-af_1t}.dt = \frac{2\pi}{a}\int_0^{af_1t} \frac{1}{1-(af_1t)}.d(af_1t) = -\frac{2\pi}{a}\log(1-af_1t) = \frac{2\pi}{a}\log\frac{1}{1-af_1t} \tag{4.61}$$

- decreasing frequency:

We have in the same way:

$$E(t) = 2\pi \int_0^t \frac{f_2}{1+af_2t}.dt = \frac{2\pi}{a}\int_0^{af_2t} \frac{1}{1-(af_2t)}.d(af_2t) = \frac{2\pi}{a}\log(1+af_2t) \tag{4.62}$$

a) Sweeping velocity

- For increasing frequencies, from (4.55):

$$f = \frac{f_1}{1-af_1t} \rightarrow \frac{df}{dt} = \frac{f_1^2.a}{(1-af_1t)^2} = \frac{f_1^2.a}{f_1^2/f_2^2} = a.f^2 = \frac{f_2-f_1}{f_1f_2t_b}.f^2 \tag{4.63}$$

From which the graph derives:

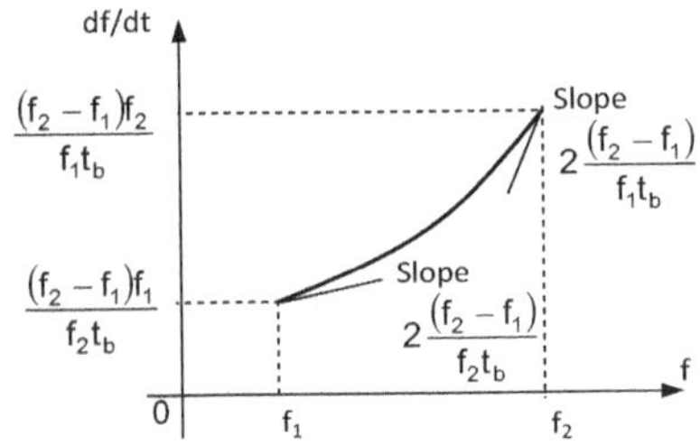

- For decreasing frequencies, from (4.59):

$$\frac{df}{dt} = -a.f^2 = -\frac{f_2-f_1}{f_1f_2t_b}.f^2 \tag{4.64}$$

The following graph is obtained:

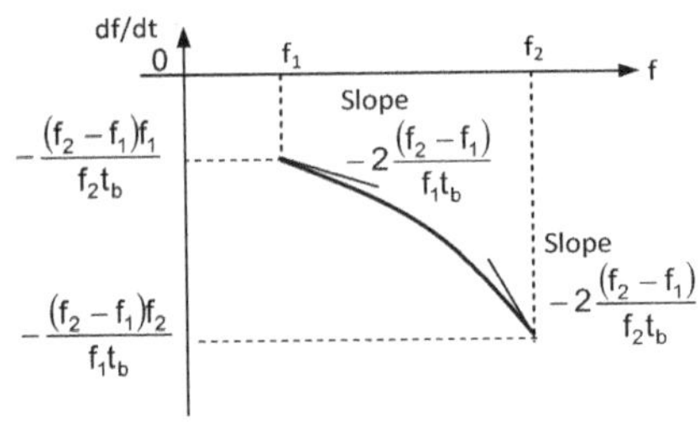

NB: A very detailed approach is given in [ref. I1] with a direct application on the damage induced on a system with 1 DOF.

4.1.2.3.5. *Influence of sweeping speed*

An extremely slow sweeping allows for the measuring and tracing of the system's transfer function of the, at 1 DOF without distortion, and getting the correct values of resonant frequency and surge.

When sweeping velocity increases, the resulting transfer function differs more and more from the actual transfer function. Deformations of the transfer function result in:

- a decrease of the maximum ΔH_m ;

- a displacement in frequency of the X-coordinate of the maximum Δf_R (resonant frequency);

- a displacement of the median axis of the curve that loses its symmetry Δf_m;

- an increase in bandwidth (interval between half power points) ΔB.

From which the shape of the following curve derives:

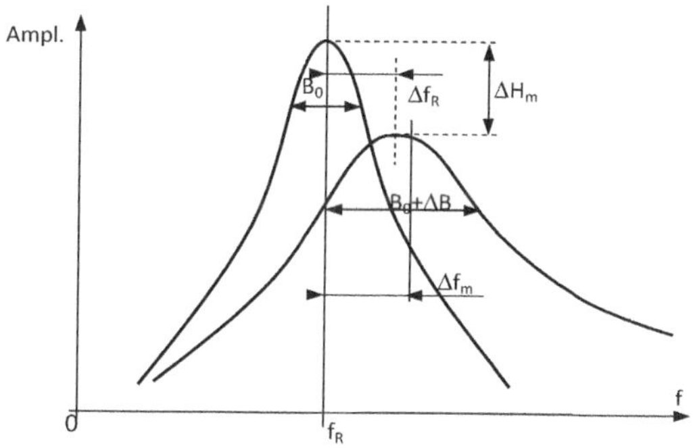

When the sweeping velocity increases again, we observe:

- First, the appearance of beats, due to interference between the free response of the mechanical system, which is relatively large after the resonance and excitation "swept sine" imposed on the system. The number and scale of these beats are even lower than their depreciation is great. The shape of the curves is the following:

- Then, if velocity increases again, everything happens as if the system was subject to a mechanical shock, the duration of the sweeping becomes small. The largest extremum of the response occurrs for $t > t_b$ (residual response observed similarly to the impulse shocks). This technique has been proposed to simulate shocks.

We note that, regardless of the type of sweeping (linear, logarithmic, hyperbolic), the response of the resonant system at 1 DOF is not the same when sweeping is going up and down. The amplitude obtained in descending sweeping is always greater than that obtained during the rising sweeping. This can be qualitatively explained by the fact that the slope seen by the excitation signal is much more steep rising than in downward sweeping.

And so, the filtering effect of the excitation signal is more significant in rising sweeping. Of course, this effect decreases with the scanning velocity. We obtain the following response, of the G (η) type, where we have the following:

- In the X-coordinate a standardized homogeneous parameter $\eta = \frac{Q^2}{f_0^2}\left(\frac{df}{dt}\right)_{f=f_0}$ under velocity.

- In the ordinate, the G ratio of the largest peak of the temporal response, whether positive or negative, with the largest peak that we would have obtained in a uniform regime, homogeneous, and with an amplitude (virtually identical curves regardless of the type of sweeping).

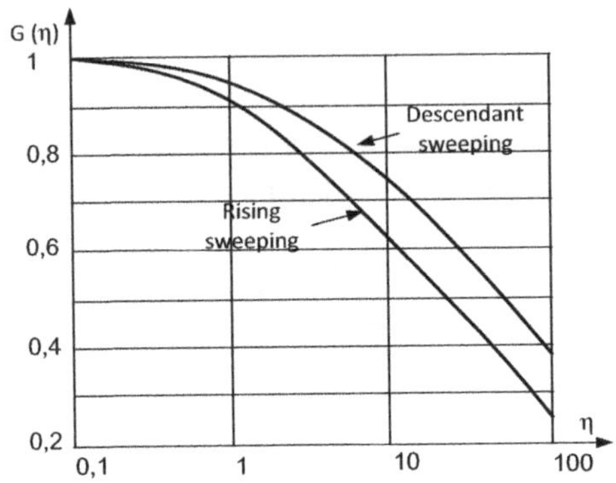

From which the importance of the sweeping velocity and sweeping direction derives.

We can explain equation (4.11) by writing: $\left|\dot{f}\right| = \frac{\Delta f^2.\pi}{\beta} \equiv K.\Delta f^2$, K < 1. And an equation of the same type is found as (4.53).

4.1.3. Conclusion

In the previous sections, we developed some concepts regarding swept sine testing. This is mainly for educational purposes, since sweeping velocities, as well as testing durations, are calculated directly by modern control consoles or are imposed by normative documents. We must remember that these types of tests are essentially useful for the characterization testing of materials, analyzed in order to identify critical frequencies (of resonance, usually reported as such for $Q \geq 10$); this is generally done at reduced levels so as not to unnecessarily fatigue the material. This information can be of assistance in the design and development domain, or can be used to build these same critical frequency endurance tests. It is important to note, as already mentioned, to the fact that it is necessary to carefully specify severity levels, as these tests may be harmful to the material, while often being little representative of the vibrations encountered during the profile of life of the material.

4.2. Vibration testing in noise or random

4.2.1. Overview

We recall that the requests met under mechanical environments are known as:

- deterministic, when their variation over time can be described by an explicit mathematical relation. We will simply say that these vibrations can be periodic (sinusoidal or complex) or non-periodic (pseudo-periodic or transitory);

- random, when the instantaneous value cannot be predicted at the moment $t + \varepsilon$ (with ε having an infinitely short duration before the time constant of the studied system), even if it is known at a moment t. This is the type of solicitation that is described in this section.

We previously saw that the study of mechanical environmental problems can be done with sinusoidal signals that lend themselves well to the calculation and decomposition techniques in Fourier series (methods used for a long time). These methods are very useful but they remain limited since such a decomposition supposes that the periodic signal and its energy is concentrated at certain frequencies.

Additionally, phenomena actually found during the life profile of materials are rarely simple and energy reading signals are generally distributed throughout a whole field of frequencies.

The description of variations based on random excitation time can be done using an explicit mathematical relation as already mentioned in the previous chapter. The record obtained in the course of a test is only a possible result (or trajectory or function sample of the process) among many others. If several tests are performed, the results, non-superimposable, constitute a series of samples as a function of a certain time interval. This sample is characterized by the number of steps, which determines the DOF of measures carried out, on average. To represent a random phenomenon, we would need, theoretically, an infinite number of samples or DOF.

Then, a problem arises: given a series of records and samples, measured at the entrance of a material, how can we:

- describe in a convenient form, random vibratory environment of entry (excitation)?
- compare two random excitations from the point of view of their severity?
- perform a calculation of mechanical resistance of a material subjected to this type of excitation?

We will outline the main methods used to answer these questions. Methods roughly fall into two categories:

- those that attempt to describe the statistical properties of the set of samples or the instantaneous values of a sample;
- those that seek to describe the effects of the sample on an electric or mechanical filter and therefore focus more on the frequency content of the signal.

The types of analysis, already mentioned in Chapter 2, are found below.

a) Analysis in frequency space

- For a signal: Spectral density of power

- For two signals: Interspectral density $\begin{cases}\text{Spectrum in coincidence (in phase)}\\ \text{Spectrum in quadrature}\end{cases}$

b) Analysis in time space

- For a signal: Autocorrelation function

- For two signals: Intercorrelation function

c) Probability analysis

- For a signal: Probability density
  Distribution function
  Number of passes at a given value (counting peaks)

- For two signals: Composed probability density, etc.

NB: The power spectral density (PSD) and the correlation contain the same amount of information as they are the Fourier transforms, both direct and inverse to one another. We have already described the PSD analysis with its main properties in Chapter 2. However, this information is presented differently. In particular, the correlation is adapted to detection problems (detection of a sine in noise).

Fortunately, a certain determination of this random quantity can be, nevertheless, done by learning from its statistical properties that allow us to find out about possible future values.

4.2.2. Concepts on random functions

We will further elaborate here the complementary and necessary reminders to properly understand the random general phenomenon and thus complete the section on PSD analysis. We will use these results by applying them to testing techniques in noise or random vibrations.

4.2.2.1. *Random process*

A random function describing the history in the time of a random phenomenon is called "sample function" or "random function" or "path of the process". The set of all function samples or possible achievements that can be the result of the same experience is called "random process" or "stochastic process". The following figure shows a random process where the number of samples was limited to 5. This number of samples also characterizes the "DOF" of the process that is used in the calculations of the analysis of the cockpit consoles.

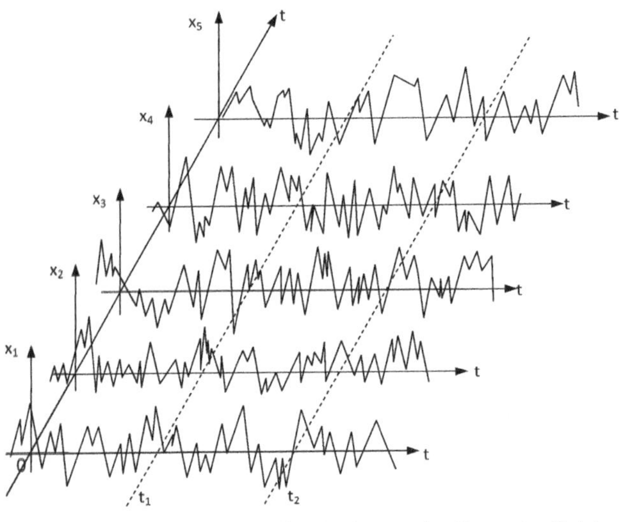

The analysis time interval is the same for each sample; it can be finished in practice (for example 10s–20s for the time of flight of a missile) or infinite in theory. All of these samples represent the possible outcomes of an experiment carried out under identical conditions.

For example, let us suppose we were to define the process that generates vibrations in a land vehicle. Initially, it is possible to imagine all the vibrations generated by any type of road on any type of vehicle, which defines a process $x_1(t)$. This process can be reduced by saying that it is a vehicle of a certain type and a second process $x_2(t)$ will be defined "all vibrations due to the imaginable road on this type of vehicle". Other subset processes can, in a similar manner, be established (the driver is careful, its velocity is limited, the vehicle is checked regularly, etc.). However, it is not possible to prevent rain or snow, whether the road is damaged or not. It is therefore difficult to find a process that provides achievements with all identical external conditions, from where the difficulty of declaring it stationary or ergodic arises, properties that we will define later on.

This aspect links the previous comments on the possibility of finding representation models which allow us to compare the severity of vibrations in terms of the damage caused on the model. This is also the current approach to testing taylorisation through *"robustness spectra"* or "*spectrum of damage by fatigue*" analysis, which is not covered in this course and is also very well developed in [ref. I1].

4.2.2.1.1. *Stationary and ergodic processes*

a) Stationarity

A random process that consists of an infinite set of sample functions can theoretically and especially hypothetically find the value of all the samples at any time, by calculating averages. If we consider the process of the preceding figure, we can define the average value (or first moment, a statistic that we will later discuss), in of the instantaneous values x_k taken at time t_1, which is equal to the sum of values x_k, taken by all samples, divided by the number of samples, which is theoretically and by definition:

$$\mu_x(t_1) = \lim_{N \to \infty} \frac{1}{N} \sum_{k=1}^{N} x_k(t_1) \tag{4.65}$$

Therefore:

- In the case where $\mu_x(t_1)$ varies with time t_1, the process is said to be "non–stationary" and its properties can only be determined, *a priori*, by calculating instantaneous averages on all the sample functions (theoretically infinite) forming the process. This is of course impossible.

- If $\mu_x(t_1)$ is invariant as a function of t_1, from the beginning of time, then the process is known as "slightly stationary". An infinite number of averages of higher orders can also be calculated ($\mu_x(t_1,t_2)$ – $\mu_x(t_1,t_2,t_3)$...) to establish a family of probability distribution functions describing the process. If all possible averages are invariant over time t, then the process is known as "stationary in the strict sense". This is of course impossible to find rigorously, and we will settle for the first two moments as we are only interested in random vibrations rather than in Gaussian laws for which stationarity of the first two moments is sufficient.

We already perceive here the difficulty of controlling consoles in noise, since it is necessary to ensure precision when it comes to measurements (and thus the sizes of sufficient samples or DOF) and thus to long periods of time, with control constraints according to the resonances of the material being tested generally, loop times.

b) Ergodicity

It is shown that the properties of a stationary process can be described in another way, by calculating averages at the time for specific samples of a whole, rather than average statistics of sample sets, that are virtually unattainable in practice. For example, let us consider in the previous figure, the fifth sample. The value of time average will be by definition:

$$\mu_x(5) = \lim_{T \to \infty} \frac{1}{T} \int_0^T x_5(t).dt \tag{4.66}$$

So, if the process is stationary and if $\mu_x(k)$ does not vary for different samples, the process is said to be "ergodic". In this case, the time average is equal to the corresponding overall average (it tends on average to) and we write:

$$\mu_x(t) = \mu_x(k) \tag{4.67}$$

NB:

(a) The stationarity of a process is not easily verifiable (there are tests to check it), because we need a great number of records (theoretically infinite). It is, in general, controlled only over one or a few samples. It is not possible to guarantee it for the future, but it is very often assumed that creative physical causes of the random phenomenon remain permanent.

(b) A stationary process is often ergodic and its properties can easily be measured from the observation of a single record for a long enough time so that, *a priori*, all outputs of amplitudes could have taken place. But this ergodicity is almost always an act of faith rather than demonstration.

(c) In practice, time averages are only used for their simplicity. For example, vibration records identified on carrier vehicles will be cut into portions that appear to be stationary in time (the frequency content being stored for a certain time). These portions are then analyzed and the result, after compilation, is considered as representative of the process.

4.2.2.1.2. *Statistical functions*

Four types of statistical functions are used to describe the basic properties of a random process. These are, average squares, probability density functions, autocorrelation functions and power spectrum density functions. As already mentioned, the autocorrelation function, which provides information in the field of time and is connected to the PSD through the FT will not be explained here, its effect on the vibrations being generally of less importance.

a) Average squares

These functions provide summary information about power or energy (also intensity) of random variables and are expressed by the following temporal equation (process supposedly ergodic):

$$\psi_x^2 = \lim_{T\to\infty} \int_0^T x^2(t).dt \tag{4.68}$$

- Also called “average squares of instantaneous values” (total average power).

- The root square (positive) of the mean square is called the “root of squares or root means square (RMS) value”, or true effective value.

- The static component is the “average value” given by:

$$\mu_x = \lim_{T\to\infty} \frac{1}{T} \int_0^T x(t).dt \tag{4.69}$$

- The dynamic component is represented by the “variance” given by:

$$V_x = \sigma_x^2 = \lim_{T\to\infty} \int_0^T [x(t) - \mu_x]^2.dt \tag{4.70}$$

- The root square (considered positive) of the variance is called “standard deviation” or average quadratic error. From previous equations, we find:

$$V_x = \sigma_x^2 = \psi_x^2 - \mu_x^2 \tag{4.71}$$

NB: σ_x therefore represents the effective value when average μ_x is zero.

Variance = average square – square value average.

b) Functions of probability density

These functions provide information about random variables in the field of amplitudes.

Let us consider the random signal in the following figure:

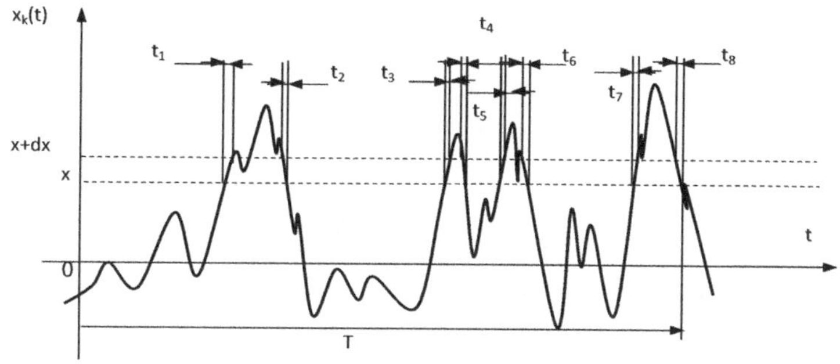

The probability that this signal has a value between x and x + dx can be estimated by ratio T_x/T, where T_x represents the total time during which x < x (t) < x + dx and T is the total observation time. If T tends to infinity, we can think that T_x/T will tend toward an exact value of probability. So:

$$\text{Prob}\{x < x(t) < x + dx\} = \lim_{T\to\infty} \frac{T_x}{T}$$

For a very small Δx, a density probability function can be established, as follows:

$$p(x) = \lim_{\substack{\Delta x\to 0\\ T\to\infty}} \frac{1}{\Delta x}\left(\lim \frac{T_x}{T}\right) \tag{4.72}$$

c) Normal law or Gaussian distribution

The normal law or Gauss law frequently intervenes in the various aspects of our daily life and in nature. If the formula is an expression that is not well known, the curve known as "bell-shaped", representing its first-order probability density, is very popular. It is indeed shown that when a size is subject to the influence of a great number of variation factors and that those are independent or almost independent from/to each other, individual measurements of this size are distributed according to the law of Gauss (a consequence of the central limit theorem and the law of large numbers).

The normal distribution is a continuous law that is of great importance in vibrations. It depends on two parameters, the mean value m (or μ_x) and the standard deviation σ or quadratic error average.

For a normal process, the probability distribution of the first order is described by the law of probability density, whose expression is:

$$p(x) = \frac{1}{\sigma\sqrt{2\pi}}.e^{-\frac{(x-m)^2}{2\sigma^2}} \tag{4.73}$$

For a stationary process, m (or μ_x) and σ are constant. If m = 0, the curve is centered and symmetrical on the y-axis. It has a maximum at the point of the X-coordinate, x = m and ordinate $y = \frac{1}{\sigma\sqrt{2\pi}}$. This curve is represented in the following figure (with m = 1 and σ = 0,5):

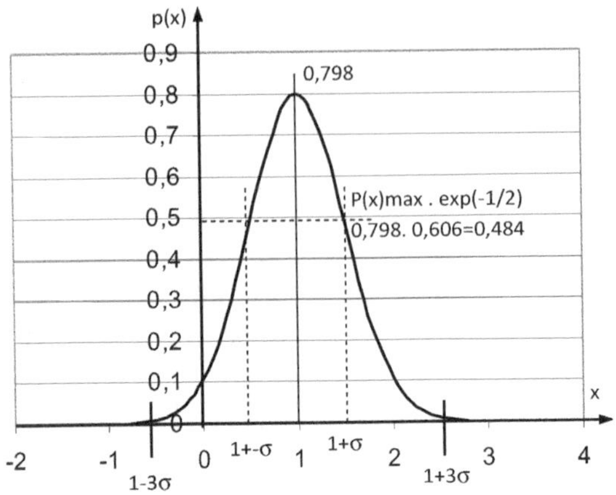

The normal or Gaussian process is the only one where higher order probability distributions can be explicitly written and studied. Concurrently to these analytical advantages, the normal process has a great importance as a model for the study of real physical processes. In particular, random vibrations will be considered stationary and of a Gaussian-type. Note that for values between $\pm 3\sigma$ (average is not taken into account), there is virtually a probability equal to 1 (integral or area under the curve). This means that in that interval, we have virtually 99% chance of finding all the amplitudes of the random process.

Moreover, since usually the average value is zero (in vibrations), we find that the standard deviation σ is called effective value by a language abuse in vibrations and specifications. In random vibration testing, we impose as a lower maximum, the limit value of $\pm 2.3\sigma$, which allows us to have a sufficient number of amplitudes with significant peaks.

d) Functions of spectral concentration of power

We will find here that most of the definitions in Chapter 2 apply to vibrations. The DSP function of a random quantity describes its composition in frequency in terms of the value of its spectral square densities and therefore its power. The mean square of a temporal sample in a frequency band between f and f +Δf can be obtained by filtering the sample with a filter band pass with stiff fronts and a narrow band by calculating the average of the squares of the signals at the output of the filter. This value approaches the true value of average squares if the observation time tends to infinity. The relation is as follows:

$$\psi_x^2(f,\Delta f) = \lim_{T\to\infty} \frac{1}{T}\int_0^T x^2(t,f,\Delta f).dt \tag{4.74}$$

where x(t, f,Δf) is the portion of x(t) in frequency band Δf from f to f+Δf. For small Δfs, a PSD function $G_x(f)$ can be defined as: $\psi_x^2(f,\Delta f) \approx G_x(f).\Delta f$, therefore:

$$G_x(f) = \lim_{\Delta f\to 0} \frac{\psi_x^2(f,\Delta f)}{\Delta f} = \lim_{\Delta f\to 0} \frac{1}{\Delta f}\left[\lim_{T\to\infty} \int_0^T x^2(t,f,\Delta f).dt\right] \tag{4.75}$$

Quantity $G_x(f)$ is always real and positive.

In terms of PSD, the value of the squares is then given by:

$$\psi_x^2(f,\Delta f) = \int_0^\infty G_x(f).df \tag{4.76}$$

NB: In certain cases, the PSD expression is carried out following the positive and negative frequencies and it is necessary to specify if taking into account a factor of 2 or not.

Thus, the value of the average squares is equal to the area under the curve representing the PSD function defined according to frequency. A typical record of a PSD can be represented by the following figure. This is the case of a wide PSD band with lines found in aeronautical standards. The corresponding tests are "sine on random vibration tests" (*sine – on – random vibration test*).

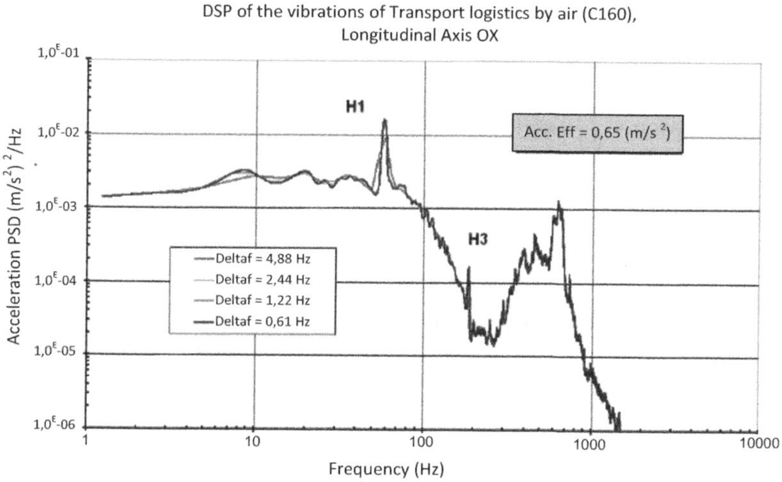

4.2.3. Random vibrations

4.2.3.1. *Overview*

We saw that random vibrations are continuous oscillatory movements whose instantaneous amplitude can be predicted on the basis of probabilities. This vibration is in a continuous spectrum of frequency whose amplitudes vary from randomly defined in statistical terms. This type of vibration is defined mathematically hereafter, assuming that the process that gives birth to it is ergodic (and is therefore stationary) and the amplitude of accelerations follow the law of Gauss, which is very close to the processes encountered in nature and especially in most of the carriers or vehicles where electronic devices are installed. Its goal is that it excites simultaneously all the frequencies of the useful spectrum.

4.2.3.2. *Probability density*

The following figure shows the recording of a random vibration based on time (in ordinates, accelerations are carried). The likelihood that an instantaneous value of acceleration is included between –a and +a is given by the integration in this interval of the Gauss law. Knowing that for an average of zero by assumption, this law means that we are much more likely to find a range near zero than of strong amplitudes.

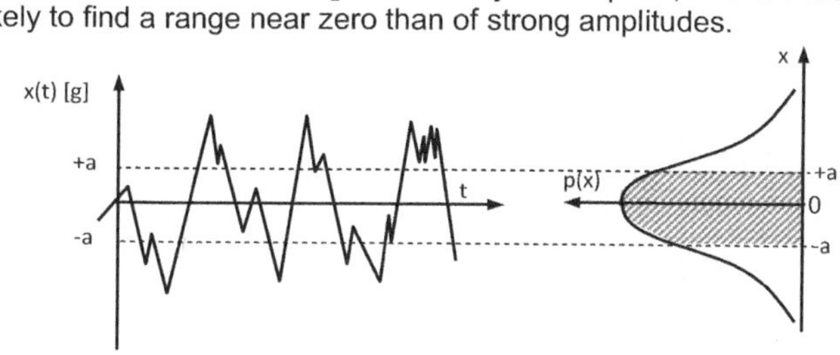

Having for the researched probability:

$$P = \int_{-a}^{a} p(x)dx = \frac{1}{\sigma\sqrt{2\pi}} \int_{-a}^{a} e^{-\frac{x^2}{2\sigma^2}}.dx \tag{4.77}$$

which is represented by the area hatched under the Gauss curve.

For memory σ (standard deviation) is the square root of the variance $V_x = \sigma_x^2 = \psi_x^2 - \mu_x^2$.

For a random vibration of average value of 0, the standard deviation σ is then equal to the effective value (or RMS), which means:

$$\sigma = \text{standard deviation} = \sqrt{\frac{1}{T}\int_0^T x^2(t).dt} = \text{effective acceleration (RMS)}$$

So, in the case of a Gaussian distribution of accelerations, the total effective acceleration of signal x(t) is equal to the standard deviation of the instantaneous amplitude of acceleration in the Gauss curve:

$$x_{eff} = \sigma \tag{4.78}$$

Let us remember that the probability of instantaneous acceleration remaining less than σ is 68%. By setting limits to 2σ, we find 95% and at 3σ it moves to 99,73%. In other words, there is a 0,27% chance that instant acceleration exceeds three times the value of effective acceleration. These settings are important because they determine the operation of steering consoles, knowing that the latter have an adjustable strict limitation to kσ, with k being generally imposed by the standards (2 or 3 in general, but it shows that it should not be less than 2,3 for a correct distribution).

NB: The curve of probability density is often standardized, i.e. the scales of the coordinates are adjusted so that the surface under the curve equals 1, which means 100% probability. By convention, in random vibration tests, levels are capped at 3σ.

4.2.3.3. Probability density of peak values

If we consider the vibrations with regards to the damage they can cause to the structures, the peak acceleration value is more important to be known than the instantaneous value. Let us suppose the signal is filtered by a circuit with a narrow band centered on frequency f_0; it results from a quasi-sinusoidal signal at a frequency f_0 with a Gaussian amplitude distribution. For such a signal, the probability that a peak acceleration has a value between a_c and c_c+ da_c is given by:

$$p(a_c).da_c \qquad \text{with:} \quad p(a_c) = \frac{a_c}{\sigma^2}.e^{-\frac{a_c^2}{2\sigma^2}} \tag{4.79}$$

This equation corresponds to the Rayleigh distribution and the corresponding curve is found below.

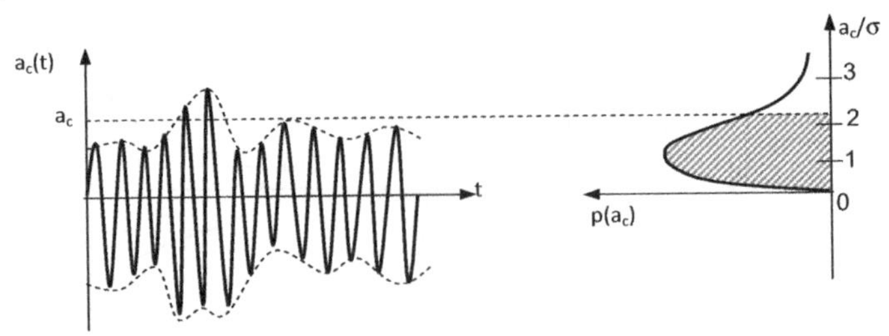

The probability that the value of the peaks is between 0 and a_c becomes:

$$\int_0^{a_c} p(a_c).da_c = \frac{1}{\sigma^2}\int_0^{a_c} a_c.e^{-\frac{a_c^2}{2\sigma^2}}.da_c \qquad (4.80)$$

This probability is equal to the area under the curve hatched. By analogy, the expression of instantaneous probability value, $p(a_c)$, is the density of probability of peak values.

4.2.3.4. *Power spectral density*

Just as the characteristics of a random vibration are described by its probability density function in the field of amplitudes and in the field of time by infinite averages of absolute value, the spectral characteristics of a random vibration are described by the measurement of its energy spectrum represented by the function of PSD. We have seen in (4.75) that the PSD is defined by equation: $G_x(f) = \lim_{\Delta f \to 0} \frac{\psi_x^2(f,\Delta f)}{\Delta f} = \lim_{\Delta f \to 0} \frac{1}{\Delta f}\left[\lim_{T\to\infty}\int_0^T x^2(t,f,\Delta f).dt\right]$. This relation expresses that the theoretical PSD is obtained by averaging the value of the mean square of greatness in a narrow frequency of bandwidth Δf over an infinite period of time by using a filter possessing an infinitely small bandwidth. In practice, it is not possible to average an infinity of filters, with an infinitely bandwidth narrow. Therefore it is necessary to estimate the PSD on the basis of finished record lengths and averaging over finite time periods with filters at finished bandwidths. Where the errors, already mentioned in Chapter 2, arise. For a random vibration, we find the definition of the PSD with the B band and with time T:

$$G_x(f_k) = \frac{1}{BT}\int_0^T x^2(t).dt \qquad (4.81)$$

where:

- $G_x(f_k)$ indicates an estimate of the PSD around f_k;
- $x^2(t)$ is the average of the square of instantaneous accelerations within the bandwidth of analysis;
- T is a finite averaging time;
- B is the effective noise bandwidth of the analysis filter.

The energy is proportional to the square of acceleration, $G_x(f_k)$, which represents the average over time of the energy available in a narrow band B, per cycle of bandwidth B. $G_x(f_k)$ is the physical quantity that will eventually cause damage to the structure subjected to a random vibration movement.

For each of the discrete frequencies of a range of reporting frequency, $G_x(f)$ could be defined by moving the central frequency f_k of the filter into the range of the useful spectrum. Curve $G_x(f)$, as traced, will perfectly define the system in an energetic manner. For example, the following figure shows a typical PSD record.

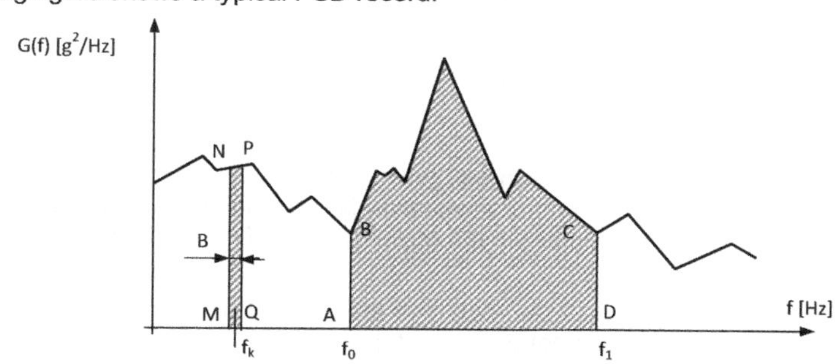

a) Relation between the effective value and the PSD

In equation (4.81), T and B are constants and we can therefore write:

$$G_x(f) = \frac{1}{B}\left[\frac{1}{T}\int_0^T x^2(t).dt\right] \tag{4.82}$$

In this equation, the expression in brackets is by definition the square of the effective value of acceleration "a" in the channel of width B. $G_x(f)$ et $g_{eff(B)}$ are therefore connected by the relation:

$$g^2_{eff(B)} = G_x(f).B \tag{4.83}$$

$a_{eff(B)}$ is represented in the previous figure by the hatched area of elementary rectangle MNPQ. Effective acceleration $g_{eff(f_1-f_0)}$ is measured by replacing B with the frequency band $\Delta f = f_1 - f_0$, and is defined by the relation:

$$g^2_{\substack{eff\\RMS}} = \int_{f_0}^{f_1} G(f).df \tag{4.84}$$

and is represented by the square root of the ABCD hatched area under the curve of the figure.

The unit used for the PSD of a random vibration under acceleration is usually g^2_{RMS} / Hz, shortened improperly to g^2/Hz, that has the dimensions of a power per unit frequency (note that the W/Hz writing = W.s = J has the dimension of energy). With g = 9,81 $m.s^{-2}$, the acceleration of gravity and, in vibration, the acceleration peak is in $m.s^{-2}$.

b) Concept of white noise

A random vibration whose PSD is maintained constant in all useful frequency bands is called "white noise" (WN), in the spectrum where we cannot distinguish a predominant frequency. By extension, the term WN is preserved when the PSD remains constant between two frequencies f_1 and f_2. From where the following graph derives:

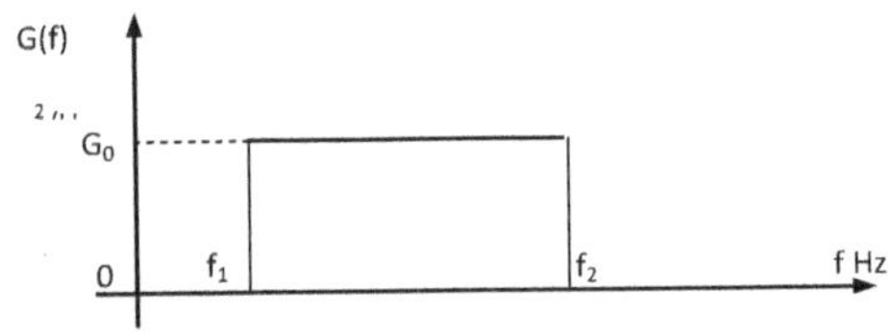

Equation (4.84) is simplified and becomes:

$$g_{RMS} = \sqrt{G_0.(f_2 - f_1)}$$

where:

- G_0 = spectral density of power in g^2/Hz;
- $(f_2 - f_1)$ = B = useful band analysis in Hz.

NB: WN is a phenomenon often seen in nature when the curve G(f) differs very little from a horizontal line. The amplitude of a random vibration is more often characterized by the RMS value of the acceleration in a frequency band given by the PSD according to the frequency in this same band.

In many standards, the ends of the rectangle are skewed (to be more realistic to a real physical acceleration, which does not abruptly stop) by slopes in dB/octave and the area of the curve becomes a trapezoid. Similarly, the actual PSD curves are often approximated by segments of straight lines of different slopes. When the noise is not white, it is said to be "pink" or in "trained spectrum".

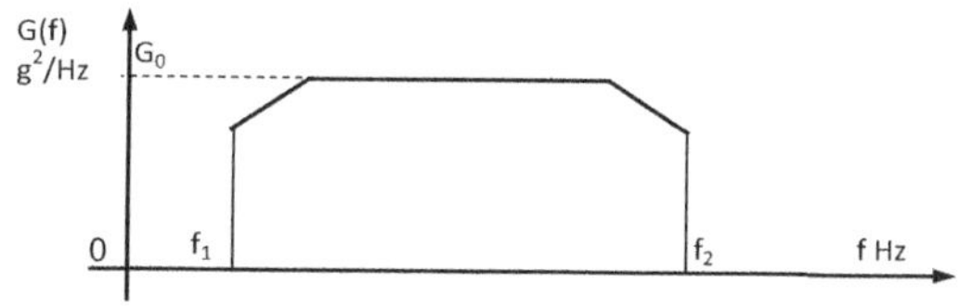

In Chapter 2, we provided a relation that allowed us to calculate the effective value of such a spectrum, taking into account the various slopes which approximated it as best as possible. The following example gives an idea of the necessary calculations.

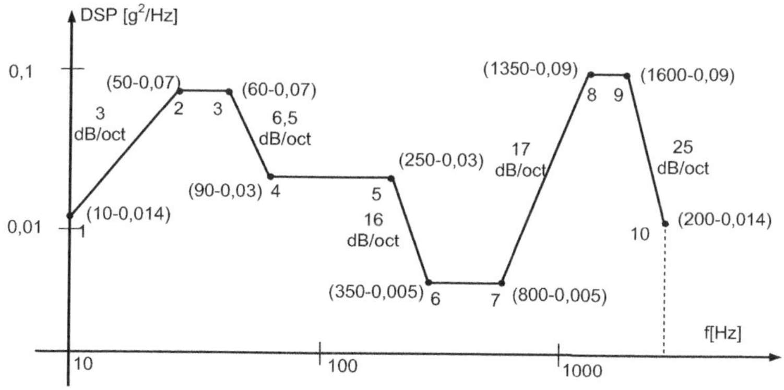

Useful relations for calculation:

- number of octaves between two frequencies f_2 and f_1: $N_{(f_1,f_2)} = \dfrac{\text{Log}(f_2/f_1)}{\text{Log}2}$;

- relationship of calculation of the effective value: $\begin{cases} g^2_{\text{eff}} = \dfrac{1}{n+1}(G_2 f_2 - G_1 f_1) & \forall n \neq 1, 0 \text{ included} \\ g^2_{\text{eff}} = G_1 f_1 . \text{Log}\dfrac{f_2}{f_1} & n = 1, \text{ or } m = 3\text{dB/octave} \end{cases}$

- n = m/3 with m = slope in dB per octave.

The calculation is made using the following table:

ref	A	B	C=A.B	D	E	F=E/3	G	H=D/G
	f	G	G.f	$Gf_{(k+1)}$-$Gf_{(k)}$	m dB/oct	n	n+1	Mean square g^2
1	10	0,014	0,14					
2	50	0,07	3,5	+3,36	+3	+1	+2	1,68
3	60	0,07	4,2	+0,7	0	0	+1	0,7
4	90	0,03	2,7	-1,5	-6,5	-2,167	-1,167	1,29
5	250	0,03	7,5	+4,8	0	0	+1	4,8
6	350	0,005	1,75	-5,75	-16	-5,333	-4,333	1,33
7	800	0,005	4	-2,25	0	0	+1	2,25
8	1 350	0,09	121,5	+117,5	+17	+5,667	+6,667	17,65
9	1 600	0,09	144	+22,5	0	0	+1	22,5
10	2 000	0,014	28,2	-115,8	-25	-8,333	-7,333	15,8
							Total of mean square	68,00 g^2
							RMS value $= \sqrt{\text{means square}}$	8,25g

NB: Of course there are calculators (VIBKIT2000 now on 64-bits) to perform this kind of operation that is done automatically with steering consoles. Table 4.2 provides a more accurate result but the calculation "by hand" is correct (8,25–8,274).

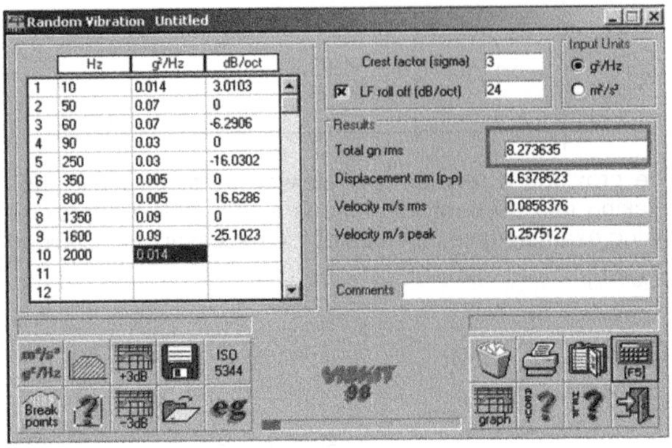

Table 4.2. *Result of the calculation of PSD with Vibkit 98*

4.3. Specific tests

We will only refer to these specific tests for information. Their detailed analysis can be done from the normative documents dealing with the subject (for example MILSTD810E – method 519 "gun fire vibration aircraft"). In this category of tests, we can classify the following tests.

a) Tests known as "shot gun"

These are tests that simulate the vibrations caused by the firing of guns mainly on aircrafts and which are generally made up of a noise spectrum formed on which added sine lines are representative of rates of gun firings (in harmonic relations). Frequencies and amplitudes are either measured (test taylorisation), or imposed by fixed standards as the MILSTD 810-method 519, previously mentioned.

An example of vibration spectrum is given below, always resulting from the same reference. The levels of severity of the spectrum of the PSD formed, as well as amplitudes of the lines, are calculated according to characteristic parameters of the aircraft, distance to guns, etc., described in tables given by the standards, as specified below.

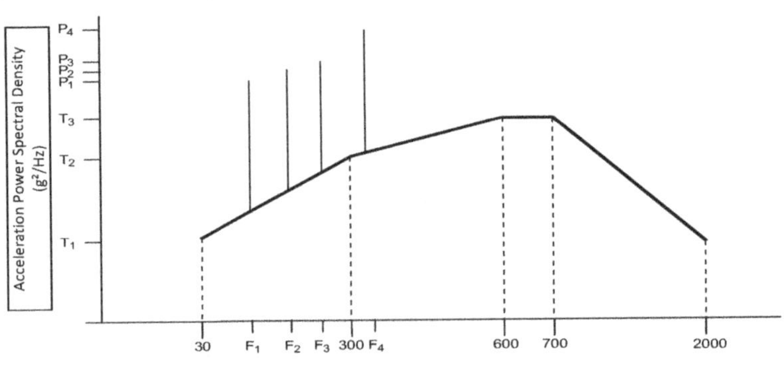

Figure 4.1. *Typical configurations of shooting guns associated with aircraft classes (MILSTD 810 standard)*

Aircraft/Pod	Gun (Quantity)	Location	Firing Rate Rnds/Min	Firing Rate Rnds/Sec	Capacity
A-4	MK12 (2)	Wing roots	1 000	16.6	100/Gun
A-7D	M61A1 (1)	Nose, left side	4 000 & 6 000	66,6 & 100	1020
A-10	GAU-8/A (1)	Nose	21 00 & 42 00	35 & 70	1175
A-37	GAU-2B/A (1)	Nose	6 000	100	1500
F-4	M61A1 (1)	Nose	4 000 & 6 000	66,6 & 100	638
F-5E	M39 (2)	Nose	3 000	50	300/Gun
F-5F	M39 (1)	Nose	3 000	50	140
F-14	M61A1 (1)	Left side of nose	4 000 & 6 000	66,6 & 100	676
F-15	M61A1 (1)	Right wing root	4 000 & 6 000	66,6 & 100	940
F-16	M61A1 (1)	Left wing root	6 000	100	510
F-18	M61A1 (1)	Top center of nose	4 000 & 6 000	66,6 & 100	570
F-111	M61A1 (1)	Underside of fuselage	5 000	83,3	2084
MIRAGE	DEFA 554		12 00 & 18 00	20 & 30	
RAFALE	DEFA 791B		2 520	42	
GEPOD 30 (GAU-8/A)	GE430 (1)	POD	2 400	40	350
SUU-11/A	GAU-2B/A (1)	POD	3 000 & 6 000	50 & 100	1 500
SUU-12/A	AN-M3 (1)	POD	1 200	19	750
SUU-16/A	M61A1 (1)	POD	6 000	100	1 200
SUU-23/A	GAU-4/A (1)	POD	6 000	100	1 200

b) Tests of vibrations on helicopters

There is a similar approach to the vibrations generated by the rotating blades on helicopters. Here is also a spectrum of PSD in which we add harmonic relations in lines, with sometimes a light sweep around the latter. An example from Stanag4370-AECTP200-247-1 "deployment on rotary wing aircraft" is given below.

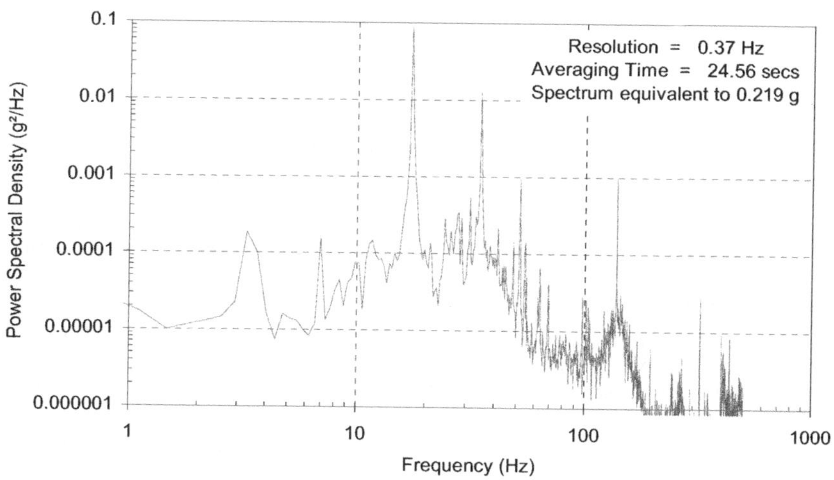

c) Acoustic tests of vibrations

The latter are supposed to simulate the vibrations caused by acoustic noise present at the launch of missiles or rockets or possibly large transport aircrafts. They differ slightly from the tests usually encountered, in the sense that they are simulated in reverberant rooms using large sirens. The severity of the test is therefore expressed in terms of sound pressure, such as the following example taken from the GAMEG13 issue 48 "acoustic vibrations".

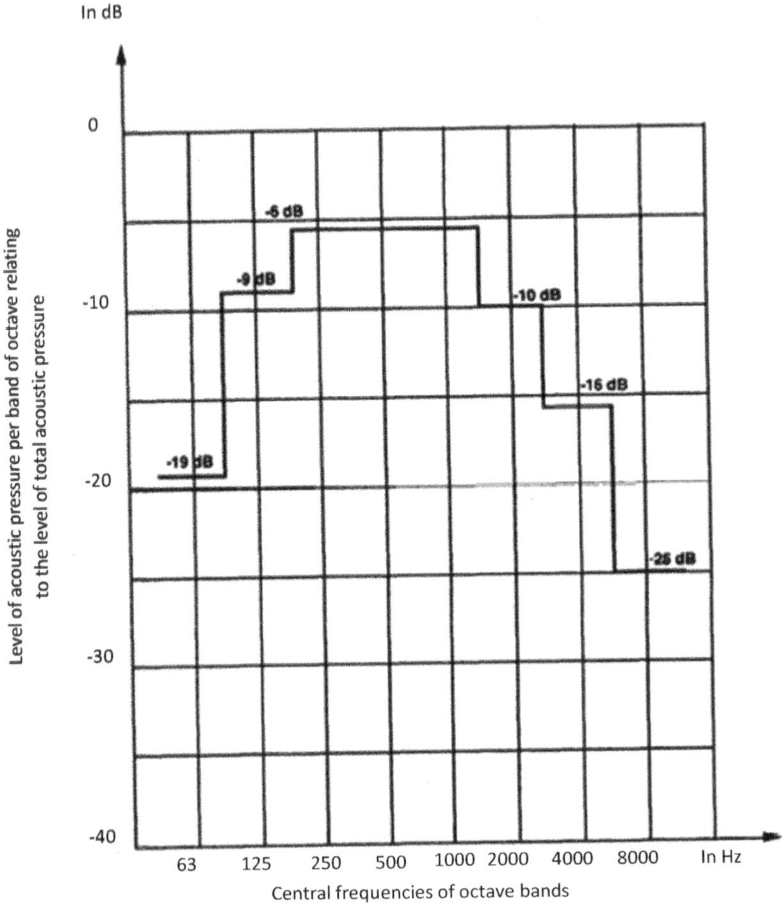

Form of the spectrum of acoustic pressure by D4 octave band for preferential severities

There are of course other particular vibrations tests (for example the combined vibrations/climatic tests), but the main ones used by large clients were mentioned here.

d) Tests of acoustic discretions

These are the vibrations generated by electronic equipment measurement tests (bays, boxes, etc.) and mechanical ones (motors, fans, etc.) conducted on carriers through the connections between the equipment and the carriers, supports, generally, hydraulic piping, ventilation, electric cables, dampers.

It is necessary to know that, indeed, these vibrations are measured up to 10 KHz for marine buildings, which must remain discrete, and at these frequencies, dampers do not remain perfect vibration insulators, because modes of propagation appear, which degrade the cut-off effect of the HF dampers. These tests are subject to defense constraints secrecy.

e) Sine on noise or noise on noise tests

They characterize, for example, the 'shot-gun' tests with regards to the sine on noise tests, and the specific tests concerning noise (of narrow band) on noise (for broadband) tests.

5

Equipment Applications

In this chapter, we will approach the main vibration sources and their impact on electronic equipment. We will do this through examining the weaknesses and behaviors of the latter, and we will suggest some very basic rules and techniques. It should be noted that despite more vibration-resistant surface-mount device (SMD) assembling editing techniques, we can still often find traditional assembly components, which justifies some sections in this chapter.

5.1. Vibration sources and effects

5.1.1. Overview

We will, of course, recall the usual main sources of vibration (primarily for military equipment). Electronic equipment shipped on board of mobile carriers (land, maritime, planes and missiles) is permanently subjected to different types of vibrations of diverse origin during their operational use. These are primarily transmitted by the structures they are attached to, and the degree of pressure ranges from the lower levels of rail transport and land vehicles, to more severe levels recorded in applications for supersonic aircrafts, rockets and missiles.

5.1.2. Vibration sources

Of course, it is better to limit vibrations from the source, but this is not always possible.

a) In land vehicles

- Soil profile.
- The engine and transmission forwarding bodies.
- The force of the wind.
- For armored vehicles, the acoustic and vibrational field of the muzzle and the force created during recoil (creations of transients), etc.

b) In marine vehicles

- The motor and transmission components.
- The hydrodynamic forces on the hull.
- The variable pressure field around the propellers.
- The movement of the waves.
- Vibrations caused by underwater explosions, etc.

c) In air vehicles

- Movement of the ground while it is rolling

1. For aircraft propellers and helicopters:

- Gas combustions and reactions of the engine's pistons.
- Defects of balancing propellers and drive shafts.
- The pressure field created around each propeller blade, etc.

2. For subsonic, supersonic aircrafts and missiles:

- The acoustic pressure of the gases' flow escaping the reactors at high speed.
- Combustion noises.
- Thermodynamic disturbances.
- Noise due to the turbulent boundary layer that runs along the aircraft or machine.
- The firing of guns and gears giving rise to complex mechanical constraints similar, in some cases, to vibration impulses.
- The intense acoustic field generated by the engine of a missile, etc.

5.1.3. Vibration effects

The main failures caused by vibrations in electronic equipment are as follows:

- Loosening of securing devices.
- Cable cords cut up.
- Defects in the welding (cold junctions and cracks).
- Broken component outputs.
- Unplugging of the components (heavy hybrid circuits, etc.);
- Contact defects in connectors, relays, supports, index cards.
- Short circuits on circuit boards, or between CI cards due to large relative movements.
- Functional failure of components.
- Misalignment of optical circuits.
- Structure ruptures due to high local constraints (overvoltage effects) or at very significant rates of fatigue.
- etc.

This list is far from thorough, but it shows that each element that makes up a piece of equipment, from the component to the primary structure, must be studied and manufactured with the greatest possible care and tested by representative tests. That is why it is advisable to perform a *"test tailoring"* approach and to use the computer facilities available (databases, calculations, and simulation) to design the selected specifications, which will form the basis for establishing a test plan that accurately represents the conditions encountered during the life of the material duration profile.

5.1.4. Notions of fatigue and damage due to vibrations

5.1.4.1. *Overview*

Under the action of repeated or alternating loads, breakage of material may occur under stresses weaker than the material's strength in static. This is known as failure under fatigue. This breakage mainly comes from microscopic cracks that appear in the material during the first cycles of charging and which progress each time it is charged. There are many curves to define the lifetime of a material because the problem is very complex. Indeed, parameters such as initial damage, corrosion, nonlinearities, geometric shapes, etc., also affect this and make the problem more difficult. We will stick to a few terms used in the current normative documents and which are still current. In our case, the relative displacement z(t) of our model at 1 DOF can then be analyzed to calculate the fatigue damage suffered.

5.1.4.2. *Wöhler curves*

The curve that is most used, linking the amplitude of the stress S to the number of N cycles to the rupture, named the S.N curve (*Stress–Number of cycles or Wöhler curve*), can be represented in analytical form by the Basquin equation as follows

$$N.S^b = C = A^b$$

Having the following classic form:

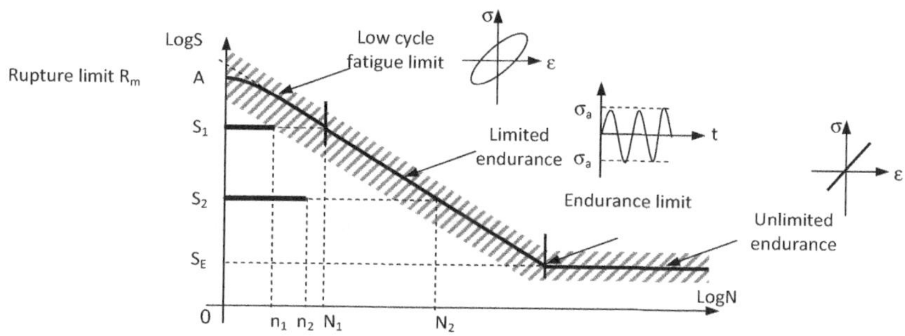

The linear aspect in the Log/Log coordinates comes because of the exponential aspect of the curves. These are of course important dispersions and are usually presented with a level of confidence of 95% (Gaussian distribution with a constant N and normal log at a constant S).

- There is a standard deviation of 5% in the plastic area ($N < 10^4$) and 8% in the endurance area ($N > 10^7$).

- A and b are characteristic constants of the material and A is the true stress at breakage.

As an indication, the parameter b can take the following values:

- aluminum alloys: 6–10;
- steels: 10–15;
- magnesium alloys: 20–25.

Normative documents call for a precise value for the b parameter, for example:

Standard	b
DEF STAN 0035 (Part 5)	5
Air 7304 (1972)	8
Air 7306	8 (but can be adapted)
MIL STD 810 F	Random: 8, Sine: 6
GAM EG 13	Free choice

In a complex material, many types of damages are possible other than fatigue damage. For electronics, when the breakage is caused by mechanical fatigue (for example, electrical contacts), the Wöhler curve is applied with a coefficient b in the order of 3 to 4. However, degradation by corrosion *(fretting)* of the electrical contacts may be more important. Therein lies the difficulty of obtaining precise quantitative results.

In logarithmic scales, this law is thus represented by a slope of –1/b, with:

$$LogS = -\frac{1}{b}.LogN + LogA$$

- For N < 10^4, the right curves because the behavior of the material becomes plastic at high levels.

- For some materials, there is a fatigue limit stress S_E below which there is no breakage regardless of the N number of cycles. At a very high frequency, the existence of this limit is questionable (results of the curves are acquired at a low frequency, in general). This is why it is advisable to extend the linear part up to 10^7 cycles.

5.1.4.3. *Linearity*

We can admit that there is a linear relation between the stress and the deformation of the following type:

$$S = K.z$$

This hypothesis in the case of vibration signals remains valid for low values of S away from the break, which are the desired operating conditions. In this equation, z represents the peak of the deformation.

5.1.4.4. *Miner's rule*

It predicts that the damage defined for a sinusoidal excitation by the ratio of the number of cycles n_i carried out at an S level to the number of N cycles in breakage is linearly cumulative regardless of stress order. Thus:

- For a S_1 given if one only carries out n_1 cycles, there is no breakage, but one admits to accumulated damage by fatigue given by: $d_1 = n_1/N_1$. Therefore, there will be breakage when $n_1 = N_1$ meaning $d_1 = 1$.

- Additionally, if we take n_1 to S_1 then n_2 cycles to S_2, we admit that partial damage is given by: $d = n_1/N_1 + n_2/N_2$. And, generalizing to i constraints levels, it becomes:

$$D = \sum_i \frac{n_i}{N_i}$$

Breakage does indeed happen when D = 1.

NB: Miner's rule is questionable. We can indeed observe a significant influence in stress application order. The lifespan is diminished when starting with strong levels with a smooth test tube (with a reversed effect for notched specimens). This is why this effect is less marked for random vibrations.

Moreover, value D = 1 at breakage should be considered an average value, with the number of cycles N itself being a statistical parameter character as previously shown. However, it is the most used rule because of its simplicity. It also allows correct comparisons between the different vibration types.

We can also note that damage D is a function of the natural frequency of the system at 1 DOF and its absorption, of parameter b and constants K and A, as well as input excitement. This leads us to the following expression [ref. I1]:

$$D = \sum_i \frac{n_i}{N_i} = \sum_i \frac{n_i.S_i^b}{A^b} = \left(\frac{K}{A}\right)^b.\sum_i n_i.z_i^b$$

We apply the term *"fatigue damage spectrum"* to the curve representing damage D based on the natural frequency of the mechanical system with 1 DOF. As absolute knowledge of D is made difficult by the lack of exact knowledge of K and A, these parameters are taken to be equal to 1 for the calculations that are then used to provide points of comparison as previously mentioned (value of D given to an arbitrary multiplying coefficient).

In case of vibrations, the damage suffered by the system during the application of a half-cycle of stress S_i is: $d_i = \frac{1}{2N_i} = \frac{S_i^b}{2C}$, and for n_i half-cycles at the level of stress S_i: $d_i = \frac{n_i}{2N_i} = \frac{n_i.S_i^b}{2C} = \frac{K^b}{2C} n_i.z_i^b$, therefore we get for total damage:

$$D = \frac{K^b}{2C}.\sum_i n_i.z_i^b$$

NB: The damage can also be calculated from expression: $d_i = \frac{S_i^b}{2C}$ relative to a half cycle, with an N_T total number of half-cycles:

$$D = \frac{1}{2C}\sum_{i=1}^{N_T} S_i^b$$

So if N_T is big:

$$E[S_i^b] = \frac{1}{N_T}.\sum_{i=1}^{N_T} S_i^b$$

Or:

$$D \approx \frac{N_T}{2C}.E[S_i^b]$$

This equation can be used according to vibration type (sine, random, etc.), through the estimation of E. In an example of the calculation of D for an excitation in PSD, we find:

$$E[D] = \frac{n^+.T}{C}.S_{eff}^b \cdot \int_{-\infty}^{+\infty} |u|^b.q(u).du$$

With: $u = \frac{z_p}{z_{eff}} = \frac{S_p}{S_{eff}}$ and $N_p = n_p^+.T =$ number of positive peaks during a T observation duration, q(z) = density of peak probability of the system stress response to the excitation during a T time.

Thus, knowing q(u), we can calculate E[D]. So, for a Gaussian excitement [ref. I1]:

$$E(D) = \frac{K^b}{C}.\frac{n_p^+}{z_{eff}} \int_{-\infty}^{+\infty} z_p^b \left\{ \frac{\sqrt{1-r^2}}{\sqrt{2\pi}} e^{-\frac{z_p^2}{2(1-r^2)z_{eff}^2}} + \frac{r.z_p}{2z_{eff}^2} e^{-\frac{z_p^2}{2z_{eff}^2}} \left[1 + \mathrm{Erf}\left(\frac{r.z_p}{z_{eff}\sqrt{2(1-r^2)}} \right) \right] \right\} dz_p$$

5.2. Electronic equipment

5.2.1. In land vehicles

Electronic equipment is becoming more and more common in automobiles (for example radio stations, computers, complex television systems, built-in electronics, etc.). In military vehicles, it is the sophisticated launch systems and missile guidance (crotales and others, for example). They come either in a compact case or in a rack drawer form. Vibration levels are generally higher. For example, values range from 2 g of 10 at 30 Hz and 0,5 g of 30 at 100 Hz.

In armored vehicles or transport, spectrums in PSD reaching 500 to 1 000 Hz and 5 g_{RMS} are, however, very common. Because of the severe shocks that occur when stowing freight cars, train equipment is excited by shock signals (a few 10 g and 10 ms), with levels of vibration being generally low (< 1 g). The following graphics illustrate this and are taken from GAMEG13 databases *"Appendix: models and environment data"*.

Tractor trailer - TRM 10 000 (20 tons)
Corrugated iron with V = 23,5 km/h

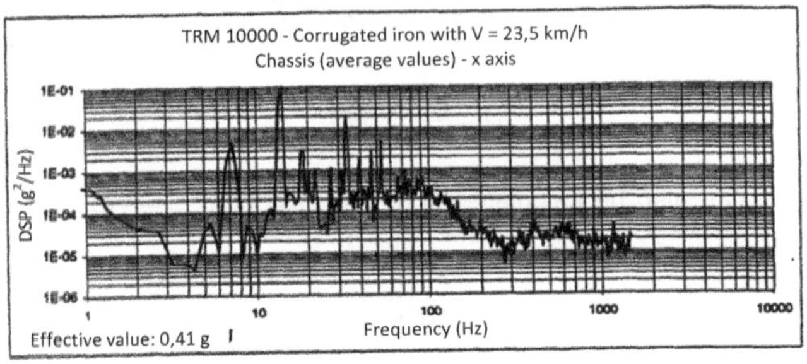

Rail borne freight carrier - SNCF Wagons type A-B-C
mediocre tracks - speeds between 80 and 130 km/h

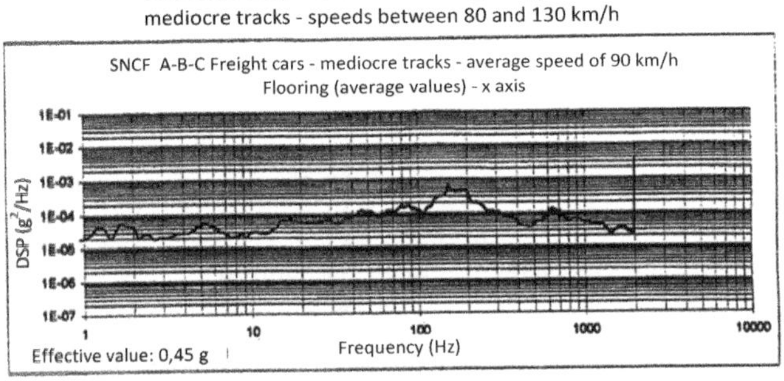

Railroad - gondola car
Shock response spectrum

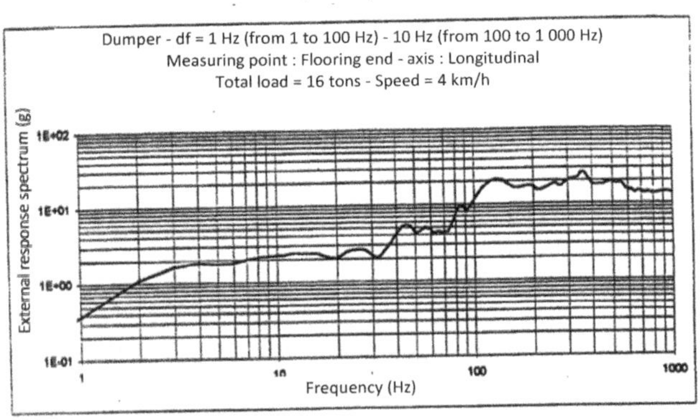

5.2.2. In ships and submarines

Electronic equipment is mostly designed in drawer, racks or cabinet forms installed in large-sized electronic bays (800 x 800 x 2 000) for weights from 50 to 500 kg. Vibration spectra are usually of low frequencies (0.1–50 Hz), with levels of the order of 1–2 gc/ = peak value and sometimes 200 Hz for the torpedoes, for example. In military vessels, shocks have a greater importance due to the underwater explosions of different origins that can occur (for example grenades, underwater mines, etc.). In submarines, on the other hand, the vibrations are generally weak and also limited to around 50 Hz. To reduce acceleration levels due to these shocks and vibrations, electronic materials are usually mounted on dampening systems or floating platforms. These resonances generally measure around 10 to 15 Hz maximum (ergodicity problem). Therefore, it is best if at least the first material modes are located past 30 Hz to respect the rule of the octave we will discuss later on.

5.2.3. In aircrafts, helicopters, missiles

The control units mounted in these carriers often have special shapes that allow them to better fit in the necessarily limited spaces.

Unlike land and naval equipment, their volumes and weights are very critical and their density varies from 0.8 to about 1.2 kg/dm^3, approximately. Its weight is between a few kilograms and 30 kg.

In helicopters, the vibratory frequency spectrum extends from 5 to 500 Hz and levels of 0,5 to 5 gc/. Toward 10 Hz, shifting may reach 5 mmc/c. In turbojet civilian aircrafts, the frequency band goes from 5 to 1 000 Hz with acceleration levels varying from 1 to 5 gc/. The strongest levels appear along the vertical axis between 100 and 400 Hz. In military supersonic aircrafts, the frequency spectrum reaches 2 000 Hz and the average level is 10 gc/.

For missiles, the lowest frequency corresponding to the first bending mode is about 3 Hz, the highest frequencies are around 5 000 Hz. Above 1 000, Hz 30 gc/ levels are commonly recorded. The vibratory environment in aircrafts and missiles is more random than periodical. Because of the high frequencies, it is impossible for these types of carriers to create resonance-free equipment. These frequencies will excite many modes, and we will have to be careful while studying, analyzing and testing these electronic systems. In case of very severe vibrations, especially during military aircraft's gunfire, where intense vibratory energies can be generated (intensity according to the material's distance to the gun barrel and its firing rate), the study of a suspension makes it possible to filter dangerous frequencies. A typical example is that of a gun sight installed on a plane which, during firing (30 mm DEFA cannon), was able to spot the gyroscope's braces that would systematically break. Measurements taken while in flight confirmed levels of 0,4 g^2/Hz in the 80 to 150 Hz band, that is overall levels of 30 g_{RMS}. A damper elastomer suspension can reduce these levels to around 0,1 g^2/Hz.

A few examples extracted from the DEFSTAN 0035-part5.

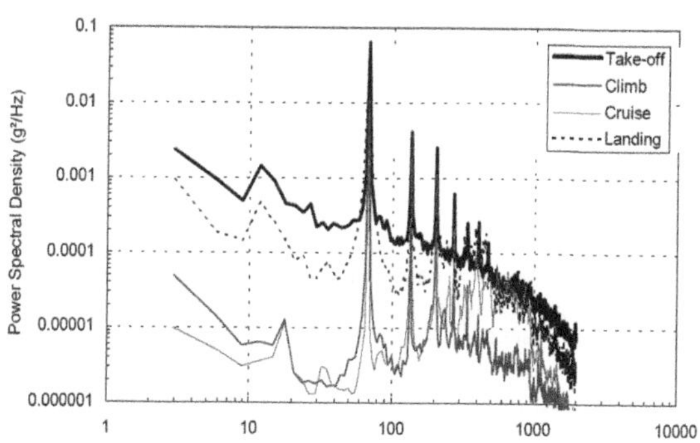

Figure 5.1. *Vibration spectra of a C130 Hercules turbo-propeller transport aircraft (data of the plane's propellers in the vertical axis)*

5.2.4. Conclusion

These data were provided for information and to illustrate our point. Obviously, it is advisable to consult any data bank available to best adapt severity levels selected for the design and tests.

5.3. Design of electronic equipment subjected to vibrations

5.3.1. Operating mode

There are several possible methodologies to develop an equipment or system depending on the complexity of the latter. The phases suggested here are common and concern classical equipment (cases, bays, racks, etc.), but there are others, of course.

5.3.1.1. *Establishment of selected specifications*

This is the most important initial part because it is the one that will determine the severities the material will have to endure. Therefore, it is necessary to have good knowledge of the environment encountered during the lifespan of the material (for example mechanical, acoustic, aerodynamic, etc.). For this, there are methods and tools already mentioned several times, which consist of using:

- a flat rate suggested by the normative documents when the lifespan is bad or unknown;

- tayloring tests from available databases, measures on site, etc. as indicated in the various normative documents, which to this day and for the most part of the normative military documents, advocate a *"test taylorisation"* approach. The development of specifications then follows a method that is currently up and running and is perfectly described in the normative military documents and in [ref. I1];

- if the equipment is conceived for one type of carrier only, there may already be a personalization process carried out by the client that leads to particular specifications (for example the mirage 2 000 specifications).

This phase is important but will not be developed in this course. The study in [ref. I1] is a good resource to assimilate this approach.

5.3.1.2. *Location, zoning and mounting system*

It is absolutely necessary to know the zoning of the carrier's material and, especially, the suggested mounting and support techniques before undertaking the study of the material. Indeed, the rigidity of the primary host structure determines the behavior of the material by its transferability at the entrance of the latter. The mounting system that is often dictated by the carrier manufacturer must be analyzed, discussed and amended, if necessary, in order to avoid significant coupling.

5.3.1.3. *Critical sets and subsets*

Each piece of equipment includes a rather large number of sets, subsets and electronic components which are critical (relays, quartz, optical viewfinders, gyroscopes, laser transmitter/receiver, etc.) It is therefore important to ensure that they are able to maintain their performance when they are subjected to high vibration stress generated by the host structures and substructures (slides, CI cards, electronic bay structure, etc.). We can then, essentially, use the following:

- the manufacturer's specifications;

- internal or external database;

- internal or external validation tests.

5.3.1.4. *Method and development stage*

Here, we usually find the following main stages:
- calculation stage;
- model stage;
- prototype stage.

a) Calculation stage

Currently, there are powerful data-processing tools that are capable of making calculations by the Finite Element Method (FEM) in statics and dynamics, for example, and this is made *a priori* before any construction, and of the powerful modal analysis tools *a priori* and *a posteriori* on the model, which makes it possible to correctly dimension the equipment. These heavy means can be completed by the calculation tools provided in the previous chapters that, on simplified models of representation (and thus subjected to restrictive hypothesis), can make correct predictions (for example we can choose to make calculations of 10% in a first pass).

A certain amount of expertise and experience is thus needed to build these models, which should make it possible to obtain measures of transfer, resonance and modes functions and to conduct behavior predictions and damage calculations by using simple material resistance tools. The point of this, is that on this level, it is easy to implement modifications on the model by changing the values of the mass, rigidity and absorption to arrive at its means. Of course, if the safety coefficients obtained are too low, appropriate means must be implemented. An example of simplified models is given below.

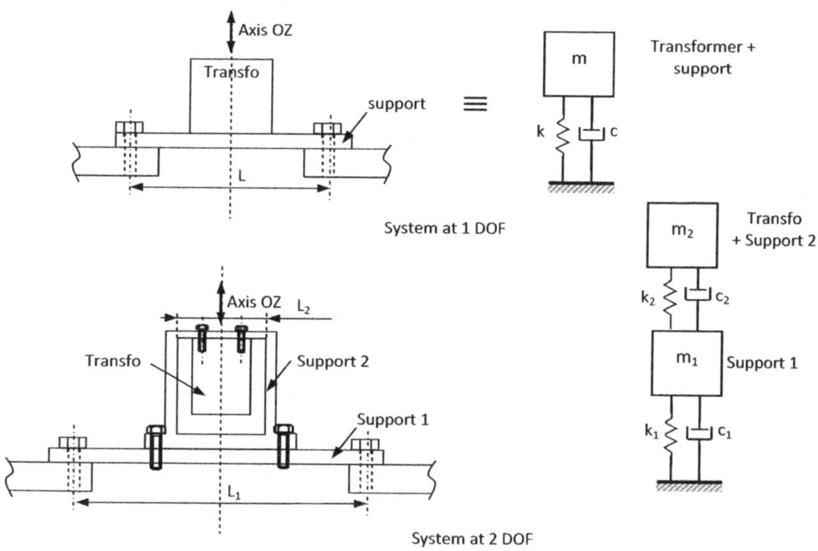

NB: Note that there is software made for calculating solid objects mounted on dampers with a large DOF number (up to 100), which allow us to make the prediction of an adopted behavior model, such as, for example, electronic bays mounted on dampers (SYMOSW of the SOCITEC company, for example). These minimum means are essential in a development office or in a test laboratory.

b) Model stage

These calculations must generally be checked on models to produce the supporting definition files. These ballasted models are made up of an assembly of metal plates and fixed masses (riveted, welded, screwed, etc.), in which finished subsets, such as micromechanics, CI cards, power supplies, etc., can be built-in. These must simulate the equipment from the point of view of masses, inertias, centers of gravity, fastenings, etc., as best as possible.

The model's sinusoidal vibration study will then make it possible to determine the resonant frequencies, as well as transferability, distortions and stresses to eventually check the calculations and modify mechanical structures, if necessary. For the suspension study, this approach is relatively simple and productive. An example is given below for the simulation of an optical support laser basket helicopter nacelle weapon. We can note the care given to the test chassis which is done so as not to disturb the model's input levels.

Figure 5.2. *ESTECH 2012 Conference – vibration and control strategies*

c) Prototype and qualification stage

The prototypes are the first operational equipment to be manufactured according to fabrication and study records, with possible alterations led by the study of models. They allow validating the manufacturing record and its implementation with the production tools. It will be tested with vibrations according to the specification conditions in order to:

- analyze and confirm the dynamic characteristics (resonant frequencies, power surges, transfer functions, etc.);

- control the behavior of the different components;

- verify its operation and its performance specifications in vibratory environment functional verification tests (FVT) during the qualification stage.

As a result of this test campaign, it is possible to still make minor mechanical and electrical modifications. Then, the reconditioned equipment will be presented in a qualification receipt. If possible, it is desirable to then carry out tests on higher levels to determine the limiting performances of the equipment and to determine its weak points. In general, this test is destructive or excessively fatigues the equipment, after which it can no longer be used.

5.3.2. Design techniques

The two main rules, based on the theory of simplified vibrations discussed in Chapter 1 and resulting from the experience gained in laboratory tests, are the following:

- Either *to stiffen* as much as possible, and within the possible limit, the stress of weight and volume of the component's supporting mechanical structures;

- or *to isolate* sensitive elements with dampers.

while respecting certain design rules.

5.3.2.1. *Stiffen the elements of the supporting mechanical structures*

The key factor here is the resonance's effect. Indeed, with resonances and frequencies of around 10 Hz to a few hundred hertz, significant distortions and displacements may occur if the transmissibilities are low and create high stress in the structures and components as well as fatigue rates that are too high. These relative distortions and displacements are the main causes of damage and failures in electronic equipment and others.

The technique is then to set the structure's resonance frequencies as high as possible so that the relative displacements generated are the lowest possible, or at the very least, compatible with acceptable fatigue rates of the materials and components. Indeed, we have seen that the relative displacement varies as the inverse of the frequency square for a given acceleration, so, all that is needed is to have sufficiently high resonant frequencies so that the relative displacements generated are below acceptable fatigue rates, with cycle numbers taken into account. The problem is then to determine these two parameters "*stress caused by relative displacement, maximum number of cycles admissible"*. One of the strategies to do this can be the following:

- If possible, stand in the Wöhler curve's asymptotic area for the corresponding stress $R_{lim;}$

- If this is not possible, determine the lifespan predicted based on the number of maximum cycles found for the real fatigue rate R_S.

The look of the Wöhler curve for a given material shall have the following general appearance.

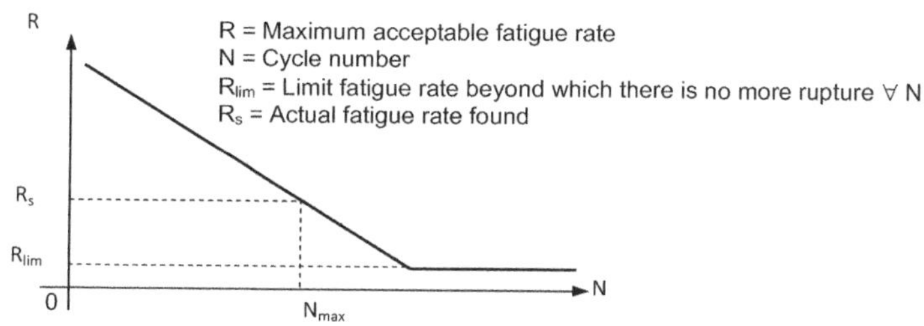

NB: Lalanne [ref. I1] details this approach in all cases of usual vibrations. The difficulty for an absolute calculation resides in the parameter determination for the calculations of N for noise or combined vibrations, as well as the determination of the maximum relative displacements for a complex component made up of different materials, or that amounts to the same determination of damage (for example an electronic card). In fact, this approach makes it possible to rather determine tools allowing the comparison of the damage caused by different mechanical signals, with the same hypothesis about the parameters used in the calculations, which allows determining the selected vibration specification severity rates. The complete approach, which remains quite complex, falls outside the scope of this course.

Thus, we can observe that there is only one way in this case, to either stiffen or strengthen the mechanical components by:

- thickening and material;
- implementation of flanges (which can also serve as thermal drains);
- choice of materials;
- modification of shapes, sections, etc.

These processes are deduced from the fact that resonant frequencies, as we have seen already, are proportional to the elasticity module E or Young's modulus of the materials, which constitute mechanical structures in moments of inertia I of their sections, and are inversely proportional to their mass per length unit m and of their length L (or their densities).

Remember that at the ends of a built-in beam, whose thickness is b, width h, length L and unit mass m, the corresponding resonant frequency of the first mode is given by equation:

$$f_n = 3{,}5.\sqrt{\frac{E.I}{m.L^4}} \tag{5.1}$$

For a rectangular beam, $I = \frac{b.h^3}{12}$. So, if h is doubled, the inertia will be multiplied by 8 and the frequency of the first mode by $\sqrt{8} \approx 2{,}8$. This simple example shows the importance of employing flanges, of the use of rectangular sections instead of square sections, and the use of materials with high elasticity modulus. The following are the natural frequencies of a few common examples of similar beams with the corresponding moments of inertia.

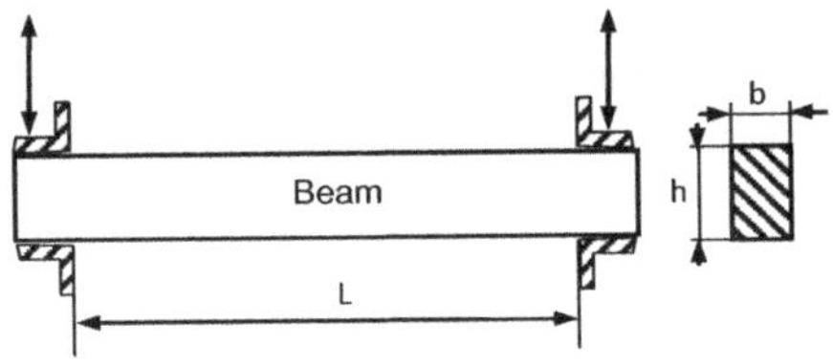

E = modulus of elasticity in N/m²
I = moment of inertia in m⁴
m = Mass unit kg/m
L = length in meters

Natural frequency $F_N = 3{,}5.\left(\frac{E.I}{m.L^4}\right)^{1/2}$

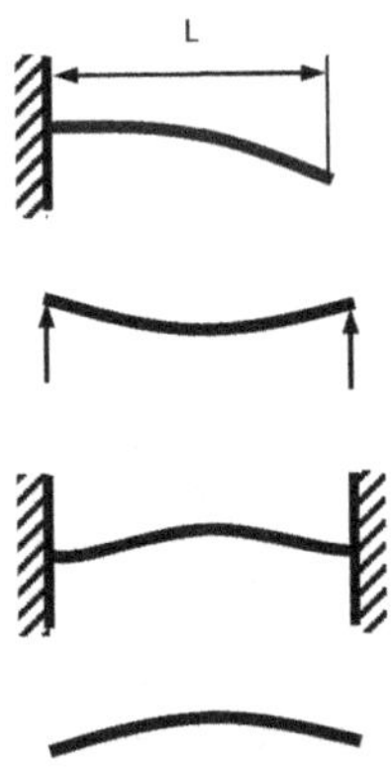

Beam built-in on one end $F_N = 0{,}56.\left(\frac{E.I}{m.L^4}\right)^{1/2}$

Beam supported at its ends $F_N = 1{,}57.\left(\frac{E.I}{m.L^4}\right)^{1/2}$

Beam built-in at its ends $F_N = 3{,}5.\left(\frac{E.I}{m.L^4}\right)^{1/2}$

Free-free beam $F_N = 3{,}5.\left(\frac{E.I}{m.L^4}\right)^{1/2}$

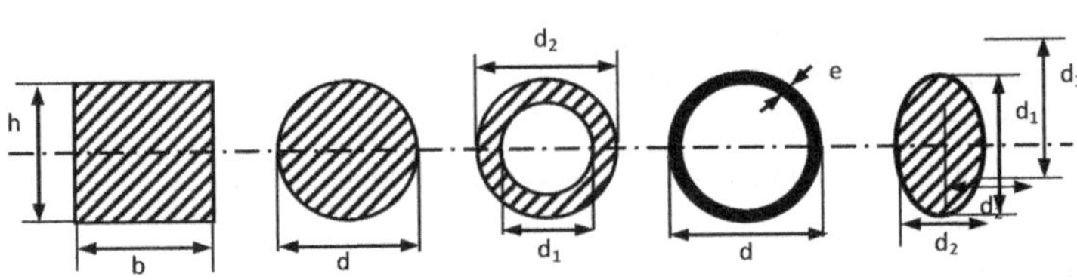

$I = \frac{bh^3}{12}$ $\quad I = \frac{\pi d^4}{64}$ $\quad I = \frac{\pi}{64}\left(d_2^2 - d_1^2\right)$ $\quad I = \frac{\pi d^3 e}{8}$ $\quad I = \frac{\pi d_2 d_1^3}{64}$

We can also consider introducing dampening as a way to reduce the Q-factor coefficients, but the resonant frequency will be slightly reduced (for example with strips of rubber, polyurethane or other products to glue or project, etc.). We will see that this dampening aspect is especially effective for low-frequency dampers that operate within a few tens of hertz. In the framework of the chapter, focus will be placed on this harmful phenomenon.

5.3.2.2. *Modes separation*

We also need to separate the flexion and torsion modes, mainly from the primary structures. The coupling of these modes significantly decreases the resonant frequency and consequently increases distortions.

5.3.2.3. *Decoupling subset modes*

It is necessary to separate the natural frequencies of each subset in order to avoid couplings, which causes significant amplitude phenomena, because power surge coefficients then multiply whenever there is a mode coincidence. To avoid this phenomenon, certain normative documents recommend using the rule of the octave.

Rule of the octave: The resonant frequency of a secondary structure must be at least two times greater than that of the primary structure. This is illustrated by the following diagram.

We can then do the following observations:

- Q_1 = 20 dB (x10) = primary structure power surge coefficient;
- Q_2 = 12,5 dB (x4,22) = uncoupled secondary structure power surge coefficient;
- Q_3 = 21.5 dB (x11,9) = power surge coefficient of the two coupled systems;
- Q_4= 2.5 dB (–10 dB) = coupled secondary structure power surge coefficient.

So, if modes coincide Q'_3 = 32,5 dB (x42,2), instead of 21,5 dB (x 11,9), representing an increase of 7,5% instead of 162%.

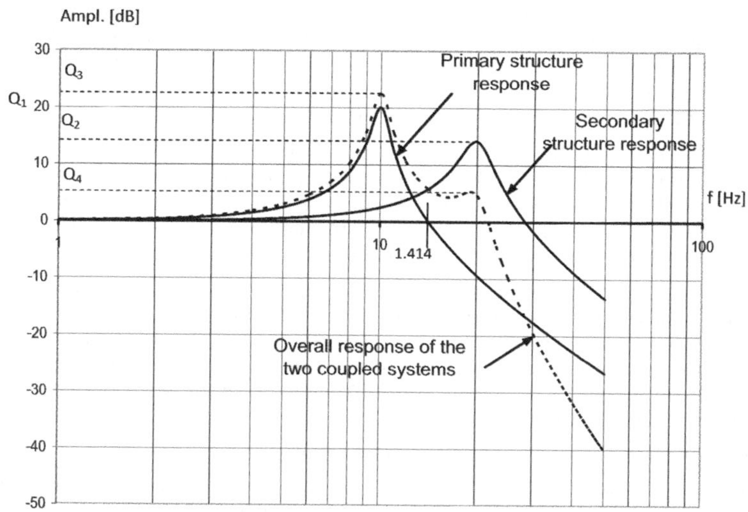

Thus, for an electronic chassis whose resonance frequency is 200 Hz, the resonance frequency of a mounted component will have to be ≥ 400 Hz, etc., for each additional DOF, which is illustrated by the following figure.

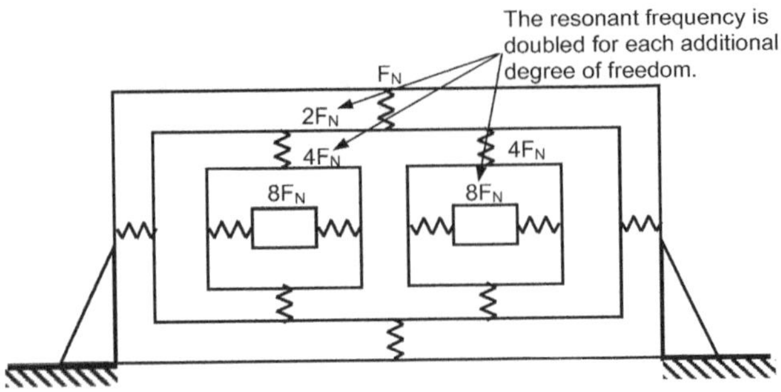

NB: Of course, this rule is not easily applied unless we know the mode values from the mass, spring, and damper models whenever they are possible (simple structures) or if we were able to measure them experimentally.

5.3.2.4. *Use of dampers*

The fallback solution when other solutions could not be implemented is to use decoupling dampening systems. This can happen in severe environments or when the component number induces multiple and distributed modes. We must then isolate all the components, either the primary structure (or the bay or electronic box's structure, for example). The principle of dampers or vibration isolators is deducted from the phenomenon of resonance of the system at 1 DOF, where from a frequency equal to $f_R.\sqrt{2}$, the transmissibility of the system becomes less than 1 and so the shock absorber reduces the entry excitation. It is therefore the shock absorber operational area. Their resonant frequencies generally lie between 5 and 50 to 100 Hz, and their Q-factor coefficient is around 3–5 at most.

The lower the decoupling frequency, the more significant the displacement rates will be. It is therefore convenient to preview the necessary displacements around the material (75 mm for marine equipment, for example). Similarly, the more significant the absorption, the less effective the insulation (less stiff slopes). The best compromise is that for which Q = 3–5, which allows good insulation twice the resonant frequency (–10 dB, rule of the octave) and more after, of course (approximately –6 dB per extra octave).

Dampers are present in the shape of metallic springs filled with shock absorber materials (steel wool, hydraulic systems or others), wire ropes (damping due to frictions of the cable strands) and elastomer elements (silicone or polyurethane based with a damping due to internal hysteresis, molecule friction.) Their architecture is almost always that of a suspension with mechanical fastenings and the three orthonormal operational axes.

However, in a suspension, the system's stability depends on the damper layout relative to the material's geometry and, especially, to its center of gravity (CoG). This is to avoid the rotating tilt phenomena and coupling between axes. They must, if possible, be located as far from each other as possible and the CoG of the suspended unit must be located in the system map or on the shock absorbers axis of symmetry, or, at least, on the axes linking them as shown in the following figures.

The following general minimum precautions must be taken:

- sufficient space should be designed around the equipment following the three directions;

- flexible systems (water, air, electric cables, etc.) installed between the equipment and the structure must have sufficient loops and support systems to avoid excessive fatigue and significant tractions (the problem of noise transmission to the hull by submarines also arises occasionally for dampers and hull connections);

- a good vibration damper (fatigue resistance) may be a bad shock damper (maximum displacement) and vice versa. Therefore, we must carefully choose the latter according to the main constraints found. Fortunately, there are good compromises with certain classes of dampers (for example, cable shock absorbers);

- there are also rack systems equipped with dampers at their base (suspended ATR racks) and other ones. A database is needed to make the best choice.

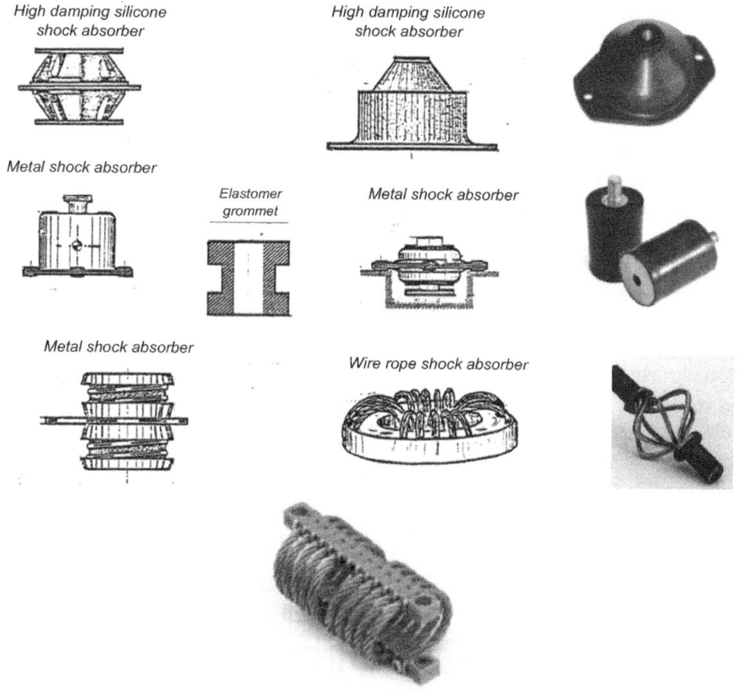

Figure 5.3. *Different types of shock absorbers*

5.3.3. Printed circuit boards

5.3.3.1. *Overview*

These are the main elements of most electronical equipment. Component carriers and electrical connections are, very often, in a vibratory environment, subject to fatigue or rupture damage, when these cards or components are excited in their resonant frequencies. We can thus find stress concentrations in output leads or the legs of the components and welds. This results in fractures, cracks and defective contacts. A good component performance is recommended and we then must:

- Estimate the distortion amplitude limits and the minimum security resonance frequencies during the design;
- Stiffen cards *a priori* or when the estimated distortions are too important;
- Ensure an effective maintenance of the cards;
- Avoid resonant coincidences (direct couplings);
- Mount the components according to confirmed methods.

5.3.3.2. *Minimum frequencies of card security*

The CI card vibrations tests study showed that the components' and welds' lifespan is related to dynamic relative displacements. Thus, in [ref. I17] (Steinberg), which achieves unanimity in this area, lifespans higher than 10^6 cycles under fatigue can be reached if the card's vibration amplitude does not exceed 0.003 times its width, that is:

$$y_{max\,c/} = 0{,}003.b \tag{5.2}$$

b = smallest side of the card (b in mm, $y_{maxc/}$in mm).

When they are excited in their resonance frequency, CI cards behave as systems with 1 DOF and the distortion's dynamic amplitude based on frequency can be determined by the following relation:

$$y_{max\,c/} = 0{,}25.g.\frac{Q}{f^2} \tag{5.3}$$

with: f = vibration frequency in Hz and Q = transmissibility in the center of the card (usually maximum amplitude critical area).

The experiment also shows that the approximate transmissibility Q, for the many and different types of cards, can be approximated starting from the resonant frequency f_n, that is to say:

$$Q = A.f_n^{1/2} \tag{5.4}$$

where A is an empirical, dimensionless constant depending on the frequency f_n and on entry accelerations levels. Therefore:

- For cards whose resonance frequency is located between 100 and 400 Hz and for levels varying between 2 and 10 g, the value of A is around 1;
- For frequencies between 50 and 100 Hz, A decreases to 0.7;
- For frequencies between 400 and 700 Hz, A is equal to 1.4.

The following figures show the possible distortions.

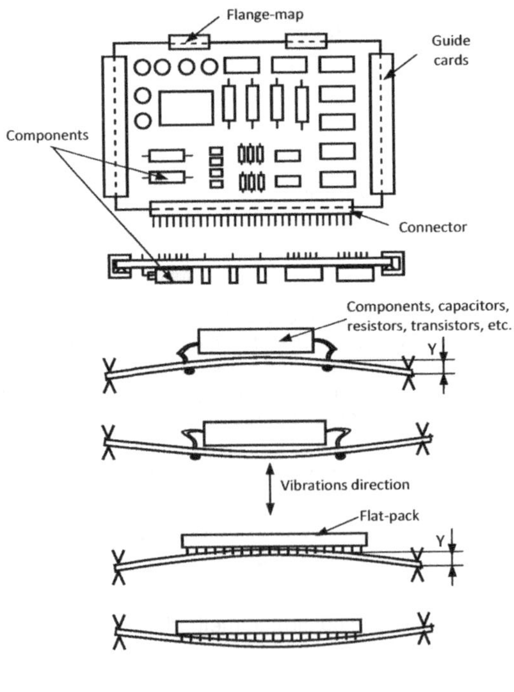

The practical formula for estimating the minimum security frequency is thus obtained by combining previous equations, that is:

$$f_{n\,min} = \left(\frac{0,25.g.A}{0,003.b} \right) \qquad (5.5)$$

Thus, if the minimum resonant frequencies calculated or measured (by vibration tests), are lower than f_{nmin}, then it will be necessary to stiffen the card. Some relations, making it possible to estimate the CI cards resonant frequencies according to their configurations, are given in Appendices C and D. A particularly useful study that is also well adapted to electronic card vibrations is presented in [ref. I17].

5.3.3.3. *Rigidifying CI cards*

There are essentially the following methods.

a) Increasing card thickness

The 1.6 mm passage, which is a commonly used thickness, to 2,4 mm, multiplies the resonance frequencies by a factor of 1,5, for example. For low-charged small dimension cards (<100 mm on the side) whose levels do not exceed 5g, for example, this solution makes it possible to obtain acceptable component distortions. A bigger thickness (3,2 mm, for example) would be more effective, but we are quickly faced with weight problems and especially manufacturing problems (hole metallization). It seems that 2,4 mm is, in most cases, an optimal thickness. Of course, there are always specific cases that should be analyzed individually.

b) Adaptation of a central fixing

This technique makes it possible to cancel the first vibration mode, where the amplitudes are largest. It is then possible to fix cards among them through a central stem, itself fixed on the primary structure, but this solution does not work for plug-in cards.

c) Usage of specific damping materials

The usage of previously mentioned plastic materials such as polyurethane foams, epoxy resins, or silicones, deposited in thick layers on the cards' surfaces dampens the resonance amplitudes and also stiffens the card. The goal of this process is to obtain very rigid cards. The difficulty then lies in repairing and changing faulty components and the need for low thermal dissipations. Or, then, it is necessary to use specific products to stimulate thermal exchanges (by phase change), but which are less effective vibration wise.

It is also possible, in the case of applications where they are short-lived (missiles, bombs, torpedoes, etc.) despite very high vibrations, to fully mold the cards in the adapted plastic products, so as to fully form entirely rigid blocks, except for the connectors.

d) Thermal drains

The already mentioned thermal drains made up of aluminum, copper or brass plates, whose role is to transfer heat energy out of the equipment, can also contribute to improving the cards' function, both in stiffness and damping (frictions between cards and drains). This technique is interesting, because it can fix two issues that may damage CI cards, and is often used.

e) Usage of stiffeners

When the previous techniques cannot be used or are insufficient, the use of simple and inexpensive stiffeners is the most practical process to rigidify the cards. The effectiveness of vibration reduction in stiffeners depends, primarily, on the form, material and fastening method.

The best technique to fasten stiffeners is to glue or weld them when it is possible to do so. If not, they can be screwed or riveted. The experiment shows that screwed stiffeners have an effectiveness of 25% compared to those welded. It can reach 50% when the fastenings are very close to one another. Due to their high modulus of elasticity, the metal stiffeners are more efficient than plastic ones; however, precautions must be taken to avoid false electrical contacts.

Stiffener sections will be selected from those that possess a high moment of inertia (in U, T, L squared or rectangular shapes). Stiffeners are parallel to modal distortion. An example of traditional stiffeners is given in the following figures.

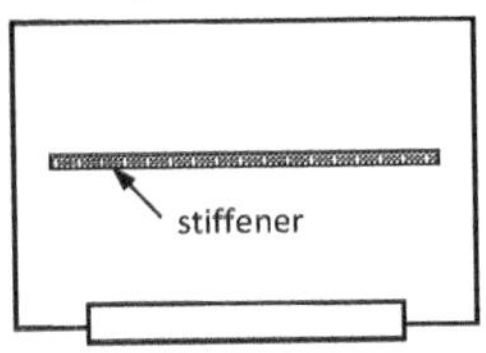

(a) A single stiffener

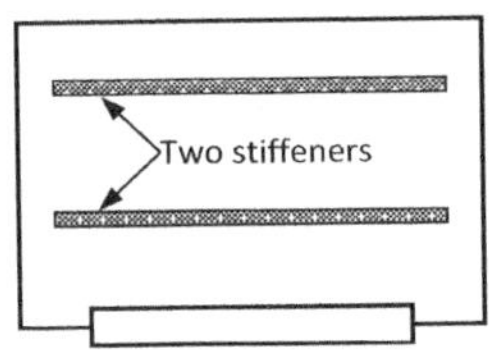

(b) Two stiffeners
(good layout)

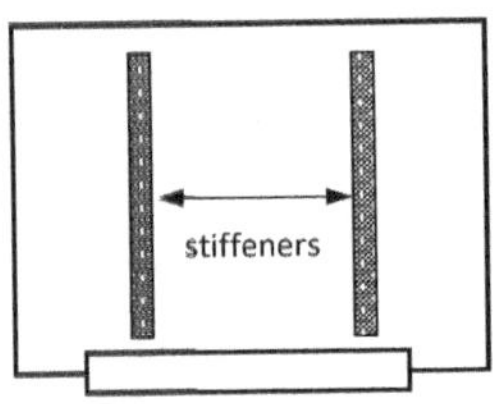

(c) Two stiffeners
(bad layout)

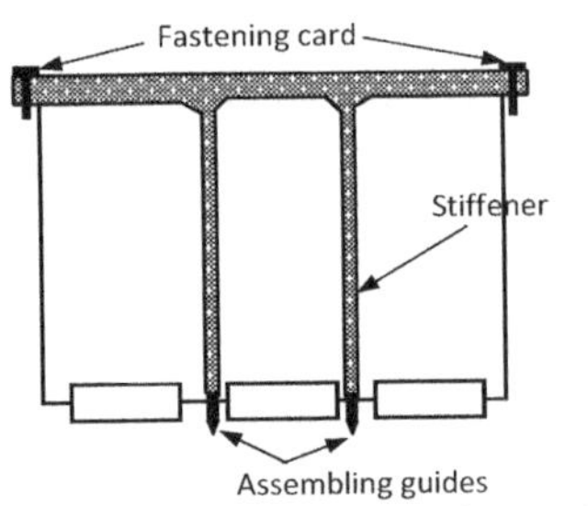

(d) Stiffeners equipped with guidance and fastening card systems

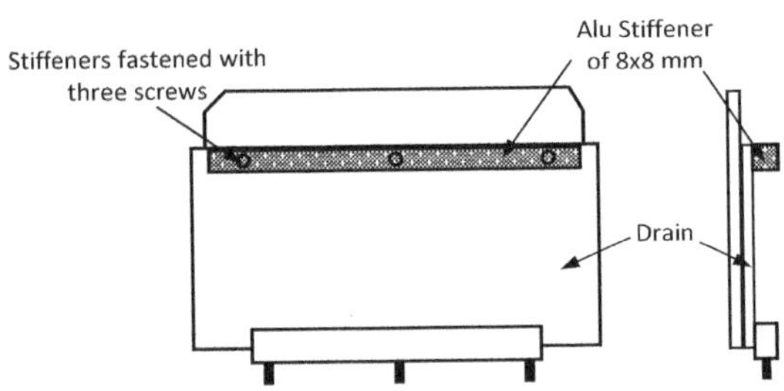

The results obtained are described in the following table.

	Resonant frequency	Peak amplitude
(1) Cards without stiffeners	138 Hz	0,6 mm
(2) Cards with stiffeners	230 Hz	0,25 mm

So, for a dimension of b = 190 mm, the amplitude at the center of the card is less than the minimum security amplitude of 0,003.190 security = 0,57 mm, thus ensuring an acceptable lifespan to the card and its components.

NB: In Appendix A, there are curves deduced from the experiment, which make it possible to determine the natural frequencies of standard cards according to the stiffener models used, without calculations.

f) Card maintenance

The resonant frequencies of thin plates excited by vibrations depend on the fastening conditions of their edges. They are maximum when the edges are built-in. Since we can compare CI cards to thin plates, when they are plugged in they would thus need to have their side edges held as strongly as possible by guide cards and their upper edges effectively restrained. The fourth edge is equipped with a connector resembling an embedding. This maintenance also reduces the risks of false contacts in the connector.

- Rigid card guides reduce dynamic deflections and stress, but then high friction forces will make card operations difficult (assembling, disassembling).

- Flexible card guides make it possible to reduce the difficulty of assembling and disassembling the cards, but they cause dynamic displacements and thus higher constraints with bumping shock risks in the metal slides.

A compromise is thus essential when choosing card guide stiffness, which behave like springs and influence the card's frequency response.

The following figures give some classic results of cards with constant thickness depending on the type of embedding (a more elaborate table is given in Appendix D):

general equation: $$F_N = \frac{B}{2\pi}\left[\frac{E.t^2}{\rho.a^4.(1-\nu^2)}\right]^{1/2}$$

where:

- E = Young modulus in N/m^2;
- t = thickness of the plates in m;
- ρ = mass density in $N.s.^2/m^4$;
- a = length of the side in m;
- ν = Poisson's ratio;
- B = coefficient function of the type card fastening.

Assembly 1
One embedded side
Three free sides

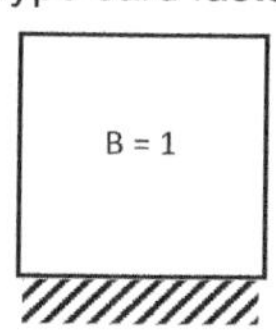

$$F_{N1} = \frac{1}{2\pi}\left[\frac{E.t^2}{\rho.a^4.(1-\nu^2)}\right]^{1/2}$$

Assembly 2
One embedded side
Three supported sides

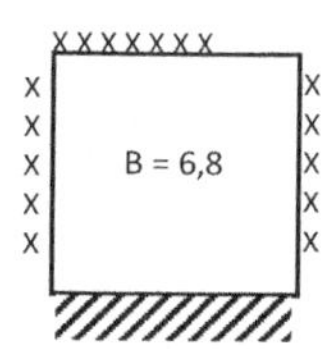

$$F_{N2} = 6,8.F_{N1}$$

Assembly 3
Three embedded sides

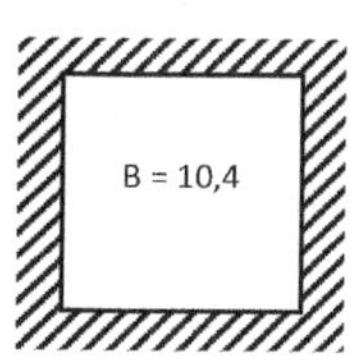

$$F_{N3} = 10,4.F_{N1}$$

$$F_{N3} = 1,5.F_{N2}$$

Many types of card guides are offered by suppliers and used in the electronics industry. The most commonly used are the following (currently, there may be others):

- Circle segment beryllium bronze "Birtcher", cards guide that allow good card functioning on their entire length of their edges and good assembling;

- Beryllium bronze card guides with undulated springs: The assembling of the cards is easy but with a lower maintenance. Stiffness is a function of the number of contacts and the thickness of the springs;

- Barbed elastomer card guides: Card maintenance is quite effective when the barb density is of least 1 per cm. Metal–metal shock bumps are avoided but maneuvering assembling and disassembling requires a major effort and the use of extractors. Quick barb wear can entail switching between cards and card guides;

- Metal corners card guides: They are mainly used with thermal drain cards to conduct heat to the structure of the cases (thermowells). Their advantages lie in the rigid bridle of the cards, which is similar to an embedment with easy card insertion and extraction;

- Card bridles: These are usually made up of elastomer and are mounted on removable straps directly placed on case lids. They must keep the top edges of the cards on the greatest length possible. They are generally made of flat edged slots or equipped with barbs. Entry chamfers must be provided in order to facilitate mounting the cards in the slots;

- For unassembled cards that are directly mounted on the walls of the materials, a sufficient number of screws must be conceived (a step of 5 cm is, technically, the maximum tolerated value). The more points there are, the more the fixation resembles an embedment and the damping created by the friction between the card and screw heads is significant.

The following diagrams provide a few common types of slides:

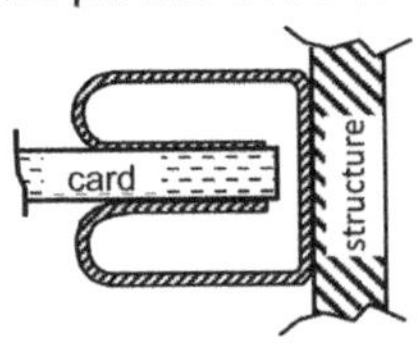

(a) Circular arcs
Material: beryllium bronze

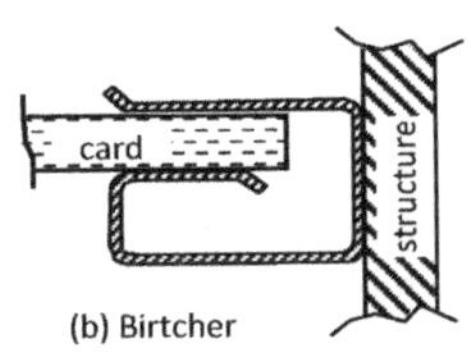

(b) Birtcher
Material: beryllium bronze

(c) Corrugated springs, material: beryllium bronze

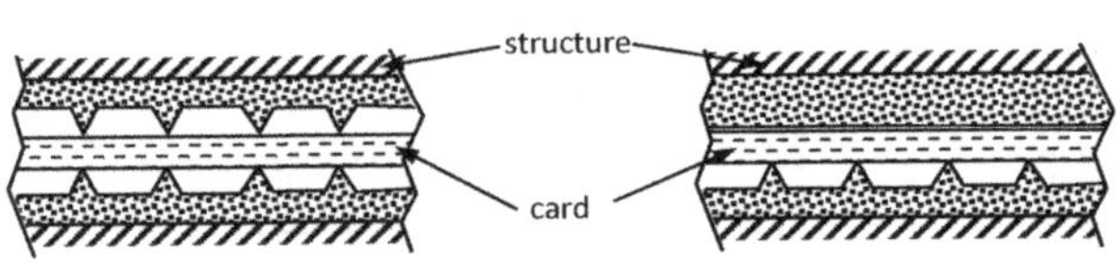

(d) Barbs, material: elastomer

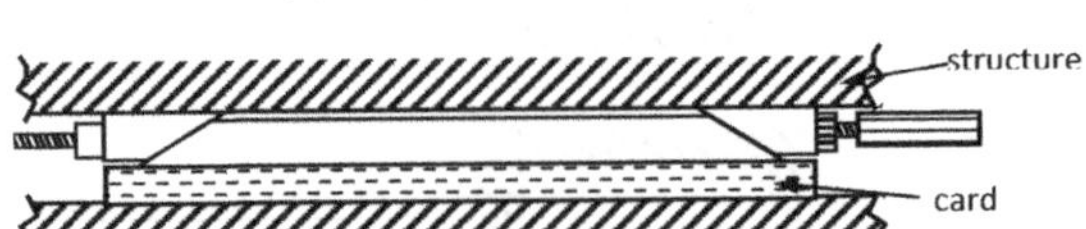

(e) Corner aluminum alloy

5.3.4. Components assembly

5.3.4.1. *Overview*

The components' capacity to withstand severe vibrations depends on many parameters including their size, their placement on the cards, their mounting methods, the cards' resonant frequencies, entry acceleration levels and duration of the environment. Damages are mostly due to dynamic stresses that are developed by the relative movements between the components, their wires or output legs, and the cards, as well as the intrinsic brittleness of their materials. The strongest stresses appear, of course, during the component and cards' resonances. Thus, resonances may appear in the components when their bodies act as masses and their outputs act as springs. They usually appear as angular oscillations whose rotation centers are located in the output welding spots (sensitive spots), as we can see in the following figures. These resonances will be mitigated by placing the component's bodies as close to the cards as possible (input of additional damping by friction). They will be lifted very high by fixing or pasting them directly on the cards.

When the cards are insufficiently rigidified, it creates complex modes in the electric outputs, which can cause fatigue breakage or false contacts in the welding. The maximum stress condition being that where the component is located in the center of the map, its body is parallel to the small side. This is due to faster distortions on the small side compared to the large side, caused by the same displacements. Of course, as previously mentioned, joining or fixing the components improves their aspect [ref. I17].

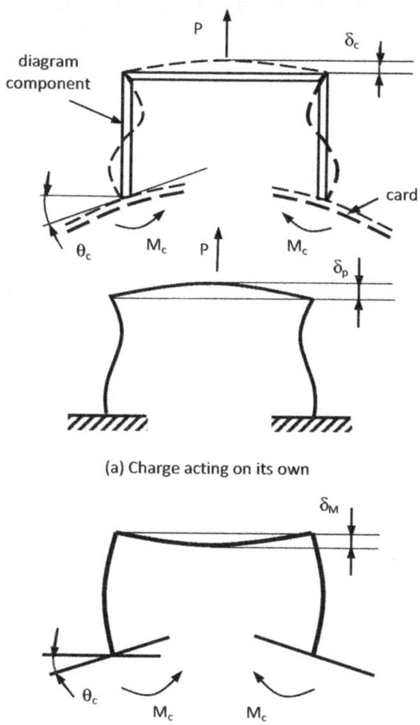

(a) Charge acting on its own

(b) Couples

Bending methods in the outputs of components mounted on cards

The most severe stress conditions appear in the fundamental card mode where distortions are the largest.

The points where the outputs are soldered to the circuit will tend to stay perpendicular to the card when it becomes distorted. The spots where the outputs are sealed to the component will also tend to stay fixed because the body of the components is more rigid than the outputs.

The result will be a complex bending mode in output wires, which roughly resembles a combination of two separate charges:

- A charge acting on its own because of the vertical displacement.

- A couple at each end of the outputs.

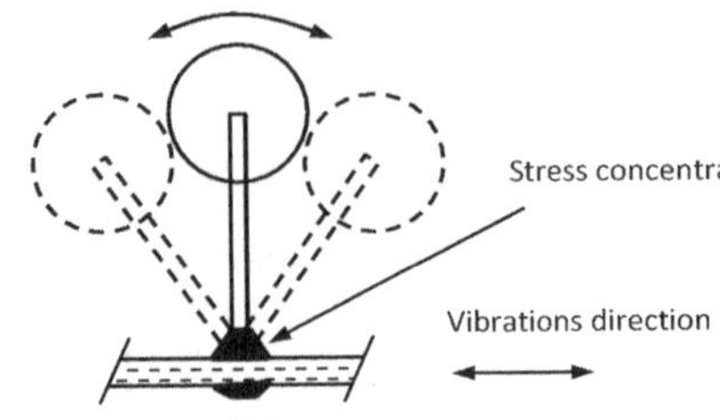

5.3.4.2. *Precautions on the assembly of components*

- Fragile components and heavy components should, preferably, be set up on the edges of the cards (near the card guide or connectors), rather than in the center.

- It is preferable to place the components' bodies parallel to the largest sides of the cards.

- Assembly of heavy components (e.g. transformers) must be made on cards rigidified by metal structures (amplifier cards, power cards, etc.). During their assembly, fragile components (ceramic cases, etc.) must be handled with precaution.

- Avoid repeated folding and unfolding of the fixing brackets.

- Output bending should be conducted at a distance of at least 1 mm less than the components bodies and with a radius of less than 1.5 times the diameter of the wires. The tools used to perform this operation should not risk damaging the wires.

- All of the components must be plated on the card except in particular cases (like power resistors). Epoxy-laminated shims must be inserted under the types of relay modules, transistors, etc., and pasted on the cards.

- Electric outputs must not be the only support of heavy and large-sized components (7 g and more). They must thus be fixed on the cards by suitable means such as metal or plastic supports, clips, epoxy adhesives, polyurethane, etc. The use of terylene wire hoops is not recommended due to the change in tightness during climatic tests.

- Outputs of components fixed by their body on the card should not be tight. Loops should be conceived in order to give flexibility to the wires and to distribute the stress in all of their lengths.

- Components equipped with glass beads crossings and ceramic glass cases will be assembled the same way.

- Modules such as integrated circuits, hybrid circuits and microprocessors, must be welded on the circuits and not inserted on special supports so as to avoid unplugging outages.

- For large modules (>48 outputs), the addition of small local stiffeners will reduce the stress supported by the outputs.

- Resistances to high energy dissipation will be fixed on relay terminals or on special metallic supports. It is possible to install them close to the cards as long as you practice cuts in them.

- The welding points must be carried out on both sides of the cards as much as possible to reduce the risk of fractures by tearing.

- The wiring cables will be fixed on the cards with a few resin drops at about 5 mm from the welding points.

- Varnishing, vacuum impregnating and coating cards and components with thin layers of epoxy resins, polyurethans or silicones *(conformal coating)*, in order to protect the circuits against oxidations and mold, helps the components bear vibrations. Coating with resin layers of 1–2 cm thick or molding parallelepipedic blocks 1–2 cm thick *(potting)* in which components and cards form compact and rigid sets allows for very severe entry vibrations without risk of mechanical failure from the components. The following figures illustrate these points.

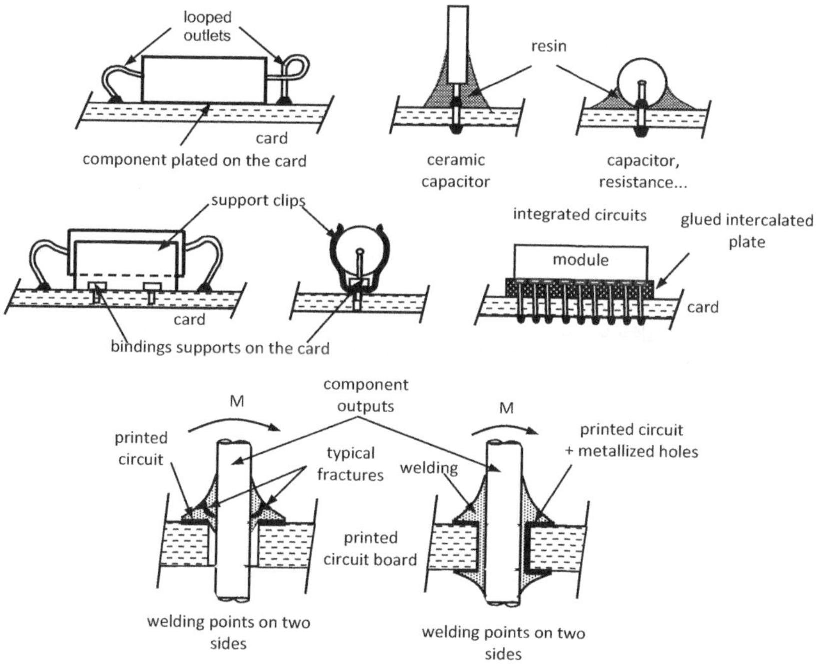

Of course, all these rules are upgradeable and strictly depend on technological developments, such as current SMD assembling that solve many of the mentioned problems. Otherwise, as sometimes specified, these rules must be compatible with other environment requirements, such as climatic and electromagnetic constraints.

5.3.5. Cases

5.3.5.1. *Overview*

Depending on the space available in the carriers, particularly on aircrafts and missiles, the electronic equipment can be produced according to various configurations.

Parallelepipedic cases and rectangular cases are generally the most used, because they are less expensive to manufacture, their assembly is simple and the plug-in modules such as the CI cards, which are the principal components, adapt well to these shapes. Made up of sheet steel, assembled sheet steel aluminum alloys (welded, bolted or riveted) or in aluminum or magnesium alloy foundry, they are fixed on carriers either directly or through shock absorbers.

In equipment where mechanical, optical or electronic systems are installed (video, cathodic, tracking mechanisms, etc.), the cases often have more complicated shapes, however, their basic structures are still parallelepipedic and rectangular. They can be equipped with cylindrical envelopes (as for missiles, bombs, targeting pods, etc.) themselves screwed or decoupled on their basic structures.

5.3.5.2. *Different assembly types*

The assembly type can affect the response characteristics of a case. The most commonly used assemblies are covered in the following sections, along with their qualities and defects.

a) Cases plugged-in directly on the carrier

The whole chassis can be assembled or removed by placing a few captive screws on the front. Interface connectors and centering fingers are assembled at the back to ensure the precision of the plug-in connectors.

- *Advantages*: Installation and fast disassembling of the case or the chassis. Easy fixing.

- *Disadvantages*: Very poor assembly type from a structural and dynamic viewpoint. The plug-in charge must pass through the case, which must be rigid enough to avoid bending. This is why a sheet aluminum case is not advised (unless its thickness is increased). The case can turn around the securing screw as long as there is only one of them. Bending modes will be coupled to the torsional modes and high transmissibility may result. On equipment, power surges of 20 were picked up on the front higher angles, for entry levels of 2 g and frequencies ranging between 150 and 200 Hz.

b) Cases plugged in on ATR racks

The ATR rack in sheet aluminum is adapted to receive airborne equipment whose casings conform to ARINC 600 standards, which defines the dimensions, assembly types, connectors, etc.

Two retractable fingers mounted on springs as well as connectors are installed at the back of the rack. A screw hinged at the front ensures the fixing of the case (or two screws for 3/4 ATR formats). It is normally equipped with shock absorbers that allow it to effectively reduce dynamic loads, as long as the shock absorbers used are carefully selected.

- *Advantages*: Quick case installation, easy fixing. Good filtering of the vibrations starting from low frequencies (# 30 Hz).

- *Disadvantages*: Usable only on cases whose only function is electronics. Significant insertion force due to the 600 series connectors used.

c) Cases fixed by fast screws 1/4 turn quick (DZUS)

The case is fixed on its front face by fast blocking screws spaced according to a standardized step and clipped on a metal snap ring.

- *Advantages*: Fast and easy installation of the case.

- *Disadvantages*: It is only used for light sheet aluminum cases. Low dynamic response in case of direct assembly on the carrier. This type of assembly is especially used for cases embedded in suspended instrument panels.

d) Cases assembled on a support dovetail

The base of the case is guided and blocked on an intermediate support (fixed to the carrier) by an assembly dovetail (in the case of gun viewfinders mounted on military aircrafts' dashboards).

- *Advantages*: Quick assembly and fixing and installation reproducibility.

- *Disadvantages*: Dynamically unstable device due to low guidance length and insufficient blockage related to the equipment's dimensions. Rotations and relatively low frequency

(# 100 Hz) resonance couplings generate forces on the equipment due to high accelerations, which are the main causes of the damages recorded on optical systems (ice, ice supports, objectives, etc.).

e) Screw fixed cases

The case is equipped with fixing brackets, generally with two at the front and two at the back, and is fixed on the carrier's structure with screws or bolts.

- *Advantages*: Effective maintenance of the case. Improved dynamic response if the fixing brackets and the sides of the case are rigid enough. This type of assembly can be reinforced with the addition of side fixings, which allow for a better distribution of dynamic loads.

- *Disadvantages*: Longer installation of the case. Need for an access point around the case.

f) Equipment assembled on missiles or bombs

The cases of this cylindrical equipment are generally fixed at the end of the machines by using flanges and screws, conical screws mounted radially, or by strapping. The effectiveness of these assembly types is a function of the number of fixing points and the stiffness of the support's surface. Significant power surges can occur for transversal vibration excitations if fixings are not regularly distributed and reinforced on the cases.

g) Equipment assembled under planes (Pods)

This equipment is fixed under military aircraft with the same devices used for bombs, specifically: two hooks and four screws jacks acting radially and according to a 60° angle from the hooks. Screw jacks have two roles: traction on the hooks and side maintenance of the Pod to cancel any rolling motion.

- *Advantages*: Easy installation. Usage of an existing system.

- *Disadvantages*: Blocking instability in relation to axis X (longitudinal axis of Pod). Sliding hazard Coupling of low frequency (#70 Hz) bending and torsion modes.

Most of the vibratory energy generated on this type of equipment comes from turbulences that sit along the front edge and the skin. The following figures illustrate these points:

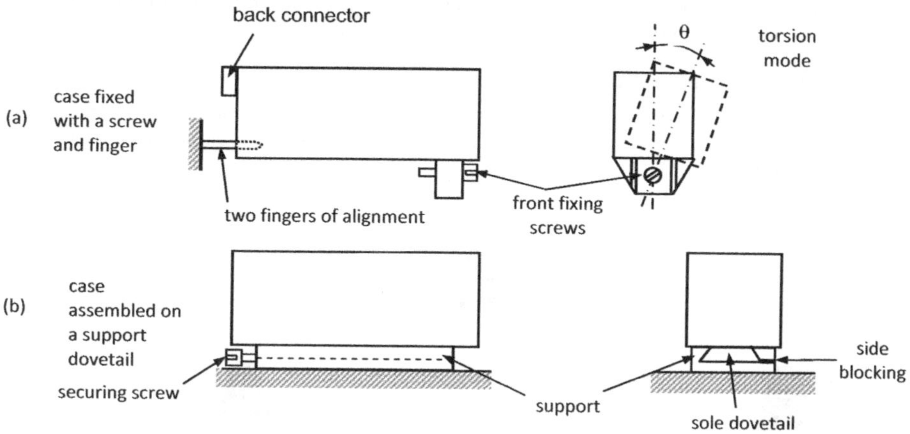

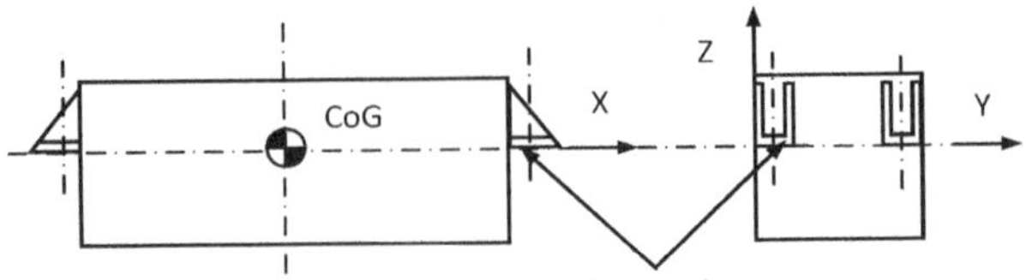

These rules, for example, can be found in the decoupling of electronic bays and seismic masses that serve to isolate the means of vibrations of buildings. With the following traditional configurations called "center of gravity assemblies".

h) Recommended configuration for electronic bays

The useful classic layout is four lower shock absorbers and two rear stabilizers in accordance with the following diagram:

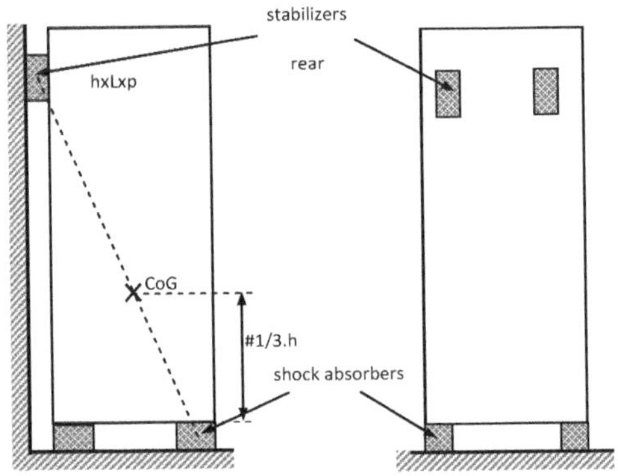

If possible, install the heavier items at the bottom of the bay (amplifiers, power supplies, transformers, etc.), so as to position the bay's CoG at about 1/3 of the height. Then, try installing the shock absorbers in a way that avoids coupling between axes. For this, align the axis joining shock absorbers and stabilizers at the far end with the CoG.

i) Recommended configuration for decoupling masses

The principle is the same. It is a CoG assembly type, with the following possible traditional diagrams; CoG is estimated by taking into account the means test weight and the average vibrated specimen weight.

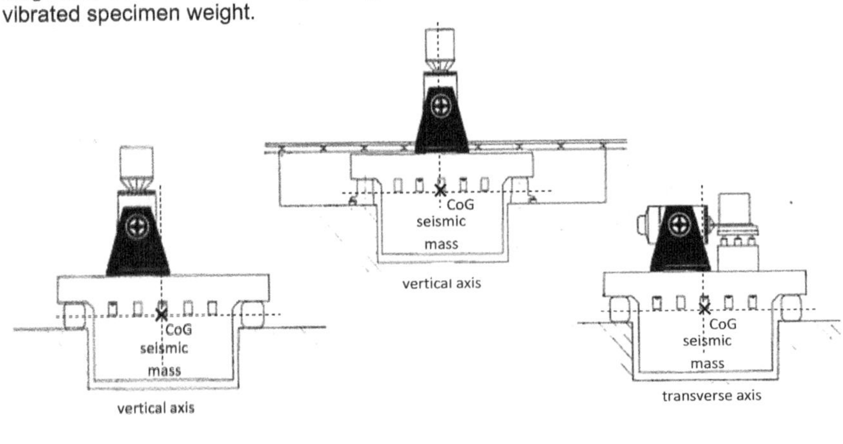

5.3.5.3. *Case design considerations*

a) General recommendations

The electronic case (or electronic bay) is the interface with the carrier that directly transfers vibrations onto the interior components from the excitations perceived on its support; it is thus a main structure. Therefore, it is necessary to provide special care to its design. It is thus advisable to analyze its dynamic behavior to identify the possible bending and torsion modes, which are very detrimental.

Indeed, torsion modes are often a source of problems in electronics components. These often forgotten modes can appear at relatively low frequencies and cause great deformations, and thus, problems.

A simplified preliminary analysis can often be carried out by regarding the case as a beam of relatively constant section and with a distributed mass, as long as the cross-section is relatively constant and if the case is supported at each end. The model used depends on the designer's experience and can give interesting orders of magnitude. Once the main modifications have been carried out, it will then be possible to refine, if necessary, with heavier calculations, such as finite element method (FEM) and modal analysis.

In general, most of the rules used for the cards can also be used for cases with some additional specifications, either:

- choose, insofar as possible, a "center of gravity" assembly type, or try to come as close as possible to it (minimization of the couplings between axes and couplings of torsion and bending modes), regardless of whether the structures are isolated by dampers or not;

- use high elasticity module materials to get the highest stiffness possible. For the record:
  - aluminum alloy modules: # 6,600 kg/mm^2 = 6,6 10^{10} Pa
  - magnesium alloy modules: # 4,500 kg/mm ^{2}= 4,5 10^{10} Pa
  - beryllium alloy modules: # 28,900 kg/mm = 28,9 10^{10} Pa
  - densities of these materials from 1,7 to 2,8;

- anticipate the greatest number of fixing points possible;

- if possible, make the cases in foundry;

- for the cases with ARINC format, format 3/4 ATR will be preferred rather than format 1/2 ATR (better fixing, weaker couplings);

- place the heavy components (transformers, optics, power supplies, power amplifiers, etc.), as low as possible and, at least, close to the edges or angles. In case of impossibility on an energetically ribbed partition;

- anticipate mounting the cases on thick and ribbed supports in order to limit high transmissibilities;

- reinforce fixings with flanges or gussets.

b) Case of sheet cases

- Increase sheets inertia with flanged edges, welded, screwed or riveted reported profiles.

- Carry out sheet foldings with curvature radii higher than these sheets' thickness. Any sharp edges are prohibited.

- Stiffen side panels with welded profiles or "sandwich"-type signs (sheet plates glued with epoxy resin). With this process, we can increase shock absorption.

- Fix sheets together with a substantial number of welding points, rivets or screws. Continuous welding is the best technique.

- Fix the lids, rigidified by flanged edges, by placing screws close to one another.

- Install reinforcement gussets in the corners.

- Rigidify motherboards whose stiffness intervenes in the case's one.

c) Case of molded cases

- Thicken and flange the structures supporting fixings.

- Rib or brace the cases' flanks.

- Anticipate fallen edges (peripheral thickening of the flanks and lids).

- Anticipate lines as thick as the panels, at least (large lines are desirable).

- Reinforce fixings by placing ribs on the sides of the cases.

- Fix the lids with a significant number of screws (parallelepipedic or cylindrical envelopes).

- Anticipate strongly ribbed foundries for secondary structures carrying heavy elements. Fix them to the main structures with a significant number of screws or bolts.

d) Means of fixing

This is a critical point that is often the reason for the observed damage and intervenes in material reliability. They must be selected based on their mechanical characteristics by exclusively taking into account the dynamic mechanical constraints. The general recommendations are the following:

- select the type of fixing in order and according to: environment, constraints, maintenance costs;

- correctly select their dimensions and positions according to the measured dynamic loads and the structures' geometry;

- anticipate a sufficient number of fixing points to better distribute stress and improve shock absorption;

- tighten screws and nuts with the recommended couples;

- choose quality steels (limiting resistance to at least 90 kg/mm^2);

- prevent any loosening of these fixings by using engraved discs, with teeth, onduflex, or by depositing Loctite on the nets;

- remove all screws or manufactured bolts, or any having sharp edges (hexagon head screws manufactured in 18/8 stainless steel for example);

- avoid using countersunk head screws, especially those that are cracked. Opt for cross-shaped head screws or other equivalents (phillys, torx, etc.).

5.3.6. Wiring

Wiring power outages are essentially translated into cut wires, short-circuits and bad contacts in the welds and the connectors. They intervene in the reliability of the equipment and, contrary to a faulty electronic card that can be fixed or changed, it is generally difficult to change, nay impossible. Hence, the card this subject needs. The following minimum rules must be followed:

- the strands must be locked alongside walls by bracing or by using a cable tie. The distance between these fastening devices should not be more than 5 cm;

- the insulated wires will be maintained using resin drops or self-adhesive fastening clips.

- the wires will be stripped with adapted tools so as not to damage the conductors and to avoid ruptures;

- the wire connections must also be cared for:
  - for welded wires: Retractable sleeves keeping the wire sheath and the contact together,
  - for crimped wires: Use systems with foots that pinch the sheath,
  - for Wrapped wires: The sheath must be coiled on the pin, at least on a length equivalent to that of a whorl. We can observe the good performance of the wrapping tin with regard to vibrations;
- Connectors must be rigidly fixed;
- The welding will be done with care by following the usual rules because they are often the cause of vibration failures.

5.3.7. Conclusion

We have scanned the main difficult points encountered during the equipment's design by taking into account any potential damage that may be caused by vibrations. Now, we have to implement provisions that allow minimizing risks under acceptable costs. This company-specific organization is an essential point that is a part of the *"quality assurance"* approach and an important issue in companies. Nor must we lose sight of the fact that the proposed rules generally fit into compliance with specifications that are always required to withstand traditional, mechanical, but also climatic and electromagnetic environments. It will thus be necessary to deal with potential conflict situations. Knowing that there are always delays and cost constraints, it is thus understood that the organization goes through the following:

- Raise awareness in the design offices about these too often ignored problems, which require minimal initial training, which can be internal in general, and an ongoing periodic training so as to maintain the skills;
- The implementation in these means of calculations and simulations adapted to the developed equipment (computer assisted design (CAD), FEM calculations, experimental modal analysis, simple software for calculations for shocks, vibrations, Electromagnectic Magnetic Compatibility (EMC), climatic, etc.);
- Publishing and implementation of design rules in all these fields as well as the follow-up and improvement of the latter according to changes in technologies;
- Sufficient test facilities to help develop the definition justification files) by allowing a traditional quality approach;
- Minimum expertise in all areas (hiring, outsourcing, training, etc.).

There are volunteer quality approaches involved in meeting these objectives, such as the ISO 9 000 (2 000 version) certification for the industrial structure and the NF EN ISO/IEC 17025 accreditation for testing and analysis laboratories, as well as all the reliability growth techniques.

Of course, it is advisable to adapt all these approaches to the needs and objectives of the company.

5.4. Study of a particular case – example of analysis of an electronic bay

The following example concerns the study of an electronic visualization bay that must comply with vibration and shocks mechanical constraints. The suggested approach puts emphasis on the aspect of vibrations and shock absorbers selection methodology [ref I18].

a) Context

Develop a new object, in a strong competitive environment, with a focus on aesthetics while satisfying operational constraints (ergonomics, environment).

b) Bay structure

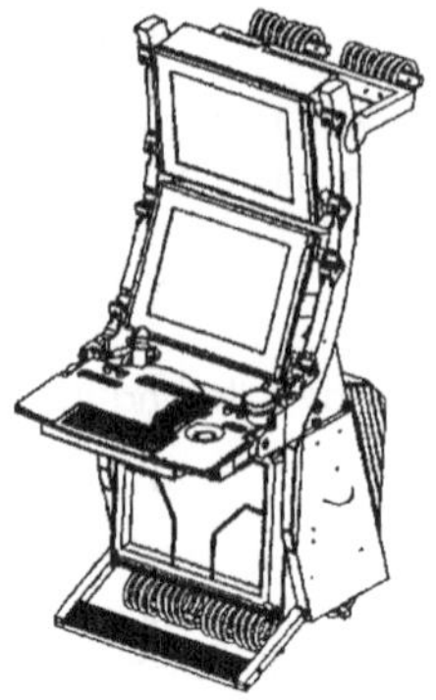

Figure 5.4. *Example of electronic visual bay*

c) Specifications

Vibrations: 1 mmc/c from 5 to 22 Hz and 1 gc/ from 22 to 50 Hz. Half-sine shock: 50 g 6 mt.

d) Criteria taken into account to define the shock absorbers

- Increased shocks or vibrations?
- Installation constraints (suspended floor, cradle, high fixation points, etc.)?
- Ergonomic constraints?
- Is there a decoupling in the system already?
- Is there a maximum residual constraint?
- Do we have control on the final mass (shelves, etc.)?
- Are the specifications clear (levels, duration, etc.)?
- Are the structure's first own modes known?

The choice of dampers is the responsibility of the design engineer that validates its design from the following results:

- chart of the CoG's residual acceleration in the three axes;
- chart of the CoG's displacement in the three axes;
- chart of the force-deformation in the connections;
- static crash;
- coupling effects (rotations, residual flexing);
- resonance and power surge (damping system mode);
- displacements at critical points (search for abutments).

Generally, this is not a service provided by damper suppliers.

e) Calculation means used

- SYMOSW software: Analysis of the dynamic behavior of mechanical systems subject to shocks or vibrations. The model is built using solids considered not deformable, and connected by links (springs, dampers, etc.). The DOF number is important, as is any direction of the shock absorbers.

- Treatment of linear and nonlinear problems.

- A postprocessing module allows for a graphic visualization of the model's responses and the model's 3D animation.

- Results may be subject to additional analysis (spectra of response to shocks, Fourier integrals, etc.).

SYMOSW is composed of four modules.

(a) Model:

- solid: 1–6 DOF;
- linear springs;
- nonlinear springs;
- understanding the "force–displacement" laws and creation of libraries;
- dampers that are viscous, dry, quadratic, etc;
- joints, ball joints, contacts, games, etc.

(b) Loading:

- static: torsor of applied forces. Imposed displacements;
- transitional: preprogrammed shocks, time signals input, library creation;
- harmonic: random sine vibration, rotating forces, oscillatory couples, etc.

(c) Analysis:

- modal;
- static: linear and nonlinear;
- transitional: linear and nonlinear (large rotations on the whole model);
- harmonic;
- shock response spectrum;
- Fourier integral.

(d) Results:

- chart of time or frequency responses (displacement, acceleration speeds) in any point of the solid, efforts in the connections, 3D animation of the model.

SymosW is also an excellent didactic tool.

f) Data to prepare for the under test equipment

- Geometry: calculation of inertias and coordinates of the CoG.
- Mass.
- Type, position, orientation of the dampers.
- DOF.
- Counter-mass and decoupling.
- Static stress.
- Shocks and vibrations of entry: profile, application direction.
- Type and calculation window.

And some subtleties: "sharing" dampers, counter-mass' "lift" force.

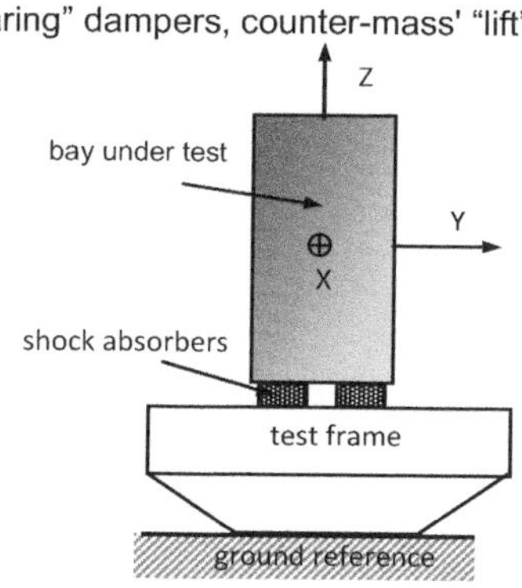

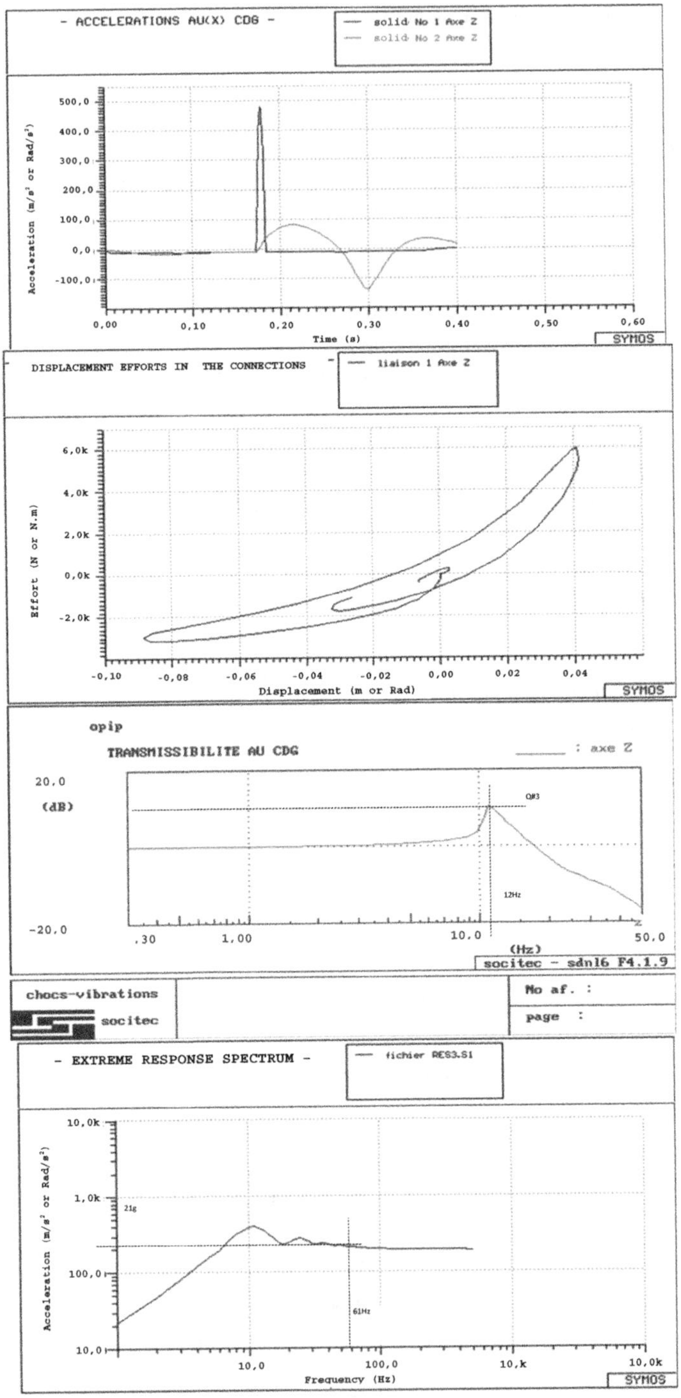

Figure 5.5. *Calculation of the response of the bay's CoG with SYMOS*

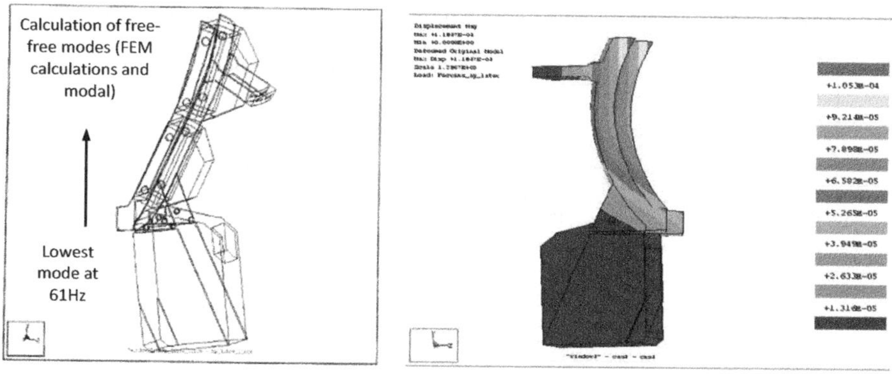

Figure 5.6. *Bay FEM calculation of the mode at 61Hz and 21g*

g) Identified difficult points

- Risk of defaults in the absence of high clamping (wish of the specifier).

h) Method used

- Definition of the applicable constraints (shocks, vibrations).
- Dimensioning of the shock absorbers (residual, ergonomics).
- Dynamic modal analysis of the structure (free–free).
- Calculation of the SRS of the residual (at the CoG of the console).
- Static calculations are equivalent (distributed acceleration, L/L structure).
- Verification of the vibration stress: residual calculation.
- Static calculation equivalent to f_R and fatigue impact.

i) Conclusions

- Shocks: In conclusion, the equivalent static loading (21 g at 60 Hz) shows in the following figure (FEM calculation) that the shock fatigue levels are held for the lowest dominating mode.

- Vibrations: The most important response level is at 12 Hz, with an acceleration given by:

$$a[g/c] = \frac{f^2.d[mmc/c]}{500} = \frac{12^2.1}{500} = 0.288 \text{ gc/ that is with Q \# 3, to \# 0.864 gc/}$$

The lowest modes of the internal components largely comply with the rule of the octave and are located well above 24 Hz.

The bay's endurance test at this frequency shows that the equipment does not tire (the excitement level and the test duration take lifespan into account).

j) Example of dimensioning of structures to random vibrations by equivalent static analysis (performance in robustness) [ref. I1]

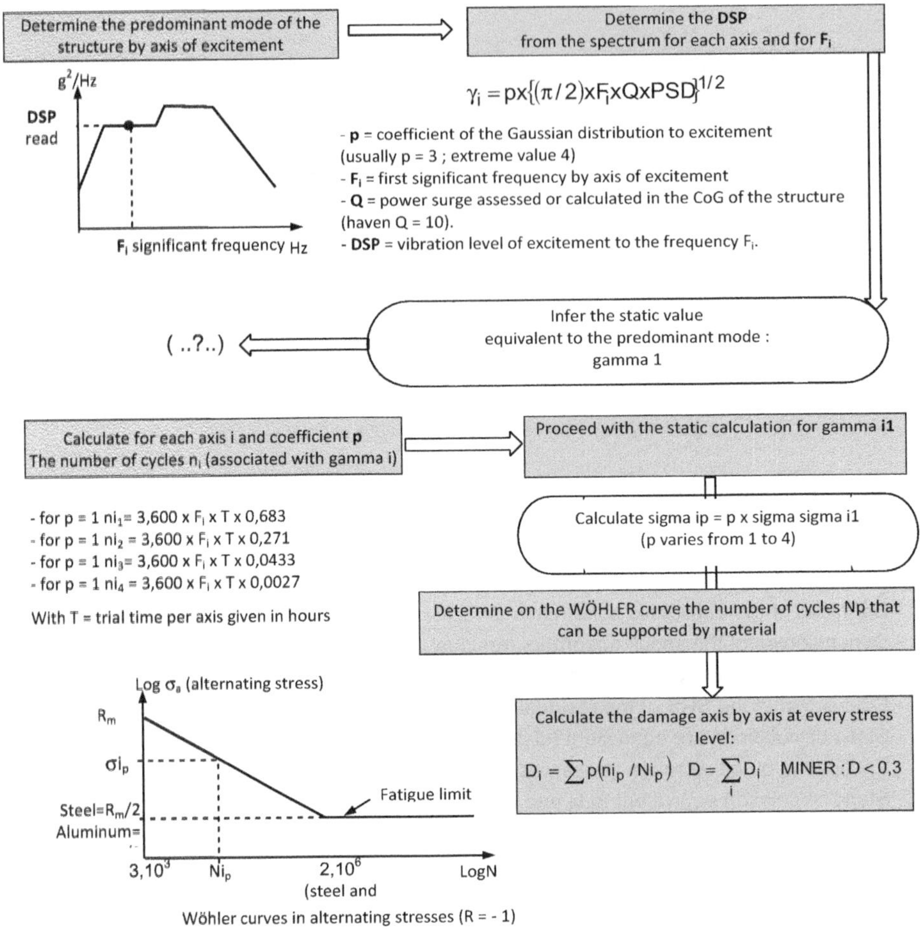

Figure 5.7. *Example of dimensioning of structures to random vibrations by equivalent static analysis (performance in robustness)*

6

Controlling Generators of Vibrations and Shocks

This chapter outlines the main control techniques for testing shocks and vibrations using modern digital consoles. Several methods and rules will be put forward, including their advantages and disadvantages.

Our aim is to present the operating principles and techniques and, with this information, analyze the optimal conditions for conducting the tests. As is always the case when using relatively closed computer systems, there are more or less correct possibilities for simulation of environments, which directly impact the quality of the test and risk abnormal deterioration of the equipment.

Of course, the fundamental principles for processing control algorithm signals are complex but these are relatively well known and are more or less unveiled by each provider, despite each of these having their own techniques that are sometimes confidential and patent protected. It is therefore essential for them to define the concrete points that we will discuss, even if they do not generally delve into the details of algorithms, at the very least mentioning the qualification tests. This is also why we cannot further discuss these techniques.

6.1. General principles

Operating the control consoles is based on the following principles.

a) Concept of an open loop

Model used. Calculation part of H(f).

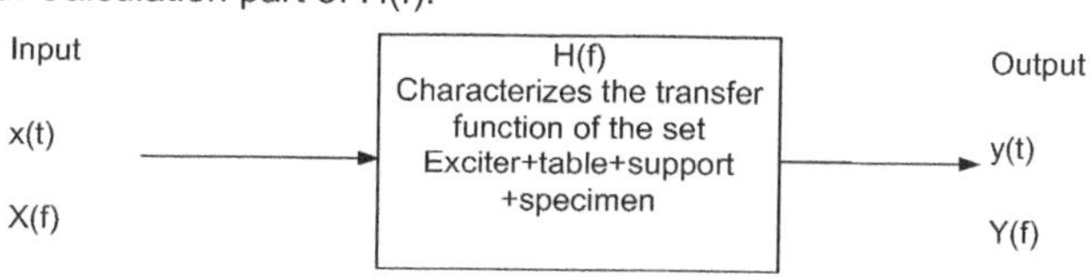

Figure 6.1. *BO model*

- X(f) = TF of x(t).
- Y(f) = TF of y(t) which must therefore be equal to the specification reference.

This principle involves calculating H(f) by a pretest, then $X(f) = \frac{Y(f)}{H(f)}$ and then the control signal x(t) by $x(t) = TF^{-1}[X(f)]$. One can immediately note that a complexity arises due to the estimation of H(f) and the calculation of X(f), which require a division that may contain zeros.

b) Concept of a closed loop

Model used. Control, calculation and correction part of x(t).

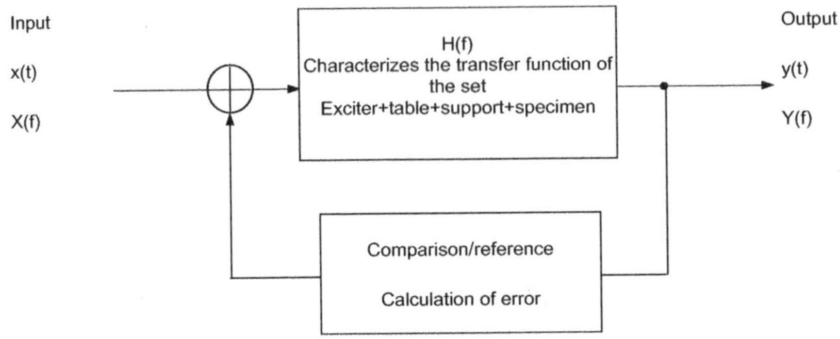

Figure 6.2. *BF model*

Control problems occur because during testing due to the nonlinearity of fixtures and possible changes in H(f), we get: $H_{pre\text{-}test}(f) \neq \Delta\, H(f)_{tests}$.

The goal is then to act on the input x(t) in case of change in the value of H(f).

Of course, here we feel the need to estimate the settings in terms of averages and statistics with concepts of associated errors. We will therefore generally have to work with a given time loop.

c) Typical problems

These include:

- the need to generate various types of complex signals:
  - sinusoidal: fixed; scanning: linearly, logarithmically, by frequency
  - noise: real, pseudo
  - shocks: known mathematical form; on the shock spectrum; created by breaking points (network of lines)

Hence, the difficulty in finding general algorithms;

- calculating the transfer function: two techniques:
  - through pretests (particularly for shocks and therefore pseudo control),
  - closed loop calculation (particularly for vibrations and therefore real time control);
- analyzing feedback signals:
  - sine: specific analysis
  - noise: Fourier analysis
  - shock: Fourier analysis + shock spectra (and transitional)
- correction algorithms, with the following common problems:
  - error calculation,
  - correction percentage,
  - correction speed.

With, of course, the intervention of the overvoltage coefficients seen in the analyzed frequency band, which affect control and the usual compromise: speed/accuracy.

6.2. Typical configuration of the equipment used

A classic block diagram is shown in Figure 6.3.

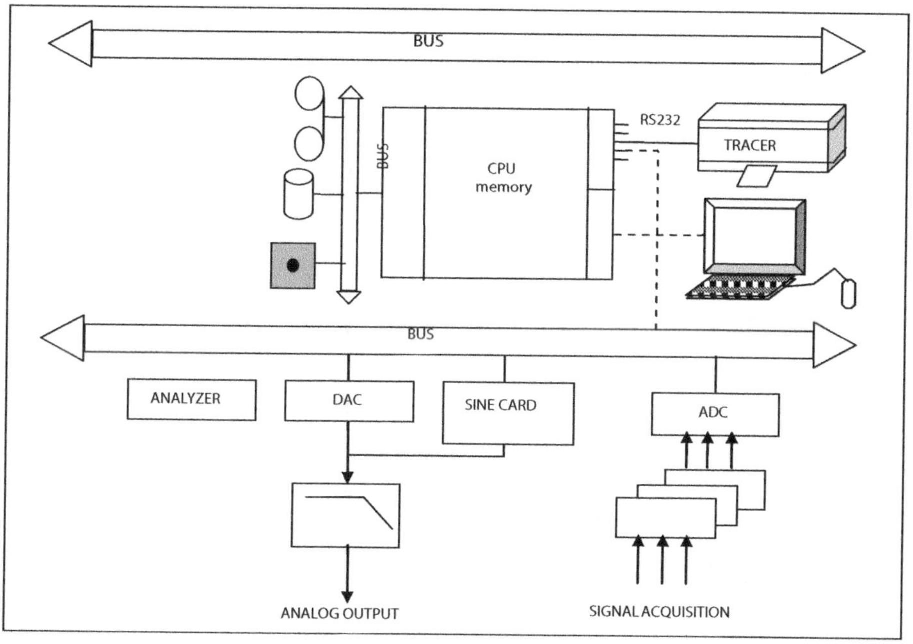

Figure 6.3. *Classic block diagram*

In other words:

- a keyboard:	input of settings
- a calculator/memory:	calculation of signals
- an analog/digital converter (ADC):	input of signals
- digital/analog converter filters (DAC):	output of signals
- a calculator + software + analyzer:	analysis of signals
- software:	corrections/control
- a graphical interface:	real-time visualization
- tracers/printers:	restitution of curves
- hard disks, Floppy disks, cartridges, tapes:	back-up and management of tests

6.3. Traceability of tests

Assistance in the development of test reports is always proposed. A folder is created that includes all the relevant information describing the test. A summary of the various stages is given below.

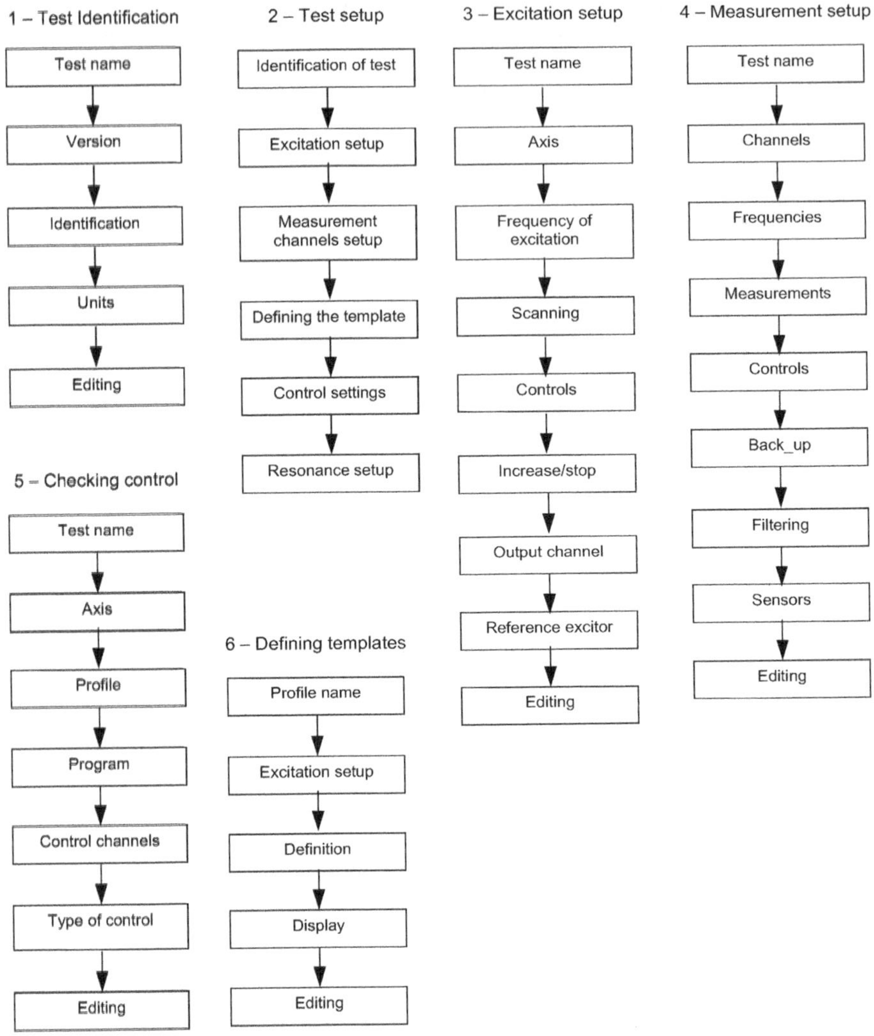

Figure 6.4. *Test identification settings*

6.4. Control in sinusoidal mode

6.4.1. Generalities

The signals on digital consoles are obtained by sampling and quantization; they are constructed in the shape of filtered staircase signals. Their distortion is based, among other things, on the number of points necessary for their construction (see Figure 6.5).

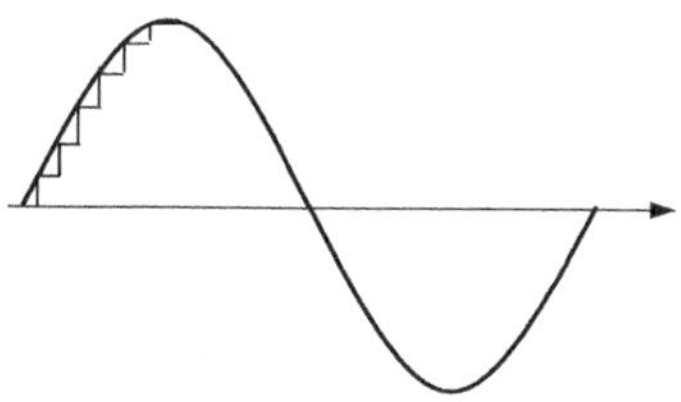

Figure 6.5. *Streamlined appearance of a reconstituted sinusoidal signal*

Meaning that:

- a large number of points for a good sine wave is needed, hence a long calculation time;
- the effect of sampling is attenuated by the output filters, but it is not deleted. Therefore, there will always be some distortion.

In general, a specific electronic card is used, which calculates the signal and converts it into an analog signal (filtering, sampling, etc.).

6.4.2. Generating a signal

Conventionally, the central system sends the calculated values to the generation card (depending on the scanning program, the velocity, the acceleration, etc. of the control signal and level error) for the following settings:

- frequency;
- amplitude;
- phase (in the case of a phase-lock).

The card calculates all the points of a period and carries out the analog conversion.

6.4.3. Signal acquisition

The principle includes that the number of acquisition points is constant for each period, and thus the sampling is not constant as a function of the period (or frequency), which optimizes the number of calculations and therefore the calculation time for each period.

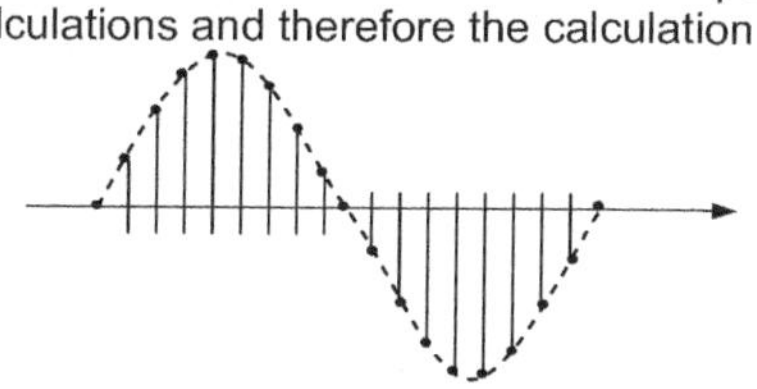

Figure 6.6. *Sampling example of a period*

Figure 6.6 is a non-constant sampling depending on frequency.

Example 6.1: If one chooses 16 points per period, for example, Shannon's theorem shows that we can "analyze" or "detect" until harmonic 7.

6.4.4. Measurements

Conventionally, the different types of sine measured parameters are:

- peak value;
- average value;
- effective value;
- filtered value.

Processing only occurs over one period or at most over a few periods.

a) Analysis of the peak value

For any entire signal analysis period, we seek the maximum value obtained.

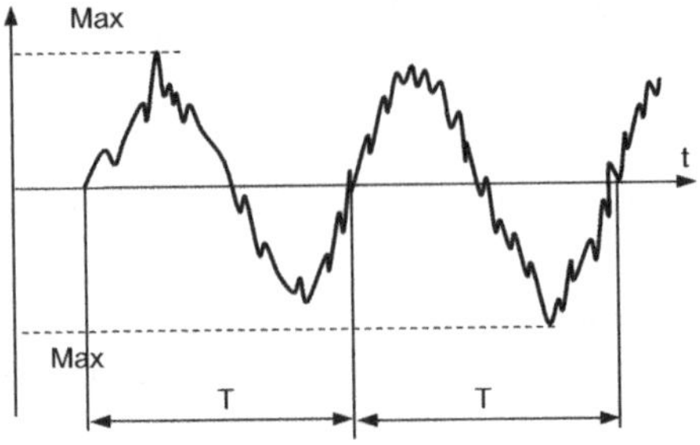

Figure 6.7. *Example of a distorted signal*

b) Analysis of the average value

If there are N points per period, the calculation gives:

$$V(\text{average}) = \overline{V} = \frac{1}{N}.\sum_{k=1}^{N}|y(t_k)| \tag{6.1}$$

We then calculate the peak value from the average value, which gives, for a sinusoidal signal:

$$V(\text{peak}) = \text{Vc}/ = \overline{V}.\frac{\pi}{2} \tag{6.2}$$

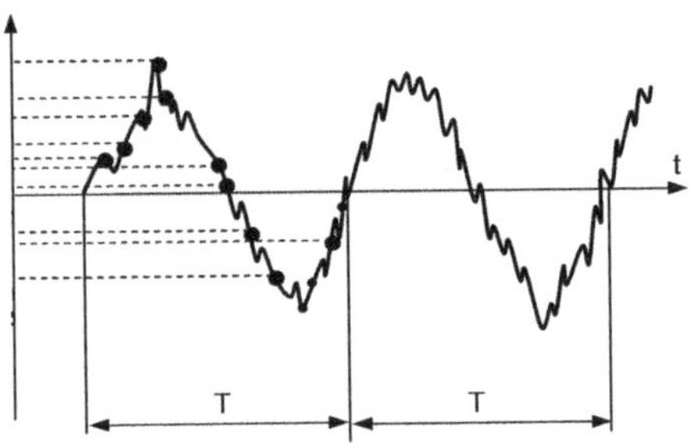

Figure 6.8. *Calculation of the peak value from the average value*

c) Analysis of the RMS

From the same curve and the same points N per period, we calculate:

$$V_{eff} = \sqrt{\frac{1}{N}\sum_{k=1}^{N} y(t_k)^2} \tag{6.3}$$

We again find the peak value from:

$$V(\text{peak}) = V_{eff}.\sqrt{2} \tag{6.4}$$

d) Analysis of the filtered value

The principle of measuring the filtered value is to calculate the correlation of the signal with the reference, over a period. In fact, this procedure allows us to:

- eliminate continuous values;

- eliminate harmonics;

- "but not eliminate noise".

The block diagram of the calculation is as follows:

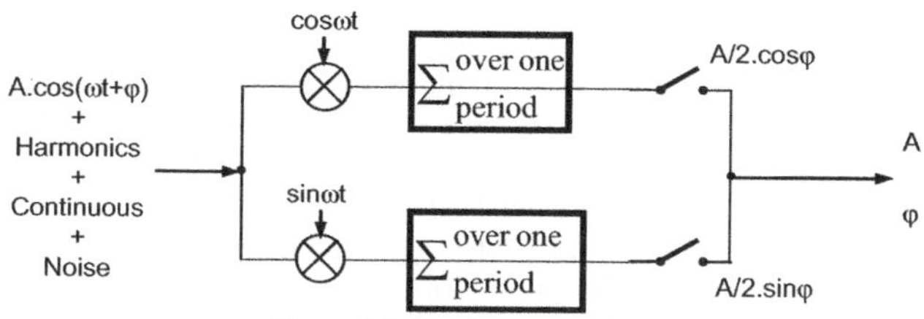

Figure 6.9. *Block diagram of filtering*

The products of the input signal with cosωt and sinωt restore a sum of signals with double frequencies of ω and cosφ and sinφ. This sum is then low-pass very low frequency (VLF) filtered (digital integration), which restores the A/2.cosφ and A/2.sinφ signals and the module then gives the values of amplitude A and phase φ (calculation module and calculation of the phase). In other words, the sinusoidal signal is reconstructed at pulse ω.

e) Possible conventional processing and strategies

This is a very important part in the operation of digital consoles, which directly impacts the quality of the vibration test. Indeed, depending on the adopted control strategy, there may be very different behaviors of the testing equipment, with potentially significant impairments or perhaps without problems. It is a point that must be carefully considered in cooperation with the test-requesting client. Therefore, we usually get:

- *an average for a control sensor*: this is a procedure carried out over several periods that eliminates noise on the control signal (assumption of white noise with Gaussian distribution with zero mean). This improves control;

- *average over multiple sensors*: the same sort of strategy as above but applied to all sensors, with the possibility of controlling relative to the average value calculated from measurements done by the different sensors. This strategy is used when the fixtures are relatively large and it is difficult to ensure homogeneity of vibrations within the scanned frequency band. We noted, in fact, that when the frequency was high, the propagation or resonance phenomena in the test fixture made it impossible to obtain a level of acceleration that was identical at all points in the tested specimen. It is also possible to weight each averaged control sensor in order to take attenuations or amplifications known in advance into account. This strategy can be risky. Indeed, let us suppose that the level to control is 2 g and we wish to control eight points on average (usual maximum possibility of control devices). We carry out the following calculation:

$$\text{average control target value 2g} \quad \text{measured value}: \ \gamma = \frac{\sum_{k=1}^{8} \gamma_k}{8}$$

The system regulates the level of excitation in order to meet the condition $\gamma = 2$ g. Let us suppose then that we are in a worst-case scenario, where seven out of eight sensors see very little or no value at the considered control frequency. The system will therefore be seeking level 2,8 = 16 g at point 8 in order to respect the set rule, which is of course invalid. We must therefore handle this possibility carefully and, for example, associate it with the following strategies. In general, an average of four control points is sufficient. In theory, this is the most realistic strategy if we want to vibrate a structure in the most homogenous manner possible and best meet the specification demands, namely vibrate at such and such level, but we have seen that it can overtest the material in particular vibration conditions;

- *stronger or weaker control*: the system chooses to control based on the strongest or weakest control sensor (at least two are required, of course). Of course, control based on the weakest is avoided and is never used because it is too dangerous (with the inevitable presence of antiresonances). Control based on the strongest, called "Notching", protects a point or a component of the testing equipment that cannot stand a certain amount of vibration. This strategy is often used when one controls across several points, but it generally undertests some selected control points. In order to maintain the desired level at a point that is experiencing overvoltage, the system is forced to decrease the levels of other points so as to respect the set; hence, the possibility of undertesting in this strategy.

It is then possible to optimize the "control strategy" approach by combining the effect of averaging and control based on the strongest. Thus, if one has eight control inputs on the console, we will assign four to average controls and four to the strongest controls in parallel. We will essentially be controlling based on the average, which is more representative, and the set of the strongest with a possibly slightly higher value than the general target allows us in order to avoid the inconvenience of overtesting as mentioned above.

6.4.5. Corrections

Correction procedures are based on the measurement of error from a setpoint, as shown in Figure 6.10:

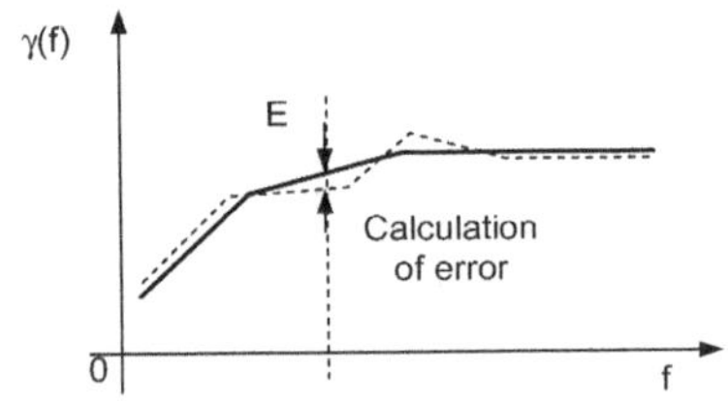

Figure 6.10. *Control error model*

where: E(error) = (reference value) – (feedback value)

The feedback value is obtained from measurements and calculations carried out on one or more controls and depends on the chosen strategy.

a) Error correction

There are usually several possibilities including:

- choosing a percentage of error to be corrected, with the option to change this setting according to the actual frequency;

- changes can be done manually;

- changes can be done through preprogramming.

b) Consideration of the transfer function

Of course, the correction is done for feedback signal measurement, which takes into account the shape of the transfer function. The goal is, for all changes of frequency f_k to f_{k+1}, to develop a control signal that is calculated from the transfer function corresponding to the frequency f_{k+1}. This can be done in two ways:

- calculation during an estimation pretest of H(f);

- calculation during a scan.

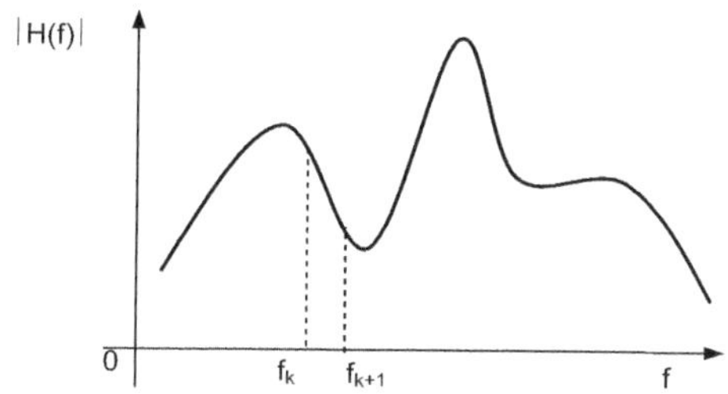

Figure 6.11. *Simplified example of a transfer function*

c) Potential weighting

Calculating the new "drive" (electric signal of the final control amplifier that controls the vibration generator) can be done through weighting and averaging the two calculations. These operations allow us to optimize control according to the overvoltage obtained in the different phases of resonance, where:

- D_1 = *drive* modified according to the measured and calculated error (weighting P_1);
- D_2 = *drive* calculated from the transfer function (weighting P_2).

With the following calculated result of drive to k+1:

$$D_{k+1} = D_1.P_1 + D_2.P_2 \tag{6.5}$$

6.4.6. Safety devices on controllers

Of course, in case of excessive setpoint overrun, there are alarms and security policies in place to avoid problems due to breakages.

a) Alarms

- These consist of + and – levels relative to the nominal set level.

- These levels can be different depending on the frequency and they are displayed on the console screen.

- If these levels are exceeded, a visual and audible signal is triggered.

b) Aborts or stops

- These are also limited by + and – from the nominal set level.

- They can also be different here depending on the frequency and are also displayed on the console screen.

- However, if these levels are exceeded, the test stops, usually with a "soft" decrease in the level to avoid excitation HF of the equipment.

6.4.7. Safety devices on the measuring channels

The concept of Notching, mentioned above, is used for measurement channels. The principle is the same: a measurement point acts with the same basis on the control set as on the control point (hence the confusion) and at the considered point, the system strives to maintain a different and specified set level. So for the Notching strategy, this is done:

- either on the control channels;
- or on the measurement channels.

The graphs in Figure 6.12 illustrate this.

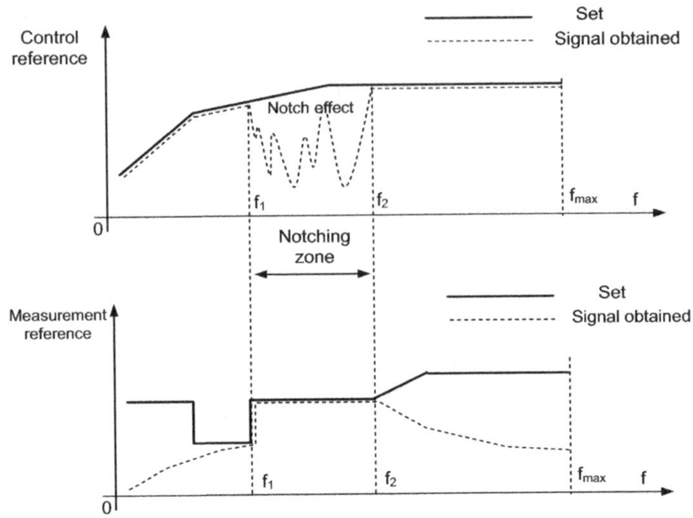

Figure 6.12. *Example of notching*

In case an alarm channel profile is exceeded, several steps could occur:

- alarm,
- stop,
- notching (or notch in the control or measurement profile).

6.4.8. Back-ups

The traceability of the test requires safeguarding the relevant test information. In general, the key information is automatically saved. This includes:

- for the drive: the transfer function;
- for the control channels: for each scan, backing up:
  - the frequency response curve for each control channel,
  - the frequency response curve of the control signal;
- for the measurement channels: for each scan, a back-up procedure performed on the measurement channel:
  - peak value,
  - average value,
  - effective value,
  - filtered value.

With the usual types of back-up:

- manual,
- automatic,
- preprogrammed.

6.4.9. Deferred time

Information restoration work can be done in deferred time, with:

- restoration of backups. Frequency response curves:
  - pilots,
  - control,
  - measurements;
- deferred procedures:

- time back-up,
- frequency back-up. Transfer functions between channels if backups are the *filtered* response curves, which give the module + the phase relative to the reference (or all the information).

6.5. Random control

6.5.1. Signal generation

There are several techniques that can be used to generate a PSD from a formed noise spectrum, which is generally a summary of the measurements carried out in the field during the life profile of the material to be tested. These often generate a PSD block of the spectrum to be created from white noise with Gaussian distribution, which is concatenated in order to restore the overall PSD. Other techniques also use inverse Fourier transforms of the same blocks to restore the PSD of each block and then the overall PSD.

a) Defining the reference

This is the PSD control spectrum as shown in Figure 6.13. Classical shape, with:

$$\Delta f = \frac{f_{max}}{N \text{ points}}$$

Calculation of the RMS value: $|g_{ref}| = \sqrt{PSD.\Delta f}$

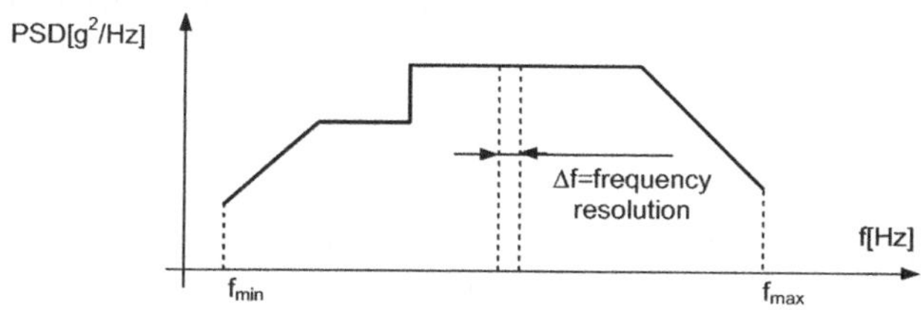

Figure 6.13. *Example of PSD*

This gives the module in effective g of the reference in the Δf band.

b) Building a calibration block

The system builds a standard noise block that will be used to form the spectrum of the desired PSD. Several techniques are possible for this.

(a) Analog

In the first analog control systems, thermal noise existing in a Zener diode was used and was polarized at its bend, as shown in Figure 6.14.

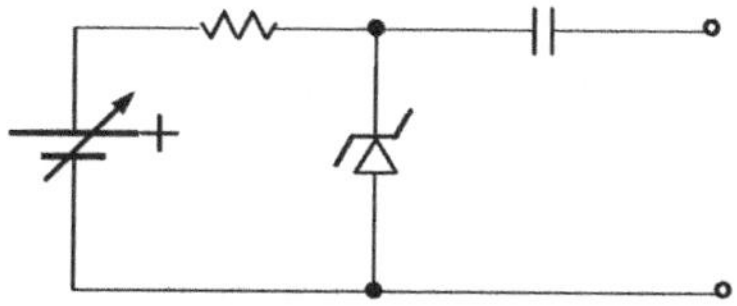

Figure 6.14. *Example of an analog BB generator*

This produces white noise (BB = uniform PSD) with a more or less Gaussian distribution. One can also use the properties of the central limit theorem by combining several white noises (at least 20) to approximate a Gaussian distribution more accurately (which is a consequence of the central limit theorem).

Other analog techniques use the jitter phenomenon caused by temperature of the phase of the sinusoidal oscillators or the metastability of a flip-flop in digital.

Next, these signals are processed using optimized Δf filters in order to generate the right effective level in each band based on the spectrum of the requested PSD.

The idea here is to have a purely random signal that is quite difficult to set, which is why these digital control consoles have been developed.

(b) Digital

In digital systems, everything is done through calculations using known relationships in order to process the input–output signal through the Fourier transform, fast Fourier transform (FFT), autospectrum and interspectrum techniques. The methods for generating random signals with Gaussian distribution that are capable of forming a PSD are from [ref. I1 Volume 3].

First of all, let us mention some reminders. Calculation techniques for white noise with Gaussian distribution use random functions built from random uniform variables that feature completely independent products (just like white noise, for which the autocorrelation function is then a Dirac, bearing in mind that practical white noise is of course limited in the bandwidth where PSD can be considered constant, a uniform law that is therefore used when nothing is known).

If r is a random variable (RV) of the uniform law on interval [a, b], its probability density is given by:

$$p(r) = \begin{cases} \dfrac{1}{b-a} & \text{for } a \le r \le b \\ 0 & \text{elsewhere} \end{cases}$$

In this case, we show that if $r \in [0, 1]$, then RV $y = a + (b - a).r$ is also uniform in the same interval with average $(b-a)/2$ and standard deviation $\sigma = (b-a)/2\sqrt{3}$.

This feature is used to generate white noise with a given Gaussian distribution of standard deviation and in a given block Δf, the right effective value.

1) Sum of sinusoids with random phases method

If the random time signal that generates the PSD of the maximum frequency f_{mx} is of duration T, one must:

- determine a time interval (step) given by: $\delta t = \dfrac{1}{f_{ech}} = \dfrac{1}{k.f_{mx}}$ (usually k = 2,6);
- choose a number M of the PSD definition points (power of 2 in order to use FFT techniques);
- define the number of calculation points of the time signal: $N = T/\delta t$.

NB: Bearing in mind that it is possible to change the parameters M and N according to the PSD to be formed, but keeping a statistical error ε_0 given by: $\dfrac{M}{N} \le \dfrac{\varepsilon_0^2}{2}$;

- define the frequency interval between two points for calculating PSD: $\Delta f = \dfrac{f_{ech}}{2M}$;
- for each point M of the PSD, it is then necessary to calculate at each time point $t = k.\delta t$ (k is the integer from 1 to N), the sinusoids with random phases of duration T with:
  - shape: $x_m(t) = x_{m_{max}}.\sin(2\pi f_m t + \varphi_m)$ where $f_m = m.\Delta f$ $(1 \le m \le M)$;

- amplitude: $x_{m_{mx}} = \sqrt{2G(f_m)\Delta f}$ where $G(f_m)$ is the value of PSD for f_m. The amplitude of the sinusoid is equal to twice the RMS value $G(f_m)\,\Delta f_m$;

- phase φ_m is a generated random phase (BB) for a distribution probability of instantaneous values of the signal as specified, which is generally Gaussian;

- then carrying out the sum of M sinusoids at each time point t.

This is a fairly large number of calculations.

The φ_m phase is obtained, for example, by the Box–Muller method, which generates RVs following a standard normal distribution (it has two forms, the basic form and the polar form).

Basic form: two independent RVs, U_1 and U_2 uniformly distributed in the interval [0, 1], the variables expressed in the following equations are then independent and follow a standard normal distribution [ref. r9] and [ref. I1]:

$$\varphi_m = \sqrt{-2\log(U_1)}.\cos(2\pi U_2) \qquad \varphi_m = \sqrt{-2\log(U_1)}.\sin(2\pi U_2)$$

Polar form: these are two independent RVs u and v are uniformly distributed over the interval [–1, 1]. Upon calculating $s = u^2 + v^2$, if s = 0 or s > 1 then the couple is rejected (it must remain within the unit circle). Otherwise, the couple is accepted and the variables of the following relationships are independent and follow a standard normal distribution:

$$Z_0 = u.\sqrt{\frac{-2\log(s)}{s}} \qquad Z_1 = v.\sqrt{\frac{-2\log(s)}{s}}$$

The polar form is faster than the basic form. Calculations are performed using tables for the functions used.

For digital and discrete, obtaining uniform RVs is done using software or hardware because of a linear feedback shift register, for example. This comes from techniques for generating binary maximum length sequences, which allow us to obtain random digital words (those that exist in the registry) with uniform distribution. These techniques are also used in error detector and corrector coding in telecommunication and encryption [ref. I19, Volume 5]. An example is given in Figure 6.15.

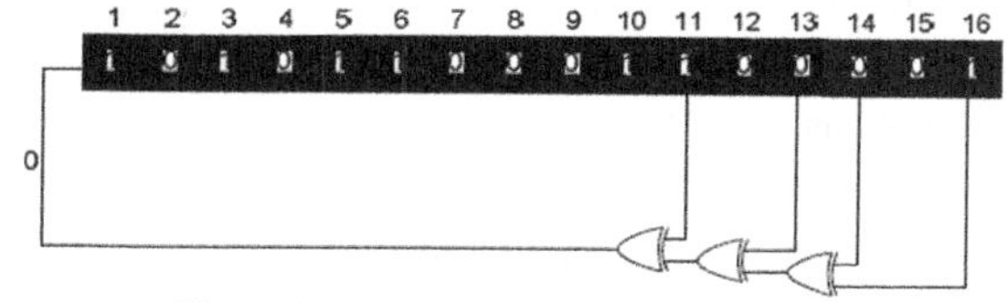

Figure 6.15. *Example of Fibonacci type LFSR*

Another technique for calculating the signal from a simpler and faster PSD is possible [ref. I1] by:

- selecting a number of blocks;

- time step: $\Delta t = \dfrac{1}{2.f_{mx}}$;

- as here we want to do an FT from a PSD that is itself calculated by direct FT, we require the amplitude and phase or the real and imaginary parts, meaning we need to calculate the real and imaginary parts. As the PSD is obtained from the squares of the amplitudes of the FT, we must consider the square root of the PSD [g] in order to create the real and imaginary parts, meaning that:

PSD_R(i) = $\sqrt{DSP(i)}.\cos(\varphi) = \sqrt{g(i)}.\cos\varphi$

PSD_I(i) = $\sqrt{DSP(i)}.\sin(\varphi) = \sqrt{g(i)}.\sin\varphi$

The index i corresponds to the actual point for n points of the starting PSD, with: $\varphi = 2\pi \frac{\text{generation of a random number}}{32{,}767}$ where, here, the random phase is obtained from a random number generator software that gives a value between 0 and 32,727. For example: Phi = 2,0*PI=rand()/32,767.0; // Random phase;

- calculating the inverse Fourier transform from PSD_R(i) and PSD_I(i);

- the acceleration signal is the one obtained with the inverse transform divided by $\sqrt{2\Delta f}$ (Δf being the PSD step).

This process is repeated in order to build the signal by concatenating the “s” signals thus obtained.

2) Inverse Fourier transform method

In this method, the PSD of the random signal x(t) of duration T is calculated from samples of K blocks of duration ΔT and by averaging the PSD of these blocks. For a given block, PSD is given by:

$$G(f) = \frac{2}{\Delta T}|X(f)|^2$$

when X(f) is the FT of this block.

Conversely, if we provide the PSD G(f), it is possible to calculate a random signal respecting the PSD from the following:

- calculate the sampling frequency $f_{ech} = 2f_{mx}$ and time step $\delta t = 1/2f_{mx}$ from the maximum frequency f_{mx} of the PSD;

- calculate the signal block of duration ΔT that is obtained by FT^{-1} from the frequency interval (step) Δf of the same PSD;

- multiply PSD G(f) by $e^{j\varphi}$, where φ is a random phase;

- calculate the FT of this time block using the relationship: $X(f) = \sqrt{\frac{\Delta T}{2} G(f).e^{j\varphi}}$;

- calculate the FT^{-1} of X(f) to obtain a complex time variable of duration ΔT. The useful signal x(t) is the real part of this complex value;

- verify the validity of the obtained signal by comparing it with the original PSD (given by a statistical error ε) by generating a signal with a total duration of $T = \frac{1}{\varepsilon^2 \Delta f}$, or in other words, repeat K times the last three steps of the calculation (where K is the integer increment of $T/\Delta T$).

We then get the following signal types:

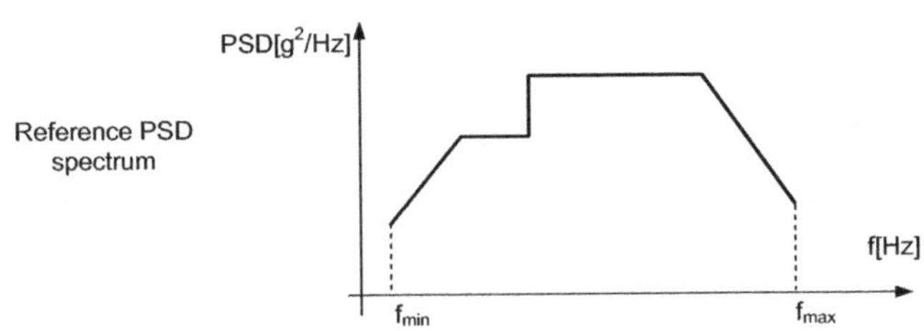

Random phase with its Nyquist plot

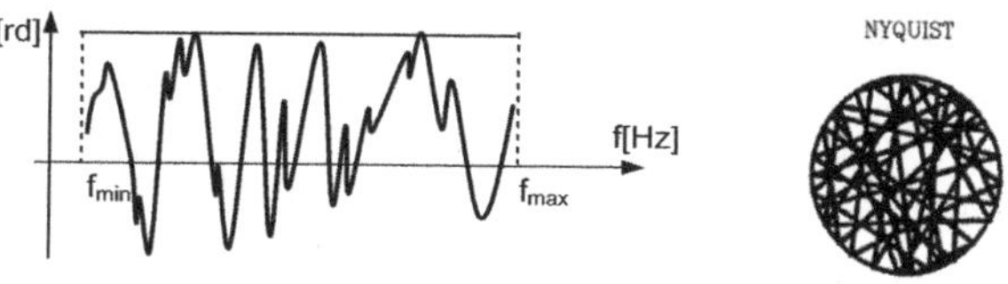

Random time calibration block

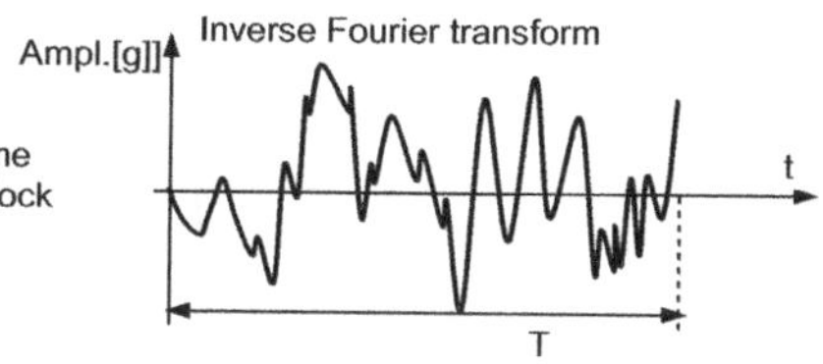

c) Generation of pseudo-noise

We start with a calibration block.

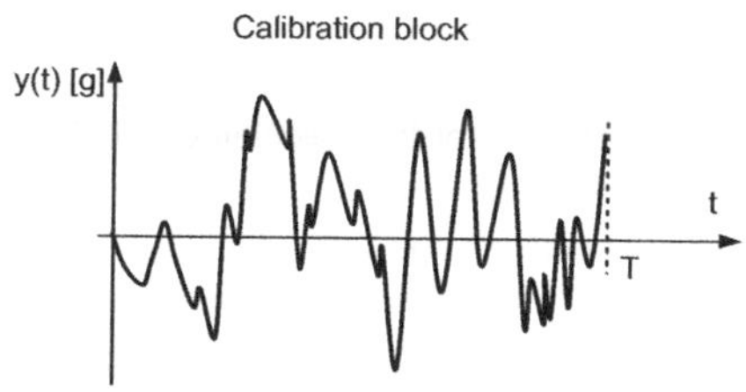

This calibration block is concatenated:

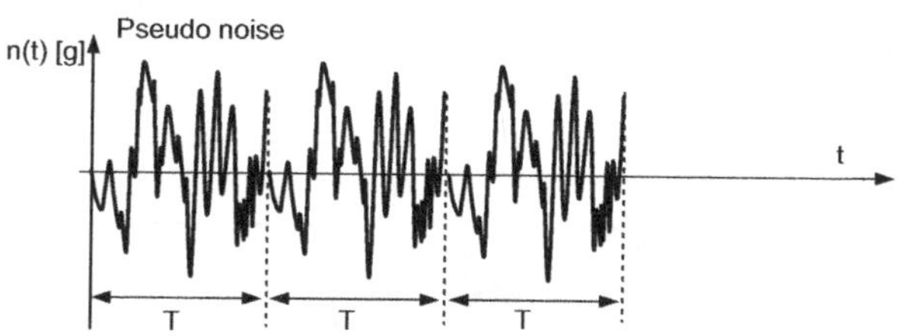

- the block is repeated equal to itself;
- it has the required PSD properties:
  - random phase;
  - required level;
- there is no energy introduced between two FFT calculation points;
- continuous noise is obtained using weighting windows (Hanning, Hamming, Blackman, etc.) for passage from one block to another in order to limit abrupt transition problems (generating HF and shocks) and secondary lobes.

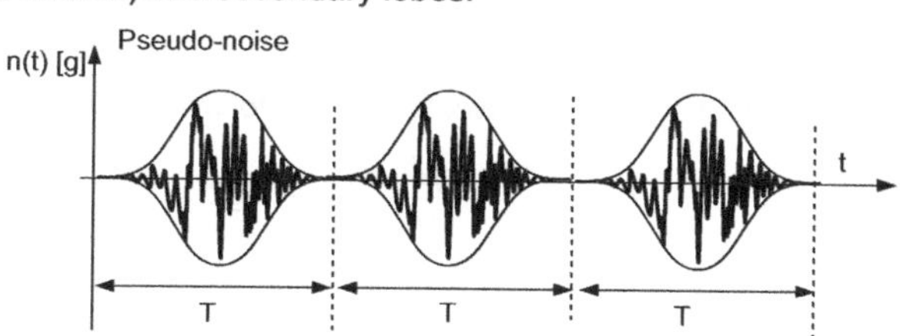

d) Generation of real noise

The procedure is the same but the blocks are different:

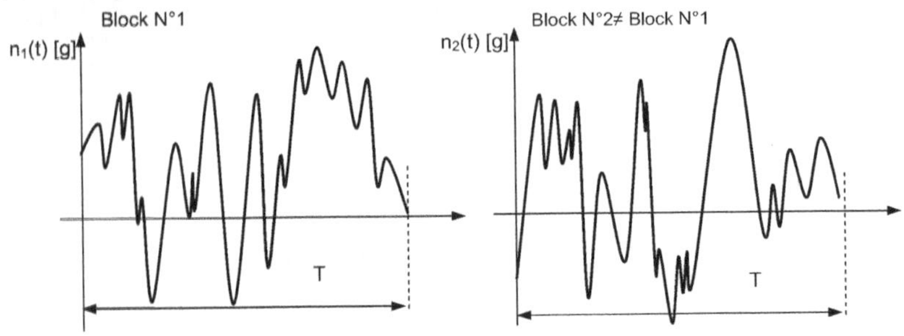

And so on...

e) Signal type

This is the rate of phases restarted from one block to another in the generated control signal.

- 0% = pseudo-noise: better accuracy of the calculated spectra, but there may be gaps in the excitation (between two lines of analysis).

- 100% = real noise: all frequencies are excited.

- If it is difficult to obtain a control spectrum corresponding to the reference when using a poorly dampened structure and real noise (100%), one can select values between 0 and 100% to ensure better accuracy.

- Recovery: the overlap between two data blocks can be used when the reference profile has a significant level in BF.

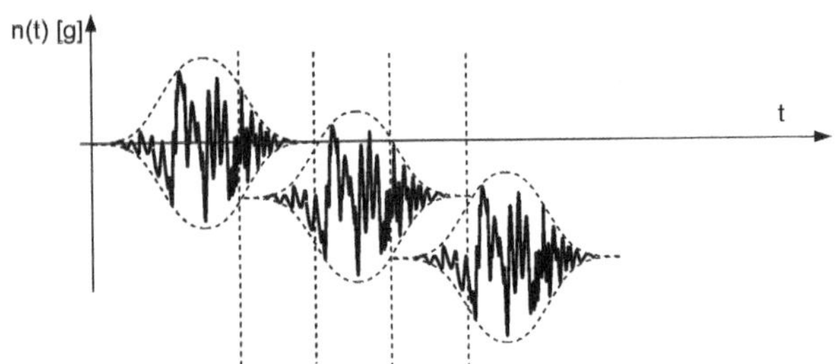

f) Characteristic parameters

Sampling frequency: for example, traditional relationships:

$f_{ech} = K.f_{max}$

K is linked: $\begin{cases} \text{To N number of converter bits} \\ \text{To slope p of the filter cutoff} \end{cases}$

The operator gives f_{max} and the system knows K from the following relationship:

$$\text{LogK} = \frac{10.\text{Log}2^N.\text{Log}2}{p} + \text{Log}2$$

It then calculates f_{ech}.

Example 6.2: N = 12 bits; p = 100 dB/octaves → K = 2.569 (usual value).

6.5.2. Acquisition of measurements

The usual block diagram is then:

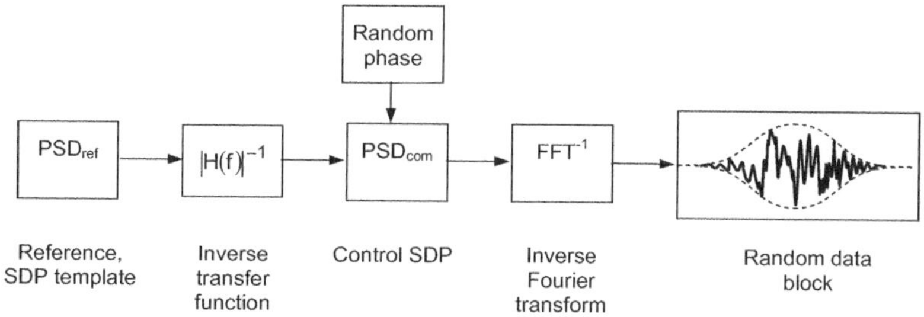

a) Frequency resolution

- N lines = number of analysis points.
- The operator indicates N lines.
- The system recognizes Δf as: $\Delta f = \frac{f_{max}}{N_{raies}}$.

REMINDER.– Let us recall here the relationships issued from Shannon's theorem:

$$\text{Shannon} \rightarrow f_{ech} = 2.f_{max} = \frac{1}{\Delta T} \qquad \Delta T = \text{sampling period}$$

$$T = 2.N_{lines}.\Delta T = \frac{2.N_{lines}}{2.f_{max}} = \frac{1}{\Delta f}$$

Therefore: $$\Delta f = \frac{1}{T}$$

b) Calculating PSD from the response of y(t)

$$\text{SDP} = \frac{\overline{Y(f).Y(f)^*}}{\Delta f} = \frac{\overline{G_{YY}(f)}}{\Delta f}$$

$\overline{G_{YY}(f)}$ = Autospectrum of the response y(t) (linear average; there are others, such as exponential average, etc.)

- Exponential average: we then use a weight coefficient P such as:

$$\overline{PSD(i)} = \overline{PSD(i-1)} + \frac{PSD(i) - \overline{PSD(i-1)}}{P}$$

$\text{NDOF} = 2.(2.P - 1)$ Nr of degrees of freedom

The aim here is to optimize the correction speed/statistical accuracy ratio by reducing the loop time as much as possible (time between two rounds of calculation and correction).

c) Calculation accuracy

If M is the number of averages, then 2.M = number of degrees of freedom (NDOF).

Statistical accuracy is then proportional to:

$$\sigma \approx \frac{1}{\sqrt{M}}$$

Therefore, accuracy is ever better with increasing NDOF.

d) Calculation of the feedback spectrum

Calculation from one or more control sensors.

- Choice of strategy:
  - min/max: controlled from the sensor with the smallest or the largest value;
  - average: controlled from the PSD from the spectra average (line by line).
- Choice of weighting coefficient: this is common to all sensors used.
- The following curves are available:
  - f(t): time function
  - F(f): Fourier transform
  - PSD: spectral density power
  - $G_{XX}(f)$: autospectrum
  - $G_{Yx}(f)$: interspectrum
  - H(f): transfer function
  - Coh: coherence

e) Surveillance

- Alarms:
  - levels are generally dependent on frequency;
  - if exceeded, there is a sound and visual cue on the screen.
- Stop:
  - levels of tolerance in general and depending on frequency;
  - several excesses on one or more noise lines are tolerated before an abort is triggered;
  - if thresholds are exceeded, then stops occur.

- Notching: a maximum level not to be exceeded can be requested for each measurement channel:
  - constant level or profile depending on the frequency;
  - actions taken: - alarm
    - stop
    - notching

f) Back-up

- In general, PSD spectra for tests, controls and measurements.
- Some control systems calculate the FRF between channels (module and phase).
- Time signals (not interesting for noise).
- These can be:
  - manual,
  - automatic (back-up rate is configurable by the operator (one at the beginning of a test, one in the middle and one at the end, for example).

g) Deferred time

(a) Restitution of back-ups:

- PSD: - tests;
  - control;
  - measurements;
- FRF between channels.

(b) Batch processing:

- time backups? (not very interesting);
- manipulations of curves:
  - correction coefficient,
  - multiplication, etc., using a calibration block,
  - integrals, derivatives,
  - superposition of results.

6.6. Shock and transient control

The shapes of available signals are then as follows.

6.6.1. Generation in temporal form

(a) Most of the usual mathematical shapes such as:

- the half-sine:

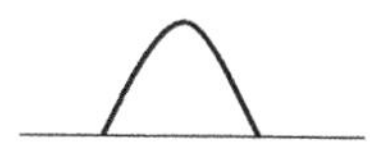

- the symmetrical or non-symmetrical triangle:

- the final peak sawtooth (FPS):

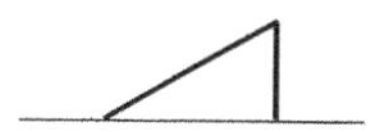

- the initial peak sawtooth:

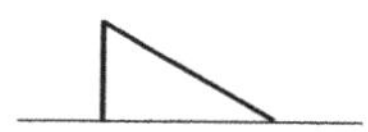

- the rectangle:

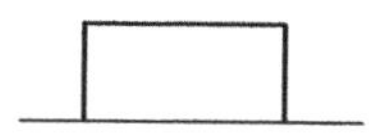

Bearing in mind that the most common are the half-sine and the FPS.

(b) The signals are made of broken points (transient) or segments from known shapes

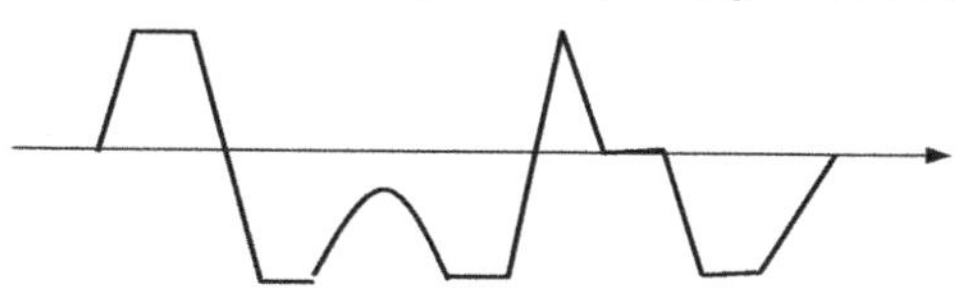

(c) Or through the use of external files

- Recorded signal
- Calculated signal
- Natural signal

(d) Optimization of conventional signals

The signals described above are not signals with zero velocity variations. But the signals generated by tests are usually physically zero velocity variations (equipment not moving before and after the shock).

We must therefore add pre- and/or postshocks (secondary lobes) to the main shape to ensure that this zero velocity variation (positive area under the acceleration equal to the negative area).

The aim of the proposed optimization is therefore:

- to ensure the speed and the zero displacement at the end of the shock;
- to reduce the maximum displacement of the test medium (particularly for exciters of electrodynamic vibrations).

The choice of secondary lobes is then out of:

- rectangular;
- single half sine;
- double half sine (a period);
- optimized in terms of minimum displacement.

With the possibility of a "pre-pulse", a "post-pulse" or both. An example of such a signal is given below:

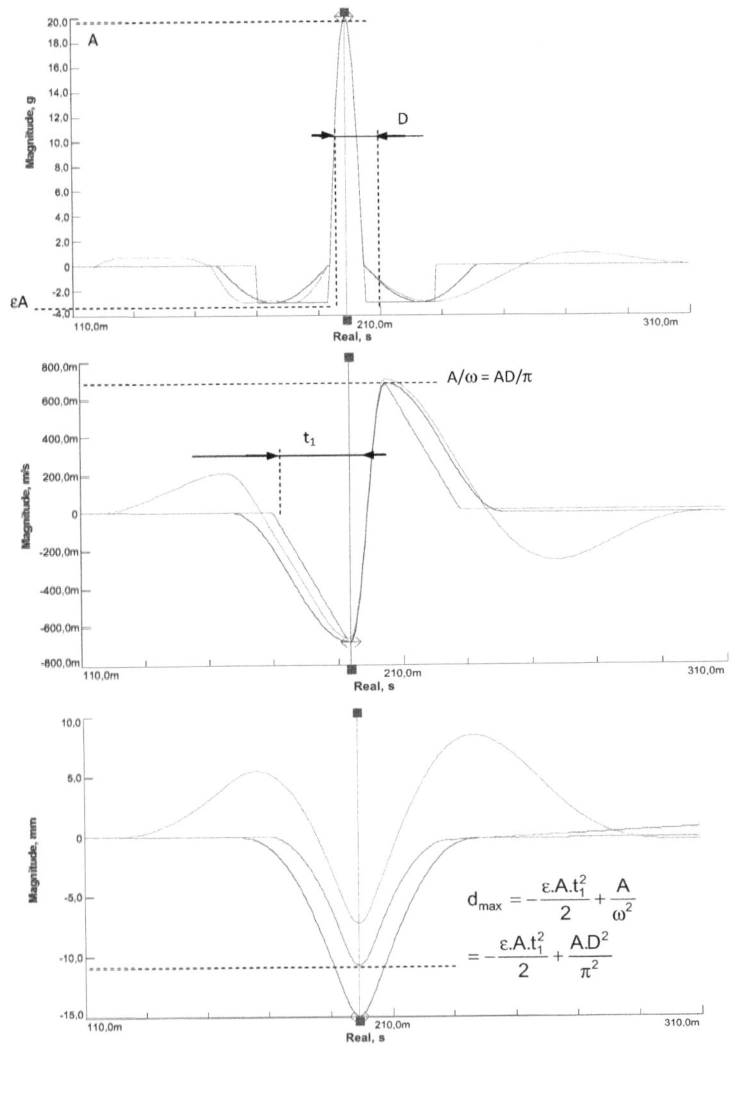

If ε is the percentage of the peak value of the main shock and if it also represents the negative peak value of the secondary lobe used, we then get, for example, a rectangular pre- and "post-pulse" for a half-sine shock (similar to a free fall shock impact) with the following relationships:

- for speed: $v_{max} = \pm\varepsilon.A.t_1 = \pm\frac{A}{\omega}$ instead of $\frac{2.A}{\omega}$. Giving a lower half maximum speed;

- for displacement: $d_{max} = -\frac{\varepsilon.A.t_1^2}{2} + \frac{A}{\omega^2} = -\frac{\varepsilon.A.t_1^2}{2} + \frac{A.D^2}{\pi^2} = \frac{A.D^2}{\pi^2}.\left(1 + \frac{1}{2\varepsilon}\right)$. So a lower displacement of the negative term and after development value, which is ever smaller with increasing ε.

This technique allows us a safety margin with return to zero speed at the end of the shock and minimizes the required speed as well as the maximum displacement required to achieve the nominal shock. Of course, the excitation signal is more distorted with increasing ε, especially its shock spectrum in VLF. Therefore, great care must be taken when implementing this method, see [ref. I18], bearing in mind that the choice of ε then depends on:

- the nature and shape of the main time signal;
- the acceptable VLF distortion based on the lowest modes of the tested material;
- maximum possibilities of testing means;
- maximum permissible error (MPE) of the test specification and all tolerances attached to the test impact (transverse, speed, distortion, etc.).

Generally, MPEs impose a value of ±20% of A and must therefore remain within this template to comply with the specification.

(e) Acquisition

The following parameters must then be defined:

- maximum frequency f_{max}:
  - selected by the operator:
  - conditions: position of filters;
    sampling frequency f_{ech};
- frequency resolution Δf:
  - imposed by digital technology: $\Delta f = \frac{1}{\text{observation window}} = \frac{1}{T}$;
- the selection criteria are then as follows:
  - to scan the shock finely and observe the harmonics, one must oversample, with:
    $\Delta t = \frac{1}{f_{ech}} = \frac{1}{k.f_{max}} \qquad k > 2;$
  - to observe the pre- and postpulses, side effects, rebounds, dampening issues, etc., one must observe (T = total duration of signal acquisition):

$$\Delta f = \frac{1}{T}$$

(f) Corrections

We have already mentioned that the generation of the command or control signal is calculated from the estimated total transfer function (exciter loaded by the specimen and its support). This estimate is in open loop before the impact test itself, or even between each impact if there are increasing levels. The procedure is as follows:

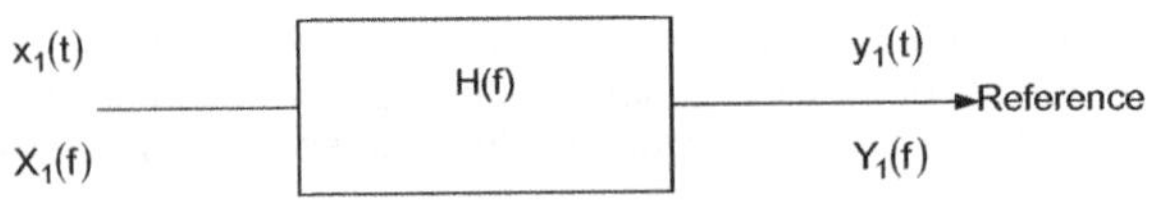

Thus, if Y(f) = the reference, FT with Y(f) = reference FFT. We then calculate the pretest:

$$\Delta[Y(f)] \rightarrow \Delta[X(f)] = H^{-1}(f).\Delta[Y(f)]$$

and: $\Delta[x(t)] = FFT^{-1}\Delta[X(f)]$

Hence, the correction: $x_2(t) = x(t) + \Delta[x(t)]$

With the possibility of only correcting a percentage "p" of the error by:

$$x_2(t) = x(t) + p.\Delta[x(t)]$$

NB: The speed of actual consoles allows correction after each shock if there are multiple shocks.

(g) Tolerances or MPE

In crash tests, tolerances (increasingly referred to as MPE) take the following shape:

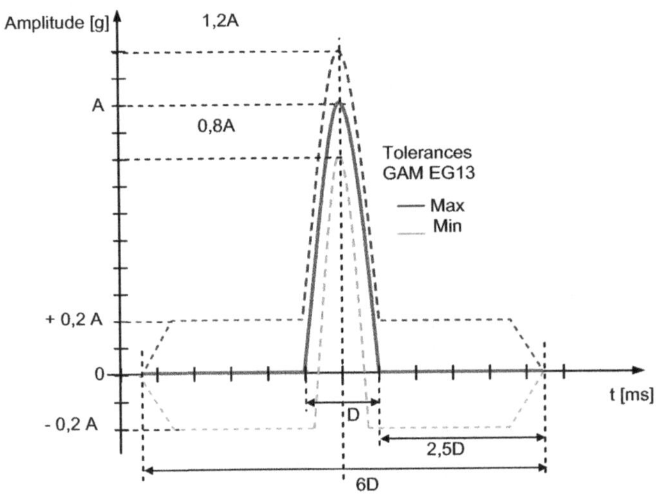

where:
D = nominal duration;
T = 6D = minimum time during which shocks must be observed if these are from vibration exciters;
——— nominal shock;
- - - - - - MPE limit.

Control feedback must be contained within these limits.

(h) Back-up

- For one or more shocks:
  - control sensor (time);
  - measurement channels (time);
  - control signal (*Drive*);
  - transfer function.

- For shaking:
  - all shocks (time);
  - one part with rate 1/N;
  - plus the previous curves.

(i) Restitution

Deferred time:
- Time back-ups (control + measurements);
- Fourier transforms: (control + measurements);
- shock spectrum: (control + measurements);
- transfer function: (module + phase):
  control channels/signal,
  channels/control,
  channels in between;
- shaking: synthetic curve, where:

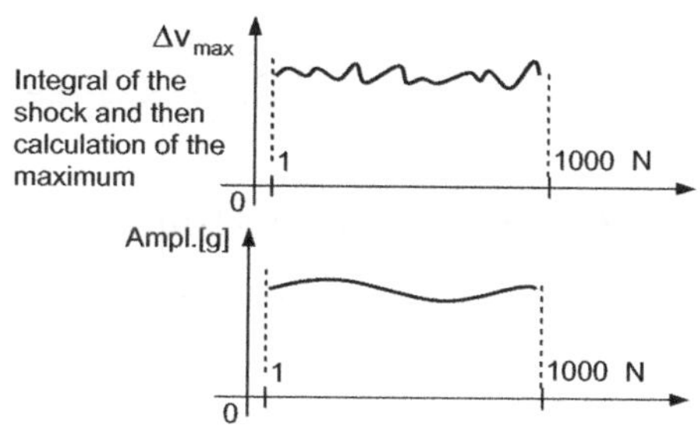

6.6.2. Generation and control in the shock spectrum

a) Definition

REMINDER.– The shock spectrum or shock response spectrum (SRS) allows us to:

- simulate the severity of a shock;
- thus if two signals have the same SRS, they have a similar severity as defined in SRS;
- this severity is measured using a representation model that was mentioned in Chapter 1, which consists of an infinite number (within the useful bandwidth) of basic systems with 1 DOF, linear and decoupled. In other terms:

$$m.\ddot{x}(t) = -c.\dot{z}(t) - k.z(t)$$

$$z(t) = x(t) - e(t)$$

$$m.\ddot{x}(t) + c.\dot{x}(t) + k.x(t) = k.e(t) + c.\dot{e}(t)$$

$$\omega_0 = \sqrt{\frac{k}{m}} = \text{own pulse}$$

$$\xi = \frac{c}{\sqrt{km}} = \text{reduced damping}$$

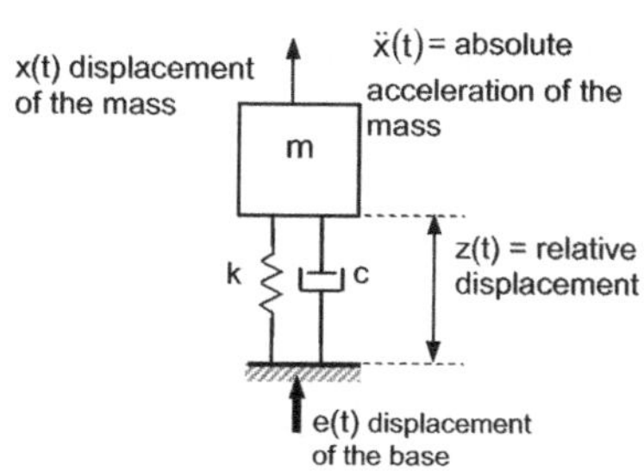

Thus, the plot of the maximum value taken by z(t) for each natural frequency of the basic systems is the SRS.

With the data: f_0; ξ; e(t), we deduce z(t). We show that for small ξ (from Q#3 or ξ# 0,166), we virtually get the following equation:

$$\ddot{x}(t) = \omega_0^2.z(t)$$

We seek the maximum of this expression, called "pseudo-acceleration", based on the useful band for f_0.

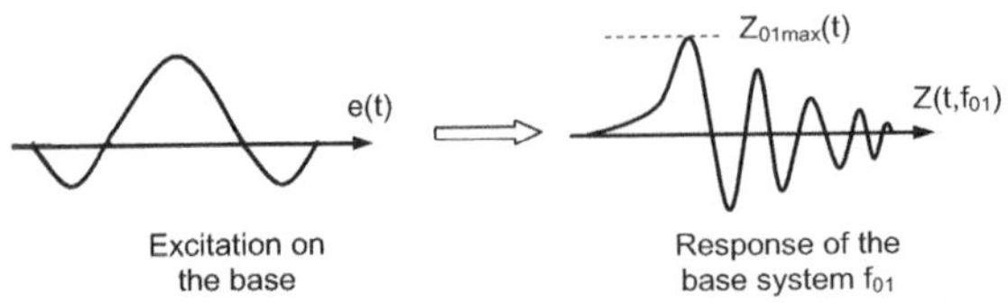

Hence, the appearance of the SRS as a standard half-sine.

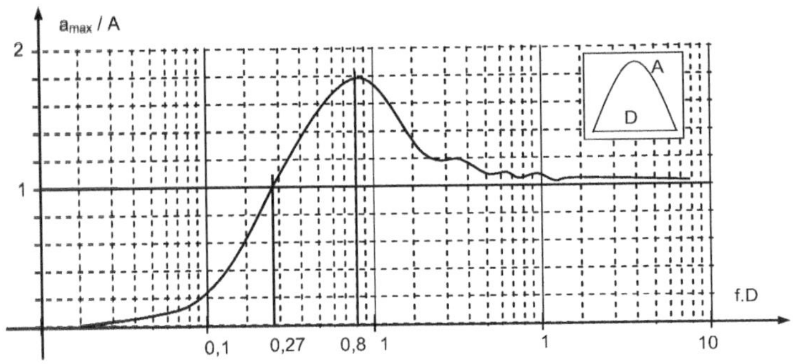

Shape of the SRS in general and for specifications.

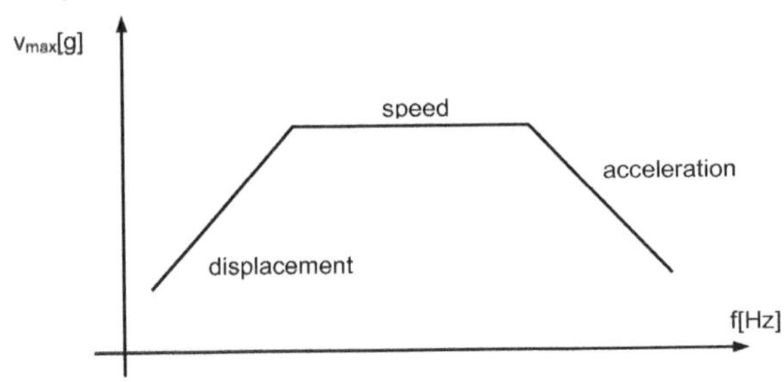

REMINDER.– SRS of a sinusoid.

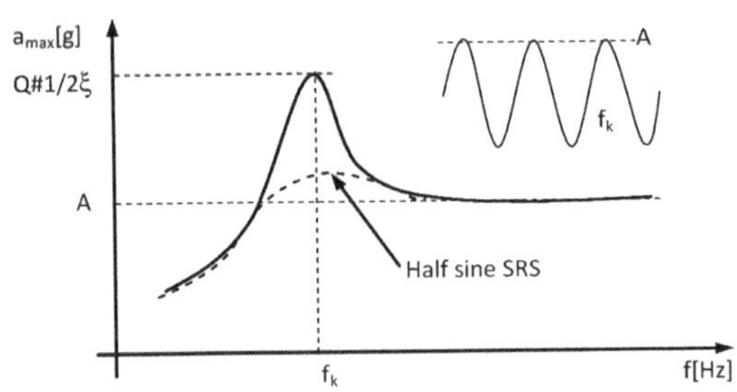

b) Key features

The interest in using the SRS lies in the following features:

- it characterizes a signal by its extreme effects;

- it is easier to interpret than a FFT for short-term signals where the means are complicated;

- it allows us to find averages or severity envelopes for different time signals;

- an analysis is possible in points per octave (speed);

- algorithms are specific;

- no concept of fatigue (practically and generally not taken into account) or different effects according to the temporal form.

c) Generation

There are several methods for building SRS. Without going into too much detail, let us imagine a method that follows the definition of the SRS by a sum of components.

(a) Sum of components

- Divided into number of components per octave in general (not Hz per Hz).
- The following shapes:
  - dampened sinusoid:

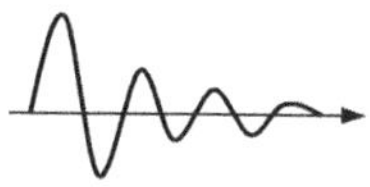

  - modulated sinusoid:

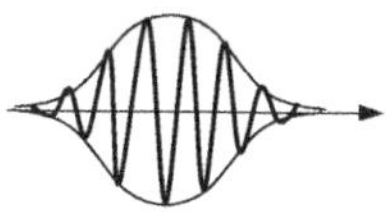

(b) Characterization of a component

It is characterized (for a dampened sinusoid) by a pseudo-frequency, dampening and phase at the beginning, or a delay (which amounts to the same thing).

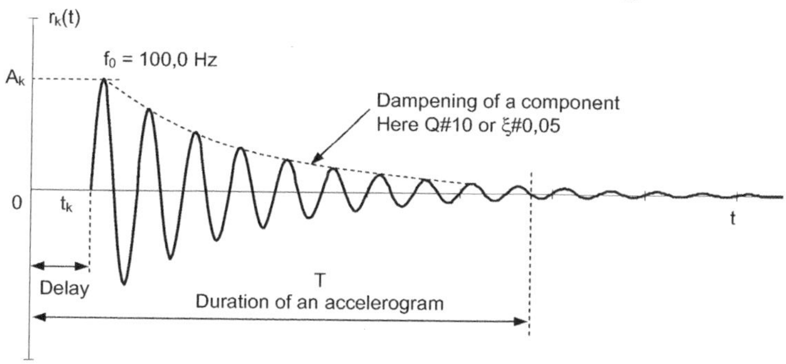

(c) Definition of construction data

Rate of the shock spectrum to be achieved:

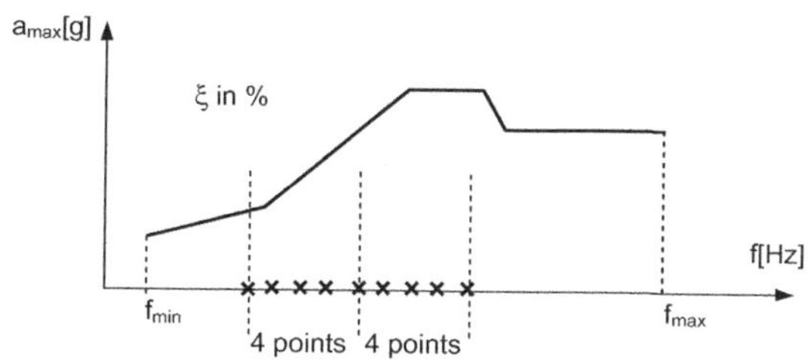

In other words, the data: f_{min}; f_{max}; number of points per octave, dampening ξ in %, component frequencies.

(d) Operator selection

The possibilities are:

- duration T of the accelerogram. In other words: $\begin{cases} T \;;\; N\,\text{points} \rightarrow \Delta t = \dfrac{T}{N}; \\ \dfrac{T}{\Delta t} = T.f_{ech} = T.k.f_{max} \end{cases}$

- delay of each component;

- dampening ξ of the dampened sinusoid, or number of periods for the modulated sinusoid;

- polarity + or –.

The algorithm should find initial levels that allow synthesis to begin, such as the complete following procedure:

1) selected parameters:
   - manually
   - automatically; with: delay draw, polarity, dampening;
2) sum of the components:
   - initial level = value of the spectrum or 1;
3) calculation of SCR and calculation of error relative to the nominal spectrum;
4) altered levels;
5) stopping the synthesis when the accuracy level is reached.

(e) Limits and deviations (errors)

The measurement is mainly done for the following two cases:

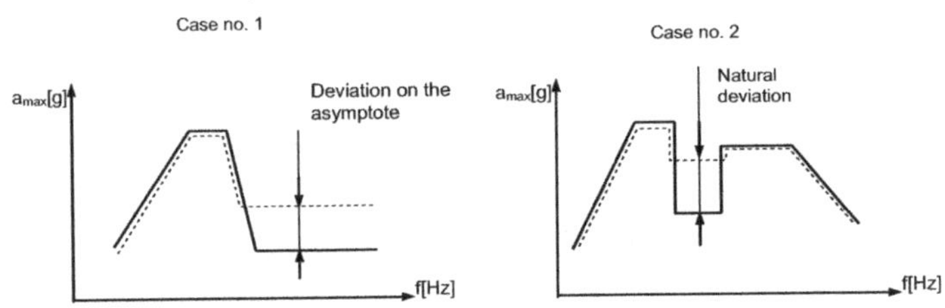

d) Control

Control is standard for all transient signals:

- either in open loop control;
- or in iterative control.

(a) Procedure

- Preliminary calculation of the total transfer function H(f) by pretest with:
  - one or more noise blocks sent (measurement of H(f) in noise);
  - the signal itself being sent at a reduced level (–XdB).
- Calculation of drive and sending.
- Calculation of error either in the frequency domain or in the time domain. With:

$\Delta Y(f) = Y_1(f) - \text{FFT(reference)}$ → Spectrum of the output signal error.

$\Delta X(f) = H^{-1}(f).\Delta Y(f)$ → Spectrum of the input signal error.

- New control signal:

$X_2(f) = X_1(f) + \Delta X(f)$ → Spectrum of the input corrected signal.

$x_2(t) = \text{FFT}^{-1}[X_2(f)]$ → Input corrected temporal excitation signal.

- Calculation parameters:
  - frequency band;
  - number of points per octave or Δf for linear calculation;
  - dampening of basic oscillations (%);
  - unit: relative displacement; pseudo-speed; pseudo-acceleration;
  - type: primary; residual; composite or (envelopes all maxima);
  - sign: positive; negative; absolute.

e) Example of processing and the spectra obtained

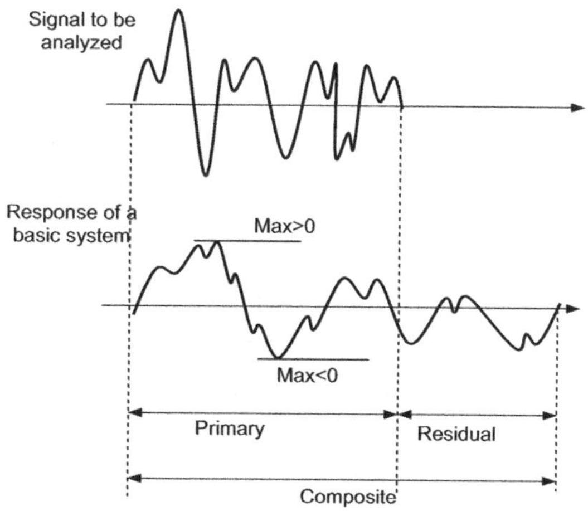

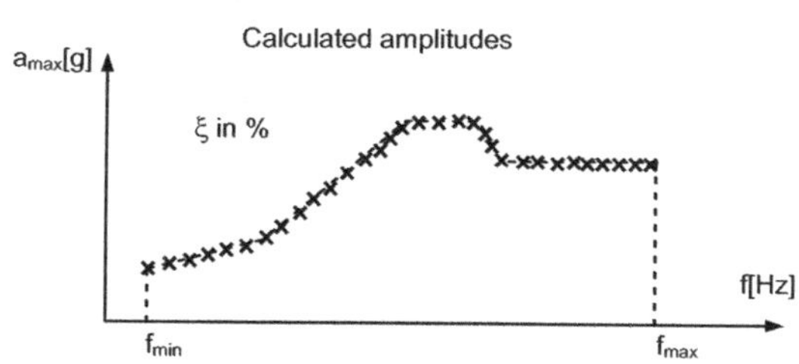

6.7. Combined vibrations control

6.7.1. Overview

Included in these types of combined vibrations is the simulation of real environments consisting of:

- broadband noise with flat or shaped profiles (white or pink noise) (1);

- superimposed narrow bands of noise with possibilities of scanning and level changes (with a given profile) (2) or sent in bursts (3);

- pure sinusoidal components with the possibility of scanning and level change (according to a given profile) (4) or sent in burst (5).

Example of the following graph models:

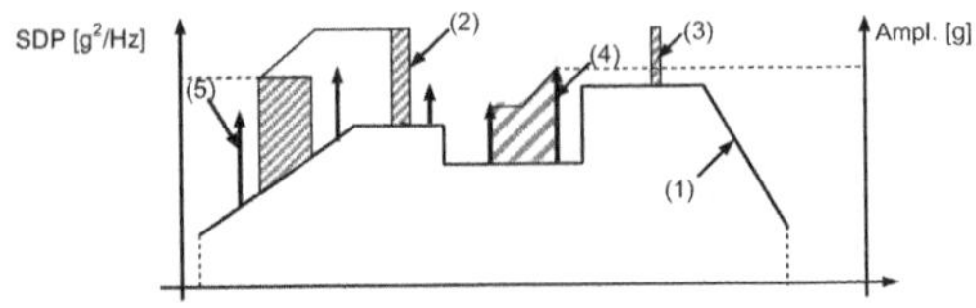

Example of normative documents that require this kind of vibrations:

RTCA DO-160: Environmental Conditions and Test Procedures for Airborne Equipment-Vibrations

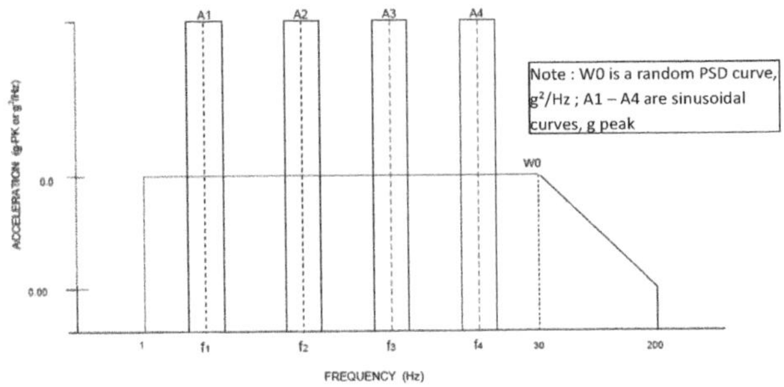

Sine-On-Random vibration Test Curve for Helicopters

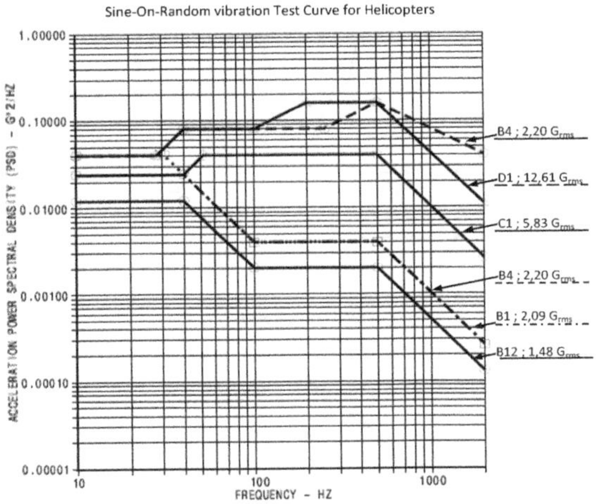

Robust Random Vibration Test Curves for Equipment Installed in Fixed Wing Aircraft with Turbojet or Turbofan Engines

6.7.2. Methods

a) Generation

- Noise: identical to the problem of noise alone, with the possibilities of:
  - a basic block while taking account of the profile;
  - possible randomization;
  - scanning frequencies or bursts.

- Sine: digital generation (or also analog for some devices) with components, with the possibilities of:
  - exact frequency;
  - level based on profile;
  - frequency sweep or bursts.

b) Measurement and control

- Separation of noise from sinusoidal signals:
    - different analysis;
    - different references;
    - g^2/Hz with a profile;
    - g with a profile.
- Method:
    - we calculate the overall spectrum of the feedback signal containing all the components;
    - the sinusoidal components are known accurately, so we can calculate the frequency spectrum and analyze them by correlation (noise extraction);
    - influence from the global spectrum is removed;
    - we can separately control:
        - the noise spectrum [g^2/Hz];
        - the sinusoidal components [g];
    - thereby using the methods presented for sine and noise.

c) Important points

To be checked depending on the type of control console used:

- possibility of mixing noise on noise and sine on noise (example of previous standards);

- possibility to scan the components according to the given profile (sine and noise);

- generating accurate sine between two lines if necessary;

- analysis of sine must be specific and presentation of components must be separate;

- analysis of noise by FFT must be done after removing the influence of sinusoidal components to actually excite the structure with the required noise spectrum.

6.7.3. Current and emerging needs

a) Overview

These mainly come from the relatively recent concept of "tayloring of tests" or "*tailoring*" by replacing the use of flat severities that are less suited to the goal. The goal is to optimize the development and testing of equipment with goals for improving the cost performance ratio. This is in conjunction with:

- the use of constraints from the real environment encountered during the life profile of the material;

- simulation of the real environment through environments that are equivalent and that may contain multiple environments encountered during the life profile of the material [see ref. l1, f5, f6, n4].

The following diagram illustrates this point:

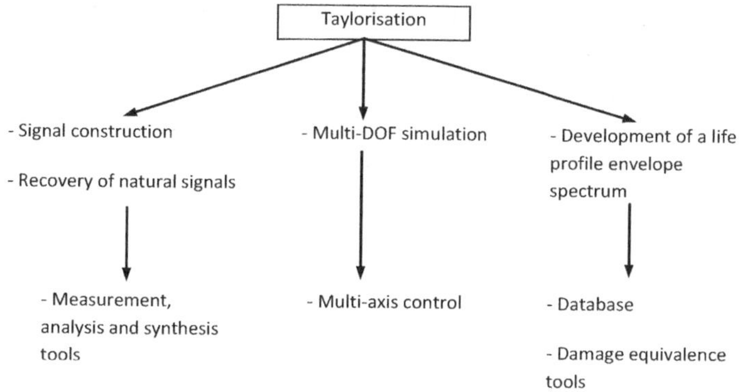

b) Tools

- Have pre- and postprocessing techniques for the development of requested test profiles.
- Easier recovery of files.
- Common file structures (avoid conversions).
- The tools must be integrated into the control system.

c) Multiaxis control

- Should be a normative standard.
- It must have the same characteristics (homogeneous systems).
- It should be flexible and open, as there are many developments in this field.
- It is desirable for it to be based on a current standard system.

Possible block diagram:

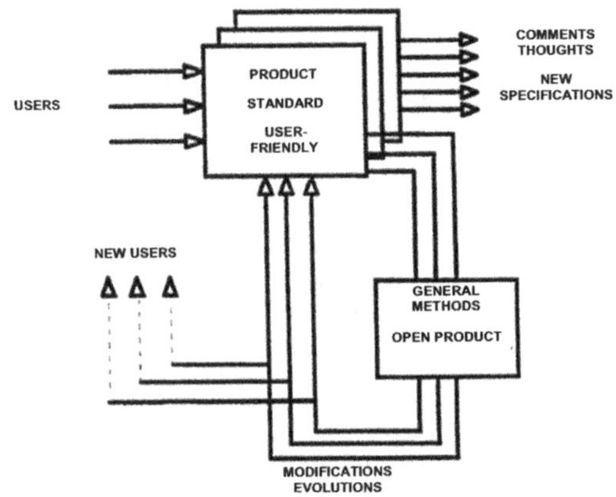

d) Tayloring

- Find real environments that correspond to the life profile of the equipment or access to databases and database management.

- Calculate the equivalent damage spectrum in order to carry out the test by containing several situations of life profile if possible (similar in general); calculations tools are therefore needed.

- The calculation can be performed by an adjunct system with the possibility of communication and easy transfer with the control system. It can optionally be integrated into the control system, but it requires significant expertise, which is where problems can arise such as for users trained to use the tools and interpretation of results. Moreover, the same profile or the same services are not concerned.

e) Integrated modal analysis

- The goal is to use the environmental simulation test to achieve modal analysis of the equipment since most of the means and tools are in place and already used.

- The advantages include:
  - not requiring additional testing;
  - not requiring specific excitatory modal analysis;
  - allowing the structure to be excited to an energy level equivalent to that in a real environment (reveals nonlinearities and makes better modes).

- The disadvantages still needing to be addressed are:
  - the different boundary conditions (built equipment);
  - measuring the force produced by the exciter, needed to obtain the generalized masses associated with each mode is not always easy.

- The proposed procedure:
  - during the test, acquire the necessary structural responses;
  - calculate transmissibilities between responses and control in acceleration;
  - use a specific smoothing algorithm for obtaining:
    - natural frequencies;
    - dampening;
    - modal shapes.

- Calculation of generalized parameters:
  - same procedure as for the control test;
  - in addition: calibration of the exciter to get coefficient Γ such that $F = \Gamma.I$ (where I = current in the moving coil, Γ = force current transformation coefficient);
  - potential calculation of the dynamic mass through:

$$M(\omega) = F_{empty}(\omega) - F_{with\ structure}(\omega)$$

6.8. Control: a few essential rules

a) Know what is actually observed

- Consideration during correction and control (position of control sensors).

- Filter positions.

- Acquisition:
  - number of points per period;
  - sampling frequency.

- Type of processing technique:
  - strategy;
  - averages (number and type);
  - real degrees of freedom.

b) Know what is being presented

- Number of frequency points.

- Rate of reload:
  - calculation;
  - display.

c) Judge the speed of the system

- Classic trap: The greater the coincidence is between the reference and control, the faster the system (false).

- Not always checked: If a system is slow to intervene and therefore to correct, some local phenomena cannot be "seen" in the case of sine. It looks as if the system was quick to correct.

- Watch out for the conditions of analysis or measurement that can slow down a system:
  - do not forget that some measurement strategies suppress or group information;
  - the more we visualize it, the more we measure and the less control we have.

d) Manage the response speed problem

- Conditions for acquisition: (T seconds)

for analysis: ($\Delta f=1/T$)
for processing: (one or more channels, type of analysis, etc.)

- Conditions for correction: weighting of a new block
accepted correction percentage.

e) Manage the accuracy/speed compromise

- Consider whether the equipment:
  - is fragile;
  - has a developmental behavior or not;
  - is linear or not.

- Consider the type of test:
  - investigation;
  - endurance;
  - environment.

f) Manage the correction speed problem

- Speediness can be dangerous (oscillations, aborts, etc.).
  - Problem of "pumping" in sine.

- Speediness can cause falsities.
  - Problem of statistical accuracy.

- Speediness can cause ineffectiveness.
  - The test never "passes".

- The concept of speed is related to old systems (# 10 years) for which the IT imposed control limits.

- Today, systems are too fast compared to the physical phenomena of vibrations or shocks. So the operator must judiciously use all accessible parameters to best manage:
  - quality;
  - safety;
  - the possibilities.

NB: Appendix J provides a summary of the common rules on control strategies.

7

Metrology of Measurement and Testing Methods

This chapter is an introduction to the measurement techniques used in testing laboratories with the associated requirements for metrology of measurements as well as the devices and means used.

We focus particularly on accelerometer measurement, which is the most commonly used technique. We will discuss the need to validate the testing facilities and estimate their uncertainties in order to provide usable test results.

As in Chapter 4, we will rely on the requirements of the standard NF EN ISO/CEI 17025, which is the main organizational document for testing and analysis laboratories in this field.

7.1. Introduction to accelerometer sensors

7.1.1. Overview

We have seen that vibration phenomena have often been characterized by the acceleration parameter (absolute acceleration generally relative the ground), which we usually express as "number of g", with g being the acceleration of gravity at ground level (g # 9,81 ms^{-2}). Indeed, through the relationship $F[daN] = m.\ddot{x}(t)[g]$, which is more simply written as: F = mg. It is then easy to calculate the forces applied to simple structures if one knows their mass.

Moreover, the levels found on equipment such as rockets or missiles are 10 to 100 times higher than those encountered in the automotive industry, for example. It is therefore necessary to carry out measurements with high dimensions.

Ultimately, these signals use a relatively wide frequency band; from 0,1 Hz to about 10 Hz for VLF signals, such as the suspended VLF platforms with frequencies of 2–10 kHz and even up to 50 kHz for signals measured in acoustic discretion or for characterizing pyrotechnic shocks.

Among all accelerometric sensors, piezoelectric sensors are the most suitable for these types of measurements. For the latter, the polarized piezoelectric element generates charges under the effect of compression or shear stress, which is substantially proportional to the force applied on the crystal and thus also proportional to the acceleration experienced by the accelerometer through an inert mass. This is why the piezoelectric sensor is the most commonly used sensor (see [ref. I18]).

7.1.2. Principle of operation

A piezoelectric accelerometer essentially consists of an inert mass (inertial mass) within the frequency band (infinitely rigid) and placed on a piezoelectric crystal of which two opposing surfaces have become conductive by depositing a conductive metal film. This is shown in the following diagram.

When an acceleration or force F_s is applied to the mass, electrical charges appear on the surfaces of the crystal. These charges, which are proportional to acceleration, are captured by the electrodes.

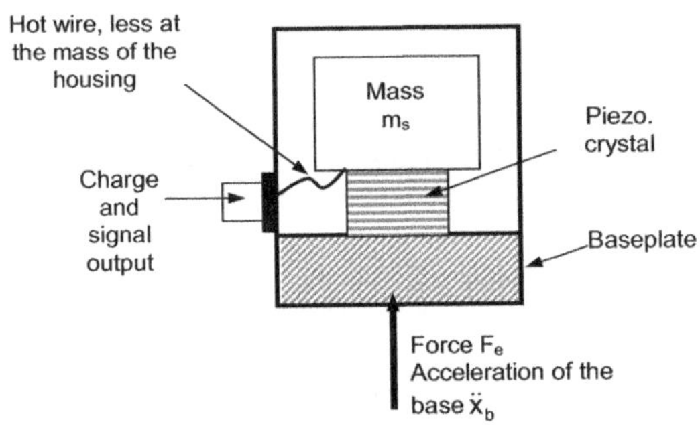

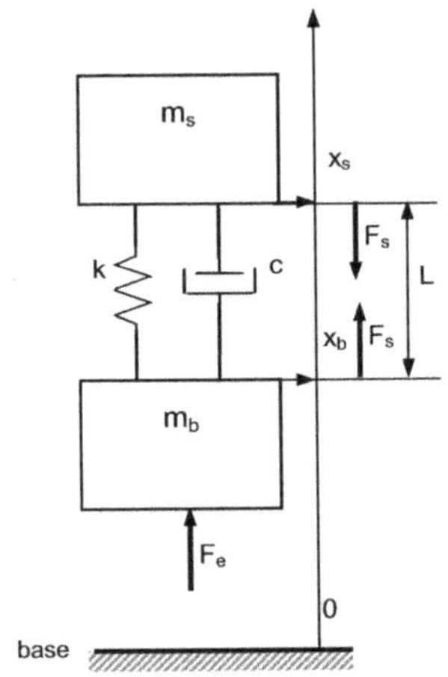

From a mechanical point of view, the above model illustrates a fundamental principle with the following assumptions:

- x_s = displacement of the inertial mass relative to the ground (absolute);
- x_b = displacement of the base relative to the ground (absolute);
- $x_s - x_b - L = r$ = relative displacement of the inertial mass relative to the housing;
- m_s = total inertial mass;
- m_b = mass of the accelerometer base;
- L = initial displacement relative to resting state;
- F_e = base excitation force;
- F_s = force of the spring on the base and the inertial mass.

The equations of motion then become (with the model notation):

$F_s = k(x_s - x_b - L)$ Spring return force

$m_b.\ddot{x}_b = F_s + F_e$ Force on the base

$m_s.\ddot{x}_s = -F_s$ Force on the inertial mass

The differential equation of motion can be written in terms of the relative motion as:

$$\ddot{x}_s - \ddot{x}_b = -\frac{F_s}{m_s} - \frac{F_s + F_e}{m_b} = -F_s.\frac{m_s + m_b}{m_s.m_b} - \frac{F_e}{m_b} = -\frac{k(x_s - x_b - L)}{\mu} - \frac{F_e}{m_b}$$

where: $\mu = \dfrac{m_s m_b}{m_s + m_b}$ = often called reduced mass and for the harmonic force: $F_e = F_0.\sin\omega t$

$$\mu.\ddot{r} = -k.r - \frac{\mu}{m_b}.F_0.\sin\omega t \tag{7.1}$$

- Thus, when the accelerometer is not excited and is free in its response, the differential equation (second term zero) is incited:

$\mu.\ddot{r} = -k.r$ if through the assumption, the relative movement is also in $r = R.\sin\omega t\,(\phi = 0)$

Giving : $-\mu R\omega^2 \sin\omega t = -kR\sin\omega t \rightarrow \omega_n^2 = \frac{k}{\mu}$

This gives the natural angular frequency of the system through:

$$\omega_n^2 = k\left(\frac{1}{m_s} + \frac{1}{m_b}\right) \tag{7.2}$$

However, if m_b is very big relative to m_s or if the base of the sensor is perfectly integrated with the structure on which it is fixed, we can then write:

$$\omega_n^2 = \frac{k}{m_s} \tag{7.3}$$

which represents the natural angular frequency of the inertial mass and also the natural angular frequency of the accelerometer, mounted correctly on the structure it is measuring. In practice, it is impossible to mount the accelerometer perfectly rigidly, but as a first approximation for a screw mounting, this is acceptable. We will see that for certain setups, this is not the case.

In forced response to a harmonic excitation on the base, we ignore c and (7.3):

$$\mu.\ddot{r} = -k.r - \frac{\mu}{m_b}.F_0.\sin\omega t \qquad \rightarrow \qquad \ddot{r} + \omega_n^2 r + \frac{F_0}{m_b}.\sin\omega t = 0 \tag{7.4}$$

As the solution is also sinusoidal, we get:

$$-\omega^2 R\sin\omega t + \omega_n^2 R\sin\omega t + \frac{F_0}{m_b}\sin\omega t = 0 \rightarrow R\left(\omega_n^2 - \omega^2\right) + \frac{F_0}{m_b} = 0 \rightarrow R = -\frac{F_0}{m_b\left(\omega_n^2 - \omega^2\right)}$$

For very low frequencies and for $\omega << \omega_n$, the value of R tends toward the constant value R_0 such that $R = -\frac{F_0}{m_b\omega_n^2}$.

If we normalize the expression for displacement R in High frequency (HF), relative to the displacement R_0 in VLF, we get:

$$A = \frac{R}{R_0} = \frac{1}{1 - \left(\frac{\omega}{\omega_n}\right)^2} \tag{7.5}$$

This result is remarkable because it shows that the displacement between the base and the inertial mass when the accelerometer is subjected to sinusoidal excitation tends toward increasing the natural frequency of the accelerometer.

This is the same for the force on the piezoelectric crystal. Thus, a typical curve is as follows:

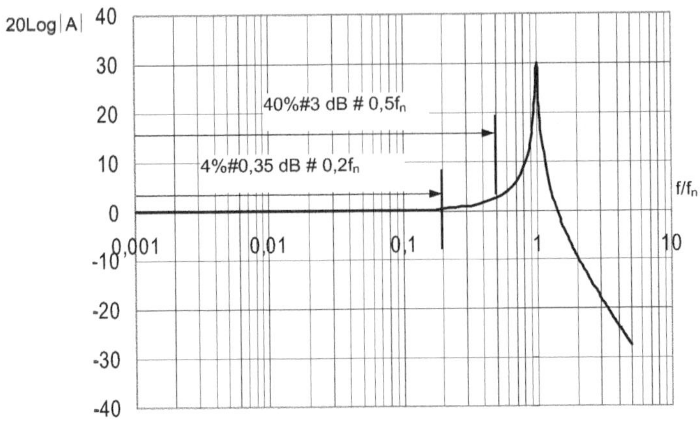

So for frequencies between VLF and 0,2 f_n, the curve is nearly flat and the sensitivity of the accelerometer is constant. Interpretation of the relationships (7.4) and (7.5) shows that the transmissibility of the assembly is practically constant up to 0,2 f_n, and therefore the force transmitted on the base or the corresponding acceleration through m_s is transmitted directly onto the crystal. Thus, for a peak amplitude of force or acceleration of the given base, the stress on the crystal will be constant up to 0,2 f_n. In theory, the signal from the sensor will show a constant acceleration peak (if its sensitivity does not vary depending on the frequency, which is usually the case).

However, an electric charge Q_C across the crystal that is proportional to the tensile compression force generated by the relative acceleration $\ddot{r}$ and thus through the mass m_s brings about the relationship (overvoltage at f_n being very strong $\ddot{r} \approx \ddot{x}_b$):

$$Q_C = K.\ddot{r} \equiv K.\ddot{x}_b \tag{7.6}$$

The accelerometer thus transforms acceleration into an electric charge. Its sensitivity is then evaluated in picocoloumb per gram (pC/g). This depends on K and the crystal type.

7.1.3. Main features

Let us recall the following main features.

a) Advantages

- It is self-generating (except for integrated electronic sensors of the ICP type, for example, which is the trademarked PCB name for IEPE (integrated electronics piezo electric), it stands for "integrated circuit piezo electric").
- It is robust (no moving parts).
- It is dynamic.
- It has a relatively wide frequency range (several 10 kHz).
- It has good stability over time.
- This is an absolute sensor and therefore relatively easy to calibrate.
- It works in all directions.
- Through integration, it allows us to obtain the velocity and displacement (easier than deriving, which always adds noise).
- It is light (a few to about 10 g).
- It is compact (1 cm^3).
- It has a good resolution (0,01 g).

b) Limitations

- The output is highly impedant because it works with picocoloumb charges (except for integrated electronic sensors).

- The Low frequency (LF) limit is linked to the electronics.

- The HF limit is related to the geometry of the sensor, as well as its fixation.

Hence, the importance of developing this type of sensor. Some examples of structures are given below (BK notes).

Principle of the DELTA SHEAR®

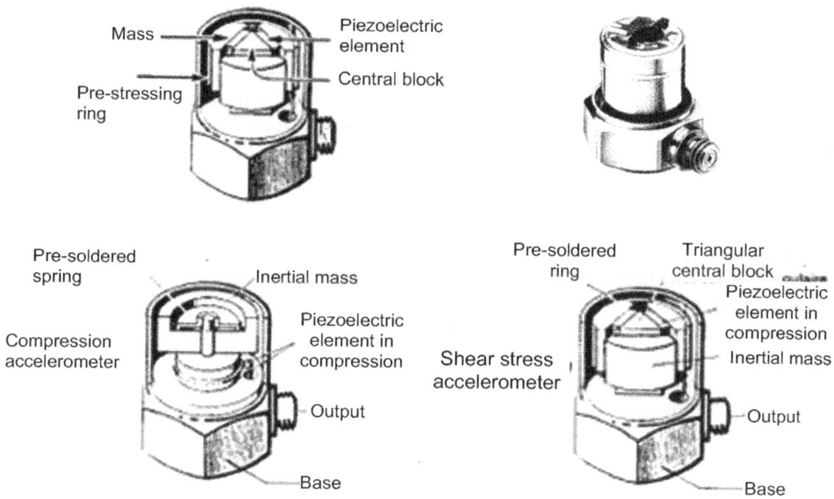

Figure 7.1. *Examples of accelerometer sensor types*

c) Frequency response

We have seen that it practically corresponds to the response of a system with 1 slightly dampened DOF. As the dampening is very weak, there is a risk of damage if one excites it at f_n. In addition, the sensitivity varies as one nears f_n. Therefore, it is generally advised to work at:

$$f_{max} = \frac{f_n}{5} \qquad (7.7)$$

In addition, in the previous calculation, we have assumed that the sensor was firmly fixed to the structure. When this assumption is challenged by the possible fixation techniques, the resonance frequency of the assembly decreases severely, through the following techniques, for example:

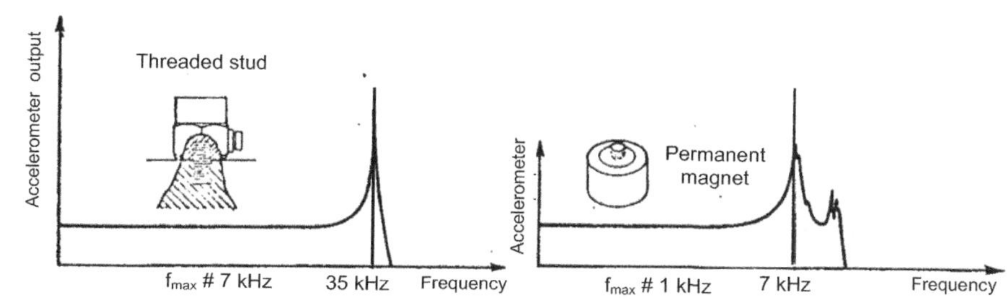

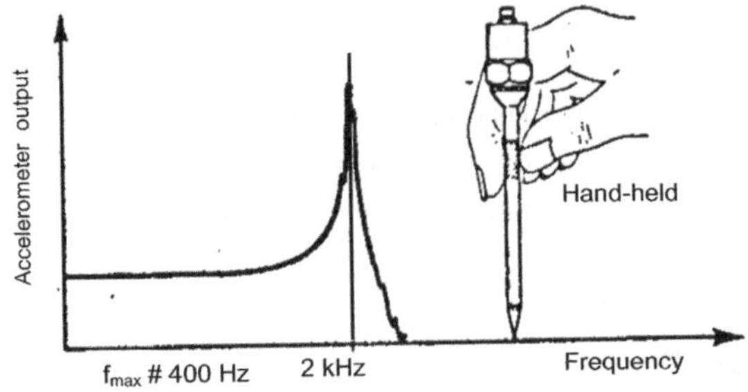

Other techniques exist, such as using adhesives with intermediate supports to prevent damage to the sensor base, sticker films, etc. In addition, to get the three orthogonal axes of vibration, we often use intermediate cubes on which the three sensors are fixed and the cube is then secured to the structure. It is therefore imperative to validate the chosen solution by a frequency analysis in order to determine the useful bandwidth of the measuring assembly, including the cube.

If possible, always preferentially use the technique of the screwed sensor.

We also discussed the fact that it was critical to excite the resonance frequency of the sensor. This is for two reasons: one concerns the deterioration of the crystal and the other that the crystal can become saturated under the effect of high frequencies generated by such shocks (jounce in the tested material) and distort the measurement (for example shocks from underwater explosions or pyrotechnic shocks). We can then use an intermediate interface to control the entire sensor + accessory cutoff frequency. This can be developed by the laboratory, where one can use filter systems, as proposed by BK for example, with the following type of structure:

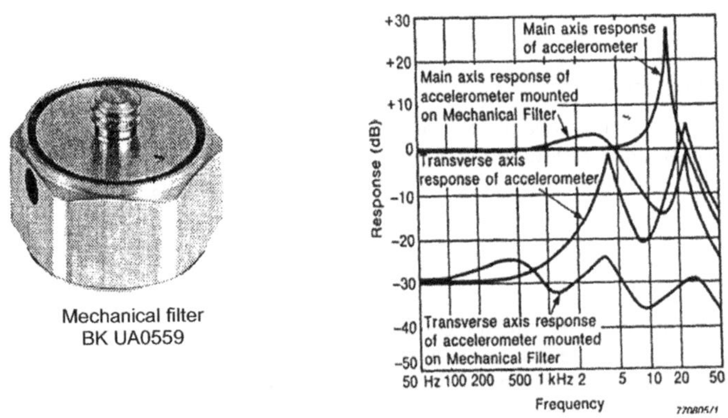

Figure 7.2. *Example of a mechanical filter*

d) Sensitivity

In the direction of the sensitive axis of the sensor, the accelerometer has a typical resolution of about 0,01 g. The following graph shows how the sensitivity of an accelerometer varies relative to the direction of the exciting force. We see that it also has a transverse sensitivity that we try to minimize, but which is far from zero and therefore needs to be taken into account in the measurements. It is usually expressed as a percentage of nominal longitudinal sensitivity.

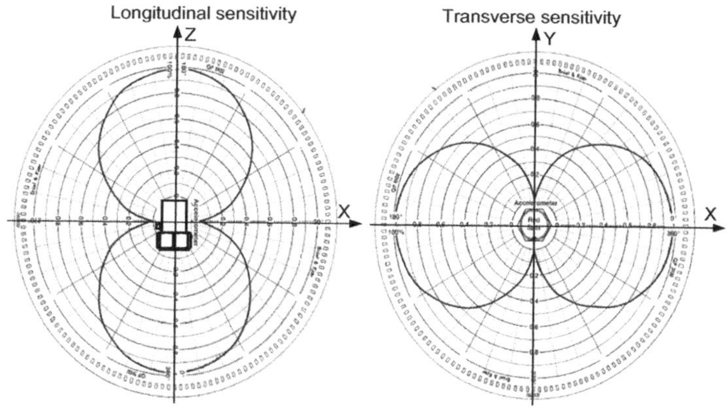

Figure 7.3. *Example of transverse sensitivity of accelerometric sensors*

So for transverse sensitivity, we see that there is a preferred direction that should be used for difficult cases.

Manufacturers usually give the maximum transverse rate along Oy. This direction is sometimes indicated by a red dot on the accelerometer housing.

Let us note that a "good" accelerometer has a maximum transverse sensitivity of 2–5%. As an example, the sensitivity of miniature accelerometers (weakest overall) varies from 2 to 10 pC/g.

In general, sensitivity tends to increase by 1% per 500–5 000 g.

The variation in sensitivity depending on temperature is typically around 0,45 pC/g per °C.

e) Environmental influence

We saw in the previous features that certain parameters depend on the environment. This is a point that must be taken into account when estimating uncertainties as influencing factors. We then get the following environments:

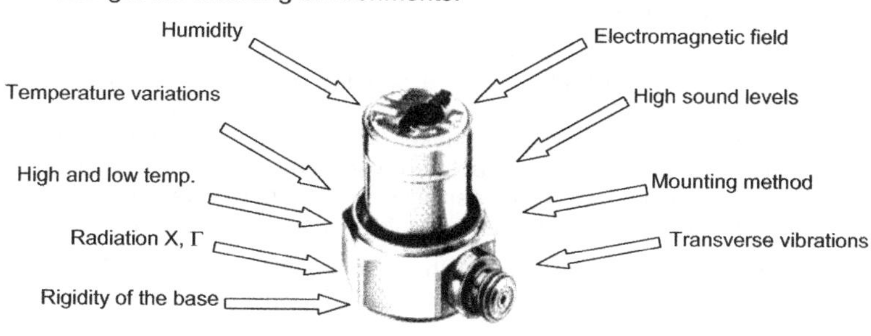

Figure 7.4. *Example of influencing factors on accelerometer sensors*

f) Reminder on accelerometer fixations

(a) Screwing

This is the most effective fixation. Unfortunately, it cannot always be used because it cannot penetrate all structures. However, for control sensors, it can be used more often (it is almost imperative) because they are often placed on the test frame near the fixation points. Some standards require screwing (acoustic discretion, for example).

(b) Adhesion

This is the most used method, particularly for measurement sensors. One should remember the following types of glues:

- Cycril (cyanoacrylate monomer), which sticks instantly but reacts badly to humidity and must be used on a smooth surface (this is the most used method because it holds well and is simple to implement);
- Hottinger, a sort of powder cement and curing agent;
- Métolux, which works on rough and awkward surfaces, humidity has no effect on it but it is not instantaneous (30 min).

There are certainly others that we have not mentioned here.

(c) Double-sided adhesive tape

This method of fixation is used on delicate materials (electron tubes, etc.). Its only advantage is that it is easy to take off and leaves no trace but this is also its disadvantage. This is used more in BF.

Let us not forget old methods that used magnet systems and manual points.

7.1.4. Calibration of accelerometer sensors

The sensitivity of sensors varies depending on the weather and other environmental parameters; therefore, it is necessary to periodically provide calibration rounds (typically once a year). This is a delicate operation and precise sensor calibration can only be done by an accredited COFRAC metrological body. Without going into too much detail, the physical techniques involve interferometric measurements of displacement by laser or measurement by reciprocity techniques (a kind of double weighing), which overcomes the sensitivity of sensor accessories.

For intense use and very critical tests (Ariane boxes, sonar, satellite, etc.), calibration is redone more often (every 3 months or right before the test). It is therefore appropriate, depending on the needs and requirements of the maximum permissible error (MPE) specifications, to allow the possibility of performing calibration operations internally at the laboratory (option offered by standard 17025, subject to validation of the method by a metrological COFRAC expert audit). This is all the more essential the bigger the set of sensors is (sometimes hundreds).

The most commonly used method involves using a secondary standard sensor, which is carefully stored and regularly calibrated by an accredited organization. It should be used in combination with the sensor to be calibrated. The accelerometer is placed back to back with the calibrated accelerometer on the specific shaker (transverse and minimum harmonic rate). To do this, the calibration sensor is manufactured specifically for this, particularly regarding its fixations, stability, robustness and sensitivity. We then compare the results of measurements and deduce the measurement and its associated uncertainty.

Typically, measurements are carried out with an uncertainty of about 3–10% in the worst cases (depending on the frequency). MPE is usually ±20%, which it is easy to observe.

A simplified example is given below. Other methods and presentations are also possible (see Appendix G).

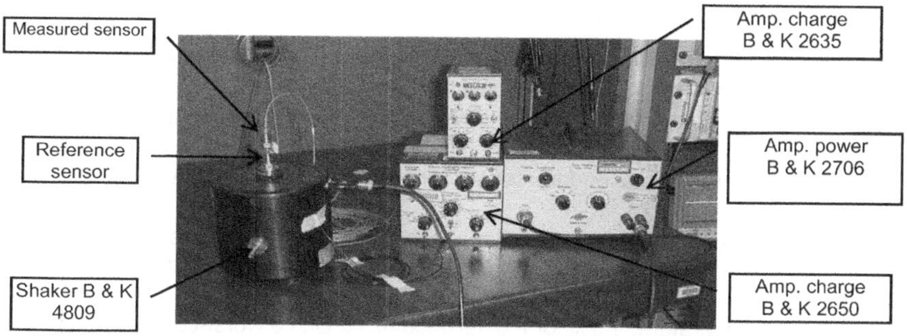

Figure 7.5. *Example of a calibration method for accelerometer sensors*

Example for determining measurement uncertainty of a accelerometer

Type : xxx	No. of série = xxx	No. of CB = Xxx

Verif date	xx/xx/xxxx	Freq.	Réf. value	Meas. value	Sensibility deviation %
N° CVM	XXX	20	1,0000	1,01	1,00
		100	1,0000	1,01	1,00
		250	1,0000	1,015	1,50
		1,000	1,0000	1,01	1,00
		2,000	1,0000	1,00	0,00

Method for calculating absolute uncertainty

- With **Ie** (ref. system uncertainty),
  we get uj(Ie) = Ie/2
- With **Ivxi** (frontal acquisition uncertainty),
  we get uj(Ivxi) = Ivxi/2
- Standard deviation **e**
  We get ua(e) = e
- Correctness error **Ej**
  Ej = avg value (meas) – avg value (ref.)
- Error due to the method
  ujmet = EMTmet/3
- Error due to influencing factors
  uh = tolerance BK4809
  utr = transverse rate BK4809

uh = tolerance BK4809

Average	1,0000	1,0090	
min	1,0000	1,0000	
max	1,0000	1,0150	
Ecart E_j = Deviation Ej = average (meas.) – average (ref.)	0,01	0,90%	
e = standard deviation on sensitivity deviation			0,55%
Ie = uncertainty on the reference system	1,5%		
Ivxi = uncertainty of the frontal acquisition	2,00%		
MPE Estimated method	3,00 %		
uh = Influencing factor tolerance (BK 4809)	1%		
utr = Influencing factor transverse (BK 4809)	0,5%		

		Validation
MPE criteria ≤	**5**	Complies
Sensitivity deviation criteria ≤	**5**	Complies
Initial value in mV/g	**10,00**	
Actual value in mV/g	**10,00**	
Measurement sensitivity deviation	**1,00**	

MPE method	e	Ie	Ivxi	Uh	Utr	uc^2 = sum (ua^2 + uj^2)
3.00	0,55	1,50	2,00	1,00	0,50	
ujmet = MPE/3	ua(e) = e	ujIe = Ie/2	ujvxi = Ivxi/2	uh	utr	uc
0.50	1,00	1,42	0,75	1,2	0,5	2,03%
Widespread uncertainty of the reference system $U = 2u_c$						**4,06%**
Absolute uncertainty Ej + |U| =						**4,96%**

NB: Let us note that in this case, it is necessary to have expertise in the field of metrology and appropriate training.

7.1.5. Accuracy of accelerometers

This depends on the following factors, which are given as examples and overlap relatively well with the calculations that were previously proposed in Excel tables:

- calibration: the absolute accuracy of the secondary reference calibration is around 1 to 1.5%;

- error due to calibration: secondary sensor with better amplification around 2% from 20 to 2 000 Hz;

- error due to frequency: the retained sensitivity is the sensitivity to specific frequencies (260 Hz or 1 000 Hz). The error introduced by keeping this sensitivity for the entire frequency range is around 2%;

- errors due to temperature: 1,5% for a difference of 10°C in a worst-case scenario;

- various influencing factors: noise level, background noise, transverse, mass of the accelerometer, etc., around 0.5%. The adhesion methods can provide much larger errors if improperly controlled for (50 and even 100%).

The aim is to obtain a combined uncertainty of about 5% in order to remain compatible with MPE specifications; otherwise, the measurement system must be reviewed by working on the quality of each element.

a) Error due to transverse sensitivity

Let us consider, for example, the structure to be vibrated below. In a vibration along the 0Z axis, let us suppose that in A, we have vibration of 10 g along 0Z and 1 g along 0X generated by the geometry of the structure. An accelerometer with 5% sensitivity placed at A then expresses:

$10g+0,5:1=10,5g$, meaning an error of $\frac{10,5-10}{10}.100=5\%$

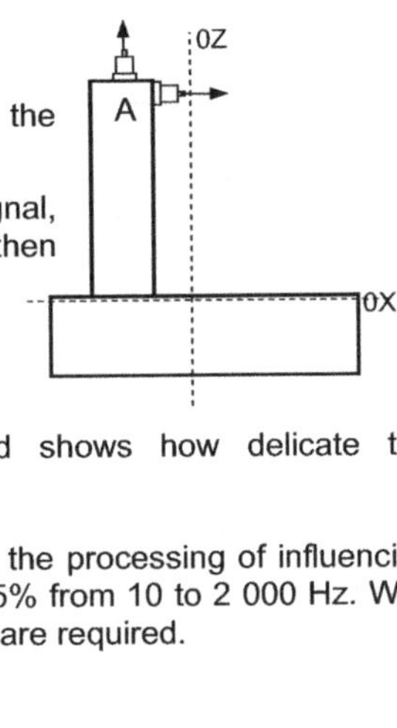

which is the error due to the transverse sensitivity of the accelerometer, on the nominal measurement along the 0Z axis.

Let us suppose that we want to measure the transverse signal, of the following structure along 0X. The accelerometer will then show:

$1g+0,05.10=1,5g$, meaning an error of $\frac{1,5-1}{1}.100=50\%$

which is substantial compared to previous errors and shows how delicate the measurement of transverse components is.

In conclusion, depending on the quality of the sensors and the processing of influencing factors, composed uncertainties typically vary between 1 and 5% from 10 to 2 000 Hz. With each stricter requirement in terms of MPE, specialized sensors are required.

7.1.6. Sensor installation rules

The main rules are as follows:

- choose the type of sensor based on need (mass, cable output type, integrated electronics) or not, isolated or not (for problems of electrical noise due to ground loops), possibilities for fixation, sensitivity following shocks or vibration, transverse, etc.. The initial choice is determinant;

- avoid unnecessary stress at the fixation if it needs to be isolated or installed with an intermediate adapter piece;

- set the measuring cable on the structure at the immediate output of the sensor (1–2 cm from the housing) usually using Scotch paper;

- check that there are no overriding influencing factors (temperature, EM field, significant acoustic noise, large transverse vibrations, etc.), and if there are, they must be taken into account in the measurement.

7.2. Measurement amplifiers

7.2.1. Overview

We have seen that the piezoelectric accelerometers have high internal impedance, so the signal needs to be amplified using a special amplifier. Two types of amplifiers can be used:

- voltage amplifiers;
- charge amplifiers or converters.

Although currently, there are increasingly robust integrated electronic sensors that do not require the use of special amplifiers, we will, nonetheless, discuss classic amplifiers in order to fully understand the operation of sensors and measurement systems. Let us note that in general, actual front-end acquisitions have sensor inputs with or without integrated electronics.

7.2.2. Equivalent circuit

Before going into the different types of amplifiers, one must determine the electrical model of the sensor in order to calculate the analytical relationships of the measurement system. We have seen that the piezoelectric sensor is a generator that produces an electrical charge across its surface when one applies a mechanical force:

$$q = K.F \qquad K = \text{piezo constant} \tag{7.8}$$

The accelerometer can then be modeled either by a current generator or by a voltage generator.

a) Voltage generator equivalent circuit (vacuum)

The open circuit voltage that appears across the model is:

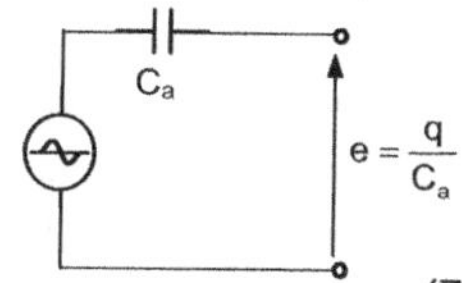

$$e = \frac{q}{C_a} \tag{7.9}$$

C_a = capacity of the accelerometer.

b) Charge (or current) generator equivalent circuit

We ignore the internal resistance R_a, which is very high (R_a > 200 000 MΩ). In practice, to collect the signal from the accelerometer, one must include a circuit that adds additional parasitic capacitance due to:

- the amplifier: C_i;
- the connecting cable: C_c;
- an amplifier input parallel resistance R_i.

The equivalent circuits become $C_t = C_c + C_i$:

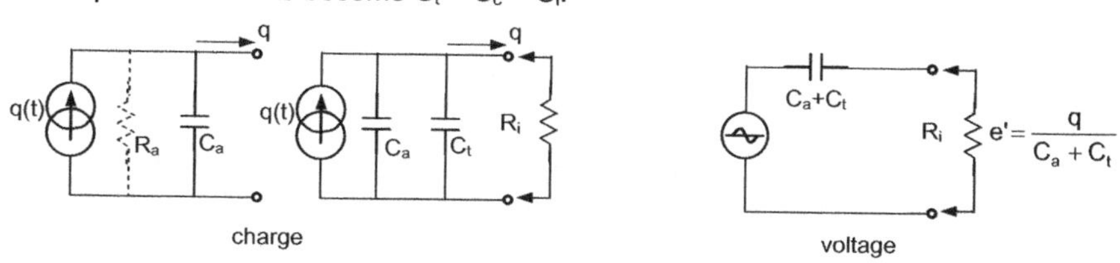

In the voltage equivalent, when the amplifier is connected, the voltage e′ that appears at the R_i terminals (the amplifier input resistance) is equal to $q/(C_a+C_t)$, if R_i is big ($>10^8\ \Omega$). We note of course that e′ is smaller than e, the voltage delivered by the vacuum equivalent generator.

More specifically, e′ during test phase measurements is different from e_{cal} during laboratory calibration, as C_t has different values, which is a disadvantage.

In contrast, the amount of charge q does not vary.

We can then calculate e′ through:

$$e' = \frac{q}{C_a + C_t} \quad (\text{where } R_i > 10^8\,\Omega)$$

e′ in V/g

q sensitivity declared by the manufacturer in pC/g

C in pF

Or:

$$e' = e_{cal} \cdot \frac{C_a + C_{cal}}{C_a + C_t}$$

E′ in mV/g

e_{cal} sensitivity in mV/g measured during calibration

C_{cal} capacitance of the cable and amplifier used for calibration in pF

Advantage: By adjusting C_i, we can get e′ constant ("standardized or normalized") for all measurement channels, although e, q and C_a vary for each sensor.

7.2.3. Voltage amplifier

The voltage amplifier is an amplifier with a classic design but with a very high input impedance, which are obtained by:

- a counter-reaction circuit (R_I #100MΩ = $10^8\Omega$);
- an electrometer tube (R_I #100000MΩ = $10^{11}\Omega$);
- field effect transistors (R_I #$10^{14}\Omega$).

Disadvantages:

- sensor sensitivity variation along the length of the connecting cable;
- multiple risks of alteration of the input internal impedance;
- low-frequency attenuation.

Indeed, the LF response is a function of the time constant RC, in which the capabilities and strengths of the accelerometer and associated electrical circuits come into play. With, for example, the following typical values:

- C_a = 300 pF;
- C_c = 200 pF;
- C_i = 10 pF;
- R = 100 MΩ.

Meaning: $\tau = RC = 100\times10^6 \times 5\times10\times10^{-12} = 5{,}1\times10^{-2}\,s \rightarrow \omega_c = \frac{1}{RC} = 19{,}6\,rd/s \rightarrow f_c = 3{,}1Hz \text{ at } -3dB$.

which already gives 5% attenuation at 10 Hz, while the tests often start at 2 or 5 Hz. This response in LF will be even worse if the input resistance is lowered and the risks are numerous throughout the manipulation (fingers on connectors, air humidity, etc.).

7.2.4. Charge converters

The charge converter, mistakenly called "charge amplifier", is the only one that is currently used in practice. This is an operational amplifier (OA) with total capacitive counter-reaction (works in Miller's amplifier). We recall the ideal OA model:

- large open loop gain (BO) and 180° phase shift;
- high input impedance (∞);
- no input current;
- low output impedance (0 Ω).

a) Reminder of the main results

The following model is used:

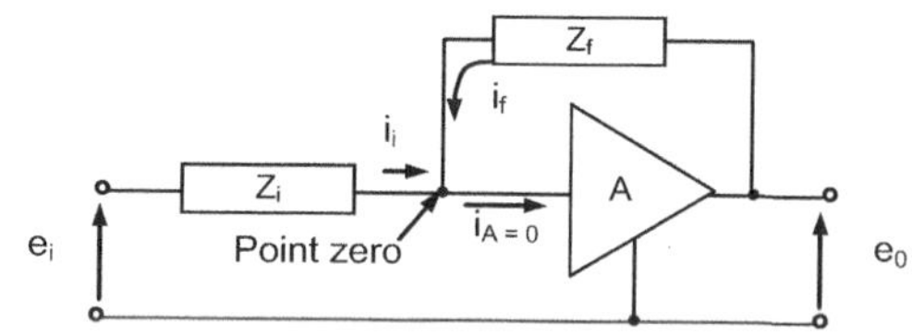

where:

- currents: $i_f + i_i + i_A = 0$ (1)

- voltages: $-\dfrac{e_0}{e_i} = \dfrac{Z_f}{Z_i} \cdot \left\{ \dfrac{1}{1 + \dfrac{1}{A} \cdot \left(1 + \dfrac{Z_f}{Z_i}\right)} \right\}$ (2)

- if A is big: $-\dfrac{e_0}{e_i} = \dfrac{Z_f}{Z_i}$ (3) (7.10)

- input impedance for zero point: $Z_{input} = \dfrac{Z_f}{1 + |A|}$ (4)

b) Applications to capacitive impedances

In the charge converter, the accelerometer itself and its cable (C_a, C_c) act as head impedance Z_i of the OA input (including the input capacitance that is already defined and that is very low relative to the others, typically 10 pF). The model is therefore:

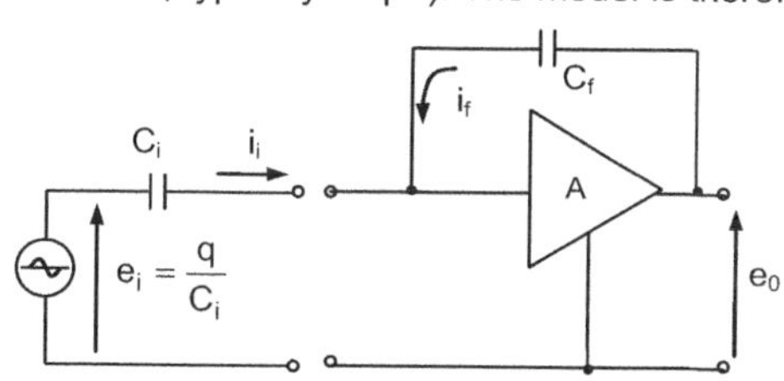

To calculate it, the voltage equivalent circuit is used so as to highlight the charge generated from the known characteristics of the accelerometer. From equation (3), we deduce:

$$e_0 = -e_i . \frac{Z_f}{Z_i} = -e_i \frac{C_i}{C_f}$$

However : $q = e_i.C_i$

Wherefrom : $e_0 = -\dfrac{q}{C_i}$

The output voltage is therefore only a function of the input charge q and counter-reaction capacitance C_i of OA. The circuit thus operates like a current and is not affected by the capacity of the head circuit. As the connecting cable between the accelerometer and the amplifier is only used to transport the charges, its length and capacity do not intervene.

c) Input equivalent impedance

Equation (4) gives:

$Z_{input} = \dfrac{Z_f}{1+|A|} = \dfrac{1}{C_f.(1+|A|)}$. The input impedance is purely capacitive and if gain A is big:

$$C_{input} = C_f.|A| \qquad (7.11)$$

Example 7.1: If A = 10; C_f = 1 000 pF, then: C_{input} = 10^7 pF = 10 μF. The accelerometer sees the amplifier as a large capacitor.

d) Conversion error

According to formulas (2) and (3), we see that accuracy of e_0 depends on the accuracy of knowledge of Z_f, Z_i, A only. The exact formula is then:

$$e_0 = -e_i.\frac{C_i}{C_f}.\left\{\frac{1}{1+\frac{1}{A}.\left(1+\frac{C_i}{C_f}\right)}\right\} \rightarrow \text{big A gives} \rightarrow e'_0 \approx -e_i.\frac{C_i}{C_f}$$

The relative error is then directly given by the following expression, regardless of whether the cable length and the gain A is taken into account in the calculation of e_0:

$$\frac{\Delta e_0}{e_0} = \frac{e_0 - e'_0}{e_0} = 1 - \frac{e'_0}{e_0} = 1 - \frac{1}{e_0}.e'_0 = 1 - \frac{1+\frac{1}{A}\left(1+\frac{C_i}{C_f}\right)}{-e_i(C_i/C_f)}.\frac{-e_iC_i}{C_f} = 1 - \left\{1+\frac{1}{A}(1+C_i/C_f)\right\}$$

$$\frac{\Delta e}{e_0} = -\frac{1+(C_i/C_f)}{A}$$

This presents two cases:

- First case: C_i # C_f. $\left|\dfrac{\Delta e_0}{e_0}\right| = \dfrac{2}{A}$. Thus, if A = 10^4 (typical), $\left|\dfrac{\Delta e_0}{e_0}\right| = 0{,}02\%$, which is a very small error;

- second case: $C_i >> C_f$. Let us then calculate C_i (# C_c) to get an error of 1% on $\left|\dfrac{\Delta e_0}{e_0}\right|$, which is an acceptable error. We get:

$$10^{-2} = \frac{1+C_i/C_f}{A} \rightarrow C_i = (10^{-2}A-1).C_f$$

With : $A = 10^4$; $C_f = 1000$ pF (typical) $\rightarrow$ $C_i \approx 100.C_f = 100\,000$ pF

Thus, for a typical cable of 100 pF/m (high enough), we can use a length of 1 km with an error of only 1%, compared to using a short 3 m cable, for example.

Hence, the interest in using this type of charge converter.

e) HF response

The charge converter behaves like an OA connected as an integrator circuit, therefore equivalent to a low-pass filter of very low cutoff frequency. We have seen that:

$$e_0(t) = -\frac{q}{C_f} = -\frac{1}{C_f}.\int i_i(t).dt \quad \text{Where: } i_i(t) = I \sin \omega t, \text{ we get:}$$

$$e_0(t) = \frac{1}{C_f}.\frac{I}{\omega}.\cos \omega t$$

Therefore, e_0 decreases as ω increases. In addition, an OA is further limited at the cutoff frequency (as for any amplifier) and beyond the cutoff frequency, its gain decreases. We usually characterize this by its product A.B (gain in bandwidth). For good quality OA, A.B > 15 MHz, which means that A = 1 at 15 MHz or A = 15 000 at 1 000 Hz and 7 500 at 2 000 Hz with an order of magnitude of 10^4.

f) LF response

The following model is used:

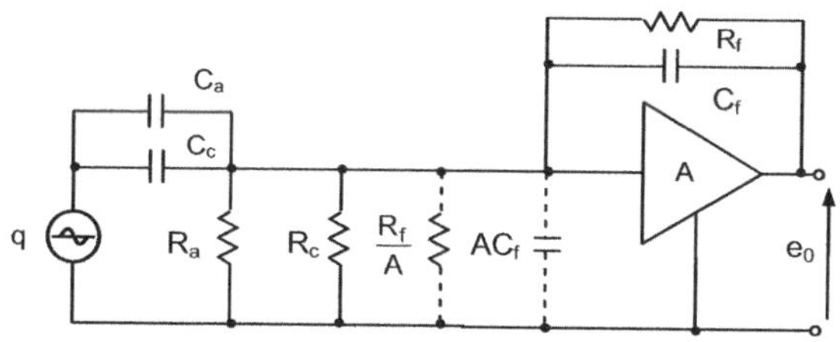

The time constant is typically given by: $\tau = \frac{1}{\sum(1/R)}.\sum C$

Where:

- C_a = accelerometer capacitance;
- C_c = cable capacitance;
- AC_f = reduced input capacitance;
- R_a = resistance of the accelerometer;
- R_c = cable resistance;
- R_f = leakage resistance of C_f;
- R_f/A = reduced input resistance.

The calculation shows that the R_f/A and AC_f parameters are paramount, indeed:

$$\frac{R_f}{A} << R_a \text{ or } R_c$$

$$AC_f >> C_a + C_c \qquad (7.12)$$

$$\text{Therefore:} \quad \tau = \frac{R_f}{A}.AC_f = R_f.C_f$$

The answer in LF only depends on qualities of the counter-reaction capacitance. If the leakage resistance R_f is infinite, C_f retains the charge and the system is pseudo continuous.

There are amplifiers with a time constant of around 1 000 s. Such amplifiers can be used to measure static forces, for example, pressure sensors, flip-flop force sensors, etc.

In practice, we limit the bandwidth in LF to overcome possible derivatives from the zero of OA in continuous by placing a high-value output capacitor.

For example:

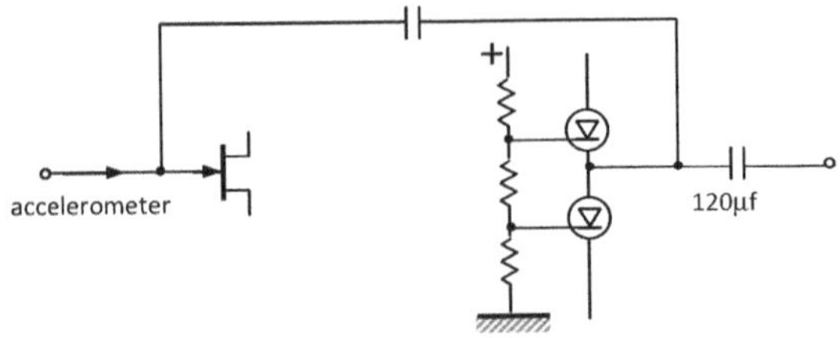

Ultimately, the practical useful band generally extends from 2 to 20 000 Hz at ±5% and the cutoff frequencies at –3 dB are around 0,1 to 1 400 kHz.

g) Background noise

It can be equated to a voltage noise e_b reduced at the input. The model is the following:

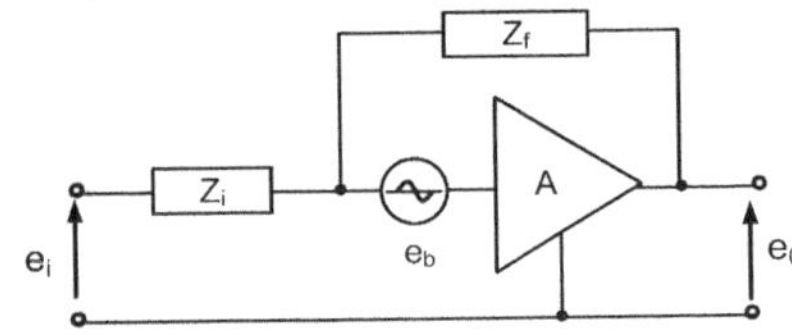

Once all calculations are done, we get:

$$-e_0 = \frac{e_i.(Z_f / Z_i) + e_b(1 + Z_f / Z_i)}{1 + \frac{1}{A}.(1 + Z_f / Z_i)} \tag{7.13}$$

If A is big and the impedances are purely capacitive (in practice), we get:

$$-e_0 = e_i.\frac{C_i}{C_f} + e_b.\left(1 + \frac{C_i}{C_f}\right) \tag{7.14}$$

The noise voltage thus acts as an input voltage. If C_i increases, e_b increases and the output noise increases, the longer the connecting cable.

h) Typical block diagram of a charge amplifier

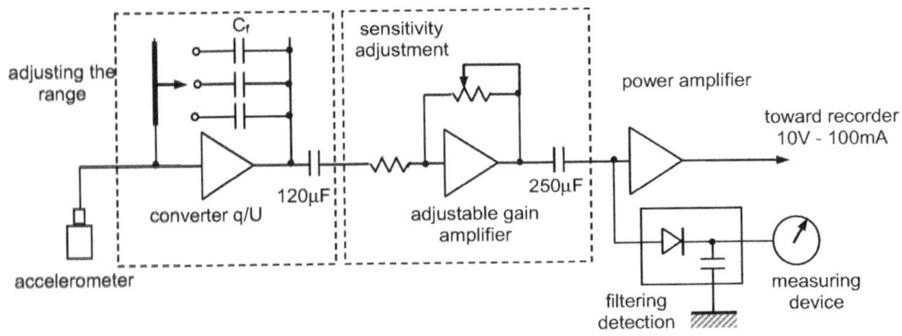

The RANGE adjustment switches the acceleration value of acceptable full scale from 1 to 3 000 g/c, to get a standard output voltage at ±RMS. The SENSITIVITY setting is a 10-turn potentiometer at 1%, calculated such that it suffices to display the sensitivity number of the sensor used on the potentiometer vernier in order to directly obtain the normalized output (there are, for example, conversion factors of around 0.2–10 mV/pC).

NB: Currently, for significant sensitivities or to measure strong shock amplitudes, one must use charge amplifiers. In other common cases, it is simpler to use integrated electronic accelerometers for which the sensitivity is factory-set once and for all (usually around 10 pC/g) and for which the output is directly normalized at the accelerometer output. The front-end acquisition input amplifiers are simplified, as well as the adjustments and costs.

Here is a diagram of a generic measurement system:

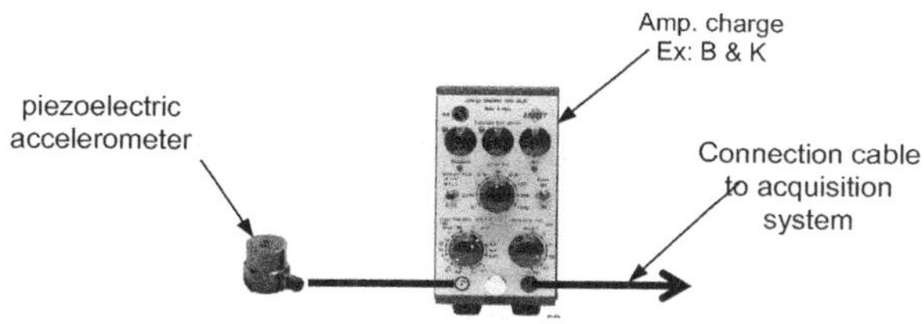

Figure 7.6. *Example of a charge amplifier*

7.3. Validation and verification of the testing means

The requirements for quality in testing laboratories impose verification and validation testing means (standard NF EN ISO/CEI 17025). These include:

- verification (or characterization) of the testing means to evaluate the different influencing factors that occur during the course of a test in order to control for them and include them in the estimation of uncertainties and in the operation validation;

- validation of tests to verify the performance of the testing means compared to those that are expected in order to meet the requirements of the standards used.

To do this, one may base themselves on the following documents (non-exhaustive and possibly outdated or amended):

NF EN ISO CEI 17025 "General requirements for the competence of testing and calibration laboratories";

COFRAC: LAB Ref. 02 "Guide for using the NF EN ISO 17025 standard in accredited laboratories or accreditation candidates";

EN ISO 9001 V2000 "Quality management systems – Model for quality assurance in design, development, production, installation and servicing";

NF X 50-142 (12/1990) Quality of tests;
RG.Asro 000 11 (03/1986) Environmental quality of tests;
RE.Asro 601 11 (02/1994) Provisions relating to quality of tests in a mechanical environment;

Theory and application Handbook by B&K November 1987. Piezoelectric Accelerometers and Vibration Preamplifiers;

Technical review B&K N° 2 1987, Trends in Accelerometer Calibration;

Mesures dynamiques Volume III Endevco 1989.

a) Example of proposed methodology for an electrodynamic exciter

(a) Required operations

- Composition of the means.

- Limitations of the means.

- Description of the annual preventive maintenance operation.
- Description of metrological operations on measurement systems;
  - calibration:
    - if calibration is carried out by an accredited laboratory, the calibration certificate is proof that standards are met;
    - if calibration is carried out by a non-accredited laboratory, the body responsible for the calibration provides a calibration report and provides evidence of its connection through the traceability of standards used during the calibration and a copy their calibration certificates;
  - verification: in this case, the body responsible for checking provides a verification certificate (report of measurements) and provides evidence of the connection through the traceability of standards used when checking and copying their calibration certificates.

NB: Calibration and verification can be done internally with an audit and validation of the method, called "metrological confirmation", for a measuring device or instrument.

- Initial characterization of the testing means and the following:
  - definition of the time frame;
  - measurement of influencing factors (table homogeneity, transverse rates, harmonic rate, etc.).

(b) Definition of the actual methodology

- Configuration of the testing means.
- Estimation of uncertainty of the test means (based on MPE standards).

NB: In terms of difficult measurements, the calculation of uncertainty does not take special cases due to subtest assemblies into account (fixture + tested material interaction): these induce resonances, transverse rate, harmonic rate and abnormal jounce.

- Establishment of the verification certificate.

NB: Of course, this example is only given for information's sake, each methodology has to be adapted to the specific needs of each testing laboratory.

7.4. Control of metrology in a testing laboratory

The aim of this section is to raise awareness in the laboratory of the importance of metrology in a testing laboratory. For this, one does not need to delve into the details on metrology control operations, which are relatively large and complex. We highlight the NF EN ISO/CEI 17025 standard. It primarily involves:

- organizational quality from a metrological point of view (responsibility, training, etc.);
- the calibration procedure (internal or external);
- the verification operations (internal or external);
- estimation of uncertainties.

a) Reminders

Metrological requirements include the concept of MPE, which was already mentioned in section 7.1.4 and is defined in the ISO/CEI 99 guide in Article 4.26:

"Maximum permissible error: error limit or extreme value of measurement error, relative to a known reference value, which is permitted by specifications or regulations for a measurement, a measuring instrument or a given measurement system". The methods of standardized tests do not generally use the term MPE, other terms may be used such as "tolerance" or "accuracy".

One should select the MPE from standards' requirements and allocate them to the different components involved in obtaining a required objective by taking the measurement uncertainties into account in the declaration of compliance.

The MPE can be used to measure the accuracy of the equipment by itself (determined by calibration) within the context of its use, taking external influencing factors into account in its implementation (installation, handling, environmental conditions, etc.). This can be denoted U_{ext}. The result of the measurement process then comprises a combination of these two terms. This then determines the accuracy of the measurement or its capability (see n5 and NF E 02-204 for more details) to declare compliance (or non-compliance) of the equipment. The capability of a measurement process is defined as the ratio between the tolerance to be checked (T: difference between specification minimum and maximum values) and the measurement uncertainty (denoted U) expressed during the verification of the object, meaning C = T/U. In the event that the requirement used is a MPE, T must be replaced by the extent of the MPE, meaning 2*MPE (or MPE in ± %).

b) Allocation of MPE

The MPE is generally either attached to the main magnitudes that are involved and which are to be achieved in the test or the magnitudes of influences as expressed above; although these are not popular, they are inevitable and therefore influence the main measurement values or behavior and/or the response of the test material.

Moreover, the values written in the standard procedures are not always directly usable for verification of testing means or measuring devices, because they can mask a measurand deduced by combining several components (different measurement or test equipment).

Each of these components must then be "allocated" the share of MPE that we consider to be the most suitable (see Appendix H).

c) Quality organization

Mainly:

- procedures (management and technical);
- responsibility (management);
- training, expertise and competence (technical);
- management of measuring equipment and test facilities;
- validation of specific methods, confirmation of standard methods.

d) Calibration and verification

- External, control of subcontracting, development of specifications.
- Internal, method validation.

Example 7.2: Simplified calibration method of accelerometer sensors (a more elaborate method is given in Appendix G).

e) Evaluation of influencing factors

(a) Measurement of harmonic rates

A sine vibration scan is performed from 10 to 2,500 Hz. Measurement of the harmonic rate is determined from the ratio (NF/F) RMS level/filtered level recorded by the reference sensor. Calculations are performed using the following relationships:

$$T_{\%} = \left(\frac{NF}{F} - 1\right) x100 \text{ and } D = \sqrt{T^2 + 200T} \text{ With D and T in } \% \rightarrow \frac{D}{100}\sqrt{\left(\frac{T}{100}\right)^2 + \frac{2T}{100}}$$

T = tolerance on the signal
NF = RMS signal value including harmonics
F = value of the filtered fundamental
D = distortion

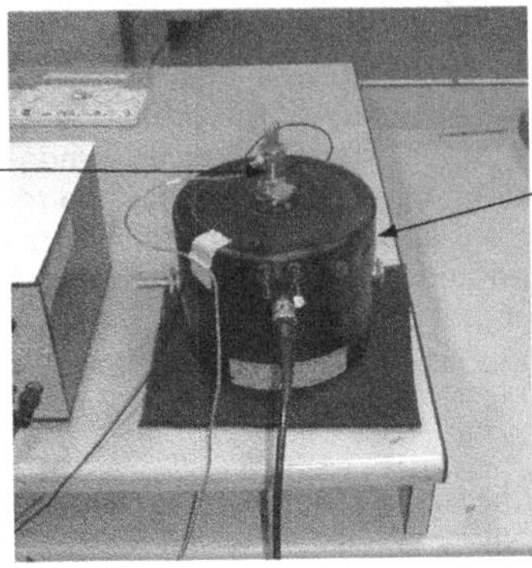

Figure 7.7. *Common testing setup*

Frequency	20	100	250	1 000	2 000
NF/F	1.0017	1.0006	0.9996	0.9988	0.9983
T%	0.17	0.06	-0.04	-0.12	-0.17

Uh (standard devitation on T)
0.14

Table 7.1. *Table of the measured values*

(b) Measurement of the transverse rate

A sine vibration scan is performed from 10 to 2 500 Hz. The measurement of the transverse rate is performed using two measuring sensors stuck to the reference sensor.

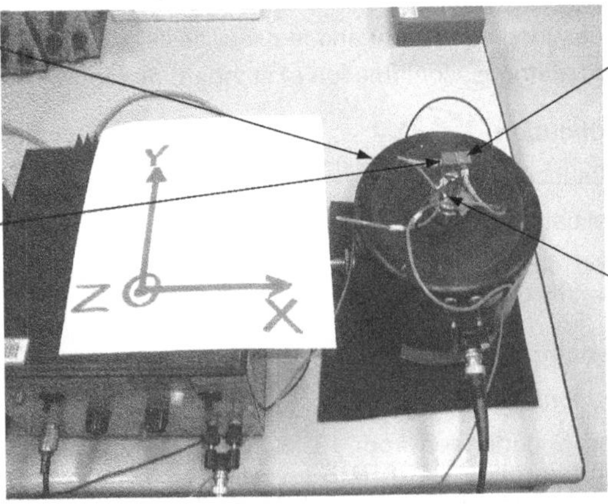

Figure 7.8. *Test setup for measuring the transverse*

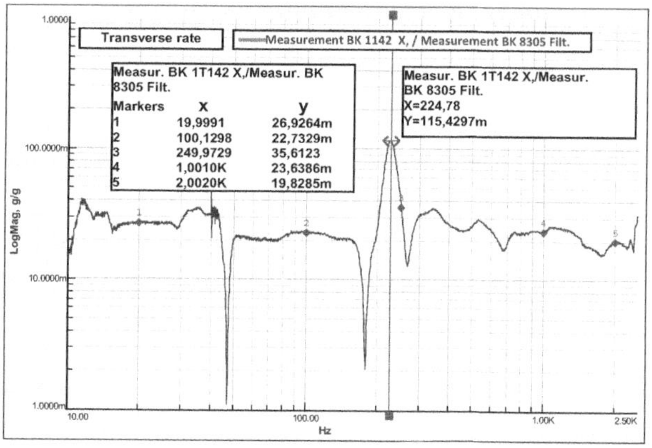

Figure 7.9. *Measurement charts: max. transverse rate on the five frequency points axis OX = 3.6%*

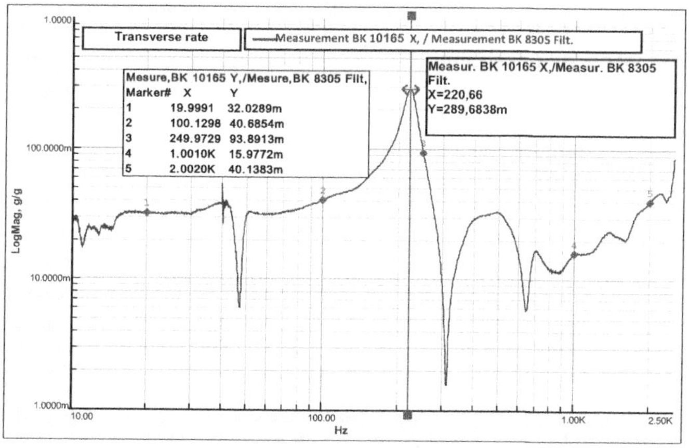

Figure 7.10. *Measurement charts: max. transverse rate on the five frequency point axis OY = 9,4%*

(c) Measuring the influence of ambient conditions

Measuring the influence of ambient conditions is carried out at two frequency points (20 Hz and 1 000 Hz). The low temperature variation is simulated using an icing bomb and high temperature variation is simulated using a hot air blower.

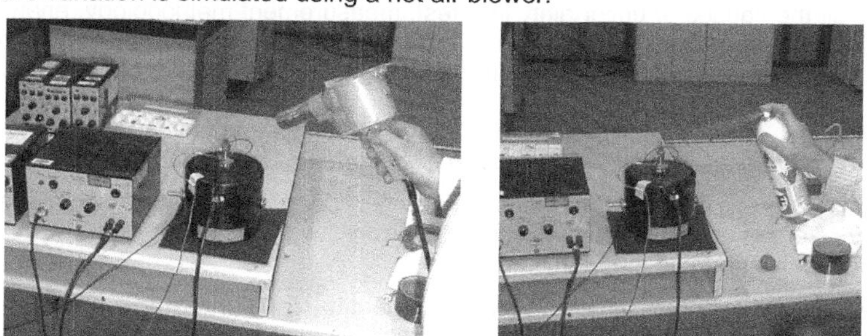

Figure 7.11. *Test setup*

Influence of temperature variations

	Verif. Date		27/03/2003			
	PShaker B&K 4809		Serial no = 1374964		N° CB = 010642	
Type of variation	Variation range	Delta T°	Frequency Hz	Reference value	Measurement value	Level variation in %
Low	20°C à 10°C	-10°C	20	2,023	2,030	0,35
			1000	1,976	1,964	-0,61
High	20°C à 30°C	10°C	20	2,023	2,065	2,08
			1000	1,957	1,967	0,51

Table 7.2. *Table of readings and determination of the amplitude variation*

The only variation in significant amplitude (2,08%) was measured at 20 Hz for $\Delta T \approx 10°$ C.

f) Estimation of uncertinaties

As an approach that is now mandatory, it requires significant expertise and therefore extensive training in this field. There is no need here to go into the details of the whole approach for estimation of uncertainty, as it is relatively complex. Here, we outline the following requirements (see Appendix G):

- procedures (management and technical);
- responsibility (management);
- training, expertise and competence (technical).

(a) Concept of uncertainty and MPE

The tests described by outsourcers in the technical specifications for all levels commonly refer to civil or military normative documents. These describe the tests based on standard norms or refer to a "personalized test" approach. In each case, there is a constraint on the tolerances of the generated and/or measured signals, regardless of whether they are temporal like the pulses from conventional shocks or described in terms of SRS.

The usual metrological term is "maximum permissible error", which expresses the extent to which the measurement must be assigned with its extended uncertainty. This requirement is extremely important and must be treated with rigor because it directly impacts on the quality of testing. This is all the more important currently, as the demand of outsourcers in terms of testing quality refers to the COFRAC accreditation of testing laboratories, which is very stringent for the measurement aspect.

Here, we briefly recall the general method for evaluating and expressing the uncertainty of measurement.

1 - Identify all the causes of uncertainty in a test, measurement methodology, analysis using the "5M" diagram for example (Figure 7.12), particularly the specific influencing factors in the concerned test (homogeneity of the measurement, sensor transverse rate, distortion, etc.):

"5M" = Material – Medium (environment) – Method – Matter – Human resources (*from the French: Matériel - Milieu (environment) - Méthode - Matière - Moyens humains*)

2 - Reduce systematic errors. This step is especially important with decreasing measurement uncertainty. To eliminate significant residual errors, we will have to use comparative measurements by changing the method.

3 - Mathematically express, if necessary, the quantity to be measured based on all the quantities on which it depends, including all corrections, which may have a significant effect

on the measurement result (for a simple measurement using a measuring device, the quantity to be measured may well be directly obtained. It is then the uncertainty of the device itself that should be taken into account. This is usually expressed in percent or ± an extent in the unit of measurement).

4 - Determine for each of these quantities the estimation of their typical value, either by measurement or through external source (physical constants, etc.).

(5) Determine for each of these quantities the value of the typical uncertainty (uncertainty of measurement expressed as a standard deviation), either by the direct type A method, which corresponds to the calculation of a standard deviation over several successive measurements, or through the direct type B method, which involves finding an estimate by referring to external information sources (metrological characteristics of measuring devices, etc.).

6 - Determine the combined uncertainty (if the quantity to be measured depends on several quantities), either from the variances of these units if they are independent or from their covariance if they are correlated.

7 - Determine the final measurement uncertainty or expanded uncertainty from the combined uncertainty based on a given confidence level (generally the accepted expansion coefficient is k = 2).

NB: For measurements during shock tests, the combined uncertainty is often achieved through a type B method. This can be done using a quadratic average across different typical uncertainties of the devices used, including their weighting coefficients and significant test influencing factors. The expanded uncertainty is then obtained by doubling the combined uncertainty (about 96% confidence). We must then meet the following relationship:

$$\mathbf{U_{expanded} = 2.U_{combined} \leq MPE}$$

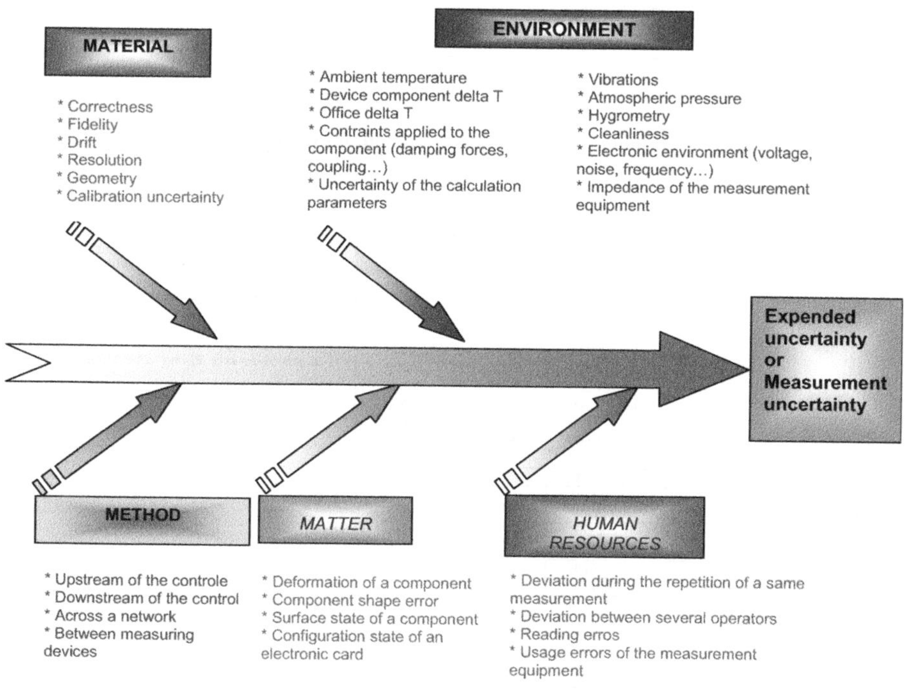

Figure 7.12. *The "5M" method*

Definition of measurement uncertainty: Uncertainty of measurement is by definition the result of the evaluation for characterizing the extent of the area within which the true value of a measurand is estimated to be, generally with a given probability.

(b) Sample calculation table

This table is for demonstration purposes of a simplified approach. Other techniques are possible (see Appendix G).

Calculating the uncertainty of a measuring system-Mechanical test

Date 09/07/2003

Table for determining the expended uncertainty of a complete measurement system			
Device - EMT - I			ujemt = EMT/3 or ujl = l/2
Accelerometer	MPE in %	3.00	1.00
	I en %		0.00
Amp, charge	MPE in %	3.00	1.00
	I in %		0.00
Méthod	MPE in %	3.00	1.00
	I in %		0.00
4	MPE in %	3.00	1.00
	I in %		0.00
5	MPE in %	0.00	0.00
	I in %		0.00
6	MPE in %	0.00	0.00
	I in %		0.00
	uc		0.00
	U = 2uc in %		**0.00**

Calculation metrhod:
The combined uncertainty uc of the complete measurement system is determined from the uncertainty of each measuring device involved. The uncertainty of each device of type B is denoted **uj** and it takes the correctness deviation of each device info account.

The calculation formula is:
uc² = u²j1 + u²j2 + u²j3 + u²j4 + ...u²jn

The expanded uncertainty of the complete measuring system is:
U = 2uc

Uncertainty uj of each device will be calculated :
- Either from the absolute uncertainty **I**, so **uj = I/2**
- Or from the MPE, so **uj = EMT/3**

Expended uncertainty =			**4%**
Criterion on MPE (control channel)=	**5.00%**	Validation =	**Complies**
Criterion on MPE (measured channel)=	**8.00%**	Validation =	**Complies**

Validation of calculation		Expected value	Validation
Calculation of the combined uncertainty **uc**		2.00	OK
Calculation of the expanded uncertainty **Ul**		4.00	OK

Control system uncertainty and acquisition of mechanical test

Testing means	**Impac 36x36**	**Serial N° 21**	**N° CB = 016644**
Acquisition system	**VXI Thoms-01**	**Serial N° US3901059**	**N° CB = 035164**

Vérif. date

Sample N°	Referencre value	Measurement value	Sensitivity deviation %
1	-21,000	-20,500	-2,38
2	-21,000	-20,300	-3,33
3	-20,900	-20,300	-2,87
4	-21,000	-20,400	-2,86
5	-21,000	-20,200	-3,81
6	-20,800	-20,400	-1,92
7	-21,000	-20,400	-2,86
8	-21,000	-20,300	-3,33
9	-20,900	-20,300	-2,87
10	-21,000	-20,300	-3,33

Absolute uncertainty calculation method:
- With **le** (reference system uncertainty),
  we get uj(le) = le/2
- With **Ihp** (spectrum analyser uncertainty),
  we get uj(Ivxi) = Ivxi/2
- Standadr deviation **e**
  we get ua(e) = e
- Correctness error **Ej**=avg value (meas.) - avg value (ref.)
- Error due to the method
  ujmet = EMTmet/3

Average	-20,96	-20,34	
min	-21,00	-20,50	
max	-20,80	-20,20	
Deviation Ej = avg (meas.) - avg (ref.)	0,62	2,96 %	
e = standard deviation on sensitivity deviation			0,54 %
le = reference chain uncertainty	1,75 %		
Ihp = spectrum analyseur uncertainty	0,58 %		
MPE estimated methode	1,00 %		

MPE method	e	le	Ihp	uc² = sum (ua² + uj²)
1,00	0,54	1,75	0,58	
ujmet = EMT/3	ua(e) = e	ujle = le/2	ujvxi = Ivxi/2	uc
0,33	0,54	0,88	0,29	1,12
Expended uncertainty on the reference system U = 2uc				**2,23 %**
Absolute uncertainty Ej + |U| =				**5,19 %**

Criterion on MPE ≤	**VALIDATION**
10	**complies**

8

Testing Means for Vibrations

This chapter accurately describes how usual vibration means operate, namely electrodynamic generators and electrohydraulic generators. An approach using models is proposed for both cases in order to go into the details of how these two tools are used, such that we can understand their limitations, advantages and disadvantages. Some commonly used terms are electrodynamic vibrator, shaker, vibration generator, electrodynamic exciter, etc.

One can loosely say that electrodynamic generators operate at a higher frequency (a few thousand hertz) but only for relatively low masses (around one ton when coupling several generators, three for triaxial vibration), and they are also limited in their displacement, which is a significant limitation during impact tests (currently, 3 inches of displacement are possible). They also have relatively pure signals.

While the latter usually work in LF and even VLF (they skip the continuous state and some go up to 1,000 Hz) with large forces related to the hydraulic part of the operation, they are less limited in their displacement (for example for earthquake tests). The signals generated are generally more distorted [ref. I2, I11, I12, I18].

8.1. Electrodynamic exciters

8.1.1. Brief description of an electrodynamic exciter

An electrodynamic exciter works with the same physical principle as a speaker. It essentially consists of a moving coil, which supports the material being tested and is traversed by a modulator current of the signal to be generated and is immersed in a magnetic field created by another coil, which is itself fixed. Of course, the system is mechanically adapted to the stress that this means allows since it tests itself during testing.

A fragmented view and a simplified view allow us to a better understand how it works (Figures 8.1 and 8.2).

8.1.1.1. *Fragmented view*

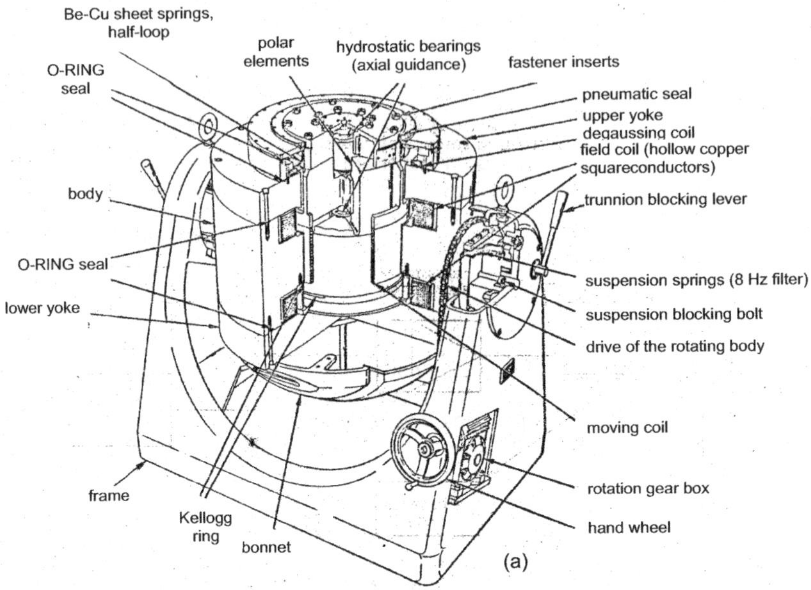

Figure 8.1. *Diagram of an electrodynamic exciter: fragmented view*

8.1.1.2. *Simplified view*

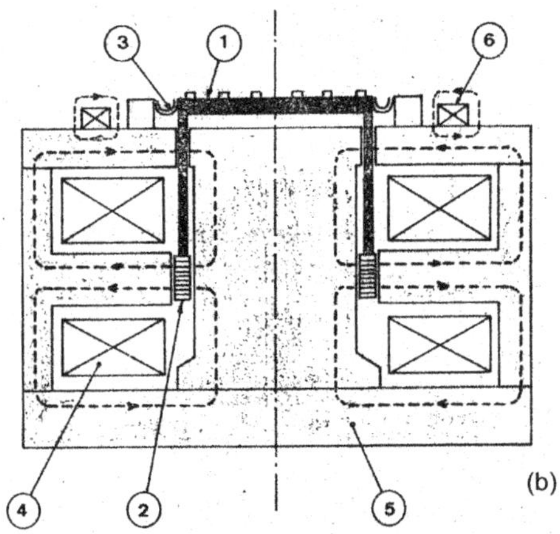

(1) - Vibrating table where specimens are fixed.
(2) - Moving coil secured to the table. The assembly is suspended by springs (3).
(4) - Field coils (fixed).
(5) - Frames
(6) - Demagnetization circuit

The induction field created by the fixed coil is constant. When the moving coil is traversed by a current i (t), an electromagnetic force is generated

$$F(t) = B.\ell.i(t)$$

and the coil moves.

Figure 8.2. *Diagram of an electrodynamic exciter: simplified view*

Figure 8.3 shows an image of a Ratier Forest VRF30 electrodynamic exciter and the usual related performances.

Figure 8.3. *Ratier Forest VRF 30 electrodynamic exciter*

Technical sheet for the VRF 30 shaker

Supply:	Remote amplifier 50 KVA - TEA 5 000
Thrust:	3 000 daN
Vibrations:	3 axes (avec table transverse)
Bandwidth:	5–2 000 Hz
Crankshaft:	
Vertical:	Moving coil weight = 30 kg
	Resonance frequency when empty = 30 87 Hz
Transverse:	Moving coil + transverse table weight = 112 kg
	Resonance frequency in a vacuum = 12 00 Hz

Vacuum and vertical performance:

Acceleration:	981 m/s² or 100 g (between 82 and 20 00 Hz)
Speed:	1,9 m/s (between 25 and 82 Hz)
Displacement:	25 mm c/c (between 5 and 25 Hz)
Max. vibrated mass:	200 kg

Frequency of exciter suspensions = 15 Hz

Control system:
Sine and random vibration
Random vibration + sine lines
Cannon shot
Shock tremor (1/2 sine, triangle, etc.)

8.1.2. Impact opportunities and performance

8.1.2.1. *Possibilities*

The advantage of this testing means lies in the following notable possibilities:

- use of the same test fixture for vibration;

- the impact test can be performed directly after (or before) the vibration test without dismantling the test frame;

- combined vibration-shock tests with a single testing means;

- optimization and "personalization" of tests by "feedback of the shock for a specified shock spectrum";

- flexibility of the procedure due to electric control and the possibility of "enslaving" the waveform by "equalizing the transfer function" (manual or digital console);

- various types of shocks are possible. Impact (with or without rebound) and pulse shocks;

- impact tests for fatigue-shaking tests, etc.

8.1.2.2. *Shock performance*

From the known sinusoidal data, we can roughly define the performance of the entire transient exciter amplifier assembly. The following recommended values are generally permitted:

- maximum displacement: 25,4 mmc/c (1") and 50,8 mmc/c (2");

- maximum peak strength: two to three times that in the sinusoidal state (which roughly corresponds to the peak maximum strength factor in the random state);

- maximum peak acceleration: two to three times that in the sinusoidal state (except for mechanical limitations).

NB: Many vibration tests that are currently in service were not initially designed to function in shock states. There is therefore a high risk of mechanical damage, in particular for the moving coil that, through its construction, has glued elements. Preventive maintenance operations must take place more frequently. For this reason, some manufacturers have adapted the design of vibration impact testing means by:

- increasing the maximum permissible track to 2" instead of 1" (3" today);

- increasing the power of the amplifier with a new moving coil design;

- increasing the opportunities for eligible overvoltage from the amplifier (because of the resonance and antiresonance frequencies, especially for specimens testing, called notch frequency);

- reducing the weight of the moving coil to move its resonance frequency to a higher frequency;

- increasing the stiffness of the moving coil by incorporating the coil into the moving component (instead of gluing it on).

The same type of action may be considered for hydraulic exciters, which due to the way they are built generally show larger displacements (especially in static state) and significant strength capacities. However, their bandwidth rarely exceeds 500 or 1,000 Hz at most, which is usually sufficient for most commonly requested tests (minimum duration of a few ms).

8.1.3. Producing an acceleration pulse on the movable table of a shaker

8.1.3.1. *Mechanical limitations*

(a) The displacement of the moving assembly must be within the limitations imposed by the manufacturer, which is ± 12,7 or ± 25,4 mmc/ relative to the middle position of the moving coil at rest.

(b) In pulse mode, it is assumed that the nominal sinusoidal acceleration level can be exceeded (usually 100 g or 1,000 ms^{-2}c/) by a factor of 2 to 3. This is generally limited to 200 g in order to best avoid the risk of deteriorating the moving coil.

(c) The displacement of the movable table in the air gap of magnetic field coils invokes the existence of a counter electromotive force that opposes the control voltage e(t) of the power amplifier. This is written as:

$$U(p) = B.\ell.\dot{X}(p) \qquad \begin{cases} B = \text{fixed magnestic induction} \\ \ell = \text{length of the winding} \\ \dot{X} = \text{speed of the mobile assembly} \end{cases}$$

Thus, the control voltage from the power amplifier is:

$$E(p) = (R + Lp).I(p) + U(p)$$

E = amplifier control voltage

R = moving coil resistor

L = moving coil inductor

U = counter electromotive force (c.e.m.f)

In other terms: $E(p) - U(p) = E(p) - B.\ell.\dot{X}(p) = (R + Lp).I(p)$ (8.1)

We see that speed is limited by the maximum control voltage E_{mx} permitted by the power amplifier. The limitations given by the manufacturer in sinusoidal state are around 2 ms^{-1}. Also in the transient state, because of the planned random operation and associated peak factor, an exceedance Δ of U due to an exceedance $\Delta\dot{X}$ of the maximum speed is tolerated; the current i(t) is then maintained for the discharge time constant of the filter capacitors of the amplifier power supply (buffer effect). In this example, we are limited to 1,9 ms^{-1} according to manufacturer data.

(d) Another constraint is imposed on speed through the movement of the movable table, which must be "strictly zero" before and after the shock (as we saw in previous chapters). We therefore get the following constraint:

$$\dot{x}(t) = \int_0^t \ddot{x}(u).du = \text{velocity at time "t"}$$

And the total duration of the shock is:

$$\Delta\dot{X} = |\dot{x}(D) - \dot{x}(0)| = \int_0^D \ddot{x}(t).dt = 0 \qquad (8.2)$$

As the integral of acceleration is zero for duration D of the shock, the share of surfaces on either side of the time axis must be equal. It is therefore not possible to have, as we have already seen, a pulse that is always in positive or negative acceleration. There will always be a pre- and/or postshock pulse that has to be respected (8.2).

This constraint also exists for conventional shock machines (freefall machines, for example). This is more significant for electrodynamic exciters due to their significant displacement limitation. Indeed, the secondary pulse, which can satisfy $\Delta\dot{x} = 0$, should not affect the profile of the shock beyond the MPE specifications, otherwise the shock signal itself will be too distorted. To achieve this:

- its amplitude must be small compared to that of the shock (maximum 20%);
- its duration may be significant;
- the generated displacement should also be significant and usually far greater than that generated by the shock itself.

The displacement is the main source of mechanical limitation for electrodynamic exciters.

(e) The moving element (moving coil) is not infinitely rigid. There is a resonance frequency of about 3,000 Hz for large exciters (3,000–15,000 daN) and around 15 kHz for smaller ones (1–5 daN), which are often used in the calibration of accelerometers.

8.1.3.2. *Electrical limitations*

- When there are connections through transformers between the power amplifier and the electrodynamic exciter, limitations can appear in low frequency (LF), which are particularly disadvantageous for impact tests. In general, these are generated by pulses that have most of their energy in LF. Most modern vibrators have a direct coupling to avoid this problem.

- The power of the amplifier is finite and therefore its thrust or force, expressed in Newtons, is also finite. As the force and maximum speed parameters are limited, the acceleration "duration" and "amplitude" will also be limited. For example, we choose 70 kVA for a thrust of 3,000 daN. So in pulse mode, we allow a transient force that is double this, so 6,000 daN.

8.1.3.3. *Maximum permissible displacement*

Load compensation systems borne by the pneumatic or electric moving coils when the amplifier is on continuously allow us to adjust the center position in order to compensate the static load of the test specimen. We can also use this option to preposition the moving coil so that we fully benefit from the available track. Thus, if one can place the table in a higher position (respectively, low), we almost double the available maximum displacement Δx and thus the product $\frac{\ddot{x}_m D^2}{\pi}$ for a half-sine and shock without rebound. For a given D, we double the maximum possible speed given as $\frac{\ddot{x}_m D}{\pi}$ and therefore the maximum possible level of acceleration.

Similarly, if one considers duration D, it is possible to double the duration of the shock for the same given maximum acceleration. All intermediate combinations are of course possible.

8.1.3.4. *Theoretical limitations*

We are able to analyze the theoretical peak performance in "acceleration" and "duration" for a half sine of the electrodynamic exciter with its maximum performance retained for the following mechanical reasons:

$$\begin{cases} \ddot{x}_{mx} = 2\,000 \text{ ms}^{-2} \\ \dot{x}_{mx} = 1{,}9 \text{ ms}^{-1} \\ F_{mx} = 3\,000 \text{ daN} \\ x_{mx} = 25{,}4 \text{ mmc/c} \end{cases}$$

These limitations are primarily based on the parameter ε and duration D of the shock. We therefore propose to draw the following networks of curves:

$A = f(D,\varepsilon)$ A in [g] D in [ms]

The study of half-sine pulses with half sine pre- and postpulses [ref. I18] and pre- and postrectangular pulses are used here in a pedagogical approach:

- for a maximum acceleration curve: $A(D) = A_{mx} = cte$

- for maximum speed: $$A(D)=\frac{\Delta\dot{X}_{mx}.\pi}{D}$$

- for maximum displacement: $$A(D)=\frac{\Delta x_{mx}.\pi^2}{D^2.\left(1+\frac{1}{2\varepsilon}\right)}$$

Hence the curve in Figure 8.4 for ε = 0,2:

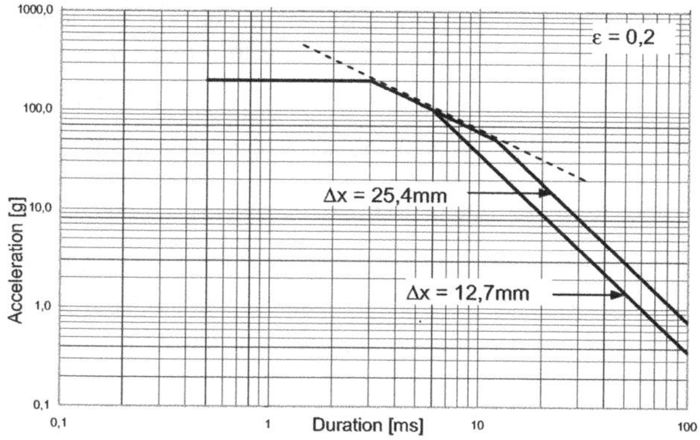

Figure 8.4. *Theoretical maximum performance for ε = 0,2*

NB: We note that, for example, for 20 ms and Δx = 25,4 mm, we hope to get 20 g, whereas with half-sine pre- and postpulses, we only get 12 g. Rectangular pre- and postpulses are therefore more effective (indeed, they have the largest surface area for a specified period, which compensates for the zero variation in speed). In reality, it is physically impossible to achieve these, we can only get close. In addition, current digital systems calculate performance automatically and even generate shock pulses with pre- and postpulses that are optimized in terms of minimum displacement of the moving coil in the exciter. We will further eloboarate on calculation "by hand" in order to better understand the phenomena.

NOTES.–

- In the field of acceleration limitation, there is a constant horizontal line (slope 0) that corresponds to the maximum acceleration performance of the exciter.

- In the field of speed limitation, we get 1/D, a slope of –6 dB/octave or slope –1.

- In the field of moving limitation, we get 1/ D^2, a slope of –12 dB /octave or slope –2.

- For a given torque (ε, D), these curves are used to determine the maximum acceleration A that we can expect to get, but in an empty moving coil (without specimen or test frame). We must then take the mass of the test specimen and its frame into account to obtain the actual value of the maximum possible acceleration, where:

m_b = mass of the moving coil in kg

m_s = mass of the specimen in kg

A_{max} = maximum possible peak acceleration of the loaded shaker

A = maximum peak acceleration of the empty shaker deducted from the previous curves

$F_{maxshock} = 3.F_{maxsine} = 9\ 000$ daN

We then get: $A.m_b = A_{max}.(m_b + m_s) \leq 3.F_{maxsine}$ (8.3)

Or: $$A_{mx} = A.\frac{m_b}{(m_b + m_s)}$$ (8.4)

Example 8.1: If the mass of the moving coil is 30 kg, can we get a half-sine 150 g, 2 ms with ε = 0,2 for a sample of 15 kg? If not, what is the maximum acceleration we can expect?

From the curve (Figure 8.3) for D = 2 ms, we read an empty moving coil acceleration of 200 g. We are therefore in the horizontal area where we are still limited by maximum possible acceleration. However, here we want 150 g with a mass of the specimen equal to 15 kg, so we must check that A_{mx} = 150 g, or $A_{max} = 200\frac{30}{30+15}$ # $133\text{ g} < 150\text{ g}$.

We cannot satisfy the requirement. The maximum possible value is 133 g.

The same question goes for a half sine of 15 g, 10 ms with ε = 0,2 and a 2" shaker.

From the curve (Figure 8.3) for D = 10 ms, we read a moving coil vacuum acceleration of 60 g. This is within the speed limit zone. We must then check that A_{mx} = 15 g, or $A_{max} = 60\frac{30}{30+15}$ # $40\text{ g} > 15\text{ g}$. It is therefore possible to satisfy the requirement and even go up to 40 g.

In addition, the force required should be compatible with maximum performance of the shaker, since: $F = (30 + 15).15 = 675 \text{ daN} = 6750 \text{ N} < 30\,000 \text{ N}$ (this should always be checked).

8.1.3.5. *Use of current control consoles*

Of course, the previous approach discussed for calculation "by hand" is done automatically in current control consoles, which allows some parameters such as maximum displacement, for example, to be more optimized. In the following curves (Figure 8.5), some results from the Data Physics console show the significant differences that may occur between the alternative strategies.

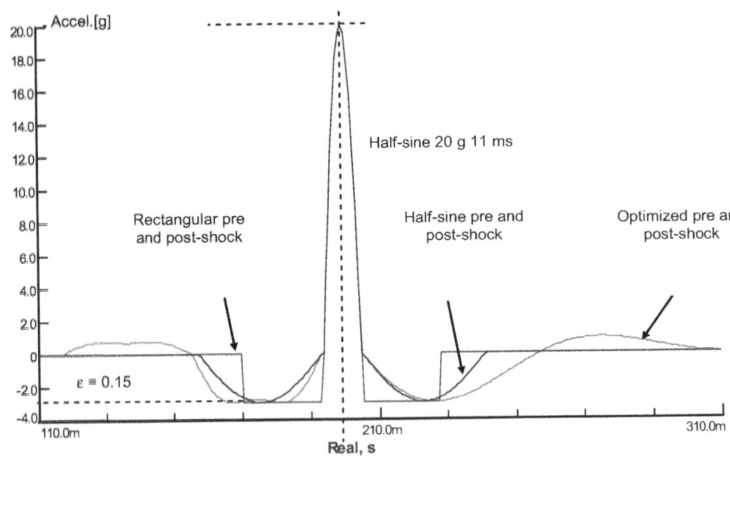

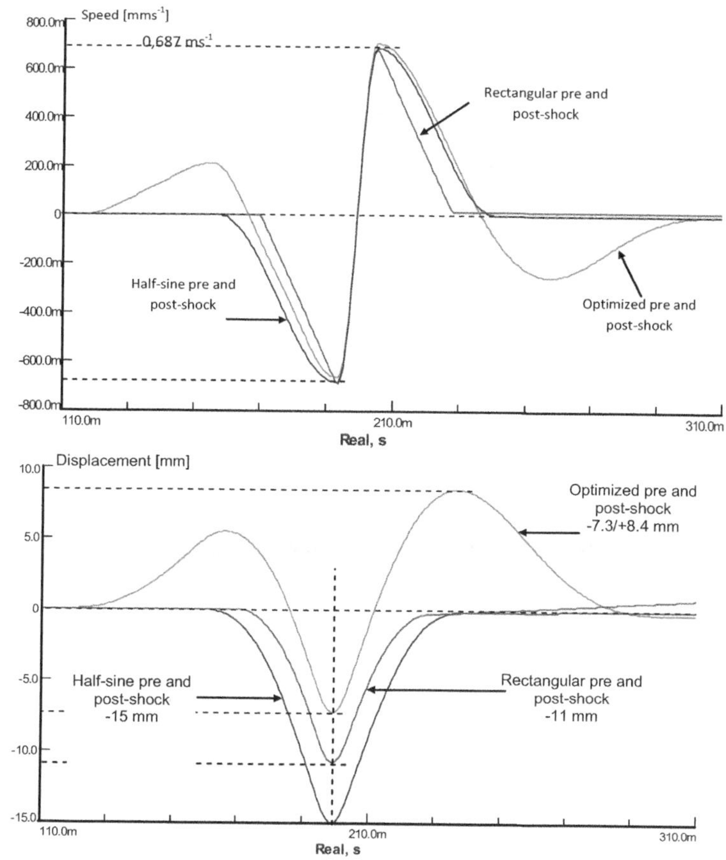

Figure 8.5. *Strategies for choice of pre and post-shock on a digital console*

NB: The impact of pre- and postpulses on the quality of the test regarding LF equipment modes still remains to be seen.

8.1.4. Modeling an electrodynamic exciter

This section is elaborated through to the end for educational purposes and in order to better understand how electrodynamic vibration exciters work. To model an electrodynamic exciter, we will first use the concepts and properties of mechanical impedance to find the mechanical model and its associated parameters. In a second step, we will determine the equivalent electrical network that will allow us to study the transfer functions that we are interested in [ref. I11 and f8].

8.1.4.1. *Mechanical model*

The mechanical model used, as well as the parameters used, is described in Figure 8.6.

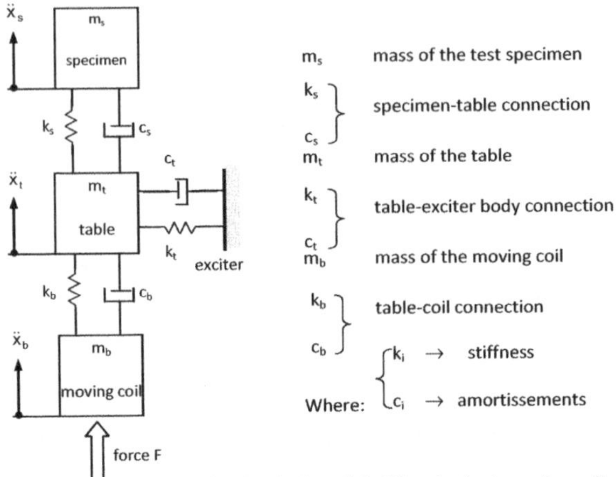

Figure 8.6. *Mechanical model of the electrodynamic exciter*

The equivalent mechanical model, which uses the concept of mechanical impedance with the strength–voltage (speed–current) analogy, is shown in Figure 8.7.

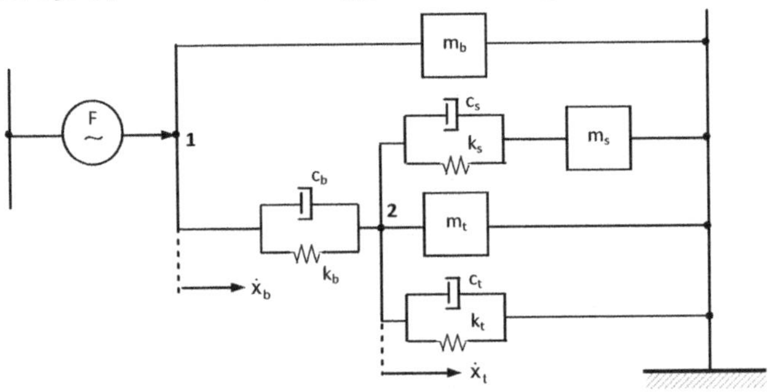

Figure 8.7. *Model in terms of mechanical impedances*

8.1.4.2. *Case of an empty table or one loaded with an inert mass*

The inert load assumption leads to an isostatic axial system where it is accepted that the specimen-table connection is infinitely rigid compared to other flexibilities that are involved in the working frequency band. The diagram is then simplified with $k_s = \infty$ and $c_s = 0$, and by including the mass m_s in m_t, the model is shown in Figure 8.8.

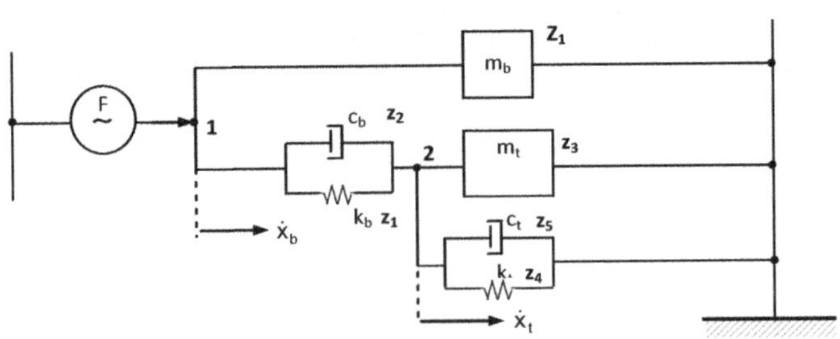

Figure 8.8. *Simplified isostatic axial model*

With the associated operational impedances:

$$Z_1 = m_b.p, \quad z_1 = \frac{k_b}{p}, \quad z_2 = c_b, \quad z_3 = m_t.p, \quad z_4 = \frac{k_t}{p}, \quad z_5 = c_t$$

The desired transfer function is one that connects the amplifier control voltage e(t) to the acceleration response of the vibrating table of the exciter $\ddot{x}(t)$. One must therefore connect force F(t) to the input e(t) and the output $\ddot{x}(t)$.

(a) We seek to evaluate the mechanical impedance Z_0 such that $Z_0 = \dfrac{F}{\dot{x}_b}$.

If z_0 is the combination of z_i (i = 1–5), then: $Z_0 = z_0 + Z_1$.

However:

$$z_0 = \frac{(z_1 + z_2).(z_3 + z_4 + z_5)}{\sum_{k=1}^{5} z_k} \tag{8.5}$$

But the concept of mechanical impedance requires the transmission of force F(t) (see section 1.4) within the series networks, due to the strength–voltage analogy used for this concept, as well as through the effect of steady state action–reaction.

Thus:

$$(1) \rightarrow \frac{F}{\dot{x}_b} = z_0 \qquad (2) \rightarrow \frac{F}{\dot{x}_t} = z_3 + z_4 + z_5$$

Therefore:

$$\frac{\dot{x}_t}{\dot{x}_b} = \frac{z_0}{(z_3 + z_4 + z_5)} \quad \rightarrow \frac{\dot{x}_t}{F} = \frac{\dot{x}_t}{\dot{x}_b}.\frac{\dot{x}_b}{F} = \frac{z_0}{(z_3 + z_4 + z_5)}.\frac{1}{Z_0} \tag{8.6}$$

And:

$$\frac{\dot{x}_t}{F} = \frac{(z_1 + z_2)}{\sum_{k=1}^{5} z_k}.\frac{1}{Z_0}$$

(b) Let us now evaluate the force generated by the electromagnetic vibrator. Laplace's law gives the moving coil of length ℓ the following force when it is in motion within magnetic induction $\vec{B}$:

$$\vec{F} = i.\vec{B}.\ell = z_0.\vec{\dot{x}}_b \quad \rightarrow \quad |F| = F = i.B.\ell = z_0.\dot{x}_b \qquad \text{where:} \qquad B.\ell = \text{cte}$$

Furthermore, the counter-electromotive force across the moving coil is given by: $u = B.\ell.\dot{x}_b$.

The control voltage becomes: $e = (R + jL\omega).i + u$

R = ohmic resistance of the moving coil = cte

L = coil inductance of the moving coil = cte

In terms of operational impedances, we get: $E(p) = (R + Lp).I(p) + B.\ell.\dot{X}_b$

Thus: $I(p) = \dfrac{z_0}{B.\ell}.\dot{X}_b(p) \quad \rightarrow \quad E(p) = (R + Lp).\dfrac{z_0}{B.\ell}.\dot{X}_b(p) + B.\ell.\dot{X}_b$

Or:

$$\frac{\dot{X}_b(p)}{E(p)} = \frac{B.\ell}{(R + Lp).z_0 + B^2.\ell^2}$$

However, by defining the sought transfer function H(f), we get:

$$H(p)=\frac{\ddot{X}_t(p)}{E(p)}=\frac{p.\dot{X}_t(p)}{E(p)} \quad \text{with zero CI}$$

Or: $$H(p)=p.\frac{\dot{X}_t(p)}{\dot{X}_b(p)}.\frac{\dot{X}_b(p)}{E(p)}=p.\frac{z_0}{(z_3+z_4+z_5)}.\frac{B.\ell}{(R+Lp).Z_0+B^2.\ell^2}$$

$$H(p)=\frac{p.B.\ell.(z_1+z_2)}{\sum_{k=1}^{5} z_k.[(R+Lp).Z_0+B^2.\ell^2]} \tag{8.7}$$

Where: $Z_0=z_0+Z_1$ we get:

$$H(p)=\frac{p.B.\ell(z_1+z_2)}{\sum_{k=1}^{5} z_k.(R+Lp)Z_1+(R+Lp)(z_1+z_2)(z_3+z_4+z_5)+B^2.\ell^2.\sum_{k=1}^{5} z_k}$$

By elaborating each term based on impedance values, we get:

$$H(p)=\frac{p^3.N_1+p^2.N_0}{p^5+p^4.D_4+p^3.D_3+p^2.D_2+p.D_1+D_0}$$

$N_1=B.\ell.c_b/m_t m_b L$

$N_0=B.\ell.k_b/m_t m_b L$

$D_4=(m_t m_b R+c_b m_b L+c_t m_b L+m_t c_b L)/m_t m_b L$

$D_3=\left(k_b m_b L+c_t m_b R+c_b m_b R+k_t m_b L+c_b m_t R+m_t k_b L+c_b c_t L+m_t B^2.\ell^2\right)/m_t m_b L$

$D_2=\left(k_b m_b R+k_t m_b R+k_b m_t R+c_b k_t L+c_b c_t R+k_b c_t L+k_t c_b L+c_b B^2.\ell^2+c_t B^2.\ell^2\right)/m_t m_b L$

$D_1=\left(c_t k_b R+k_b k_t L+c_b k_t R+k_b B^2.\ell^2+k_t B^2.\ell^2\right)/m_t m_b L$

$D_0=k_t k_b R/m_t m_b L$

These expressions can be simplified by intially ignoring the damping terms of the table and the moving coil, which are very low due to the purpose-built "stiffness" that guides their design, including:

$c_t \# c_b \# 0$

Hence, the approximate relationship of the desired transfer function:

$$H(p)=\frac{p^2.\frac{k_b B.\ell}{m_t m_b L}}{p^5+\frac{R}{L}.p^4+\left(\frac{k_b}{m_t}+\frac{k_b}{m_b}+\frac{k_t}{m_t}+\frac{B^2.\ell^2}{m_b L}\right).p^3+\frac{R}{L}\left(\frac{k_b+k_t}{m_t}+\frac{k_b}{m_b}\right).p^2 +\left[\frac{k_b k_t}{m_b m_t}+\frac{B^2.\ell^2.(k_b+k_t)}{m_t m_b L}\right].p+\frac{R k_b k_t}{m_t m_b L}} \tag{8.8}$$

In general, when one takes into account the damping terms and as it is a physically achievable system, the denominator can be put into a more practical form, namely:

$$H(p)=\frac{p^2.\frac{k_b B.\ell}{m_t m_b L}}{(p+\omega_1)\left(p^2+2\xi_e\omega_e p+\omega_e^2\right)\left(p^2+2\xi_b\omega_b p+\omega_b^2\right)} \tag{8.9}$$

This latter relationship allows us to evaluate analytical expressions of characteristic frequencies $f_i = \frac{\omega_i}{2\pi}$. Indeed, as the rigorous calculation is complex and adds nothing to our problem, some realistic approximations will allow us to tackle these sought-after relationships with annotations from the following physical experimental findings:

- the damping terms have little influence on the position of resonance frequencies, including the moving coil;

- we will show that the above-mentioned characteristic frequencies correspond to the following resonance frequencies:

$$\begin{cases} f_1 = \frac{\omega_1}{2\pi} = \text{Resonance frequency of the table-exciter body connection} \\ \text{or resonance frequency of the suspension} \\ f_e = \frac{\omega_e}{2\pi} = \text{Electric resonance frequency of the moving coil} \\ f_b = \frac{\omega_b}{2\pi} = \text{Axial resonance frequency of the moving coil} \end{cases}$$

Where: $\omega_b \geq \omega_e \geq \omega_1$

- the relative values of coefficients in p terms of the denominator give the following results (see digital applications): coefficients of:

p^5 # 1; p^4 # 10^3; p^3 # 10^9; p^2 # 10^{10}; p # 10^{15}; cte # 10^{14}

- moreover, from a rigidity standpoint, we still get:

$k_b \geq k_t$, of several orders of magnitude

The expression of the transfer function (8.8) is further simplified by:

$$H(p) \# \frac{p^2 . \frac{k_b.B.\ell}{m_t m_b L}}{(p+\omega_1)(p^2+\omega_e^2)(p^2+\omega_b^2)} \tag{8.10a}$$

One is then able to calculate values for different characteristic frequencies depending on the mechanical and electrical parameters of the model.

The approximations do not directly give ω_1 from these relationships. Due to the significant differences between the characteristic frequencies and therefore their strong decoupling, ω_1 is calculated directly from the total mass $m_e = m_b + m_t$ and stiffness k_t through:

$$\omega_1 = \sqrt{\frac{k_t}{m_e}} \qquad f_1 = \frac{1}{2\pi}\sqrt{\frac{k_t}{(m_b+m_t)}}$$

After processing the denominator, we get:

$$D(p) = p^5 + p^4\omega_1 + p^3(\omega_e^2+\omega_b^2) + p^2\omega_1(\omega_e^2+\omega_b^2) + p\omega_e^2\omega_b^2 + \omega_1\omega_e^2\omega_b^2$$

By identifying and noting that the coefficient of the term in p^2 and the constant term are not independent from the previous terms and by taking the comments made on the relative values of various parameters into account, we get the following relationships:

$$\omega_e^2\omega_b^2 = \left[\frac{k_b k_t}{m_t m_b} + \frac{B^2.\ell^2.(k_b + k_t)}{m_t m_b L}\right] \# \frac{B^2.\ell^2.k_b}{m_t m_b L} \quad (8.10b)$$

$$\omega_e^2 + \omega_b^2 = \left(\frac{k_b}{m_t} + \frac{k_b}{m_b} + \frac{k_t}{m_t} + \frac{B^2.\ell^2}{m_b L}\right) \# \frac{k_b(m_b + m_t).L + B^2.\ell^2.m_t}{m_t m_b L} \quad (8.10c)$$

with (8.10b) and (8.10c) and by making P = (1) and S = (3), the solutions are given by the equation:

$$X^2 - SX + P = 0 \qquad \rightarrow \qquad X_{1,2} = \frac{S \pm \sqrt{S^2 - 4P}}{2}$$

We want the smallest root ω_e^2 because the calculation is simplest here.

$$\omega_e^2 = \frac{S - \left[S^2\left(1 - \frac{4P}{S^2}\right)\right]^{1/2}}{2}$$

However, in AN, we find that: $\frac{4P}{S^2} \# \frac{10^{15}}{10^{9x2}} = 10^{-3} \quad \rightarrow \quad \left(1 - \frac{4P}{S^2}\right)^{1/2} \# \left[1 - \frac{1}{2}\left(\frac{4P}{S^2}\right)\right]$

Therefore:

$$\omega_e^2 = \frac{1}{2}\left[S - S\left(1 - \frac{2P}{S^2}\right)\right] = \frac{P}{S} = \frac{B^2.\ell^2.k_b m_t m_b L}{[k_b(m_t + m_b)L + B^2.\ell^2 m_t].m_t m_b L} = \frac{B^2.\ell^2}{[(m_t + m_b)L + (B^2.\ell^2.m_t / k_b)]}$$

But the $(B^2.\ell^2.m_t / k_b)$ term is very small, so we can write: $\omega_e^2 = \frac{B^2.\ell^2}{[(m_t + m_b)L]}$.

And from (1), we get: $\omega_b^2 = \frac{k_b(m_t + m_b)}{m_t.m_b}$.

Relationships close to the resonance frequencies are then:

- the resonance frequency of the suspension of the table plus moving coil assembly (table-exciter body connection):

$$\omega_1 = \sqrt{\frac{k_t}{m_e}} \qquad f_1 = \frac{1}{2\pi}\sqrt{\frac{k_t}{(m_b + m_t)}} \quad (8.11)$$

- an electrical resonance frequency in series between the moving coil inductor and the capacitor due to the table-coil masses, which is very damped by its electrical resistance (the frequency varies between 50 and 300 Hz depending on the excitation power):

$$f_e = \frac{\omega_e}{2\pi} \qquad f_e = \frac{1}{2\pi}\frac{B.\ell}{[(m_t + m_b)L]^{1/2}} \quad (8.12)$$

- the axial mechanical resonance frequency of the moving element (frequency variable from 1500 to 20 000 Hz depending on the excitation power):

$$f_b = \frac{\omega_b}{2\pi} \qquad f_b = \frac{1}{2\pi}\left[\frac{k_b(m_t + m_b)}{m_t m_b}\right]^{1/2} \quad (8.13)$$

From a module perspective, we get:

$$|H(p)|_{p=j\omega} = \frac{|N(\omega)|}{|D(\omega)|} = \frac{R_N(\omega)}{\left[R_D^2(\omega) + I_D^2(\omega)\right]^{1/2}}$$

Where: $D(\omega) = R_D(\omega) + jI_D(\omega)$

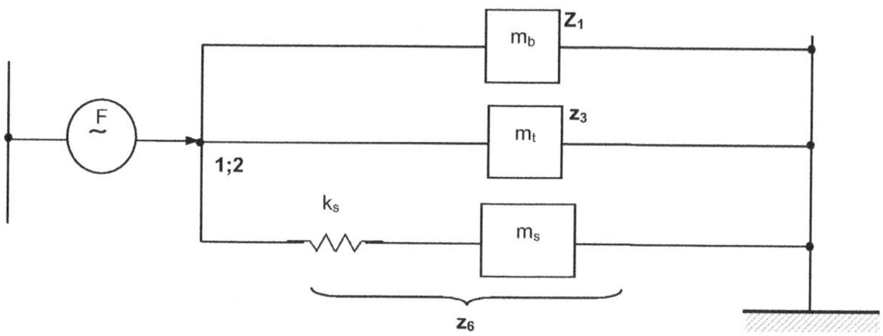

Figure 8.10. *Simplified model loaded with a resonant system with 1 DOF*

NB: This is a diagram that directly gives the relationship (8.19) of $H_s(p)$, of course.

Hence $H_s(p)$, which can be written in the general form:

$$H_s(p) = \frac{\frac{B.\ell}{L(m_t + m_b)} p\left(p^2 + \omega_n^2\right)}{\left(p^2 + 2\xi'_e\,\omega'_e\,p + \omega'^2_e\right)\left(p^2 + 2\xi_p\omega_p p + \omega_p^2\right)} \tag{8.19}$$

Where:

- ω_n and ω_p respectively characterize the "antiresonance and resonance" of the tested specimen;
- ω'_e is the new electric resonance value, modified by the presence of the specimen;
- ω_s and ω_b are ignored by inaccuracies due to previous calculations.

The evaluation of the analytical expressions of characteristic frequencies is here also done from the reduced expression of $H_s(p)$, where:

$$H_s(p) \# \frac{\frac{B.\ell}{L\ (m_t + m_b\ )} p\left(p^2 + \omega_n^2\right)}{\left(p^2 + \omega'^2_e\right)\left(p^2 + \omega_p^2\right)} \tag{8.20}$$

After developing and eventually identifying each element term by term, we get:

$$\frac{B^2.\ell^2.k_s}{Lm_s(m_t + m_b)} = \omega'^2_e.\omega_p^2 = P$$

$$k_s\left[\frac{1}{m_s} + \frac{1}{(m_t + m_b)}\right] + \frac{B^2.\ell^2}{L(m_t + m_b)} = \omega'^2_e + \omega_p^2 = S$$

$$\omega_n^2 = \frac{k_s}{m_s}$$

Wherefrom we immediately get the antiresonance value:

$$f_n = \frac{\omega_n}{2\pi} = \frac{1}{2\pi}\sqrt{\frac{k_s}{m_s}} \tag{8.21}$$

NB: Frequency f_n is sometimes called notch frequency; its importance becomes apparent when calculating the dynamic force required for the exciter to maintain the desired level of vibration of the specimen at this frequency. This is because its mechanical impedance and its apparent mass become very large.

To elaborate on $H_s(p)$ and calculate the characteristic frequencies that we are interested in, it is more convenient to consider the following two cases, which are functions of the relative positions of f_n and f_e:

$$f_n < f_e \quad \text{or} \quad f_e > f_n$$

Let us remember that:

Non-loaded case: $f_e = \dfrac{1}{2\pi}\dfrac{B.\ell}{[L(m_t+m_b)]^{1/2}} = \dfrac{1}{2\pi}\dfrac{1}{\left[\dfrac{L}{B^2.\ell^2}(m_t+m_b)\right]} \rightarrow f'_e$ loaded case

The expressions f'_e and f_p are then calculated in the same way as in the previous chapter. These are also the solutions for the following system: $X^2 - SX + P = 0$. Similar considerations lead to the same simplifications and after further elaboration, the smaller of the two is the solution:

$$X_1 \# \frac{P}{S} = \frac{B^2.\ell^2.k_s}{Lm_s(m_b+m_t)} \cdot \frac{Lm_s(m_b+m_t)}{L(m_b+m_t)k_s + k_sLm_s + B^2.\ell^2.m_s}$$

$$X_1 = \frac{1}{\dfrac{L}{B^2.\ell^2}(m_t+m_b+m_s) + \dfrac{m_s}{k_s}}$$

We deduce that:

$$X_2 = \frac{k_s(m_t+m_b+m_s)}{m_s(m_t+m_b)} + \frac{B^2.\ell^2}{L(m_t+m_b)}$$

Where: $\dfrac{X_2}{X_1} \# \dfrac{Lk_s(m_t+m_b+m_s)^2}{B^2.\ell^2 m_s(m_t+m_b)} > 1 \quad \rightarrow \quad X_2 > X_1$

Case 1: $f_n > f'_e$

As we obviously have:

$\dfrac{1}{2\pi}(X_2)^{1/2} > f_n = \dfrac{1}{2\pi}\left(\dfrac{k_s}{m_s}\right)^{1/2}$ and therefore > f'_e for case 1. We are then mandatorily in a case where: $\dfrac{1}{2\pi}(X_2)^{1/2} \equiv f_p$ and $\dfrac{1}{2\pi}(X_1)^{1/2} \equiv f'_e$

Or: $f_p > f_n > f'_e$

Hence, the respective values of the sought characteristic frequencies:

$$f'_e = \frac{1}{2\pi}\frac{1}{\left[\dfrac{L}{B^2.\ell^2}(m_t+m_b+m_s) + \dfrac{m_s}{k_s}\right]^{1/2}} \tag{8.22}$$

NB: Let us note that f_e was modified by the appearance of m_s in the total mass and the intervention of $\dfrac{m_s}{k_s}$. In this case, we indeed find that $f'_e < f_e$.

$$f_p = \frac{1}{2\pi}\left[\frac{Lk_s(m_t+m_b+m_s) + m_sB^2.\ell^2}{Lm_s(m_t+m_b)}\right]^{1/2} \tag{8.23}$$

NB: We observe an antiresonance for f_n and resonance at f_n for f_p. The shape of the transfer function $|H(j\omega)|$ is shown in Figure 8.11.

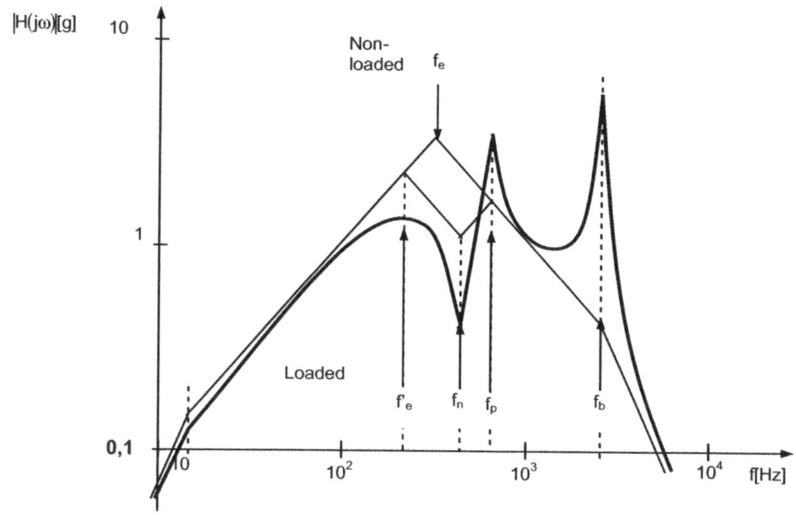

Figure 8.11. *Case for resonant case $f_p > f_n > f'_e$*

NB: Amplitudes at different characteristic frequencies are of course limited by their respective damping (not included in the calculations).

Case 2: $f_n < f'_e$

This is the opposite case. The relationships are:

$$f_p = \frac{1}{2\pi} \frac{1}{\left[\frac{L}{B^2.\ell^2}(m_t + m_b + m_s) + \frac{m_s}{k_s} \right]^{1/2}} \tag{8.24}$$

$$f'_e = \frac{1}{2\pi} \left[\frac{Lk_s(m_t + m_b + m_s) + m_s B^2.\ell^2}{Lm_s(m_t + m_b)} \right]^{1/2} \tag{8.25}$$

Or:

$$f'_e > f_n > f_p$$

However, in this case, we can show that f_p is very close to f_n by its lower value. Indeed, one can write:

$$f_p = \frac{1}{2\pi}\left(\frac{k_s}{m_s}\right)^{1/2} \frac{1}{\left[1 + \frac{k_s L(m_t + m_b + m_s)}{m_s B^2.\ell^2}\right]^{1/2}}$$

But in practice: $\frac{k_s L(m_t + m_b + m_s)}{m_s B^2.\ell^2} \# \frac{k_s L}{B^2.\ell^2} \# k_s.10^{-10}$, so for a general maximum value of a prominent mode for ks # 108, for example, we are well within 1% that f_p < and # f_n.

In addition, damping introduced by resistance R is predominant in this area and as the speeds in the masses $(m_t + m_b)$ and m_s are in phase, we simply observe an antiresonance for f_n, hence the rate $|H(j\omega)|$ in case (Figure 8.12).

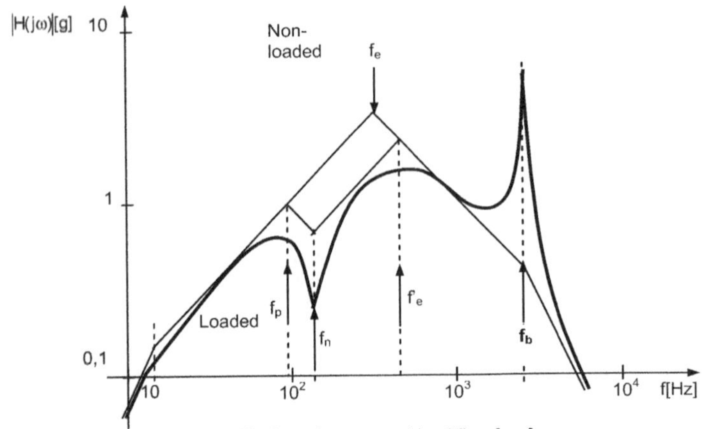

Figure 8.12. *Case for resonant load $f'_e > f_n > f_p$*

8.1.4.4. *Influence of the resonant load on H(p)*

If we now calculate the transfer function H(p) of the empty exciter with the same assumptions as those used in the calculation of $H_s(p)$ to compare them, we get:

$$\omega_s \# 0 \qquad \text{and} \qquad \omega_b \# \infty$$

$$H(p) = \frac{p^2 \dfrac{k_b B.\ell}{m_t m_b L}}{p\left(p^2 + 2\xi_e\omega_e p + \omega_e^2\right)\omega_b^2\left(\dfrac{p^2}{\omega_b^2} + \dfrac{2\xi_b}{\omega_b} + 1\right)} \# \frac{p\dfrac{k_b B.\ell}{m_t m_b L}}{\left(p^2 + 2\xi_e\omega_e p + \omega_e^2\right)\omega_b^2} = \frac{p\dfrac{B.\ell}{L(m_t + m_b)}}{\left(p^2 + 2\xi_e\omega_e p + \omega_e^2\right)}$$

So by taking the damping into account:

$$\frac{H_s(p)}{H(p)} = \frac{\left(p^2 + \omega_n^2\right)}{\left(p^2 + 2\xi_p\omega_p p + \omega_p^2\right)} \cdot \frac{\left(p^2 + 2\xi_e\omega_e p + \omega_e^2\right)}{\left(p^2 + 2\xi'_e\,\omega'_e\,p + \omega'^2_e\right)} \qquad (8.26)$$

However, the term $\dfrac{\left(p^2 + 2\xi_e\omega_e p + \omega_e^2\right)}{\left(p^2 + 2\xi'_e\,\omega'_e\,p + \omega'^2_e\right)} \# 1$ simply reflects a displacement of the electrical resonance frequency due to the presence of the resonant load, which therefore does not change the asymptotic and actual shape of the empty transfer function.

This shows that the presence of a resonant load with 1 DOF modifies the empty transfer function H(p) by a correction factor $H_c(p)$, which only depends on the pulses ω_n and ω_p such that we generally get:

$$H_c(p) = \frac{p^2 + \omega_n^2}{p^2 + 2\xi_p\omega_p p + \omega_p^2} \rightarrow H_c(p) = \frac{p^2 + 2\xi_n\omega_n p + \omega_n^2}{p^2 + 2\xi_p\omega_p p + \omega_p^2}$$

(ξ_n = damping of the specimen)

Such that:

$$H_c(j2\pi f) = \left(\frac{f_n^2}{f_p^2}\right) \cdot \frac{1 - \left(\dfrac{f}{f_n}\right)^2 + j2\xi_n\left(\dfrac{f}{f_n}\right)}{1 - \left(\dfrac{f}{f_p}\right)^2 + j2\xi_p\left(\dfrac{f}{f_p}\right)} \qquad (8.27)$$

NB: We should note that $H_c(p)$ is a general formula where the parameters ω_e, ω_p, ξ_e, ξ_p can be of any value that enables $H_c(p)$ to represent the different cases possible.

The rate of the transfer function module $|H_c(j\omega)|$ graph is evaluated for both previous characteristic cases in Figure 8.13, where:

case $f_n > f_p$ (1) and $f_p > f_n$ (2)

$$\begin{cases} \xi_n = 0,1 \\ \xi_p = 0,05 \\ (1) \rightarrow f_n = 100\text{Hz} \quad \text{and} \quad f_p = 70,7\text{Hz} \\ (2) \rightarrow f_p = 100\text{Hz} \quad \text{and} \quad f_n = 70,7\text{Hz} \end{cases}$$

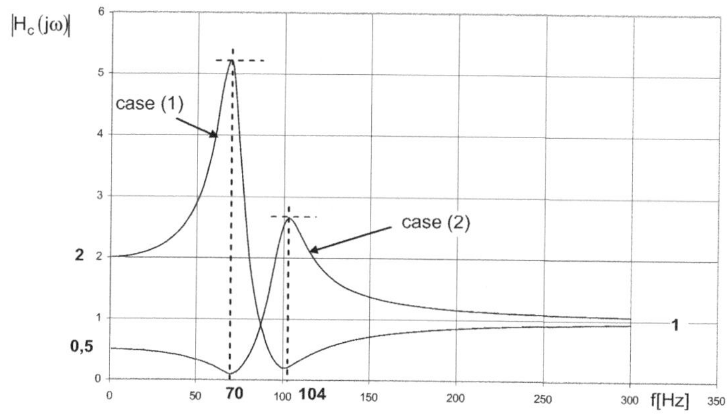

Figure 8.13. *Empty correction factor H(p) for the resonant load*

NB: We then point out the following annotations:

$$\begin{cases} f \rightarrow 0 \quad |H_c(j\omega)| \rightarrow \dfrac{f_n^2}{f_p^2} \\ f \rightarrow \infty \quad |H_c(j\omega)| \rightarrow 1 \end{cases}$$

Moreover, we note that "f_p" (resonance) depends on the complete system, while "f_n" (antiresonance) is characteristic of the tested specimen.

8.1.5. Evaluation of the modeling parameters

8.1.5.1. *Experimental settings*

NB: The measurement methods used are inspired from the NF 90,210 September 1978 standard: "Electrodynamic testing means for the generation of vibrations. Method for describing the features".

8.1.5.1.1. *Natural frequency for the suspension of the moving element f_1*

This is determined by the first minimum of the curve of the moving coil current $i_b(f)$ when it is servocontrolled at constant speed $\dot{x}(t) = \text{cte}$. As the table is not loaded and the suspensions of the exciter are unlocked (in fact, at this frequency, mechanical impedance is minimal and since it is at constant velocity, the force and hence the necessary current is minimal). We then get the following results:

- $i_b(f)$ = current in the moving coil measured using a current transformer (or amperometric clamp) set to 300;
- $\dot{x}(t)$ = speed setting at 22 cms^{-1} in the band [5 Hz, 500 Hz].

We find that:

$$f_1 \# 18.5 \text{ Hz} \tag{8.28}$$

8.1.5.1.2. *Axial resonance frequency f_b of the moving coil*

Analyzing the curves in an open loop (OL), non-loaded shaker gives the following value as the mean value f_b:

$$f_b \# 3085 \text{ Hz} \tag{8.29}$$

8.1.5.1.3. *Conventional mass m_e of the moving component*

It includes the mass of the vibrating table and that of the moving coil, which is: ($m_e = m_t + m_b$).

Where:
$$m_e = m_i.\frac{1}{\left(\frac{\ddot{x}_1}{\ddot{x}_2}-1\right).\left(1-\frac{f_1^2}{f_2^2}\right)} \tag{8.30}$$

m_e = conventional mass of the moving component

m_i = inert load at the working frequency (see below)

$\ddot{x}_1$ = acceleration of the vacuum table, measured for current I at frequency f, with:

$$3f_1 < f < \frac{f_b}{3} \quad \rightarrow \quad f = 200\text{Hz}$$

$\ddot{x}_2$ = acceleration of the table loaded by m_i, measured for the same current I at the same frequency f

Measurements in Table 8.1 give:

I [A_{eff}]	$\ddot{x}_1$ vacuum [g]	$\ddot{x}_2$ loaded [g]	m_e [kg]
193,8	20	6,45	30,73
394,2	40	12,8	30,38
591,9	60	19,4	30,84

Table 8.1. *Conventional mass measurements*

Such that the average value of m_e is:

$$m_e \# 30{,}65 \text{ kg} \tag{8.31}$$

8.1.5.1.4. *Transformation ratio Γ*

This is the ratio between force F and generated current I in the moving coil. We measure the acceleration $\ddot{x}$ of the vacuum table, for current I, at frequency f for which I and $\ddot{x}$ are roughly constant.

However: $F = B.\ell.I = m\,.\ddot{x} = \Gamma.I$ where: $\Gamma \equiv B.\ell$

So:

$$\Gamma = \frac{m_e.\ddot{x}}{I} \qquad \begin{cases} m_e \text{ in kg} \\ \ddot{x} \text{ in peak ms}^{-2} \\ I \text{ in peak A} \end{cases}$$

In this case, we choose f = 200 Hz (electrical resonance), so we get:

$$\Gamma \# 21{,}59 \text{ NA}^{-1} \tag{8.32}$$

8.1.5.1.5. *Stiffness k_t of the suspension of the moving component*

At the resonance frequency f_1 of the suspension of the moving component, the frequency is low enough (18.5 Hz) that the following relationship is verified in steady state:

$$k_t.x(t) = m_e.\ddot{x}(t) \qquad \rightarrow \qquad k_t = \frac{m_e.\ddot{x}(t)}{x(t)}$$

In sinusoidal steady state: $k_t = \omega_1^2.m_e$

Hence the average value for stiffness:

$$k_t \# 0{,}414 \times 10^6 \text{ Nm}^{-1} \tag{8.33}$$

8.1.5.1.6. *Electrical resonance frequency f_e of the moving coil*

It is determined by analyzing the phase of the curve $\frac{\ddot{x}(t)}{e(t)} = \frac{\text{table acceleration}}{\text{control voltage}}$ (more accurate measurement than amplitude due to strong damping by rotating the phase at the electrical resonance). The resulting value is:

$$f_e \# 294 \text{ Hz} \tag{8.34}$$

8.1.5.1.7. *Ohmic resistance R of the moving coil*

This is given by the manufacturer and has been verified with a milliohmmeter. It is around:

$$R \# 6{,}9 \times 10^{-3} \text{ ohm} \tag{8.35}$$

8.1.5.2. *Theoretical parameters*

8.1.5.2.1. *Mass of the table m_t and mass moving coil m_b*

By replacing ω_b with its value in the simplified relationship of H(p), we get the relationship (8.15), namely:

$$|H(p)|_{p=j\omega_b} = \omega_b \frac{m_b}{m_t.B.\ell} \qquad \text{or :} \qquad \frac{m_b}{m_t} = \frac{|H(j\omega_b)|.B.\ell}{\omega_b}$$

To go back to ratio $\frac{m_b}{m_t}$, one must determine the magnitude of $|H(j\omega_b)|$. The latter is done through an OL test with a constant coil voltage set at 200 Hz and equal to 0,707 V_{eff} or 1 V_c in order to avoid amplitudes that are too high during measurement at f_b. The measurement of the transfer function module (or of the amplitude) at the frequency f_b then gives:

$$|H(j2\pi f_b)| = 32 g_{eff} / V_{eff} \ \# \ 46 g_c / V_c \rightarrow \qquad |H(j2\pi f_b)| = 46 * 9{,}81 = 451 \ \text{ms}^{-2} / V_c$$

So with the previously established measurements:

$$\frac{m_b}{m_t} \# 0{,}502 \tag{8.36}$$

Like: $m_e = m_t + m_b \# 30{,}65$

We deduce that: $$\begin{cases} m_t \# 20{,}41 \text{ kg} \\ m_b \# 10{,}24 \text{ kg} \end{cases} \quad (8.37)$$

8.1.5.2.2. *Inductance L of the moving coil*

It is derived from the relationship that gives ω_e (8.12), where:

$$\omega_e^2 \# \frac{B^2.\ell^2}{L(m_t + m_b)} \quad \rightarrow \quad L \# \frac{B^2.\ell^2}{\omega_e^2.(m_t + m_b)}$$

So with the previously established values:

$$L \# 4.46 \times 10^{-6} \text{ H} \quad (8.38)$$

8.1.5.2.3. *Stiffness k_b of the coil-table connection*

This is derived from the relationship that gives ω_b (8.13), where:

$$\omega_b^2 \# \frac{k_b(m_t + m_b)}{m_t m_b} \quad \rightarrow \quad k_b = \frac{m_t m_b \omega_b^2}{(m_t + m_b)}$$

So with the previously established values:

$$k_b \# 2{,}56 \times 10^9 \text{ Nm}^{-1} \quad (8.39)$$

8.1.5.2.4. *Parameters derived from the comparison of curves* $\frac{\dot{x}(t)}{e_b(t)}$ and $\frac{i_b(t)}{\dot{x}(t)}$

This involves plotting the theoretical curve obtained from an electromechanical model. Next, comparing it with the experimental plot, we realign the theoretical parameters to best match the corresponding curves.

The "direct" comparison has to be done first on a curve of type $\frac{\dot{x}(f)}{e_b(f)}$ for the parameters associated with ω_b in HF, so the change in speed for the table relative to the input voltage and second, on a curve of type $\frac{i_b(f)}{\dot{x}(f)}$ for parameters associated with ω_1 in LF, so the evolution of the current in the moving coil depending on the speed of the table.

In addition, the calculation and electronic network analysis programs that are available (for example, Spice) provide access to the transfer functions in terms of voltages (and therefore, speed) or intensity (force). The selected electromechanical model is the following:

$$\begin{cases} \text{Force:} & F(t)\ [\text{N}] & \longleftrightarrow & \text{Intensity } i(t)\ [\text{A}] \\ \text{Speed:} & \dot{x}(t)\ [\text{ms}^{-1}] & \longleftrightarrow & \text{Voltage } e(t)\ [\text{V}] \\ \text{Mass:} & m\ [\text{kg} \equiv \text{daN}] & \longleftrightarrow & \text{Capacitor } [\text{F}] \\ \text{Dampening:} & c\ [\text{Nm}^{-1}\text{s}^{-1}] & \longleftrightarrow & \text{Conductor } \frac{1}{R}\ [\Omega^{-1}] \\ \text{Stiffness:} & k^{-1}\ [\text{N}^{-1}\text{m}] & \longleftrightarrow & \text{Inductor } [\text{L}] \end{cases}$$

Hence, the equivalent circuit diagram is obtained (Figure 8.14).

NB: Resistor R_1 is an actual damping resistor of 1 Ω, installed in parallel on the circuit of the moving coil to dampen its electrical resonance.

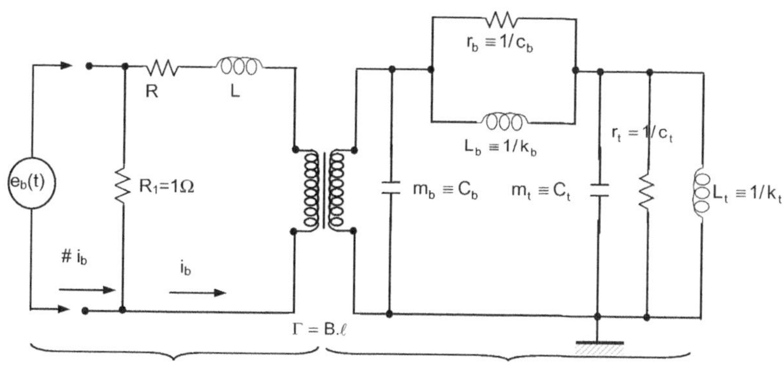

Figure 8.14. *Mechanical model equivalent circuit diagram*

8.1.5.2.5. *Scaling factor problem*

If one wishes to move from a mechanical model to an electrical model (or vice versa), the problem of "dimensions of corresponding elements" arises, as well as that of "relative values of the units used" and the technological achievement of these elements. Indeed, if the parameters are not properly corrected, we get homothetic curves with component values or unrealistic frequencies.

We therefore use a "scale factor" to carry out the necessary weightings, which takes different values depending on the type of analogy used. The value of the scaling factor is only "arbitrary" for the independent homologous elements, namely in the chosen strength–intensity analogy, the following elements:

$$m = S_1.C \ , \ \dot{x}(t) = S_2.e(t) \ , \ t_m = S_3.t_e$$

NB: t_m and t_e are the time intervals where homologous mechanical and electrical events appear (resonance, antiresonance, etc.).

We define the following scale factors:

$$\begin{cases} f_m = S_4.f_e & \text{where:} & S_4 = \dfrac{1}{S_3} \\ F(t) = \dfrac{m.\dot{x}(t)}{t_m} \equiv S_5.i(t) & \text{where:} & S_5 = \dfrac{S_1.S_2}{S_3} \\ c = \dfrac{F(t)}{\dot{x}(t)} \equiv S_6.G = \dfrac{S_6}{R} & \text{where:} & S_6 = \dfrac{S_5}{S_2} = \dfrac{S_1}{S_3} \\ \dfrac{1}{k} = \dfrac{x(t)}{F(t)} = \dfrac{\dot{x}(t).t_m}{F(t)} \equiv S_7.L & \text{where:} & S_7 = \dfrac{S_1.S_3}{S_5} = \dfrac{S_3^2}{S_1} \end{cases} \tag{8.40}$$

8.1.5.3. *Digital settings*

From the values estimated in the previous sections and in order to get realistic values for the electrical components, the following scale factors are chosen for digital implementation:

$$\begin{cases} t_m = t_e \rightarrow f_m = f_e \ \rightarrow & S_3 = S_4 = 1 \text{ (arbitrary), choice of real time} \\ S_2 = 1 \text{ and } S_2 = 10^6 \rightarrow & \text{(arbitrary), choice of realistic digital values} \end{cases}$$

Hence:

$S_5 = 10^6$; $S_6 = 10^6$; $S_7 = 10^{-6}$ (8.41)

The parameter and electrical component values are then:

$$\begin{cases} \Gamma \# 21{,}59 \times 10^4 & & \\ R \# 6{,}9 \times 10^{-3}\ [\Omega] & & \\ L \# 4{,}46 \times 10^{-6}\ [H] & & f_1 \# 18{,}5\ [Hz] \\ L_b \# 455 \times 10^{-6}\ [H] \rightarrow & k_b & f_e \# 3085\ [Hz] \\ L_t \# 2{,}41\ [H] \quad \rightarrow & k_t & f_b \# 294\ [Hz] \\ c_b \# 7{,}95\ \mu F \quad \rightarrow & m_b & \\ c_t \# 22{,}7\ \mu F \quad \rightarrow & m_t & \end{cases} \quad (8.42)$$

The damping terms still need to be estimated; these are the corresponding conductor values (or resistors in parallel) from overvoltage coefficients of usual measurements. We then get:

$$Q \# \frac{1}{2\xi} = \frac{c_c}{2c} = \frac{2.\sqrt{km}}{2c} \quad \rightarrow \quad c \# \frac{\sqrt{km}}{Q} \quad \rightarrow \quad r \# \frac{Q}{\sqrt{km}}$$

For c_b, we put Q_b # 20 into the first estimation (very overstretched for the usual curves).

Such that: $$r_b \# S_6 . \frac{Q_b}{\sqrt{k_b . m_b}} \# 76\ [\Omega]$$

For c_t, we put Qt # 7 into the first estimation (damped by Beryllium friction blades), such that:

$$r_t \# S_6 . \frac{Q_t}{\sqrt{k_t m_t}} \# 2280\ [\Omega]$$

8.1.5.3.1. *Model used to assess* $\frac{\dot{x}(f)}{e_b(f)}$ and $\frac{i_b(f)}{\dot{x}(f)}$

The selected diagram is that in Figure 8.15.

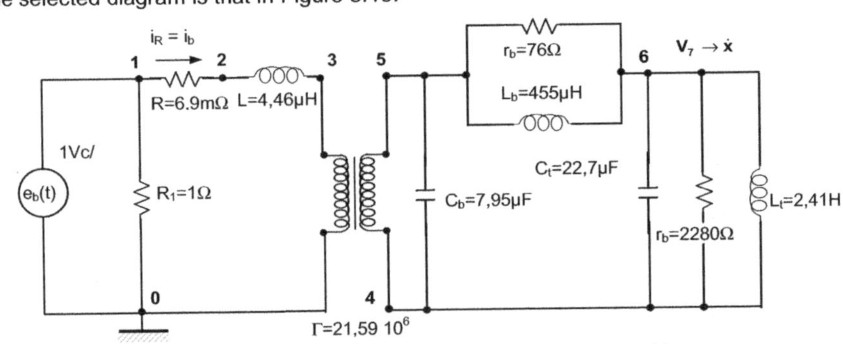

Figure 8.15. *Electrical model for assessing* $\frac{\dot{x}(f)}{e_b(f)}$ and $\frac{i_b(f)}{\dot{x}(f)}$

The iterations from the model using Spice simulation software (demonstrator) on alternative different values of r_b and r_t make it possible for us to determine the value of

- piston stroke: 254 mm
- supply pressure: 170 bars
- online pressure loss: 42 bars

8.2.2.2. *The servo valve*

This is the heart of the system. It is a kind of hydraulic amplifier that alternately directs pressurized oil to the cylinder in each hydrostatic bearing from a low control current (it plays the role of the classic Darlington circuit, which attacks the power levels in a power amplifier that is simulated here by the cylinder).

There are several principles of operation and several servo valves. The most common are the following (Figures 8.21–8.23):

- "alternating nozzle" systems;
- "electrodynamic shakers" or "paddle" systems.

The alternating nozzle system is relatively fragile. It is limited in frequency and clogs easily, which requires good oil filtering (2 µm or less), which creates maintenance problems.

Systems with electrodynamic shakers and paddles are more reliable and they accept oil filtering at only 10 µm and can operate at up to a few hundred or 1000 Hz. Furthermore, they can be maintained in "fibrillation" (a few kilohertz, typically 2 kHz) by an auxiliary oscillator, which prevents clogging, reduces friction and prevents the effects of seals blocking by dry friction. These are very commonly used.

8.2.3. Hydraulic operation equations

8.2.3.1. *Overview*

8.2.3.1.1. *Limiting speed strength*

The equations governing the operation of a hydraulic exciter involve pressure and fluid flow, which the hydraulic power supply can provide. In a cylinder, the predominant strength is provided by:

$$F = P.S \qquad \begin{cases} P = \text{cylinder supply pressure} \\ S = \text{cylinder surface} \end{cases}$$

The lifting speed at pressure P is given by, as a first approximation and ignoring losses:

$V = \dfrac{Q}{S}$, Q = oil flow supply to cylinder. We deduce the following relationships:

$$\frac{Q}{V} = \frac{F}{P} \qquad \rightarrow \qquad F.V = Q.P \tag{8.43}$$

In a hydraulic exciter, the "force × speed" product of the cylinder depends on the "flow × pressure" product of the hydraulic supply. The latter being a direct function of the electric drive power of the hydraulic pumps.

In reality, the limitations of the exciter are always a result of the flow limitations of the servo valve, which limits the flow of fluid into the cylinder and also generates significant load losses. Thus, for a given servo valve opening "a", pressure loss increases with the square of the flow (this is a nonlinear relationship), so:

$$\Delta P = \frac{1}{k}.\left(\frac{Q}{a}\right)^2 \tag{8.44}$$

k = constant related to the servo valve.

The remaining useful pressure for actuating the cylinder is then given by ($P - \Delta P$). This means that from a certain speed (linked to the flow), the dynamic thrust force becomes very weak and practically goes to zero.

In summary, we note that:

- *The supply pressure limits the force of the exciter.*
- *The flow of the servo valve limits the speed for a given force.*

8.2.3.1.2. *Limiting the servo valve*

Another one of the exciter's limitations is given by the maximum working frequency, which mainly depends on:

- *The natural frequency of the servo valve.*
- *Fluid compressibility.*

We solve the servo valve natural frequency problem by increasing the stiffness of the return springs in the control moving coil and by making the moving equipment as light as possible. One can also attempt to make the servo valve work beyond its resonance frequency.

The servo valve driver is in constant motion (to get a given flow rate) and, as a first approximation, we write the equation of the force required for motion as:

$$F = m.\omega^2.|x| \qquad (8.45)$$

x = peak servo valve piston displacement;
m = total mass in motion.

Moreover, since: $F = B.\ell.I$

We get: $$I = \frac{m.|x|}{B.\ell}.\omega^2 \qquad (8.46)$$

Knowing that the Joule losses are the square of the intensity, we get: $P \equiv f(\omega^4)$. Thus, we very quickly reach the temperature limits permissible within the servo valve. This is why the design of the servo valve is the most paramount consideration for the performance of a hydraulic exciter. For example, the resonance frequency values of good quality servo valves are in the order of hundreds of hertz. In Figure 8.24, we see the classical frequency response curves.

Let us note that it is possible to put the servo valves in parallel to increase transfer rates (for higher costs).

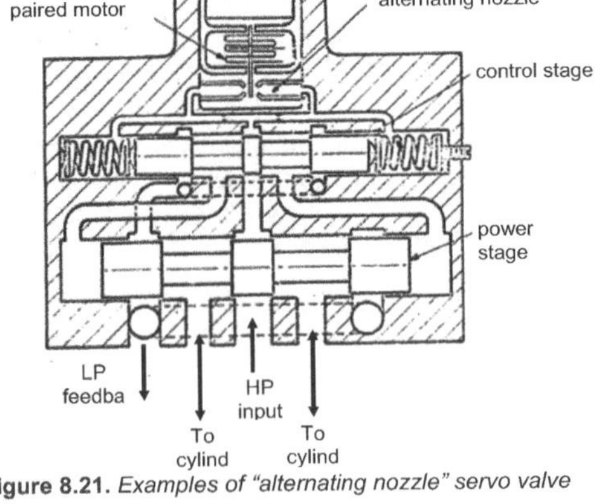

Figure 8.21. *Examples of "alternating nozzle" servo valve*

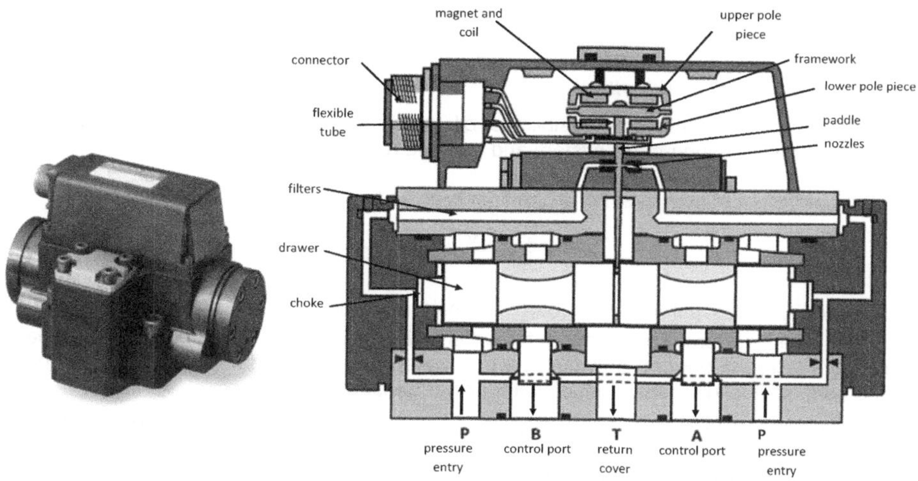

Figure 8.22. *Example of MOOG "paddles" servo valve*

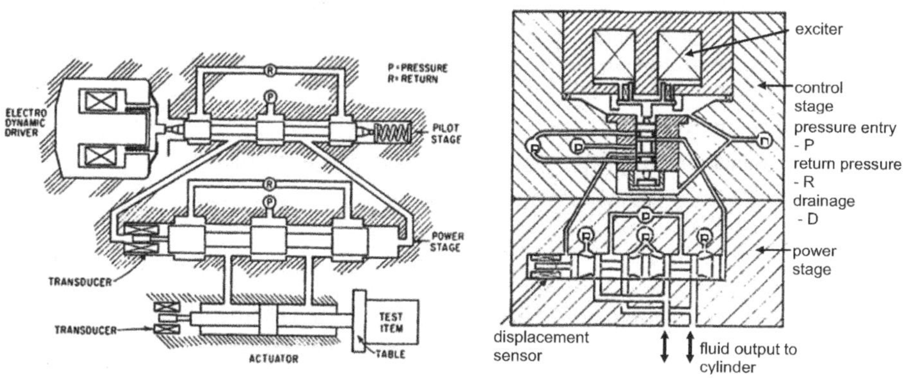

Figure 8.23. *Two examples of electrodynamic exciter servo valves*

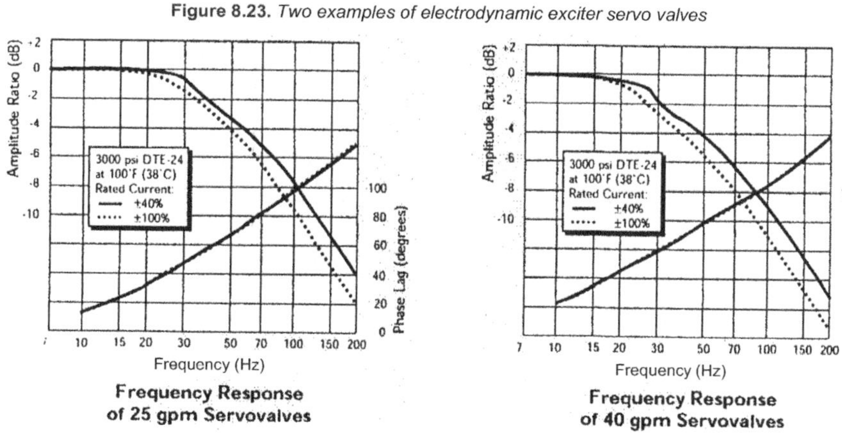

Figure 8.24. *Usual servo valve response curves*

NB: We note that there is a rapid drop in performance of the servo valve depending on the frequency, which is a limitation of hydraulic systems, as mentioned in the introduction. It is slightly compensated by the fact that in HF, flow rates are lower.

8.2.3.1.3. *Resonance of the oil column*

Another limitation is the compressibility of oil, which acts as a spring by creating, with the load mass thrust, a spring-mass system and thus the resonance frequency.

NB: The coefficient of compressibility of a fluid is given by the following relationship:

$$\frac{1}{\chi}=\frac{1}{\Delta P}.\frac{\Delta V}{V}[\text{bars}^{-1}] \tag{8.47}$$

We often use the parameter χ [bar], which is equivalent to pressure, also denoted B throughout the literature. For example, for water, χ is 23500 bar at 25 °C or about 23,5 bars of pressure increase per cm^3 of water in 1 L of water. That of standard hydraulic oil is about 17 000 bars, so it is lower than that of water (lower density).

Thus, at a given frequency, the system resonates but because of the viscosity of the oil, damping occurs to give a low coefficient, which is why it is possible to "work beyond" the resonance frequency of the oil column (typically 100–300 Hz). Beyond this resonance, the cylinder is no longer in phase with the excitement, an excess of power must be supplied in order to compensate for the filtering effect that appears after resonance.

There are empirical formulas that give the resonance frequency of the oil column depending on the type of servo valve, cylinder, the mass to push and the amplitude of the vibratory displacement. According to MB Electronics, this involves:

f_r = resonance frequency in [Hz] K = constant dependent on the servo valve

F = maximum cylinder force in [N] M = total mass in motion in [kg]

x = peak displacement of the cylinder piston in [m]

$$f_r = K.\left(\frac{F}{M.x}\right)^{1/2} \tag{8.48}$$

In Figure 8.25, we see the shape of a typical limit curve for a hydraulic exciter.

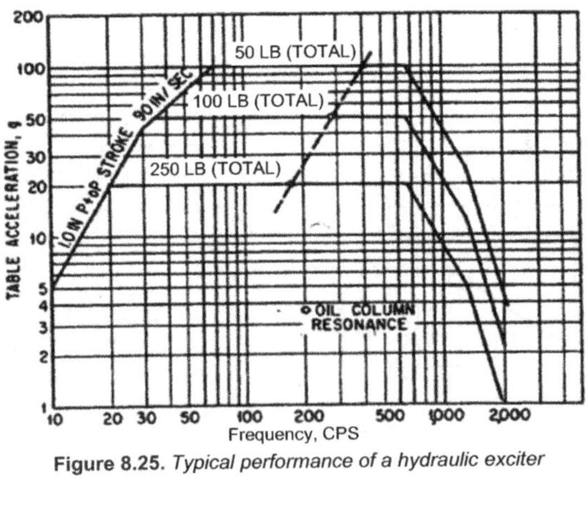

Figure 8.25. *Typical performance of a hydraulic exciter*

8.2.3.1.4. *Limiting frequency*

In LF, the vibration amplitude is limited by the maximum peak to peak displacement of the piston at values from a few hertz to a few tens of hertz, depending on the type of exciter. There is also a maximum speed limit due to the maximum throughput that the servo valve can provide. We are also limited in force and therefore for a total vibrated mass, we are limited in maximum possible acceleration. At higher frequencies, the dynamic performance of the servo valve is the limiting factor and the HF rate servo valve response curves in Figure 8.24 express this.

In Figure 8.26, we see yet another example of a performance curve for an MB Electronics cylinder of 250 kN, so of high power, with a maximum stroke of 65 mmc/c, an HV80 servo valve (maximum flow rate of 270 L/mn at 175 bar), loaded with an inert mass of 5 tons. The maximum useful track c/c of the cylinder is 2.5 inch or #63,5 mm.

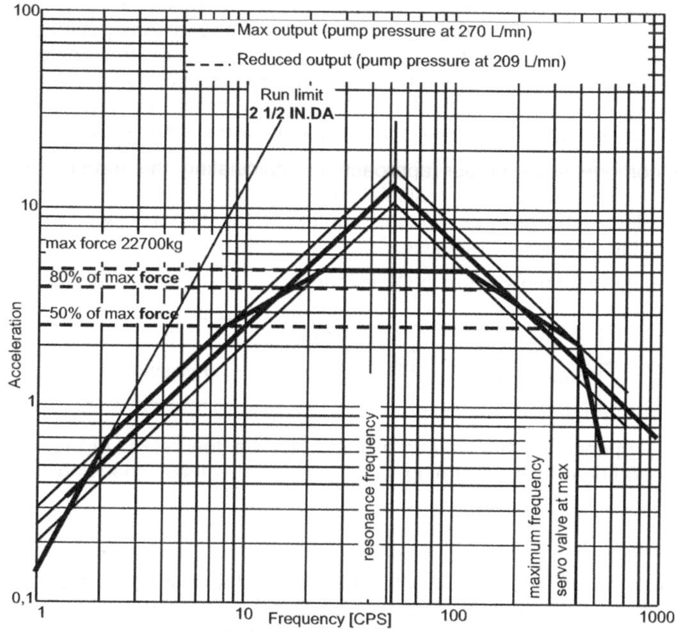

Figure 8.26. *Theoretical curve performance*

NB: USA units 1 GPM = 227.7 L/mn.

8.2.3.1.5. *Static positioning*

The displacement sensor, which is usually located in the lower end of the cylinder, allows:

- the cylinder to be placed in any static position in its possible course;
- feedback at constant amplitude or constant displacement.

The same goes for the displacement sensor in the servo valve, which can be positioned in the same way, and the servo valve piston, for recentering and for feedback during displacement.

8.2.3.1.6. *Hydraulic supply*

The hydraulic power supply, which provides "hydraulic power" in terms of flow and pressure, can use positive displacement pumps at a constant rate or variable displacement pumps that save energy. For example, one can achieve the following performances:

Force × maximum speed product [$kNms^{-1}$]	Flow below 175 bars [$Lmin^{-1}$]	Electric power required [kVA]
9	30	10
57	190	75
115	400	125
250	850	240

8.2.3.2. *Transfer function of hydraulic exciters*

8.2.3.2.1. *Overview*

The equations that describe the precise operation of hydraulic systems are usually extremely difficult to implement. They depend on physical phenomena governed by nonlinear relationships (losses of localized or dispersed load, laminar or turbulent flow, etc.). Fortunately, it is usually possible to linearize the relationships around operating points, which allows for a fairly precise approach to the involved phenomena. An introduction to the techniques and the operation of hydraulic generators is given in Appendix A.

To remain homogeneous in our approach for calculating the transfer function of an electrodynamic exciter, we propose calculating the transfer function of the input hydraulic exciter e(t) = "electrical signal for voltage control of the servo valve" and output x(t) = "displacement of the cylinder piston exciter".

The functional diagram of the system in OL is then:

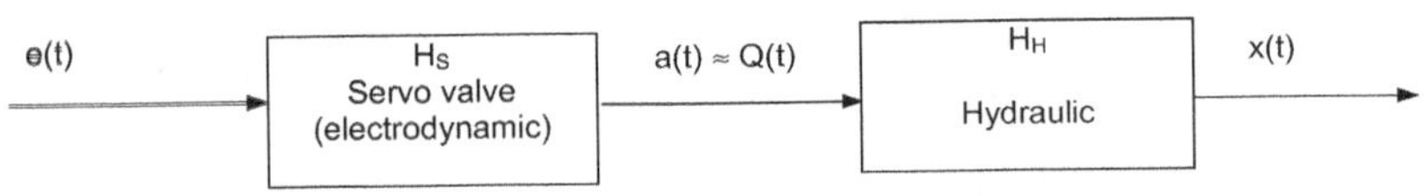

We will see (Appendix A) that we can assume that the flow rate Q of the servo valve in a chamber is proportional to the opening "a" of the latter without changing the accuracy of results.

8.2.3.2.2. *Transfer function in OL*

In Appendix A, we showed that after linearization and simplification, the transfer function of the servo cylinder in OL could be expressed with sufficient accuracy to enable us to study the performance of the system through the following relationship:

$$H_1(p) = H_S(p).H_H(p) = \frac{1/B.\ell}{(p+\omega_S)\left(1+2\xi_e\dfrac{p}{\omega_e}+\dfrac{p^2}{\omega_S^2}\right)}\cdot\frac{\omega_f/\omega_c}{\dfrac{p}{\omega_c}\left[1+2(\xi_1+\xi_2)\dfrac{p}{\omega_c}+\dfrac{p^2}{\omega_c^2}\right]} \tag{8.49}$$

Where:

$\omega_f = \dfrac{K_2}{S} = \dfrac{\xi\pi D}{S}\sqrt{\dfrac{P_e}{\rho}}$ = speed gain in an OL system;

$\omega_c = \sqrt{\dfrac{2BS^2}{VM}}$ = natural frequency of the piston on its oil cushions;

$\xi_1 = \dfrac{\omega_c}{2}.\dfrac{f_1.M}{2.S^2}$ = damping coefficient due to leakage from the cylinder piston;

$\xi_2 = \frac{\omega_c}{2} \cdot \frac{Q_e}{P_e} \cdot \frac{M}{S^2}$ = damping coefficient due to valve leaks;

ω_e = electrical resonance of the coil = $\frac{B\ell}{[L(m_t + m_b)]^{1/2}}$;

ξ_e = electrical damping;

ω_s = resonance of the valve-servo valve body connector.

NB: In the present case, we are of course interested in the input voltage, but we are particularly interested in an output cylinder acceleration because of feedback acceleration on the support table of the tested material. By multiplying by p or p², we can find transfer functions in velocity and acceleration, which means:

$$\frac{\dot{X}(p)}{E(p)} \quad \text{and} \quad \frac{\ddot{X}(p)}{E(p)}$$

Of course, the transfer function in OL (8.50) does not give the properties for operation of the servo cylinder as a vibration exciter. Nevertheless, it initializes the study of system performance based on its intrinsic parameters, for example, the study of its stability.

8.2.3.2.3. ***Example of a cylinder stability analysis***

The stability condition is calculated from the transfer function in OL. Its rigorous calculation is complex and is beyond our scope, however, an approach can be considered where one analyzes the "absolute" stability condition that applies to the transfer function in OL, with the following classical constraint to abide by:

$$|H(p)| < 1 \text{ (or 0 dB in dB)} \quad \text{for} \quad \omega \geq \omega_c$$

Thus, in the loop transfer function (feedback system), the return loop system will have a gain <1 with a phase ≥ –180 ° (phase opposition with the input signal); it will be stable.

NB: If the transfer function is calculated by only taking into account the natural damping ξ_v that is due to the viscous friction of cylinder f_v (typically small because of their construction), the transfer function obtained is the same as (8.49), one just needs to replace (ξ_1 + ξ_2) by $\xi_v = \frac{\omega_c}{2} \frac{Vf_v}{2BS^2}$ in the relationship. The stability condition is the same, so $\xi_v > \frac{1}{2}\frac{\omega_f}{\omega_c}$. If in the latter ω_c and ξ_v are replaced by their values, we get: $f_v > M.\omega_f$, which is a very difficult condition to meet since f_v is small compared to $M.\omega_f$. This is why controlled leakage in the cylinder and the servo valve was introduced into the calculation.

If one considers the cylinder, the simplified transfer function in OL then gives:

For: $\omega = \omega \rightarrow \quad H_H(j\omega_c) = \frac{\omega_f}{-\omega_c 2(\xi_1 + \xi_2)} \rightarrow |H_H(j\omega_c)| = \frac{\omega_f}{\omega_c 2(\xi_1 + \xi_2)}$

And the condition becomes: $(\xi_1 + \xi_2) > \frac{1}{2}\frac{\omega_f}{\omega_c}$ (8.50)

Thus, we draw tables giving each value of the ratio $\frac{\omega_f}{\omega_c}$ the value of (ξ_1 + ξ_2) required such that the overvoltage coefficient Q_R does not exceed 2.3 dB, which is generally the maximum accepted value that does not damage the cylinder.

We therefore note that the condition rests on the damping coefficients ξ_1 and ξ_2 and stability is more assured when these are big (including improving stability through control of leaks in the cylinder and the servo valve). ω_c must have the highest possible value (it is usually very difficult to play on this parameter).

Hydraulic systems are required for managing this significant issue because the absolute stability condition is generally not enough here (as it is very difficult to increase ξ_1 and ξ_2) and one must define a margin of stability on the overvoltage factor, for which we should usually not exceed the critical value of 1,3 (2,3 dB).

Indeed:

- the resonant pulse ω_c is almost intangible;

- the performance pulse is often imposed by the servo valve manufacturer;

- the increased leakage technique degrades the maximum performance of the cylinder and has a cost in the form of non-useful throughput (higher consumption), thus cannot be used loosely. Indeed, these solutions are not costless and generally, the advantage of the cylinder leak is clear and is usually used alone.

We will therefore have to use secondary negative feedback returns from the servo valve and the cylinder in order to improve the stability conditions, in addition to corrective systems:

- either as artificial shock absorbers;

- or as a low-pass filter that is adapted between the cutoff and resonance frequencies;

- or as a phase corrector system (phase advance), intended to prevent a resonance peak around –180°.

NB: In terms of detecting the acceleration of the cylinder as a secondary return is the most effective way, its implementation is rather more delicate.

A block diagram of a possible general transfer function is:

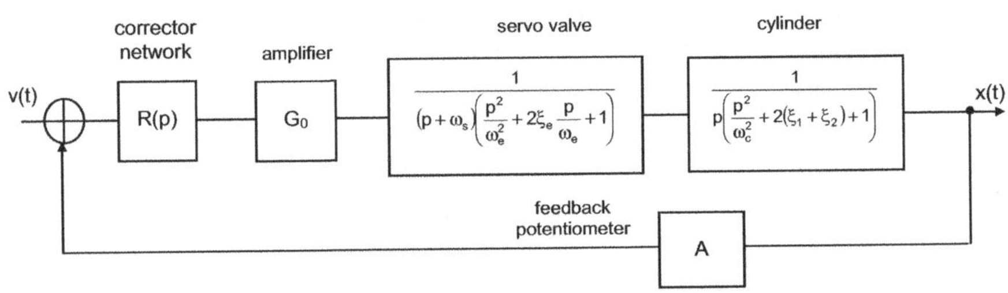

This only includes cylinder position feedback from the feedback potentiometer. For a hydraulic exciter, this function is primarily used to position the cylinder at the midpoint of operation. One must therefore add negative feedback control at the acceleration sensor located on the vibrating table in order to control the cylinder undergoing vibrations or shocks. This negative feedback control will use the remaining input jack on the card.

Let us note that proportional, integration, derivation control occurs between the input summator and the servo valve current amplifier control. One then gets the simplified block diagram in Figure 8.27.

An example of a feedback card (G122-202A1 "servo-amplifier" Moog map) is given in Figure 8.27.

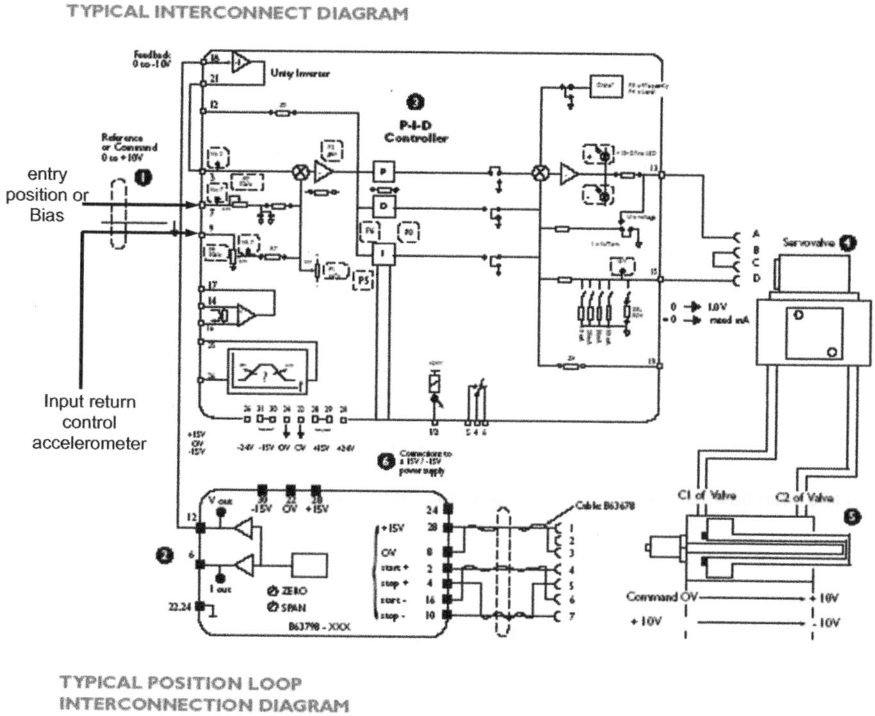

Figure 8.27. *Example of a feedback control card*

8.2.3.2.4. *Performance limitations*

- At low frequencies, performance is limited by the cylinder stroke.

- At mid frequencies, speed is limited by the maximum throughput of the setup.

- At high frequencies, performance is limited by acceleration first through the maximum allowable stress, then by the effects of hydraulic natural frequency.

In impact tests, it is essentially the flow of the servo valve that limits performance and the high distortion rate when there is significant displacement.

8.2.3.3. *Digital application*

It helps to have an idea of the magnitudes encountered.

- Cylinder force:	120 kN
- Supply pressure:	200 bars
- Frequency range:	0,1–50 Hz
- Piston + specimen mass:	1550 kg at a maximum acceleration of 80 ms^{-2}
- Maximum possible stroke of the piston:	$\pm 103 \times 10^{-3}$ m
- maximum speed of the piston:	1,56 ms^{-1}
- Useful surface of the piston ring:	90×10^{-4} m^2
- Chamber diameter:	146.5×10^{-3} m
- Piston output diameter:	100×10^{-3} m

8.2.3.3.1. *Natural frequency of a loaded cylinder*

The natural frequency of the loaded cylinder is given by:

$$f_c = \frac{1}{2\pi}\sqrt{\frac{2BS^2}{VM}} \;\#\; 65{,}5\text{Hz}$$

$B \# 15\,000\ 10^5\ \text{Nm}^{-2}$ or Pa

S useful $\# 90\ 10^{-4}\ \text{m}^2$

V useful $\# 927\ 10^{-6}\ \text{m}^3$

$M = 1550\text{kg}$

8.2.3.3.2. *Maximum instantaneous flow and servo valve opening*

- The maximum instantaneous flow Q_{max} will be equal to:

$$Q_{max} = v_{max}.S \quad \rightarrow \quad Q_{max} = 1{,}56 \times 90\ 10^{-4} \;\#\; 140{,}4 \times 10^{-4}\text{m}^3.\text{s}^{-1}$$

$$Q_{max} \;\#\; 843\text{Lmn}^{-1}$$

which is a high flow rate and the servo valve has the dimensions for this.

- Opening "a" necessary to the distributor for maximum flow: relationship (1.11) in Appendix A, which gives the flow in a cylinder chamber:

$$Q_1 = K.a.\sqrt{\frac{P_e}{2}}.\sqrt{1 - \text{sgn}(x)\frac{M}{P_e S}.\frac{d^2x}{dt^2}} = K_1\sqrt{1 - \frac{\text{sgn}(x).M}{P_e S}.\frac{d^2x}{dt^2}}$$

$$K_1 = K.a.\sqrt{\frac{P_e}{2}} = \xi\pi D\sqrt{\frac{2}{\rho}}a\sqrt{\frac{P_e}{2}} = \xi\pi Da\sqrt{\frac{P_e}{\rho}}$$

However, the flow rate is at its maximum when the velocity is maximum and therefore acceleration is zero. We deduce the following relationship, which gives the maximum flow rate according to the opening "a":

$$Q_{max} = K.a.\sqrt{\frac{P_e}{2}} \quad \rightarrow \quad a_{max} = \frac{Q_{max}}{K}\sqrt{\frac{2}{P_e}}$$

where $K = 6{,}15 \times 10^{-3}$ gain of the servo valve, given by the manufacturer.

Therefore:

$$a_{max} = \frac{140{,}4 \times 10^{-4}}{6{,}15 \times 10^{-3}}\sqrt{\frac{2}{200 \times 10^5}} \;\#\; 0{,}722 \times 10^{-3}\text{m}$$

NB: Let us note the low aperture value for maximum flow rate relative to the external passage holes of the servo valve. Servo valve technology (usually lined drawers) causes the opening "a" to not be commensurate with the diameter of the evacuation holes to the cylinder.

8.2.3.3.3. *Power*

Pressure check P_e.

$$P_e = \frac{3}{2}\frac{F}{S} = \frac{3}{2}.\frac{120 \times 10^3}{90 \times 10^{-4}} \;\#\; 200 \times 10^5\text{Pa} = 200\text{bars}$$

8.2.3.3.4. *Place of transfer*

By applying the relationships in Appendix A in conjunction with the manufacturer's data, we get the following values:

- for a loaded cylinder:

$\omega_c \# 411 \text{ rd s}^{-1}$ $\omega_f = \dfrac{K_2}{S} = \dfrac{K}{S}\sqrt{\dfrac{P_e}{2}} \# 2160 \text{ rd s}^{-1}$ $\dfrac{\omega_f}{\omega_c} = 5,26$

$q_0 \# 5,8 \text{ l mn}^{-1} = 9,8 \times 10^{-5} \text{m}^3\text{s}^{-1}$ and leak f $\# 8,4 \times 10^{-13} \text{m}^3 \text{ s}^{-1} = 5 \times 10^{-8} \text{L mn}^{-1}$

$$\begin{cases} \xi_1 = 1,49 \times 10^{-3} \\ \xi_2 = 0,0342 \end{cases}$$

- for an empty cylinder:

$\omega_c \# 2080 \text{ rd s}^{-1}$ $\dfrac{\omega_f}{\omega_c} \# 1$

$$\begin{cases} \xi_1 = 0,27 \times 10^{-3} \\ \xi_2 = 0,608 \times 10^{-2} \end{cases}$$

These values allow us to draw the cylinder transfer function.

8.2.3.4. *Harmonic distortion*

This focuses on the acceleration of the cylinder piston (second derivative of displacement). The overall harmonic rate measured for the speed and acceleration signals can have very large values (50% and over). This is particularly the case for frequency ranges around 10–80 Hz. In most cases, the main cause of distortion is odd harmonics (the highest of which is the third). This disproportionately affects impact tests in particular, since this distortion increases with nonlinearity of the transfer functions and with strokes that then become longer.

The specifications impose given distortion rates and it is imperative to abide by these (example of GAM EG13: when vibrating, the gap between the overall signal and the filtered signal must not exceed ±5% on the effective values and for shocks. This means staying within tolerances of ±20% for the amplitudes of the time or SRS).

Nonlinearities of flow (see flow equation) and the influence of joints (pipe fittings, etc.) penalize the distortion even more.

8.2.4. Use in impact tests

The means for creating hydraulic vibrations can also be used in impact tests. The interest lies within the maximum possible forces and heavy displacements; which allow us to carry out long duration shock pulses for large masses (for example a half sine of 20 g 20 ms for 1 ton). Furthermore, for the most conventional shock pulse, the useful frequency band is not very high and is compatible with the performance of hydraulic exciters.

In pulse mode, the "main limitation" is given by the "instantaneous flow rate" that the servo valve can pass, which is necessary to get the large velocity variations that are required to get the desired shapes of shock pulses. Generally, the hydraulic unit is not capable of providing the necessary instantaneous flow rate. We overcome this disadvantage by using accumulators, which act as energy reservoirs (similar to a reservoir capacitor in power supplies).

Example 8.2: We wish to carry out a half-sine shock by pulse on a total mass of 1 ton, with the following properties: A = 15 g, D = 20 ms, ε_{max} = 0,2. This is done using a hydraulic excitor. The piston surface area is 58 cm^2. We must determine the dimensions for the test.

First, we select a shock with a half-sine post-shock (pulse). We must, of course, check the impact of the distortion in LF in relation to the modes of components tested for this type of shock. We get:

$$\Delta\dot{X} = 0$$

$$|X_{mx}| = \frac{AD^2}{\pi}\left(1 + \frac{1}{\varepsilon}\right) \text{ \# } 112{,}4 \text{ mm}$$

$$|\dot{X}_{mx}| = \frac{2AD}{\pi} \text{ \# } 1{,}874 \text{ ms}^{-1}$$

- *Regarding the necessary force*, it takes place for the maximum dynamic force (if we accept not gaining in force in pulse mode as for electrodynamic systems):

F_{mx} = 1 000 x 15 x 9,81 # 147150N # 147kN

Hence the value for necessary pressure:

$$P_{mx} = \frac{F_{mx}}{S} = \frac{147150}{9{,}81x58} \text{ \# } 258\text{Bars}$$

- *Regarding the flow rate*, we get:

$$\Delta\text{volume} = \text{maximum stroke x surface} = 0{,}58x1{,}124 \text{ \# } 0{,}652\text{l in } 2x25 + 20 = 70\text{ms}$$

So:
$$\text{Flow} = \frac{0{,}652}{70 \text{ x } 10^{-3}} \text{ x } 60 \text{ \# } 559\text{lmn}^{-1}$$

We see the importance of instantaneous flow, which is necessary for calculating the accumulator that is able to achieve the instantaneous flow and particularly, the servo valve that is able to pass it through (one can also put several servo valves in parallel, as has been previously suggested).

This simplified calculation allows us to see the importance of the instantaneous flow rates required to carry out impact tests on hydraulic cylinders. This is also the main difficulty needing to be solved in order to use this particular means for impact tests.

Figure 8.28 shows a test with a half sine wave on a Fairey hydraulic exciter. The total impacted mass is 655 kg (max flow rate 200 l).

One can notice the effects of distortions on the signal, as previously mentioned.

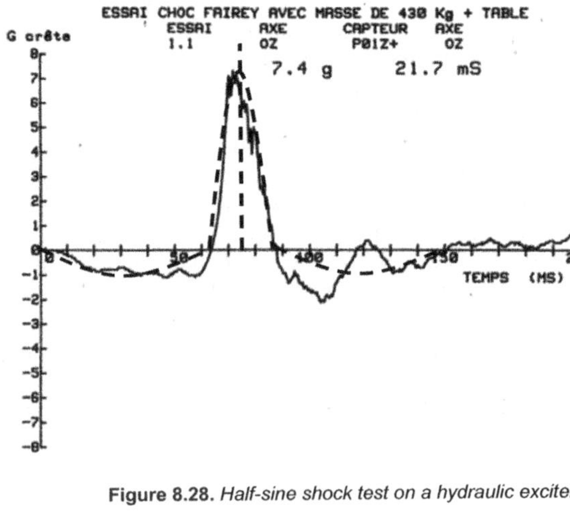

Figure 8.28. *Half-sine shock test on a hydraulic exciter*

8.2.5. Conclusion

The advantages of hydraulic exciters compared to electrodynamics in terms of maximum force developed, maximum possible stroke, robustness and operation in VLF should be weighted with the problems of stability and harmonic distortion that can be encountered in impact tests.

Nevertheless, with current technology, it is possible to use them in impact tests by taking special precautions and using properly-sized accumulators, but the system must be adapted for this. One should note that there are cylinders of significant force that are capable of generating vibrations of up to about 1 000 Hz due to progress in the development of rapid servo valves.

The approach we discussed has allowed us to understand the main problems associated with their use but the specific operation of each system must be analyzed in terms of the technologies used for servo valves and actuators and in terms of electronic transfer function correction technologies. Accurate calculations are generally more complex and belong with experts.

Conclusion

In conclusion to this lesson on vibrations, we make the following observations:

- The phenomenon of vibration is particularly damaging due to the aspect of "fatigue" that it physically imposes on the structures submitted to it. This can cause breakages that may occur at random times and possibly far away from locally severe but not destructive vibrations. In addition, it may predispose the structure to damage caused by other environmental factors (shocks, climate, corrosion, etc.). This is why special attention is usually required during the development of the equipment; the mechanical environment is the aspect that is most considered by technicians in research units and testing laboratory, along with the climatic environment.

- The physical approach is relatively complex and the tools for creating the behavior models used for prediction purposes require heavy computer tools that are designed for specialists (modal analysis, finite element, mechanical impedance, etc.). Nevertheless, we have seen that a few simple design rules allow us to avoid the most common problems. From a simulation point of view, we have also seen that the means used in vibration systems and the associated control console systems are relatively sophisticated and complex in their development and use. The existence of different types of vibrations makes them more difficult to implement and requires great expertise.

It appears that the usual recommendations for better understanding and processing this type of environment mainly include:

- guaranteeing the expertise and competence of staff involved through initial and continuous training sessions that are designed for the requirements of the design and testing structure;

- having a database on the topic that is sufficiently comprehensive and well updated (bibliography, standards, journals, articles, etc.);

- adapting the means for testing and analyzing the laboratory requirements and ensuring a thorough understanding of their possibilities and limitations. Validation operations and confirmation of the discussed means and test methods are used to initialize this approach;

- having a qualitative approach that must be approved by management and that guarantees customers the quality of the proposed testing services. The latter can be through a laboratory accreditation for the standard NF EN ISO/CEI 17025, or an equivalent standard in other countries which is a voluntary process but which is increasingly becoming a requirement for large contractors vis-à-vis testing laboratories and subcontractor analysis. In addition to the contractual aspect of the standard, subcontractor analysis usually brings a "state of mind" to the staff that is favorable for good ownership of the qualitative requirements for laboratory studies and tests.

- knowing that ultimately, the guarantee for good quality of service is based on the guarantee of finance and investment for the appropriate testing means, staff and machinery. This is essentially justified by the added value brought about by the proposed studies and tests.

Appendices

Appendix A

Fundamental Laws of Hydraulics – Application to the Study of Hydraulic Vibration Generators

Let us briefly discuss the basic principles of hydraulics in order to introduce its application to the study of the hydraulic exciters from section 8.2 [ref. I15].

A.1. Law of flow of a fluid

The fundamental relationship is given by the Bernoulli equation, which describes the law of flow in a tube of fluid (derived from the law of dynamics $\vec{F} = m.\vec{\gamma}$ or the corresponding thermodynamic relationships). It connects the fluid flow velocity to pressures that appear due to internal (hydrostatic) and external (generally, dynamics) forces.

NB: Let us note that the existence of the notion of pressure is due to the fact that all fluids are compressible (which is the definition of a fluid. Sand, for example, is not). For this reason, pressure is exerted in all directions but there is also a possible propagation phenomenon through the concept of a "sound wave".

Therefore, for a tube element ds, we get dx that makes angle α with the vertical axis and a perfect fluid (without viscosity), whereby:

- P_s = static (or hydrostatic) pressure of the element [$Pa = N.m^{-2}$] due to the weight of the height of the liquid;

- ρ = density of the liquid (mass per unit volume [$kg.m^{-3}$]). This is assumed to be constant for poorly compressible fluids;

- g = gravitational acceleration [$m.s^{-2}$];

- β = load factor ($\beta.g$ is the acceleration that the circuit is subjected to, this is the case for airplanes and systems subjected to a centrifugal force);

- h = distance from the element to a reference plane H that is perpendicular to the direction of acceleration βg [m]. This could be a horizontal plane if it only involves gravitational acceleration;

- α = angle of the velocity v with a plane that is perpendicular to plane H.

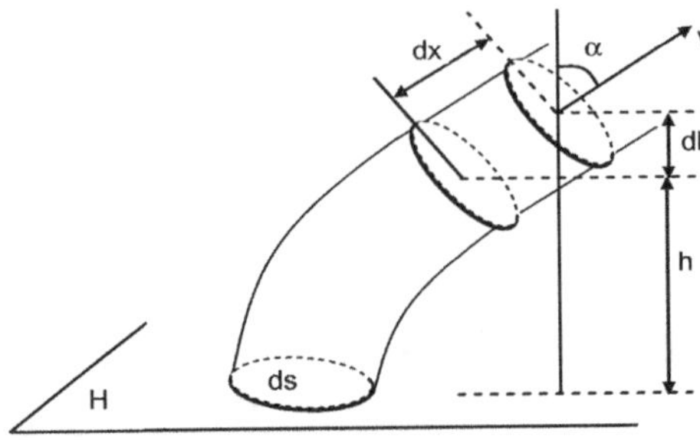

Figure A.1. *Element in a tube of liquid*

The pressure variation in element dx is given by the sum of the variation under static pressure and the weight of the liquid element due to the acceleration the circuit is subjected to:

$$dP = dP_S + \rho.\beta.g.ds.dx.\cos\alpha$$

In terms of antagonistic forces:

$$-dP_S.ds - \rho.\beta.g.ds.dx.\cos\alpha = -\frac{dP_S}{dx}.dx.ds - \rho.\beta.g.ds.dx.\cos\alpha$$

Forces oppose the movement of the fluid.

Applying the law of dynamics, we get v = velocity in the direction of movement:

$$-\frac{dP_s}{dx}dx.ds - \rho.\beta.g.ds.dx.\cos\alpha = dm.\frac{d^2x}{dt^2} = \rho.ds.dx.\frac{dv}{dt}$$

$$\text{where}: dx.\cos\alpha = dh \text{ et } \quad \frac{dv}{dt} = \frac{dv}{dx}\frac{dx}{dt} = \frac{dv}{dx}.v \quad \text{we get}:$$

$$-\frac{dP_s}{dx}dx.ds - \rho.\beta.g.ds.dh = \rho.ds.dx.\frac{dv}{dt} \rightarrow dP_s + \rho.\beta.g.dh + \rho.v.dv = 0$$

ρ remains constant in the calculation element. Thus, by integrating the equation, we get:

$$P_S + \rho.g.\beta.h + \frac{\rho}{2}v^2 = cte = P_t$$

The constant of the second member is called the "total pressure".

As there will always be energy dissipation through friction (concept of pressure loss), the total pressure decreases between two sections and so we get the generalized Bernoulli relationship:

$$P_{t1} - P_{t2} = \Delta P_t \equiv \Delta P = \Delta P_S + \rho.\beta.g.\Delta h + \frac{\rho}{2}\left(v_1^2 - v_2^2\right)$$

NB: Let us note that the term for the square of speed is directly linked to the compressibility of the fluid. This is called "dynamic pressure" and it occurs due to the effect of moving fluid.

In general, $\rho.\beta.g.\Delta h$ is negligible compared to the other terms and the relationship is reduced to:

$$\Delta P = \Delta P_s + \frac{\rho}{2}\left(v_1^2 - v_2^2\right) \qquad (A.1)$$

This relationship expresses the fact that the total pressure load loss between two points in a fluid thread is equal to the algebraic sum of the static pressure loss and the dynamic pressure loss.

If the velocity is high, which will be the case for us, we ignore ΔP_s and get the following relationship, which is generally used (and is also the result of thermodynamic considerations), if the initial velocity is zero:

$$\Delta P = \frac{\rho}{2} v^2 \tag{A.2}$$

NB: As velocity is difficult to measure, we express this relationship in a more convenient form:

where: $Q = V.S$

$$\Delta P = \frac{\rho}{2} \cdot \frac{Q^2}{S^2} \tag{A.3}$$

This relationship only makes sense for a perfect fluid (without viscosity); the situation is very different for a real fluid. Nevertheless, it is used to formulate load losses within the elements of hydraulic systems by comparison with this formula and by including the characteristic element coefficients (shape, material, etc.) (see section A.3).

A.2. Notion of viscosity

"Viscosity" μ is an important property for characterizing an oil. It creates a delaying force within the fluid ("flow" is the inverse property) with the delaying force element:

$$dF = \mu.ds.\frac{dv}{dx}$$

A.2.1. Dynamic (or absolute) viscosity μ

We define "dynamic (or absolute) viscosity μ" by the following relationship:

$$\mu = \frac{dF}{ds.\frac{dv}{dx}} \tag{A.4}$$

where $\frac{dv}{dx}$ is the velocity gradient = growth rate per unit length, where:

$$[\mu] = M.L^{-1}.T^{-1}$$

In MKSA units (SI), the unit of dynamic viscosity is the Pascal.second [Pa.s] or kilogram per meter per second [$kg.m^{-1}.s^{-1}$].

Indeed: $1.Pa = 1.N.m^{-2}$; $1.N = 1.kg.m.s^{-2} \rightarrow Pa.s = kg.m^{-1}.s^{-1}$.

The old CGS unit for dynamic viscosity is the "poise": $1.Po = 1.g.cm^{-1}.s^{-1}$

In practice, tables give the dynamic viscosity in centipoise [cPo] = $10^{-2}.g.cm^{-1}.s^{-1}$

For example, the dynamic viscosity of water at 20 °C is about 1 cPo.

Another old unit is "the Poiseuille Pl", which is equal to: 1 Pl = 10.Po.

Like $1.kg_{p,f}$ # 10N, we therefore get the same measurement in Pl and in Pa.s.

A.2.2. Kinematic viscosity ν

Similarly, we define the notion of "kinematic viscosity ν" as:

$$\nu = \frac{\text{absolute viscosity}}{\text{density}} = \frac{\mu}{\rho} \quad \text{(A.5)}$$

This is homogeneous with:

$$[\nu] = L^2.T^{-1}$$

It is therefore obtained through dividing the dynamic viscosity by the density and is expressed in $m^2.s^{-1}$.

In MKSA units (SI), 1 unit MKSA = $1.m^2.s^{-1} = 10^{-6}$ cSt = 10^{-4} St.

The old CGS unit for kinematic viscosity is the "stokes": 1.St = $1.cm^2.s^{-1}$.

In practice, tables give kinematic viscosity in centistokes 1.cSt =$1.mm^2.s^{-1}$.

Some other old viscosity units are provided for information purposes.

(a) In France

- The Engler degree [E^0]: ratio of the flow time for 200 cm^3 of the concerned liquid to the flow time of 200 cm^3 of water through an opening of 2,8 mm in the Engler viscometer:

$$E^0 = \frac{\text{time of liquid at t°}}{\text{time of water at 20°C}}$$

For liquids where $E^0 > 4$, we can apply the following conversion formula: η # 6,584 E^0

For machine oils, if the specific weight is assumed to be 0,9, we get:

$$\nu = \frac{\eta}{\rho} \# \frac{6{,}584}{0{,}9}.E^0 \# 7{,}315.E^0 \quad [cSk]$$

- The Barbey degree [B^0]: measure of the fluidity of a liquid at t°, which is expressed by the number of cm^3 of liquid flowing during 1 h in the viscometer in question.

(b) In England

- The Redwood second (No. 1 or commercial type) [R"]: an oil with density 0,915 for which 50 cm flow in 535" at 15.5r°C is said to have a viscosity of 100 R" seconds:

$$R" = \frac{\text{flow time of the liquid at t° x 100}}{535} \text{x} \frac{\text{density of the liquid at t°}}{0.915}$$

For oils, we get the following correspondence:

$$\nu = 0.260.R" - \frac{179}{R"} \quad [cSk] \qquad 34 < R" > 100$$

$$\nu = 0.274.R" - \frac{50}{R"} \quad [cSk] \qquad R" > 100$$

(c) In the United States

- Saybolt seconds (universal type) [S"]: viscosity in Saybolt seconds expresses the number of seconds required for 60 cm^3 of liquid to flow.

Water at 15.5° has a viscosity of 30 S".

We get the following correspondence:

$$v = 0{,}226.S" - \frac{195}{S"} \text{ [cSk]} \qquad 32 < S" < 100VABC$$

$$v = 0{,}220.S" - \frac{135}{S"} \text{ [cSk]} \qquad R" > 100$$

A.2.3. Variation of viscosity with temperature

We say that:

$$\mu = \mu_0.e^{-\lambda(T-T_0)} \rightarrow \frac{\Delta\mu}{\mu_0} = \lambda\frac{\Delta T}{T_0}$$

The μ_0 and λ parameters depend on fluid in question through $\lambda = \frac{E}{RT} = \frac{\text{activation energy}}{RT}$.
The relationship shows that viscosity decreases with temperature (quite significantly).

A.2.4. Variation of viscosity as a function of pressure

We say that:

$$\frac{\Delta\mu}{\mu_0} = \varepsilon.\frac{\Delta P}{P}$$

The ε property depends on the fluid in question. The relationship shows that viscosity increases with pressure. Thus, if ε = 0,003, an oil that goes from 1 bar to 50 bars will have a viscosity variation of $\frac{\Delta\mu}{\mu_0} = 0{,}003 \times 50 = 0{,}15$ or 15%.

The viscosity of a mixture of two oils of viscosity μ_1 and μ_2 at the same temperature is around (Arrhenius' law):

$$\mu = p_1.\mu_1 + p_2.\mu_2$$

p_1 and p_2 are the volume proportions with $p_1 + p_2 = 1$.

A.3. Reynolds number and pressure loss in the piping

When considering the trajectory of particles in a fluid net, there are two types of trajectory. One is regular and parallel to the axis of the net and the flow is said to be "laminar". The other is disorganized and is said to be "turbulent". The distinction between these two operating modes is done with the "Reynolds number" (dimensionless) by:

$$R = \frac{vD}{\nu} \qquad \text{(A.6)}$$

v = average velocity in m/s (or cm/s);
D = nominal diameter in m (or cm);
ν = kinematic viscosity in $m^2.s^{-1}$ or (St).

We therefore accept that:

R < 1 000, flow is laminar;
R > 2 300, flow is turbulent.

The value of R_c = 2,300 is called the critical Reynolds number. Between the two values, calculations are more complicated.

- Regarding pressure loss, there are distributed pressure losses (in the tubes and pipes) and localized pressure losses (in the restraint and connecting bodies).

"As turbulent flows are the most frequent", we link the loss to the "dynamic pressure" through:

$$\Delta P = \xi.\rho.\frac{v^2}{2} = \xi.\frac{\rho}{2}.\frac{Q^2}{S^2} \tag{A.7}$$

v = Q/S = average velocity in the reference section;
Q = volume flow rate;
S = reference section;
ξ = pressure loss coefficient in the element in question.

NB: Relationship (A.7) can also be expressed based on the "specific weight" through $\gamma = \rho.g$.

There are charts and tables that give the values of ξ based on the various elements encountered. There are also various empirical formulas for calculating losses based on the states and elements.

For example, for distributed pressure losses, we use the following relationships:

Let:
$$\xi = \lambda\frac{L}{D}$$

λ = distributed pressure loss coefficient;
L = length of the tube;
D = tube diameter.

Thus:
$$\Delta P = \lambda.\frac{L}{D}.\frac{\rho}{2}.\frac{Q^2}{S^2} \tag{A.8}$$

A.4. Typical force equations

All hydraulic system calculations, particularly transfer function calculations, are done from typical force equations (involving pressure P) and typical flow equations (which also involve P) by removing P from the two.

The equations for force express the equilibrium of moving parts under the action of the forces to which they are subjected. We get:

- forces that are dependent on the position of the movable element (cylinder), which are mainly due to stiffness "k" of type –k.x (proportional to x), to the viscous friction "c" of type -$c\frac{dx}{dt}$;

- forces that are dependent on external function parameters (pressure, flow, high acceleration, etc.) F_e. The general relationship then takes the following classic appearance if M is the mass of the moving part:

$$F_e = k.x + c\frac{dx}{dt} + M\frac{d^2x}{dt^2} \tag{A.9}$$

NB: Generally, it is possible to ignore k and c in the relationships in order to simplify things. But beware, depending on the systems, this not always possible.

A.5. Typical flow equations

These express the conservation of flow/mass $D_i = \rho.Q_i$. If, within a capacitor, a mass flow rate D_i arrives and mass flow rate D_j leaves again, the retention relationship becomes:

$$\sum_i D_i - \sum_j D_j = \frac{dM}{dt} = \rho.\frac{dV}{dt} + V.\frac{d\rho}{dt}$$

For a given mass M and a volume V: $\frac{d\rho}{\rho} = -\frac{dV}{V}$ $\rightarrow$ $B = \frac{dP}{d\rho/\rho}$

B = coefficient of compressibility and therefore:

$d\rho = \rho.\frac{dP}{B}$, in practice ρ # ρ_0 (value for P = 0)

We get:

$$\sum_i Q_i - \sum_j Q_j = \frac{dV}{dt} + \frac{V}{B}.\frac{dP}{dt} \quad \text{(A.10)}$$

which includes the term for flow rate variation $\sum_i Q_i - \sum_j Q_j$, the term for change in volume dV/dt and the term for compressibility $\frac{V}{B}.\frac{dP}{dt}$.

Application to hydraulic cylinders

One can schematically represent how the electrodynamic exciter cylinder works, as shown in Figure A.2.

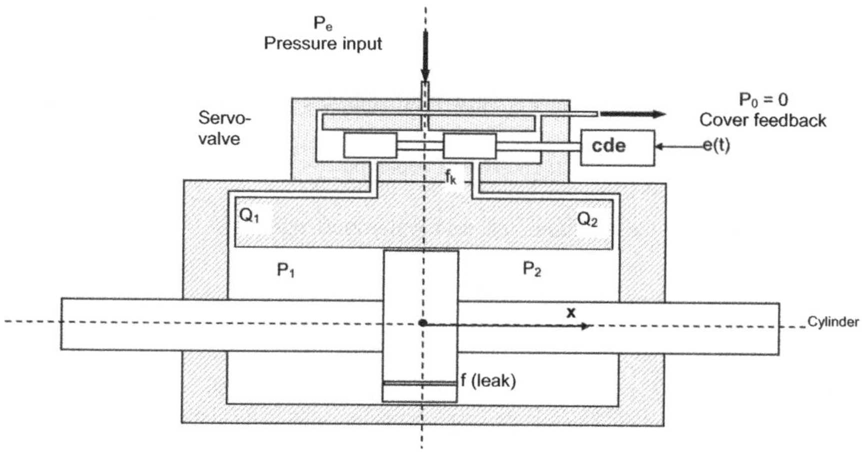

Figure A.2. *Cylinder block diagram*

Hypothetically, the servo valve is symmetrical, with no threshold or recovery. Thus, at the rate of leakage close to the cylinder and/or the servo valve (which can be ignored initially if they are not wanted or controlled for in order to improve the stability of the system), the flow in each chamber of a double effect cylinder is given by the equality of flow supplied by the servo valve and the flow rate consumed by the cylinder, which is:

$$Q_1 = S\frac{dx}{dt} + \frac{V}{B}\frac{dP_1}{dt}; \qquad Q_2 = S\frac{dx}{dt} - \frac{V}{B}\frac{dP_2}{dt}$$

where:

P_1 and P_2 = pressure in chambers 1 and 2 of the cylinder

At rest: $\rightarrow P_1 = P_2 = \frac{P_e}{2}$ (identical loss of pressure in each opening by symmetry)

Similarly: $Q_1 = - Q_2$, if we sum $Q_1 + Q_2$ and integrate it, we get:

$$P_1 + P_2 = cte = P_e$$

where: P_e = effective pressure or servo valve input pressure.

NB: This result expresses that by symmetry of the cylinder, pressures P_1 and P_2 are also symmetrical and are always less than or equal to $P_e/2$.
Thus, assuming symmetry of the cylinder, we can always write:

$$P_1 = \frac{P_e}{2} + \frac{P_1 - P_2}{2} \quad \text{and} \quad P_2 = \frac{P_e}{2} - \frac{P_1 - P_2}{2}$$

The pressure loss at the servo valve distributor level is given by:

$$\Delta P = P_e - P_1 = \frac{P_e}{2} - \frac{P_1 - P_2}{2} = P_e - P_2 = \frac{P_e}{2} + \frac{P_1 - P_2}{2}$$

$$\text{Such that:} \quad \Delta P = \frac{P_e}{2} - \text{sgn}(x)\frac{P_1 - P_2}{2}$$

sgn(x) = sign of x (x = displacement of the piston).

NB: Let us note that if P_1 # P_2 (either small displacements around 0 or linearization), the pressure loss at an opening is given by $\Delta P = \frac{P_e}{2}$. If we wish to take into account the changes in pressures P_1 and P_2 in the two cylinder chambers, we must consider that the flow rates are proportional to the square root of the pressure differential that is actually applied in the chambers and not to $\frac{P_e}{2}$, depending on the case:

- $P_e - P_1$ in chamber 1 and P_2 in chamber 2 (opening a > 0 chamber 1 in connection with P_e, chamber 2 at the cover);
- P_1 in chamber 1 and $P_e - P_2$ in chamber 2 (opening a < 0 chamber 2 in connection with P_e, chamber 1 at the cover).

The flow of fluid through the distributor follows the Bernouilli equation:

$$\Delta P = \rho \frac{v^2}{2}$$

As the opening at the distributor level is "a" and its diameter is D, we get:

$$Q = vS = v\pi Da \rightarrow \quad v^2 = \frac{Q}{\pi Da} = \frac{2\Delta P}{\rho} \quad \rightarrow \quad Q^2 = (\pi Da)^2 \frac{2}{\rho} \Delta P$$

By introducing the edge coefficient ξ and setting: $K = \xi \pi D \sqrt{\frac{2}{\rho}}$, we get, knowing that generally flow is proportional to the square root of the pressure loss and opening "a":

$$Q_1 = K.a.\sqrt{\Delta P} = K.a.\sqrt{\frac{P_e}{2} - \text{sgn}(x)\frac{P_1 - P_2}{2}}$$

The next step is to connect this rate to the differential equation of motion of the piston using the typical equations of force (A.9) (here, we can ignore c and k compared to the other parameters in the equation of forces).

NB: By assumption (linearity and symmetry), we define "gain K for each chamber or gain of the servo valve" as: $K_1 = Q_1/a$, $K_2 = Q_2/a$ (where $K_1 = K_2 = K$, which are opposed and constant). This is set for a pressure loss through an opening equal to $P_e/2$.

$$\text{So } F = M.\gamma = M.\frac{d^2x}{dt^2} = S.(P_1 - P_2) \quad \rightarrow \quad P_1 - P_2 = \frac{M}{S} \cdot \frac{d^2x}{dt^2}$$

$$\text{Thus:} \quad Q_1 = K.a.\sqrt{\frac{P_e}{2}} \cdot \sqrt{1 - \text{sgn}(x)\frac{M}{P_e S} \cdot \frac{d^2x}{dt^2}} = K_1 \sqrt{1 - \frac{\text{sgn}(x).M}{P_e S} \cdot \frac{d^2x}{dt^2}} \qquad \text{(A.11)}$$

$$K_1 = K.a.\sqrt{\frac{P_e}{2}} = \xi\pi D\sqrt{\frac{P_e}{\rho}}.a$$

NB: We then note that an element of distortion is introduced into the operation of the cylinder, since the flow rate is not strictly proportional to the distributor opening. The disturbance term is proportional to the load M and acceleration γ, so at the level at which we work when the cylinder is in vibration.

A.6. Power available on the piston

The available power is given by the product of flow through the pressure. We therefore get, for a piston alternation in a chamber:

$$W(P_1 - P_2) = (P_1 - P_2).Q_1 = (P_1 - P_2).K.a.\sqrt{\frac{P_e}{2} - \text{sgn}(x)\frac{P_1 - P_2}{2}}$$

W is at its maximum if $\dfrac{dW}{d(P_1 - P_2)} = 0$

Such that: $$P_e = \frac{3}{2}(P_1 - P_2) = \frac{3}{2}\frac{F}{S} \qquad \text{(A.12)}$$

It is therefore possible to know the maximum dynamic force F available for a given pressure P_e.

A.7. Study of transfer functions

We seek the transfer function that connects the input voltage e(t) of the servo valve motor (electrodynamic motor by assumption) to the displacement x(t) of the cylinder. Knowing that the opening a(t) of the servo valve is also connected by the transfer function of the motor to its control voltage e(t), we get:

$$H(p) = \frac{X(p)}{E(p)} = \frac{X(p)}{A(p)}.\frac{A(p)}{E(p)} = H_H(p).H_S(p)$$

$H_H(p)$ = hydraulic transfer function (cylinder);
$H_S(p)$ = electrodynamic transfer function (servo valve).

NB: This transfer function consists of two cascaded transfer functions and must include feedback loops to constitute a servo system with typical exciter hydraulic vibrations.

A.7.1. Hydraulic transfer function

The flow in a chamber is the sum of the following rates:

- that due to the displacement of the piston: $S\dfrac{dx}{dt}$
- that due to leakage of the cylinder: $f.P_k$ in chamber k
- that due to the compressibility of the oil: $\dfrac{V}{B}.\dfrac{dP_k}{dt}$ in chamber k
- that due to the distributor leakage: $f_k = 2\dfrac{q_0}{P_e}\Delta P_k$ in chamber k

With q_0 = leakage rate of the leakage section s_0 under $\Delta P = P_e/2$. In the neutral position (a = 0), if we assume that the distributor is perfectly symmetrical on all four passage sections (which hypothetically vary linearly with the relative linear distributor opening "a"), the leakage rate is proportional to q_0 through the relative pressure share that exists in every chamber, including:

$$q_0.\frac{P_e+P_1-P_2}{P_e}=2\frac{q_0}{P_e}\Delta P_1 \text{ (chamber 1) ; } q_0.\frac{P_e+P_2-P_1}{P_e}=2\frac{q_0}{P_e}\Delta P_2\text{(chamber 2)}$$

Based on previous results, we get:

$$Q_1=-Q_2=K_1.\sqrt{1-\frac{\text{sgn}(x).M}{S.P_e}.\frac{d^2x}{dt^2}}$$

$$\text{where:} \qquad K=\xi\pi D\sqrt{\frac{2}{\rho}} \quad ; \qquad K_1=K.a.\sqrt{\frac{P_e}{2}}=K_2.a=\xi\pi D\sqrt{\frac{P_e}{\rho}}.a$$

The flow term that only takes distributor leaks into account is:

$$Q_1=K_1\sqrt{1-\frac{\text{sgn}(x).M.d^2x}{S.P_e.dt^2}}=S.\frac{dx}{dt}+f.P_1+\frac{V}{B}.\frac{dP_1}{dt}+2\frac{q_0}{P_e}\Delta P_1$$

$$Q_2=-K_1\sqrt{1-\frac{\text{sgn}(x).M.d^2x}{S.P_e.dt^2}}=-S.\frac{dx}{dt}+f.P_2+\frac{V}{B}.\frac{dP_2}{dt}+2\frac{q_0}{P_e}\Delta P_2$$

By differentiating both sides, we get:

$$2K_1\sqrt{1-\frac{\text{sgn}(x).M.d^2x}{S.P_e.dt^2}}=2S\frac{dx}{dt}+f(P_1-P_2)+2\frac{q_0}{P_e}(\Delta P_1-\Delta P_2)+\frac{V}{B}\frac{d(P_1-P_2)}{dt}$$

$$\text{Similarly:} \qquad \Delta P_1-\Delta P_2=P_1-P_2=\frac{F}{S}=\frac{M}{S}\frac{d^2x}{dt^2}$$

$$\text{We get:} \qquad 2K_1\sqrt{1-\frac{\text{sgn}(x).M.d^2x}{S.P_e.dt^2}}=2S\frac{dx}{dt}+f.\frac{M}{S}\frac{d^2x}{dt^2}+2\frac{q_0}{P_e}\frac{M}{S}\frac{d^2x}{dt^2}+\frac{V}{B}\frac{M}{S}\frac{d^3x}{dt^3}$$

With the Laplace transform, we get $K_1=K_2.A(p)$:

$$K_2.A(p).\sqrt{1-\frac{.M.}{S.P_e}.p^2.X(p)}=S.p.X(p)+\left(\frac{f}{2}+\frac{q_0}{P_e}\right).\frac{M}{S}.p^2.X(p)+\frac{V}{2.B}\frac{M}{S}.p^3.X(p)$$

$$\text{So:} \quad H_H(p)=\frac{X(p)}{A(p)}=\frac{K_2.\sqrt{1-\frac{.M.}{S.P_e}.p^2.X(p)}}{S.p\left[1+\left(\frac{f}{2}+\frac{q_0}{P_e}\right).\frac{M}{S^2}.p+\frac{V}{2.B}\frac{M}{S^2}.p^2\right]} \qquad \text{(A.13)}$$

NB: sgn(x) is then directly given here by the sign of X(p) and therefore x(t) by definition.

This transfer function is not linear as we predicted because of the radical (A.11). In hydraulics, we usually include the following notations:

$$\omega_f=\frac{K_2}{S}=\frac{\xi\pi D}{S}\sqrt{\frac{P_e}{\rho}} = \text{velocity gain in an open loop system}$$

$$\omega_C=\sqrt{\frac{2BS^2}{VM}} = \text{natural frequency of the piston on its oil cushions}$$

$$\xi_1=\frac{\omega_C}{2}.\frac{f_1.M}{2.S^2} = \text{damping coefficient due to leakage from the cylinder piston}$$

$$\xi_2=\frac{\omega_C}{2}.\frac{Q_e}{P_e}.\frac{M}{S^2} = \text{damping coefficient due to valve leaks}$$

The transfer function then becomes:

$$H_H(p) = \frac{\omega_f.\sqrt{1-\frac{.M.}{S.P_e}.p^2.X(p)}}{p.\left[1+\frac{2}{\omega_c}(\xi_1+\xi_2).p+\frac{1}{\omega_c^2}p^2\right]} \qquad \text{(A.14)}$$

NB: It is possible to linearize this relationship about an operating point.

A.7.2. Principle of linearization techniques

These derive from the properties of differential relationships applied to functions. We replace y = f(x) in $(x, y)_0$ with $\Delta y = (dy/dx).\Delta x$. Let us recall the concept of the differential of a function from the following relationship:

Let the function of two variables z = f (x, y), thus: $dz = dx.\frac{\partial f(x,y)}{\partial x} + dy.\frac{\partial f(x,y)}{\partial y}$

Example of linearization of a sum: $z = a.x^n + b.y^m = z_1 + z_2$ (linearization)

$$dz = dx.\left(nax^{n-1}\right) + dy.\left(mby^{m-1}\right) = \frac{dx}{x}nax^n + \frac{dy}{y}mby^m$$

$$\frac{dz}{z} = \frac{dx}{x}\frac{nax^n}{ax^n+by^m} + \frac{dy}{y}\frac{mby^m}{ax^n+by^m},$$ around the point $z_0 = z_{10} + z_{20} = f(x_0, y_0)$

$$\frac{dz}{z_0} = \frac{dz_1}{z_{10}}\frac{z_{10}}{z_0} + \frac{dz_2}{z_{20}}\frac{z_{20}}{z_0} = \frac{dx}{x_0}\frac{nax_0^n}{ax_0^n+by_0^m} + \frac{dy}{y}\frac{mby_0^m}{ax_0^n+by_0^m} = \frac{dz_1}{z_{10}+z_{20}} + \frac{dz_2}{z_{10}+z_{20}}$$

Similarly, for the linearization of a product: $z = x^n.y^m = z_1.z_2$ (linearization)

Except around x = 0 and y = 0, we get:

$$\frac{dz}{z_0} = n\frac{dx}{x_0} + m\frac{dy}{y_0}$$

Example A.1: Linearization of an exponential:

$$y = a^x \quad \rightarrow \quad dy = a^x.(\text{Log}a).dx \ \left(\text{avec} : a^x = e^{x\text{Log}a}\right)$$

$$y_0 = a^{x_0} \quad \rightarrow \quad \frac{dy}{y_0} = (\text{Log}a).dx$$

Example A.2: Linearization of the product $z = x.y^{1/2}$:

$$dz = dx.y^{1/2} + x.\frac{1}{2}.y^{-1/2}.dy \rightarrow \frac{dz}{z} = \frac{dx}{x} + \frac{1}{2}.\frac{dy}{y} \rightarrow \text{en } (x_0, y_0) \qquad \frac{dz}{z_0} = \frac{dx}{x_0} + \frac{1}{2}.\frac{dy}{y_0}$$

Example A.3: Linearization of rates of leakage in the linear distributor (servo valve).

The block diagram is then shown in Figure A.3.

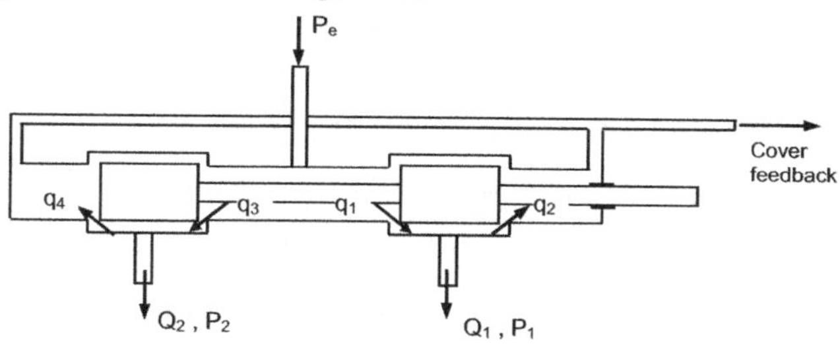

Figure A.3. *Schematic diagram of the servo valve*

We use the expressions and assumptions from the previous leak distributor. It is assumed that the four sections s_k vary linearly with a. Let a_0 be the value of the opening "a" that cancels out s_2 and s_3 ($-a_0$ then cancels out s_1 and s_4). As $q_1 = q_2$ and $q_3 = q_4$, the flow consumed at rest with these assumptions (neutral position = 0) is equal to $2q_0$. And we get:

$$s_1 = s_4 = s_0\left(1+\frac{a}{a_0}\right) \qquad \text{and} \qquad s_2 = s_3 = s_0\left(1-\frac{a}{a_0}\right)$$

Linearization of flow rates gives:

$q_1 = s_1.\sqrt{\Delta P} = s_1.\sqrt{P_e - P_1}$, which is proportional to the opening and the square root of the pressure loss or a product followed by a power, such that:

$$\frac{dq_1}{q_0} = \frac{ds_1}{s_0} + \frac{d(P_e - P_1)^{1/2}}{(P_e - P_1)_0} \equiv \frac{q_1}{q_0} = \frac{s_1}{s_0} + \frac{d(P_e - P_1)^{1/2}}{(P_e - P_1)_0} = 1 + \frac{a}{a_0} + \frac{1}{2}\frac{\Delta(P_e - P_1)}{(P_e - P_1)_0}$$

$$q_1 = q_0\left[1+\frac{a}{a_0}+\frac{1}{2}\frac{\Delta(P_e - P_1)}{(P_e - P_1)_0}\right] \quad \text{similarly:} \quad q_2 = s_2.\sqrt{P_1} = q_0\left[1-\frac{a}{a_0}+\frac{1}{2}\frac{\Delta(P_1)}{(P_1)_0}\right]$$

$$q_3 = s_3.\sqrt{P_e - P_2} = q_0\left[1-\frac{a}{a_0}+\frac{1}{2}\frac{\Delta(P_e - P_2)}{(P_e - P_2)_0}\right] \quad \text{and} \quad q_4 = s_2.\sqrt{P_2} = q_0\left[1+\frac{a}{a_0}+\frac{1}{2}\frac{\Delta(P_2)}{(P_2)_0}\right]$$

For small movements around the operating point, P_1 and P_2 are neighbors of $P_e/2$ and $(P_e - P_1)_0$, P_{10}, $(P_e - P_2)_0$, P_{20} are also similar to $P_e/2$.

Moreover: $\Delta(P_e - P_1) = -\Delta P_1$ and $\Delta(P_e - P_2) = -\Delta P_2$

This allows us to write the rates as:

$$Q_1 = (\text{useful flow rate - leakage rate}) = q_1 - q_2 = q_0\left[2\frac{a}{a_0} - 2\frac{\Delta P_1}{P_e}\right]$$

$$Q_2 = q_3 - q_4 = q_0\left[2\frac{a}{a_0} - 2\frac{\Delta P_2}{P_e}\right] = 2q_0\frac{a}{a_0} - 2q_0\frac{\Delta P_2}{P_e} = \text{useful flow rate - leakage rate}$$

And we find that the distributor leak rate is given by:

$$2q_0\frac{\Delta P_k}{P_e}$$

Example A.4: Linearization of the typical flow equation.

We have seen that its expression is a function of the square root of pressure loss (A.3), such that:

$$Q = K.\sqrt{\Delta P} = K\sqrt{P_e - P}$$

By applying the linearization relationships to the power, the product and the sum, we get K considered as a variable and $K_0 \approx K$ (proportionality factor):

$$\frac{Q}{Q_0} = 1.\frac{K}{K} + \frac{1}{2}\left[\frac{d(P_e - P)}{(P_e - P)_0}\right] = 1 + \frac{1}{2}\frac{\Delta P_e}{(P_{e0} - P_0)} - \frac{1}{2}\frac{\Delta P}{(P_{e0} - P_0)}$$

Such that:

$$Q = Q_0.\left[1 + \frac{1}{2}\frac{\Delta P_e}{(P_{e0} - P_0)} - \frac{1}{2}\frac{\Delta P}{(P_{e0} - P_0)}\right] \qquad \text{(A.15)}$$

NB: It is this linear relationship, which is associated with relationships (A.9) and (A.10), that allows us access to most hydraulic system transfer functions.

A.7.3. Linearization of the transfer function

The expression (A.14) can be directly linearized by only retaining the linear part of the expression of the numerator. That amounts to linearizing the operation of the servo valve based on the assumption that the flow rate Q is proportional to the opening "a". It has been shown that this simplification does not diminish the accuracy of the final result. The calculation can then be carried out with sufficient accuracy by writing flow rate relationships in the form $Q = K.a$.

Thus, as part of the dynamic study of the servo control (gain setting, search for closed-loop resonant frequency, consideration of stability, etc.), one can use the following simplified relationship:

$$H_H(p) = \frac{\omega_f}{p.\left[1 + \frac{2}{\omega_c}(\xi_1 + \xi_2).p + \frac{1}{\omega_c^2}p^2\right]}$$

So:

$$H_H\left(j\frac{\omega}{\omega_c}\right) = \frac{\frac{\omega_f}{\omega_c}}{j\frac{\omega}{\omega_c}\left[1 + 2j(\xi_1 + \xi_2)\frac{\omega}{\omega_c} - \frac{\omega^2}{\omega_c^2}\right]} \quad \text{(A.16)}$$

Appendix B

Study of the Basic Model with Damping

If we add a damping term to the basic model, the differential equations are modified by the introduction of an additional restoring force that occurs due to the damper. In general, we obtain (Figure B.1):

$$m\frac{d^2x(t)}{dt^2} = -k.z(t) - c.\dot{z}(t); \qquad z(t) = x(t) - u(t)$$

$$m\ddot{x}(t) + c.\dot{x}(t) + k.x(t) = k.u(t) + c.\dot{u}(t)$$

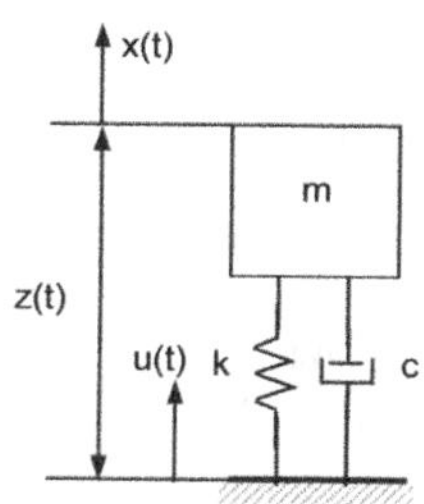

Figure B.1. *Basic model with damping*

If:

$\omega_0 = \left(\frac{k}{m}\right)^{1/2}$ = non-damped natural frequency.

$\omega_n = \omega_0\left(1-\xi^2\right)^{1/2}$ = natural frequency, step response.

$\xi = \frac{c}{2(km)^{1/2}}$ = reduced damping.

$c_c = 2(km)^{1/2}$ = critical damping.

$\omega_r = \omega_0\left(1-2\xi^2\right)^{1/2}$ = resonance frequency.

$Q = \frac{1}{2\xi\left(1-\xi^2\right)^{1/2}} \# \frac{1}{2\xi}$ = overvoltage coefficient.

The general excitement–response equation becomes:

$$\ddot{r}(t) + 2\xi\omega_0\dot{r}(t) + \omega_0^2 r(t) = 2\xi\omega_0\dot{e}(t) + \omega_0^2 e(t)$$

$$\uparrow \qquad\qquad \uparrow$$

additional terms

This describes a generalized second-order system in the presence of a derivative of excitement $\dot{e}(t)$ in the second member, and thus the existence of transmission zero and the possibility of having the same excess for $\xi > 1$. With the Laplace transform, we obtain the overall response:

$$R_g(p) = E(p).\frac{2\xi\omega_0 p + \omega_0^2}{p^2 + 2\xi\omega_0 p + \omega_0^2} + \frac{p.r(0^+) + \dot{r}(0^+) + 2\xi\omega_0 - 2\xi\omega_0.e(0^+)}{p^2 + 2\xi\omega_0 p + \omega_0^2}$$

The second term characterizes the initial conditions and is not involved in calculating the pulse response of the basic damped system. Thus, we are only interested in the first term where:

$$R(p) = E(p).\frac{2\xi\omega_0 p + \omega_0^2}{p^2 + 2\xi\omega_0 p + \omega_0^2}$$

H(p) is obtained for E(p) = 1, such that:

$$H(p) = \frac{2\xi\omega_0 p + \omega_0^2}{p^2 + 2\xi\omega_0 p + \omega_0^2} \quad \text{(B.1)}$$

Depending on the sign of the denominator determinant $\Delta' = \omega_0^2(\xi^2 - 1)$, several cases are possible. The one that interests us the most, which must optimize a non-damped case, is the one with weak ξ, where ξ <0.5. So Δ' <0 and we then set: $(\Delta')^{1/2} = j\omega_0(1-\xi^2)^{1/2}$

These are the two conjugated complex poles (Figure B.2):

$$p_1 = -\xi\omega_0 + j\omega_0(1-\xi^2)^{1/2}$$

$$p_2 = p_1^* = -\xi\omega_0 - j\omega_0(1-\xi^2)^{1/2}$$

Let :

$$T_1 = -\frac{1}{p_1} \quad \text{and} \quad T_2 = -\frac{1}{p_2}$$

Thus :

$$T_1 T_2 = \frac{1}{\xi\omega_0 - j\omega_0(1-\xi^2)^{1/2}} \quad \text{and} \quad T_1 - T_2 = 2j\frac{(1-\xi^2)^{1/2}}{\omega_0}$$

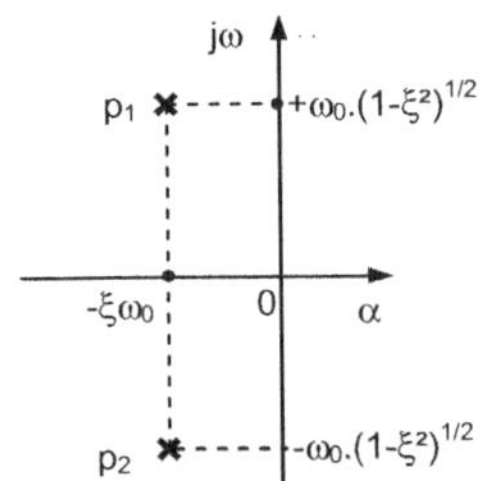

Figure B.2. *Position of the poles*

We then get:

$$H(p) = \frac{T_1 T_2 2\xi\omega_0 p}{(pT_1+1)(pT_2+1)} + \frac{T_1 T_2 \omega_0^2}{(pT_1+1)(pT_2+1)}$$

From the usual tables, this directly provides the percussive response:

$$h(t) = \frac{2\xi\omega_0}{T_1 - T_2}\left(T_1.e^{-t/T_2} - T_2.e^{-t/T_1}\right) + \frac{T_1 T_2 \omega_0^2}{T_1 - T_2}\left(e^{-t/T_1} - e^{-t/T_2}\right)$$

Hence, the expressions:

$$h(t) = \frac{2\xi\omega_0^2}{2j(1-\xi^2)^{1/2}}\left[\frac{\xi + j(1-\xi^2)^{1/2}}{\omega_0}.e^{-t/T_2} - \frac{\xi - j(1-\xi^2)^{1/2}}{\omega_0}.e^{-t/T_1}\right] + \frac{\omega_0}{2j(1-\xi^2)^{1/2}}.\left(e^{-t/T_1} - e^{-t/T_2}\right)$$

$$h(t) = \frac{1}{2j(1-\xi^2)^{1/2}}\left\{\left[\omega_0(2\xi^2 - 1) + j2\xi\omega_0(1-\xi^2)^{1/2}\right].e^{-t/T_2}\right.$$

$$\left. -\left[\omega_0(2\xi^2 - 1) - j2\xi\omega_0(1-\xi^2)^{1/2}\right].e^{-t/T_1}\right\}$$

$$h(t) = \frac{1}{2j(1-\xi^2)^{1/2}}.[\omega_0(2\xi^2 - 1).e^{-\xi\omega_0 t}.e^{-j\omega_0(1-\xi^2)^{1/2}.t} + j2\xi\omega_0(1-\xi^2)^{1/2}.e^{-\xi\omega_0 t}.e^{-j\omega_0(1-\xi^2)^{1/2}.t}$$

$$-\ \omega_0(2\xi^2 - 1).e^{-\xi\omega_0 t}.e^{+j\omega_0(1-\xi^2)^{1/2}.t} + j2\xi\omega_0(1-\xi^2)^{1/2}.e^{-\xi\omega_0 t}.e^{+j\omega_0(1-\xi^2)^{1/2}.t}]$$

$$h(t) = 2\xi\omega_0.e^{-\xi\omega_0 t}.\cos\omega_0\left(1-\xi^2\right)^{1/2}.t + \frac{\omega_0\left(1-2\xi^2\right)}{\left(1-\xi^2\right)^{1/2}}.e^{-\xi\omega_0 t}.\sin\omega_0\left(1-\xi^2\right)^{1/2}.t$$

$$h(t) = \frac{\omega_0.e^{-\xi\omega_0 t}}{\left(1-\xi^2\right)^{1/2}}\left[2\xi\left(1-\xi^2\right)^{1/2}.\cos\omega_r t + \left(1-2\xi^2\right)\sin\omega_r t\right]$$

If: $\begin{cases}\sin\varphi = 2\xi\left(1-\xi^2\right)^{1/2} \\ \cos\varphi = \left(1-2\xi^2\right)\end{cases}$ true if $\xi < 1$

Then: $$\begin{cases} h(t) = \dfrac{\omega_0}{\left(1-\xi^2\right)^{1/2}} e^{-\xi\omega_0 t}\left\{\sin\left[\omega_0\left(1-\xi^2\right)^{1/2}.t + \varphi\right]\right\} \\ \varphi = \operatorname{arctg}\dfrac{2\xi\left(1-\xi^2\right)^{1/2}}{1-2\xi^2}, \qquad \omega_r = \omega_0\left(1-2\xi^2\right)^{1/2} \end{cases} \tag{B.2}$$

The frequency response of such a system is given directly from H(p) (B.1) with:

$$\left|H(p)\right|^2_{p=j\omega} = \frac{1+\left(2\xi\dfrac{\omega}{\omega_0}\right)^2}{\left(1-\left(\dfrac{\omega}{\omega_0}\right)^2\right)^2 + \left(2\xi\dfrac{\omega}{\omega_0}\right)^2} \tag{B.3}$$

NB: For $\xi = 0$, we get H(p) and h(t) from the non-damped system.

Relationship (B.3) can also be verified from the FT of h(t) or (B.2) by definition. But here, we must take into account the fact that the system is causal and that h(t) only starts at t = 0, such that:

$$H(\omega) = TF[h(t).\gamma(t)] = K.TF\left[e^{-\alpha t}.\gamma(t).\sin[\omega_r t + \varphi]\right]$$

Where: $\begin{cases} K^2 = \dfrac{\omega_0^2}{\left(1-\xi^2\right)} \\ \alpha = \xi\omega_0 \\ \gamma(t) = \text{unit step} \end{cases}$

Thus: $H(\omega) = K.TF\left[e^{-\alpha t}.\gamma(t)\right] * TF[\sin[\omega_r t + \varphi]]$

$$TF\left[e^{-\alpha t}.\gamma(t)\right] = \int_{-\infty}^{\infty} e^{-\alpha t}.\gamma(t).e^{-2\pi j f t}.dt = 0 + \int_0^{\infty} e^{-(\alpha + j\omega)t}.dt = \frac{1}{\alpha + j\omega}$$

$$TF[\sin[\omega_r t + \varphi]] = \frac{1}{2j}.e^{j\varphi}.\delta(\omega - \omega_r) - \frac{1}{2j}.e^{-j\varphi}.\delta(\omega + \omega_r)$$

Thus: $$H(\omega) = \frac{K}{2j}.\frac{e^{j\varphi}[\alpha + j(\omega + \omega_r)] - e^{-j\varphi}[\alpha + j(\omega - \omega_r)]}{\left(\alpha^2 + \omega_r^2 - \omega^2\right) + 2j\alpha\omega}$$

This goes along with the shifting properties from the Dirac convolution product $\delta(\omega \pm \omega_r)$. By replacing $e^{\pm j\varphi} = \cos\varphi \pm j\sin\varphi$, we get:

$H(\omega) = K\dfrac{\omega_r\cos\varphi + \alpha\sin\varphi + j\omega\sin\varphi}{\left(\alpha^2 + \omega_r^2 - \omega^2\right) + 2j\alpha\omega}$, by replacing $\cos\varphi$, $\sin\varphi$, α with their values, we get:

$$H(\omega)=K.\frac{\omega_0\left(1-\xi^2\right)^{1/2}\left(1-2\xi^2\right)+\xi\omega_0 2\xi\left(1-\xi^2\right)^{1/2}+j2\xi\left(1-\xi^2\right)^{1/2}\omega}{\left(\omega_0^2-\omega^2\right)+j2\xi\omega_0\omega}$$

Either for the square of the module:

$$\left|H(\omega)\right|^2=K^2\frac{\omega_0^2\left(1-\xi^2\right)+4\xi^2\left(1-\xi^2\right)\omega^2}{\left(\omega_0^2-\omega^2\right)^2+\left(2\xi\omega_0\omega\right)^2}=\frac{\omega_0^2.\omega_0^2\left[1+\left(2\xi\frac{\omega}{\omega_0}\right)\right]}{\omega_0^4\left(1-\left(\frac{\omega}{\omega_0}\right)^2\right)^2+\omega_0^4.\left(2\xi\frac{\omega}{\omega_0}\right)^2}$$

Or the same expression with:

for the module :
$$H(\omega)=\frac{1+\left(2\xi\frac{\omega}{\omega_0}\right)^2}{\left(1-\left(\frac{\omega}{\omega_0}\right)^2\right)^2+\left(2\xi\frac{\omega}{\omega_0}\right)^2}$$

and for the phase :
$$\Phi(\omega)=\text{arctg}\left(2\xi\frac{\omega}{\omega_0}\right)-\text{arctg}\frac{\left(2\xi\frac{\omega}{\omega_0}\right)}{\left(1-\left(\frac{\omega}{\omega_0}\right)^2\right)}$$

NB: The reduced coordinate curves are shown in Figure B.3.

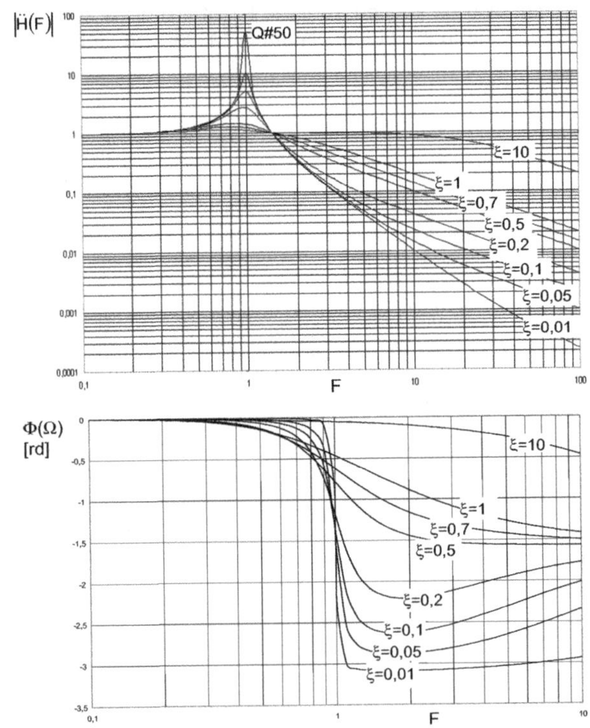

Figure B.3. *Amplitude and phase of the basic system*

Appendix C

Natural Frequencies of 1,6 mm Thick Cards Equipped with Stiffeners

The following curves give the resonance frequencies of printed circuit cards of thickness 1.6 mm equipped with stiffeners. The load of each card is # 9 gr/cm^2. This load includes the weight of the card, the stiffeners and the components. The three-dimensional cards are 100 × 150, 150 × 230 and 200 × 300. The curves give the frequencies for 1, 2, or 3 stiffeners of 1,6 and 2,4 mm thickness in stratified epoxy laminate and in steel, glued or screwed. The stiffeners are mounted in parallel to the longer side of the cards.

CARD 100 x 150 mm
(weight 135 g)

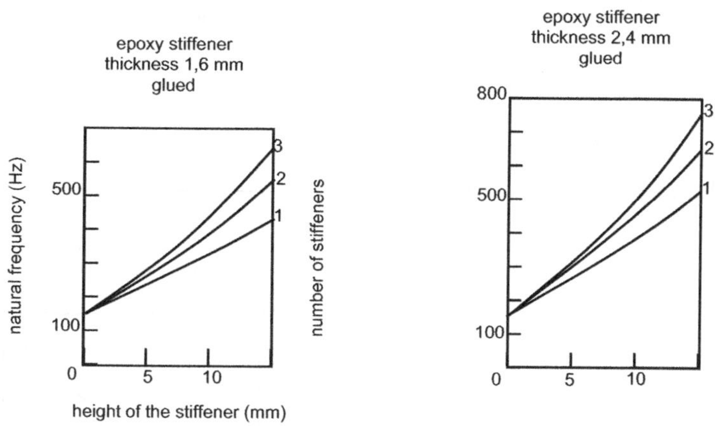

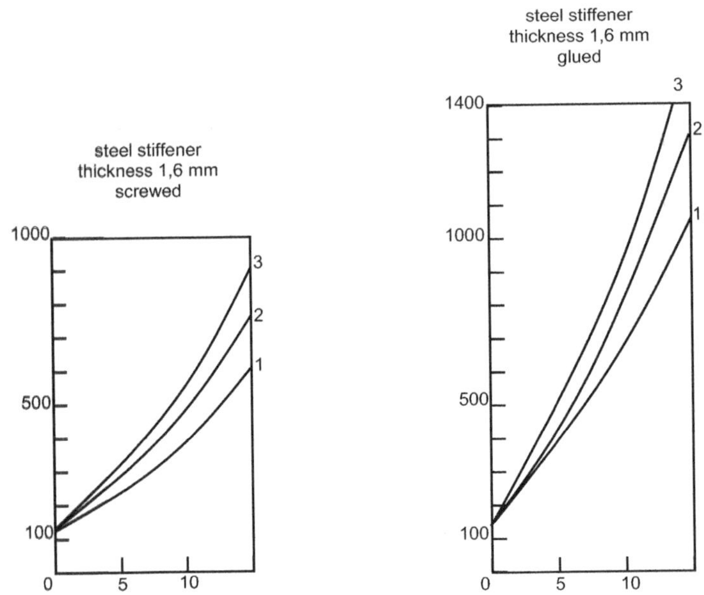

CARD 150 x 230 mm
(weight 300g)

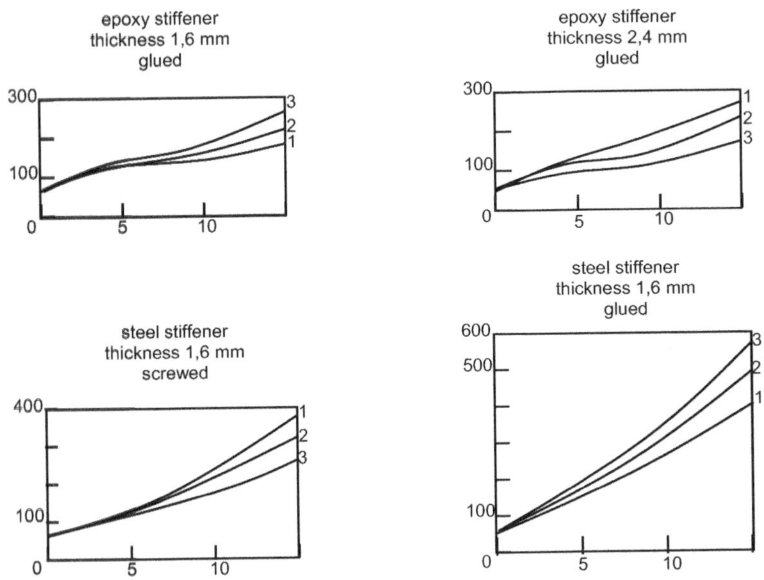

CARD 200 x 300 mm
(weight 550 g)

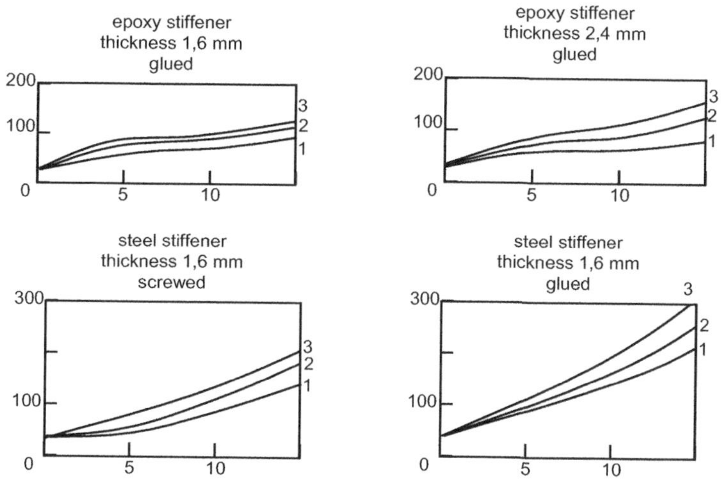

NB: For cards with the same dimensions as those shown above but for which the weights differ, their resonance frequencies can be determined by the following relationship:

$$f_2 = f_1 . \sqrt{\frac{P_1}{P_2}}$$

where:

- f_2 = frequency of the card with a different weight;
- f_1 = frequency of the reference card;
- P_1 = weight of the reference card;
- P_2 = weight of the card with a different weight.

Appendix D

Resonance Frequencies of IC Cards Depending on the Mounting Conditions

This appendix contains the resonance frequencies of boards based on the installation conditions of edges (the relationships are given here with the usual geometric parameters):

- a = length of the board in m;
- b = width of the board in m;
- ρ = mass/unit of surface in kg/m^2;
- $D = \dfrac{E.h^3}{12(1-\mu^2)}$ = stiffness factor in N.m;

where:

- E = elastic module of the epoxy laminate in N/m^2;
- h = thickness of the board in m;
- μ = Poisson coefficient (# 0.12).

free edge | supported edge | fixed edge (encased)

Equation	Board
$f_a = \frac{\pi}{8} \cdot \left(\frac{D}{\rho}\right)^{1/2} \cdot \left(\frac{1}{a^2} + \frac{1}{b^2}\right)$	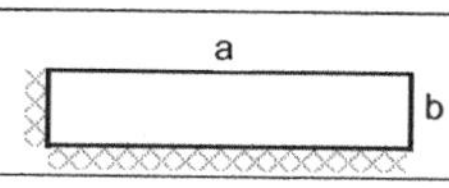
$f_a = \frac{\pi}{2} \cdot \left(\frac{D}{\rho}\right)^{1/2} \cdot \left(\frac{1}{4.a^2} + \frac{1}{b^2}\right)$	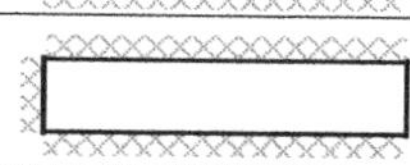
$f_a = \frac{\pi}{2} \cdot \left(\frac{D}{\rho}\right)^{1/2} \cdot \left(\frac{1}{a^2} + \frac{1}{b^2}\right)$	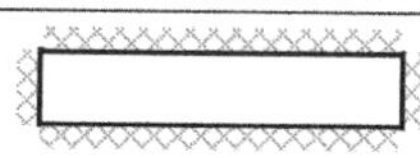
$f_a = \frac{\pi}{5{,}42} \cdot \left\{\frac{D}{\rho} \cdot \left(\frac{1}{a^4} + \frac{3{,}2}{a^2b^2} + \frac{1}{b^4}\right)\right\}^{1/2}$	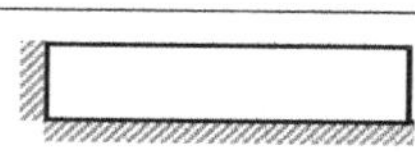
$f_a = \frac{\pi}{3} \cdot \left\{\frac{D}{\rho} \cdot \left(\frac{0{,}75}{a^4} + \frac{2}{a^2b^2} + \frac{12}{b^4}\right)\right\}^{1/2}$	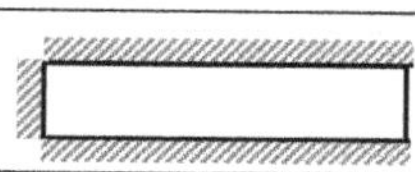
$f_a = \frac{\pi}{1{,}5} \cdot \left\{\frac{D}{\rho} \cdot \left(\frac{3}{a^4} + \frac{2}{a^2b^2} + \frac{3}{b^4}\right)\right\}^{1/2}$	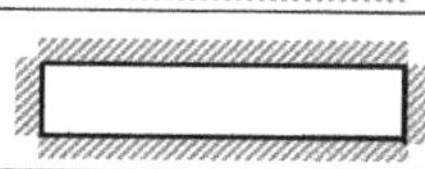

$f_a = \frac{0,78.\pi}{a^2}.\left\{\frac{D}{\rho}\right\}^{1/2}$	
$f_a = \frac{\pi}{2.a^2}.\left\{\frac{D}{\rho}\right\}^{1/2}$	
$f_a = \frac{\pi}{1,74}.\left\{\frac{D}{\rho}.\left(\frac{4}{a^4}+\frac{1}{2.a^2b^2}+\frac{1}{64.b^4}\right)\right\}^{1/2}$	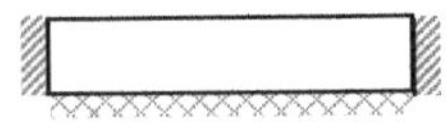
$f_a = \frac{\pi}{2}.\left\{\frac{D}{\rho}.\left(\frac{0,127}{a^4}+\frac{0,20}{a^2b^2}\right)\right\}^{1/2}$	
$f_a = \frac{\pi}{2}.\left\{\frac{D}{\rho}.\left(\frac{1}{a^4}+\frac{0,608}{a^2b^2}+\frac{0,126}{b^4}\right)\right\}^{1/2}$	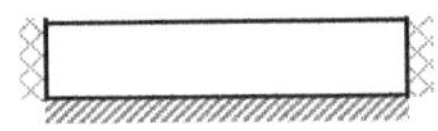
$f_a = \frac{\pi}{2}.\left\{\frac{D}{\rho}.\left(\frac{2,45}{a^4}+\frac{2,90}{a^2b^2}+\frac{5,13}{b^4}\right)\right\}^{1/2}$	
$f_a = \frac{\pi}{2}.\left\{\frac{D}{\rho}.\left(\frac{0,127}{a^4}+\frac{0,707}{a^2b^2}+\frac{2,44}{b^4}\right)\right\}^{1/2}$	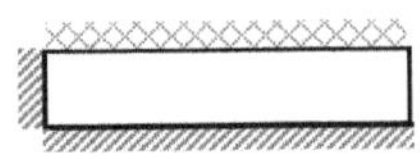
$f_a = \frac{\pi}{2}.\left\{\frac{D}{\rho}.\left(\frac{2,45}{a^4}+\frac{2,68}{a^2b^2}+\frac{2,45}{b^4}\right)\right\}^{1/2}$	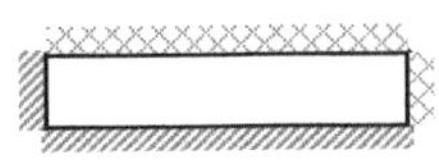
$f_a = \frac{\pi}{2}.\left\{\frac{D}{\rho}.\left(\frac{2,45}{a^4}+\frac{2,32}{a^2b^2}+\frac{1}{b^4}\right)\right\}^{1/2}$	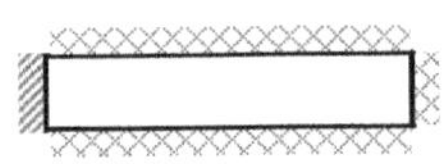
$f_a = \frac{4,50}{\pi.a^2}.\left\{\frac{D}{\rho}\right\}^{1/2}$	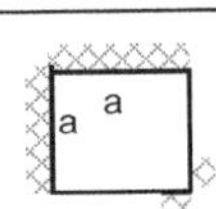
$f_a = \frac{1,13}{a^2}.\left\{\frac{D}{\rho}\right\}^{1/2}$	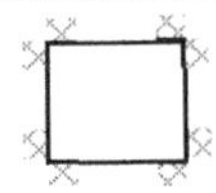

$f_a = \frac{\pi}{3,46}\cdot\left\{\frac{D}{\rho}\cdot\left(\frac{16}{a^4}+\frac{8}{a^2b^2}+\frac{3}{b^4}\right)\right\}^{1/2}$	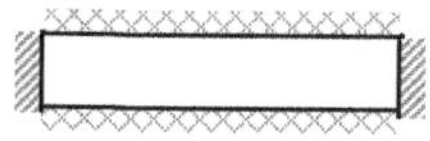
$f_a = \frac{\pi}{2}\cdot\left\{\frac{D}{\rho}\cdot\left(\frac{2,08}{a^2b^2}\right)\right\}^{1/2}$	
$f_a = \frac{0,56}{a^2}\cdot\left\{\frac{D}{\rho}\right\}^{1/2}$	
$f_a = \frac{3,55}{a^2}\cdot\left\{\frac{D}{\rho}\right\}^{1/2}$	

Appendix E

Concept of Cepstrum

The first definition of Cepstrum dates back to 1963. It was as follows:

"The spectrum of the power of the logarithm of the spectrum of the power" of the function in question.

This tool was created to analyze signals generated by earthquakes, especially in order to obtain the arrival time of an echo in a composite signal. The authors showed that the effect of a delayed echo manifests itself as a ripple in the logarithm of the spectrum. The frequency of this ripple can then easily be determined by calculating the spectrum of the logarithm spectrum (amplitude) or it is manifested by a peak. In fact, the logarithm function has the feature of accumulating terms in the product as harmonic lines of periodic signals and thus highlighting this aspect of the noise wave.

The terms used are then somewhat different from the usual terms that we know. As the "frequency" unit of this undulation is time, to avoid confusion, the authors back then called it "quefrency".

This definition was therefore translated into the following relationship:

$$C(\tau) = \left|FFT\{Log F_{xx}(f)\}\right|^2 \tag{E.1}$$

where FFT is the fast Fourier transform and the time signal spectrum $f_x(t)$ is given by:

$$F_{xx}(f) = \left|FFT\{f_x(t)\}\right|^2 \tag{E.2}$$

The definition that is commonly used today and that is implemented in measuring devices is:

$$C_p(\tau) = FFT^{-1}\{Log F_{xx}(f)\} \tag{E.3}$$

This tool is especially used for detecting periodic structures in the spectrum or for studying signals that contain harmonics (for example in the diagnosis of rotating machines), echoes and reflected signals (seismology, engine noise, radar, sonar, etc.), as well as in the analysis of speech, which mainly consists of non-stationary signals.

One can assess its usefulness from the figure below, which is taken from a document by Bruël and Kjaer and that illustrates the detection of harmonic lines that occur due to the misalignment of bearings in an engine.

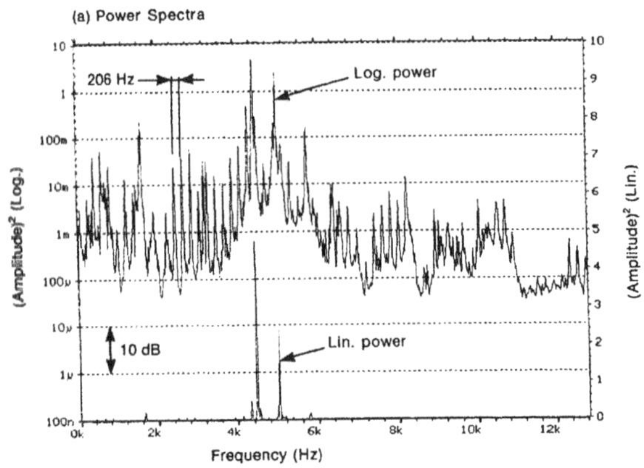

Figure E.1. *Spectral density of the power to be analyzed*

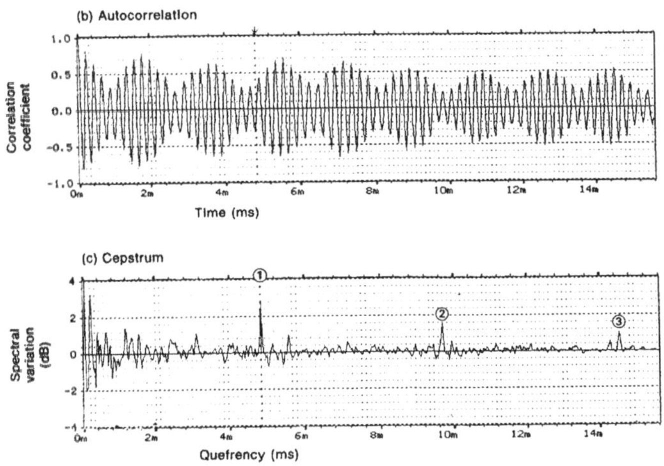

Figure E.2. *Comparison of analysis by correlation and spectrum*

One may note:

- the influence of the linear and logarithmic representation of the power spectral density. We note that in the linear state, the interesting lines at 206 Hz completely disappear and in logarithmic state, they are not clear relative to other adjacent lines such as harmonic lines (incoherent noise was eliminated by the averaging effect of the Spectrum Log function);

- the representation of the autocorrelation function in linear state does not provide any extra information;

- however, in the logarithmic representation, the Cepstrum causes the lines in (1), (2) and (3) to appear, which are in harmonic relationships since multiples of 4,85 ms = 1/206 Hz.

Hence, the interest in using this type of signal.

NB: Purely for informative purposes, let us note that BK sells two types of analyzers that carry out this kind of processing, the 2032 and 2034 (there are other and more recent ones).

Appendix F

Tolerances on Vibration Fixtures

According to E.J. Klee and W. Dustin, *Quantized goals for design of vibration and shock test fixtures*.

1- Test materials	2 – Permissible resonances	3 – Transversal responses	4 – Percentage of the levels between control point and the fixing point
Components or elements $M \leq 2{,}5$ kg $L \leq 25$ cm	None below 1 000 Hz Three resonances allowed above 1000 Hz with Q limited to 5	Below the excitement level up to 2 000 Hz	±20% up to 999 Hz ±50% from 1 000 to 2 000 Hz
Small equipment $M \leq 7$ kg $L \leq 40$ cm	None below 1 000 Hz Two resonances allowed above 1 000 Hz with Q limited to 5	Below the excitement level up to 2 000 Hz	±3% up to 999 Hz ±100% from 1 000 to 2 000 Hz
Equipment $M \leq 25$ kg $L \leq 60$ cm	None below 800 Hz Four resonances allowed between 800 Hz and 1 500 Hz with Q limited to 6 Three resonances allowed between 1 500 Hz and 2 000 Hz with Q limited to 8	Below the excitement level up to 1 000 Hz Less than twice the excitation level above 1 000 Hz in a limited frequency band	±50% up to 999 Hz ±100% from 1000 Hz
Sub-assemblies $M \leq 250$ kg $L \leq 125$ cm	None below 500 Hz Two resonances allowed between 500 Hz and 1 000 Hz with Q limited to 6 Three resonances allowed between 1 000 Hz and 2 000 Hz with Q limited to 8	Below the excitement level up to 500 Hz Less than twice the excitation level from 500 to 1 000 Hz Less than 2. 5 times the excitation level from 1 000 to 2 000 Hz in a limited frequency band	±50% up to 499 Hz ±100% from 500 to 999 Hz From ±150 to ±200% from 1 000 to 2 000 Hz
Large sub-assemblies $M \leq 250$ kg $L \leq 160$ cm	None below 150 Hz One resonance allowed between 150 Hz and 300 Hz with Q limited to 3 or 4 Three resonances allowed between 300 Hz and 1 000 Hz with Q limited to 5 Five resonances allowed between 1 000 Hz and 2 000 Hz with Q limited to 10	Less than 1. 5 times the excitation level up to 300 Hz Less than three times the excitation level from 300 to 1 000 Hz Less than four times the excitation level from 1 000 to 2 000 Hz in a limited frequency band	±50% up to 399 Hz From ±100 to ±200% from 400 to 2 000 Hz

Unless otherwise specified from the test applicant, we can draw on the tolerances shown in the table above.

- Column 1 classifies materials to be tested according to their mass (M) or their largest dimension (L).

NB: For large equipment, it is possible that the test requirements cannot be fully met by conventional means because the behavior of the fixture resonance and the equipment to be tested (by dynamic test material-fixture interaction) cannot be controlled for. It is therefore important that the testing program specifies the settings and that these are used.

- Column 2 shows:
  - the desirable frequency band without resonance;
  - the number of resonances tolerated in the rest of the frequency range;
  - the magnitude of the dynamic amplification coefficient (Q).

- Column 3 provides the magnitude of tolerable parasitic transverse movements (with limited bandwidth) relative to the desired excitement level in the test axle (loaded fixture).

- Column 4 gives the tolerated percentage between, on the one hand, the level of control sensor(s) and, on the other hand, the levels collected at the time of resonance at different fixation points of the material to be tested.

Appendix G

Determining Measurement Uncertainty – Example of a Calibration Method for Accelerometer Sensors

G.1. Overview

Chapter 7 already introduced the concept of uncertainty; so in this appendix, we will show some of the relationships that allow us to determine it. Obviously, the first step is to determine all the causes of uncertainty and the influencing factors, which can be performed using the 5M ((material, matter, means (human resources), medium (environment), method) chart given in the seven steps in Chapter 7 (or through other techniques).

First, one must respect the following recommendations:

- wherever possible, favor a *direct type A measurement*, which provides an uncertainty with limited mathematical development (average calculations and standard deviations are done from direct statistical measurements on the relevant parameters);

- *direct type B measurement* leads to a minimal number of calculations but the uncertainty thus determined requires external sources of information and is generally greater than type A uncertainty. It is used once the specified tolerance of the measurement to be carried out is large enough with respect to the measurements that were already carried out;

- *indirect measurement* consists of the most mathematical steps and is potentially the riskiest method (forgetting to consider a significant influencing factor, forgetting to take correlation or dependency between variables into account). This involves determining a combined uncertainty from relatively complex mathematical relationships. So we only use this if it is the only possible method.

NB: We do not advise using these methods without a strong background in metrology or without seeking prior support from specialists.

If appropriate, we favor an indirect measurement method based on direct type A measurements, which significantly reduces the mathematical developments and the risk of errors;

- in order to measure 1 m at a micron accuracy, one must specify the ground temperature in the laboratory (25 °C), the atmospheric pressure in the laboratory (101325 Pa), the method for supporting the meter; uncertainty calculations must consider the corrections linked to these parameters and the measurement method has to be of sufficient accuracy, which is not necessary for measuring 1 m at a millimeter accuracy.

NB: This should prompt us to not unnecessarily determine narrow dimensional tolerance intervals.

The general approach for estimating measurement uncertainty (Application of the procedure described in the GUM, Chapter 8) determined influencing factors:

Step 1: calculating the measurement result:

- definition of the measurement Y;
- analysis of the measurement process;
- writing the mathematical model that describes the measurement process.

Step 2: calculating standard uncertainties $u(x_i)$:

- estimation of standard uncertainties on the model input variables:
  - using statistical techniques (type A evaluation),
  - through knowledge, experience (type B evaluation),
  - using an indirect method through an analytical relationship that describes the measurement process $Y = f(X_1, X_2,\ldots, X_N)$;
- estimation of covariance $u(x_i, x_j)$.

Step 3: propagation of uncertainty:

- using the law of propagation of uncertainty;
- definition of sensitivity coefficients $\frac{\partial f}{\partial x_i}$;
- estimation of combined uncertainty $u_c(y)$.

Step 4: expressing the measurement result and its associated uncertainty.

G.1.1. Reminders on mathematics

- If x is a random variable (RV):

E(x) or <x> or $\bar{x}$, or μ, is the expectation of x (statistical average);

V(x) is the variance of x and σ(x) is the standard deviation [where $V(x) = \sigma^2(x)$];

$\partial_i y(x)$ is the partial derivative of the function y with respect to the ith variable of the measurement sample or vector $x = (x_1, x_2, \ldots. x_i \ldots.)$. Thus:

$$\partial_i y(x) = \frac{\partial y}{\partial x_i}(x)$$

- A function of RVs, y = y(x) is itself an RV. If the uncertainties are small, the limited development of variance of y in the first order around the average value μ of x_i is a good estimate for the variance of y, according to the following diagram, such that:

$$y(x) = y(\mu) + \sum_i (x_i - \mu)\partial_i y$$

Ignoring higher order terms in the expression, we get:

$$V[y(x)] = \sum_i \sum_j \partial_i y . \partial_j y . V_{ij}(x)$$

If x is independent:

$$V[y(x)] = \sum_i (\partial_i y)^2 . V_{ii}(x)$$

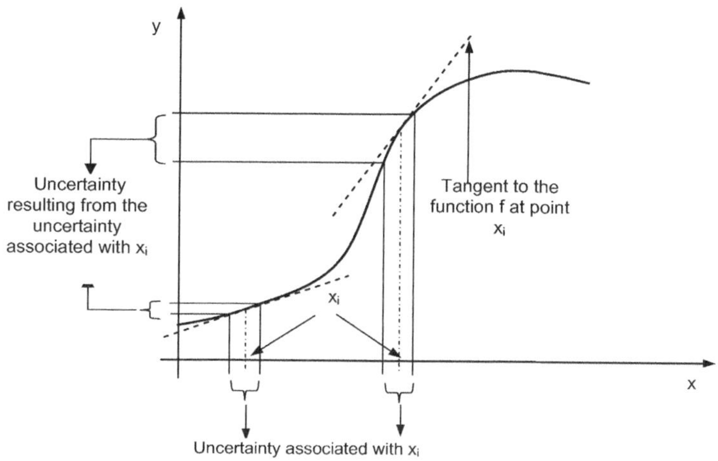

- Combined standard uncertainty:

- when uncertainty comes from several independent sources of uncertainty, we evaluate (as specified later in section G.2.2) the uncertainty u_1, u_2,....u_n that occurs due to each source for each of these sources. The uncertainty of the input variable is then evaluated by a combined uncertainty u that is determined by the relationship (standard quadratic sum):

$$u^2 = \sum_i u_i^2$$

- if the uncertainty comes from several parameters that are linked by a mathematical relationship, we apply the law of propagation of uncertainty with the following relationship in the case of independent influencing quantities:

$$u_c^2(y) = \sum_{i=1}^{n} \left(\frac{\partial f}{\partial x_i} \right)^2 .u^2(x_i)$$

The combined standard uncertainty of the measurement result y, designated by $u_c(y)$ and taken to represent the estimated standard deviation of the result, is the positive square root of the estimated variance $u_c^2(y)$ obtained from the previous equation.

If the influencing factors are dependent (there is a non-zero correlation between them), the relationship becomes (this will be explained later):

$$u_c^2(y) = \sum_{i=1}^{n} \left(\frac{\partial f}{\partial x_i} \right)^2 .u^2(x_i) + 2\sum_{i=1}^{N-1} \sum_{j=i+1}^{N} \frac{\partial f}{\partial x_i} \frac{\partial f}{\partial x_j} u(x_i, x_j)$$

These equations are, as stated before, based on a first order approximation of Taylor series of $Y = f(x_1, x_2,..., x_n)$ and is called the *law of propagation of uncertainty*. The partial derivatives are often referred to as the *sensitivity of coefficients* and $u(x_i, x_j)$ is the estimated covariance matrix associated with x_i and x_j.

The expanded uncertainty U is then obtained by multiplying the combined standard uncertainty u by a coverage factor k with:

- k = 2 for a confidence interval of # 95%;

- k = 3 for a confidence interval of # 99%.

NB1: The correlation terms in the above relationship can often be omitted by assigning the correlated terms, if any, to a same number i. This approach significantly simplifies the process.

NB2: We assume that the actual number of DOF is large enough (>10) in order to assume that the combined uncertainty approximates to the normal distribution (see GUM 1993, section G.6.6).

In statistics, the DOF is the number of random values that cannot be determined or fixed by an equation (including equations for statistical tests) to approach the concept of independence and DOF in modal analysis or SRS model systems with a decoupled DOF. Thus, the number of DOF can also be interpreted as equal to the number of observations minus the number of relationships (or number of parameters to estimate) between these observations. For example, if one seeks two numbers for which the sum is 12, neither number can be directly determined by the simple equation x + y = 12. Indeed, if one is arbitrarily chosen, the other is automatically determined. So we have two RVs but only 1 DOF. A problem is solved (it has a unique solution) when the number of DOF is 0.

If we consider line y = ax + b, we must estimate two parameters (a, b). In order for the problem to have a unique solution, it must fulfill the condition DOF = 0, so we need two observations. One observation at point 1 determines $y_1 = ax_1 + b$ where an infinite number of lines occur (DOF = 1 observation – 2 parameters to estimate < 0), therefore no single solution is found (an infinite number of lines pass through this point). A second observation $y_2 = ax_2 + b$ solves the problem with a single solution (2 observations – 2 parameters to estimate = 0). There can only be one line that passes through both points.

G.1.2. Reminders on statistics

If the measurement of a quantity under conditions that are as constant as possible is repeated n times, we get n individual results that are slightly different from one another, due to random events that are either obtained from the measuring device or from the operator or even from of the quantity itself. We assume that the values $x_1, x_2, ..., x_n$ are a sample of a population in the statistical sense of the term.

This population, which consists of all the results that we would obtain if we repeated the measurement an infinite number of times, has a distribution law characterized by a position parameter; which is the expectation E(x), and a dispersion parameter, which is the standard deviation σ(x).

The expectation E(x) cannot be deduced from measurements but, however, an estimator can be determined from n results $x_1, x_2, ..., x_n$, (unbiased): $\overline{m_x}$

$$\overline{m_x} = \frac{1}{n}.\sum_{i=1}^{n} x_i$$

NB: Unbiased means the average of the averages tends toward the average, by definition: (Bias $(\hat{\alpha}) = E[\hat{\alpha}] - \hat{\alpha}$, which is therefore an error on estimate $\hat{\alpha}$).

The standard deviation of the population σ(x) or its variance V(x) = $[\sigma(x)]^2$ cannot be directly deduced from the measurements, but, however, an estimator can be determined from the sample of n results $x_1, x_2, ..., x_n$; a variance estimator is $V(\overline{m_x})$ given by:

- for a large sample size, n > 100:

$$V\left(\overline{m_x}\right) = \sigma^2(x) = \frac{1}{n}.\sum_{i=1}^{n}\left(x_i - \overline{m_x}\right)^2$$

which is generally more easily estimated by:

$$V(\overline{m_x}) = \left(\frac{1}{n}\cdot\sum_{i=1}^{n} x_i^2\right) - \overline{m_x}^2$$

- for a small sample size, n <100 (we divide by n – 1, which removes the bias of variance):

$$V(\overline{m_x}) = \sigma^2(x) = \frac{1}{n-1}\cdot\sum_{i=1}^{n}(x_i - \overline{m_x})^2$$

which is generally more easily estimated by:

$$V(\overline{m_x}) = \frac{1}{n-1}\cdot\left[\left(\sum_{i=1}^{n} x_i^2\right) - n.\overline{m_x}^2\right]$$

NB1: It is recommended (even required for a small sample size n < 30) to carry out prior control of the normal distribution in order to eliminate sample outliers that might not be the result of random phenomena. For a non-normal sample, we can increase the sample size or eliminate outliers.

NB2: If the environmental conditions change or if a recalibration of the measuring device is carried out during n measurements, it is necessary to apply the corrections of systematic errors to the results of measurements $x_1, x_2, \ldots, x_n$ before estimating the average m_x and variance $V\overline{m_x}$.

NB3: If the measurement sample is renewable in the manner of an N-sample information stream of n_j measurements $x_{1j}, x_{2j}, \ldots, x_{nj}$ (real-time calculations, production monitoring by control cards, monitoring the quality of a product, etc.) and if each sample is representative of the population, it is then possible:

- to group all of these results and determine the average: $\overline{m_x} = \overline{m_{xij}}$ and variance $V\overline{m_x} = V\overline{m_{xij}}$ if we have all x_{ij} measurement results;

- to determine the average $\overline{m_x} = \overline{m_{xij}}$ if we only have averages $\overline{m}_{xj}$ of variances $V(\overline{m_{xi}})$ for each of the N samples and n_j number of measurements for sample j, as follows:

- if the samples are large (n > 100):

$$V(\overline{m_x}) = \frac{\sum_{j=1}^{N}\left[n_j\cdot\left\{V(\overline{m_{xj}}) + \overline{m_{xj}}^2\right\}\right]}{\sum_{j=1}^{N} n_j} - \overline{m_x}^2$$

and the estimator of variance $V(\overline{m_x}) = V\overline{m_{xij}}$ of the population of measurements;

- if the samples are small in size (n <100) but significant (n > 10) and the total population is large (Σn_j > 100):

$$V(\overline{m_x}) = \frac{\sum_{j=1}^{N}\left[(n_j - 1).V(\overline{m_{xj}}) + n_j\cdot\overline{m_{xj}}^2\right]}{\sum_{j=1}^{N} n_j} - \overline{m_x}^2$$

- if the samples are small in size ($n < 100$) but significant ($n > 10$) and the total population size is small ($\Sigma n_j < 100$):

$$V\left(\overline{m_x}\right) = \frac{\sum_{j=1}^{N}\left[\left(n_j - 1\right).V\left(\overline{m_{xj}}\right) + n_j.\overline{m_{xj}}^2\right]}{\left(\sum_{j=1}^{N} n_j\right) - 1} - \frac{\sum_{j=1}^{N} n_j}{\left(\sum_{j=1}^{N} n_j\right) - 1}.\overline{m_x}^2$$

G.2. Direct measurement

G.2.1. Principle

Direct measurement is a measurement method in which the value of the quantity to be measured is directly obtained by measuring with a measuring device, rather than through measuring other quantities that are functionally related to the quantity being measured. This measurement method is still direct if additional measurements need to be carried out in order to determine the values of influencing quantities in order to ascertain the relevant corrections (reduction of systematic errors, correction due to environmental conditions upon calibration).

G.2.2. Method

G.2.2.1. *Reduction of systematic errors*

A systematic error is a component of measurement uncertainty that remains constant over several measurements or that varies in a predictable way. Systematic errors (known or unknown) keep to well-defined causes. If we can identify the physical laws linking them to their causes and determine the intensity of the causes at the time of measurement, they must be reduced by correcting the average and the uncertainty. The most common systematic errors are:

(a) Reading error

This depends on the quantization of the measuring device (the difference between two possible interpretations, as this gap may be smaller than the resolution).

(b) Repeatability

This is the ability of a measurement process to repeat the same measurement. In general, it is evaluated by a type A method and the experimental standard deviation can then be estimated fairly simply.

(c) Correction for the calibration of a measuring device

The measuring device has a calibration sheet that specifies the correction $\overline{c_e}$ that should be applied to the device, as well as uncertainty $\overline{u_{c_e}}$ for this correction. We then correct values in the following manner.

If $\overline{m_x}$ is the measuring result of quantity x (that are average values here) and $\overline{u_{m_x}}$ is the standard uncertainty, we get:

$\bar{\mu} = \overline{m_x} + \overline{c_e}$ and $u_{\bar{\mu}} = \sqrt{u_{\overline{m_x}}^2 + u_{\overline{C_e}}^2}$ (square value of the two parameters).

(d) Correction for environmental conditions upon calibration

The measuring device indication depends on the experimental conditions and whether its calibration was performed in a reference environment (temperature, atmospheric pressure, etc.). However, we accurately know the corrections that should be applied to the measurement in the form of a function of influential environmental factors in order to establish the measurement environment.

If we consider several influencing variables $q_1, q_2, ..., q_n$, the measurement of each of these quantities introduces its own uncertainty. If $F(\overline{m_x}, q_1, q_2, ..., q_n)$ corresponds to the function that represents the relationship between the indication of the measuring device and the various influencing quantities, the corrected values are:

$$u_{\bar{\mu}} = \sqrt{\left(\frac{\partial F}{\partial \overline{m_x}}\right)^2 u_{\overline{m_x}}^2 + \left(\frac{\partial F}{\partial \overline{q_1}}\right)^2 u_{\overline{q1}}^2 + \left(\frac{\partial F}{\partial \overline{q_2}}\right)^2 u_{\overline{q2}}^2 + \cdots + \left(\frac{\partial F}{\partial \overline{q_n}}\right)^2 u_{\overline{qn}}^2}$$

This relationship assumes that the influencing variables are independent (correlation coefficient r = 0 or close to 0). If the same influencing factor intervenes explicitly or implicitly in different places of function F, it becomes reasonable to group all terms involving this variable before applying the formula.

If dependencies appear among the influencing variables, it is then possible to determine a resulting uncertainty, as shown below (the calculation principles are similar to those for the uncertainty of the indirect measurement, defined later):

$$u_{\bar{\mu}} = \sqrt{\left(\frac{\partial F}{\partial \overline{m_x}}\right)^2 u_{\overline{m_x}}^2 + \left(\frac{\partial F}{\partial \overline{q_1}}\right)^2 u_{\overline{q1}}^2 + \left(\frac{\partial F}{\partial \overline{q_2}}\right)^2 u_{\overline{q2}}^2 + \cdots + \left(\frac{\partial F}{\partial \overline{q_n}}\right)^2 u_{\overline{qn}}^2 + 2 \cdot \sum_{i-1}^{n-1}\left(\frac{\partial F}{\partial \overline{qi}}\right) \cdot \left(\sum_{j=i}^{n}\left(\frac{\partial F}{\partial \overline{qj}}\right)(\pm 1) u_{\overline{qi}}^2 \cdot u_{\overline{qj}}^2\right)}$$

Thus, after determining the sign of r (change in the same direction => r > 0, changes in the opposite direction => r < 0), we calculate the correction terms where r = +1 or r = –1, depending on each case. We will only retain these terms if they are positive in order to remain within a conservative framework.

Example G.1: To measure length L, a caliper is used. The calibration sheet includes the correction that should be added to the measurement $\overline{c_e} = 0{,}02$ mm and the corresponding standard uncertainty $u_{\overline{c_e}} = 10^{-3}$ mm, which corresponds to a variance of 10^{-6} mm²; the calibration was performed at a temperature T = 28°C difference from t_0. On the other hand, we know that the caliper temperature rule undergoes an expansion that depends on the linear expansion coefficient a = 17,10^{-6}/°C (stainless steel). With a measured length L_x, we estimate L through:

$$L = F(L_x, T) = L_x + L_x.\alpha(T - T_0)$$

By also taking the systematic calibration error into account, we get the average value of L according to L_x = 150 mm:

$$\bar{L} = \overline{L_x}.\left[1 + \alpha.\left(\bar{T} - T_0\right)\right] + \overline{c_e}$$

$$\bar{L} = 150 \text{ x}\left[1 + 17\text{x}10^{-6} \text{ x}(28 - 20)\right] + 0{,}02$$

$$\bar{L} = 150{,}0404 \text{ mm}$$

The uncertainty that occurs due to the correction of this measurement, with a standard uncertainty of L_x equal to 0.015 mm and a standard uncertainty equal to 0.1°C, is estimated by:

$$u_{\bar{L}} = \sqrt{\left(\frac{\partial F(L_x, T)}{\partial L_x}\right)^2 . u_{\overline{L_x}}^2 + \left(\frac{\partial F(L_x, T)}{\partial t}\right)^2 . u_{\bar{t}}^2 + u_{\overline{c_e}}^2}$$

$$u_{\bar{L}} = \sqrt{\left[1 + \alpha.\left(\bar{T} - T_0\right)\right]^2 \cdot u_{\overline{Lx}}^2 + \left(\alpha.\overline{L_x}\right)^2 \cdot u_{\bar{t}}^2 + u_{\overline{ce}}^2}$$

$$u_{\bar{L}} = \sqrt{\left[1+17.10^{-6}.(28-20)\right]^2 \cdot 225.10^{-6} + \left(17.10^{-6}.150\right)^2 \cdot 10^{-2} + 10^{-6}}$$

$$u_{\bar{L}} = 0{,}01528 \quad \text{mm}$$

Thus: $\bar{L} = 150{,}0404 \pm 2x0{,}01528 \text{ mm} = 150{,}0404 \pm 0{,}03056 \text{ mm}$

NB: The value of L thus measured was 28°C. We would need other corrections to determine a value of L at another temperature.

G.2.2.2. *Standard uncertainty*

Under the umbrella of the type A direct measurement method, the uncertainty that occurs due to the measurement method is then:

$$u_A = \sqrt{V(\overline{m_x})}$$

Variance is calculated by varying the number of appropriate parameters during testing (operators, equipment, environment, etc.).

Be careful, however, of the number of observations (measurements) which, if too small (less than 10) can lead to standard deviations that are not significant despite the previous relationships. If it is not possible to repeat the measurements (for example cost or availability problem), this component should be evaluated by a type B method, taking the experience of operators responsible for measurements into account, for example.

NB: Very often, we can consider that the read error contributes to repeatability. Indeed, for a type A evaluation, each observation that allows us to estimate the standard deviation of repeatability is affected by a read error that occurs in each measurement. Nevertheless, one can imagine that the quantization of the measuring device is very large compared to other possible variations that contribute to repeatability. Thus, the standard deviation of repeatability is 0 and does not take the read error into account.

By considering the principle for which the reading is involved in repeatability, there is a risk of forgetting it in some cases, especially when it is significant. We could, for example, estimate that if the standard deviation of repeatability is not 0, then it "contains" the read error, but this is not common practice.

Interoperator reproducibility (also called intermediate precision) consists of varying one of the factors from the measurement process in order to evaluate the impact of such factors on the uncertainty. Experience shows that when the operator actually engages in the process, it often plays a significant role in uncertainty and consequently, the different operators can achieve significantly different measurement results (for example, tangent to a line, detecting a color change, measurement efforts, approach, etc.). It is therefore often necessary to evaluate this feature in order to estimate a representative measurement uncertainty.

NB: Similarly, one could consider evaluating the interdevice reproducibility (by varying the measuring device), the interenvironment reproducibility (by varying measuring conditions), the intermethod reproducibility (by varying the measurement method), etc.

This principle is used in the context of the ISO 5725 standard series for collaborative trials. Measurement uncertainty is assessed by a type A "super method" in which repeatability is loosely defined as the possible intralaboratory variation (by varying the operators in the laboratory) and the interlaboratory reproducibility that evaluates dispersion between laboratories (and possibly between methods). Getting to the measurement uncertainty of the process in question, we should check that there is no bias within the process.

Apart from these near-systematic causes, measurement processes are also influenced by other environmental causes to the extent that they are often sensitive to environmental parameters (temperature, atmospheric pressure, gravitational acceleration, humidity, etc.). Obviously, knowing the physical principles related to the measurement as well as the interactions between environmental parameters and characteristics of the measured object is essential. Our bibliography provides information on this subject (law of material expansion, Archimedes' buoyancy, etc.). Naturally, the operators that usually carry out the measurements often know what influencing factors interact within the process (for example, in summer, the results are always better and when it rains, they're not as good) and can effectively guide investigations.

Other causes may be from measurement devices that, regardless of what they may be, are always imperfect. Several types of errors can coexist within a single device: accuracy or indication errors (for which we rarely know the systematic part that could be corrected for from the randomness that ultimately always contributes to the uncertainty of the process), which are common to all devices and specific errors within the technology of the device itself (hysteresis, parallel keys, etc.). These various causes of error from the device can be assessed during calibration and then be displayed on the calibration documents (certificates).

The next phase involves evaluating the various standard deviations that correspond to the identified causes, to eventually calculate the combined standard deviation of the measurement process (usually by quadratic sum). This summary may, for example, be in the following form and will be discussed later:

Causes of uncertainty	Maximum error	Law of distribution	Standard deviation
Factor 1	A	Normal (100%)	A/3
Factor 2	C	Normal (95%)	$C/\sqrt{2}$
Factor 3	D	Uniform	$D/\sqrt{3}$
Factor 4	J	Sine arc derivative	$J/\sqrt{2}$
–	–	–	–

The estimation of measurement uncertainty therefore requires estimating the deviations of influencing factors that are involved in the measurement process. Some may be valued on the basis of experimental observations (we can vary one factor and observe its influence on the measurement results): in this case, one then means type A uncertainty. The other standard deviations can be assessed on the basis of experience, bibliography, scientific knowledge, other diverse and varied sources of information, etc. In this case, the standard deviations are not directly accessible (because they are not "physical" representations). It is therefore necessary to go through the maximum errors that the factor in question can produce and their frequency of occurrence (the distribution law of the factor in question). This is called type B evaluation, as described below.

G.3. Direct type B measurement method

Whenever possible, it is best to use the type A direct method, which is the only one that takes into account the standard uncertainty that is statistically estimated from the repeatability of relevant measurements.

If the measurement is unique or if the sample of individual measurement results is too small ($n < 10$) and does not allow us to statistically process it, this analytical approach is based on the inventory and analysis of various elementary uncertainties that intervene when measuring the quantity in question. The expectation E(x) can be deduced from measurements but from the n < 10 results x_1, x_2, ...x_n, we can deduce the quantity of an estimator $\overline{m_x}$ from:

$$\overline{m_x} = \frac{1}{n}.\sum_{i=1}^{n} x_i$$

At worst $\overline{m_x}$ = x, if only one measurement is carried out.

The assessment of type B uncertainties is not based on basic statistics. As each measurement is a special case, the search for type B uncertainties must be based on a thorough analysis of the method, the procedure and the measuring devices used.

To do this, one must identify all the causes of elementary uncertainties that could occur. This can be done using the aforementioned 5M chart and determining the values of elementary or standard uncertainty.

The estimation of these standard uncertainties is as follows:

- if $e_j$max is the maximum extent of the potential error (information from various sources), the value u_j that characterizes the type B uncertainty due to $e_j$max is:

$$u_j = \frac{e_j \max}{6}$$

- if I_{cal} is the calibration uncertainty of the measuring equipment provided in a calibration certificate, the value u_j that characterizes the type B uncertainty due to the calibration of the equipment is:

$$u_j = \frac{I_{cal}}{2}$$

- if MPE is the maximum permissible error in the metrological verification of the measuring equipment, the value u_j that characterizes the type B uncertainty due to the verification of this equipment (the CEI/ISO 99 recommends a capability of 4) is:

$$u_j = \frac{MPE}{3}$$

- if *resol* is the resolution of a digital measuring device, the value u_j that characterizes the type B uncertainty due to the verification of this equipment is:

- for an ADC rectangle: $u_j = \frac{resol}{2\sqrt{3}}$

- for an ADC triangle: $u_j = \frac{resol}{\sqrt{3}}$

For other types, we refer the reader to the previous table.

We then estimate a kind of combined uncertainty for the quantity to be measured $u_c(\overline{m_x})$ by:

$$u_c\left(\overline{m_x}\right) = \sqrt{\sum_{j=1}^{n} u_j^2}$$

Example G.2: We use the documentation from a measuring device to determine the standard uncertainty:

- if the calibration uncertainty is mentioned, for example I_{cal} = 0.1, we will directly use a standard uncertainty associated with the calibration with the following value:

$$u_{calibration} = \frac{I_{cal}}{2} = 0{,}05$$

- if the verification tolerance is mentioned as an interval, by convention we will consider that value of the extent of this interval $e_x max$ corresponds to six times the standard uncertainty, so the following value:

$$u_{verification} = \left(\frac{e_x \max}{6}\right)$$

- if the verification tolerance is ± 3%, then the extent of the $e_x max$ interval is 6% and the uncertainty $u_{verification}$ is 1%.

NB: The above relationship (square root of the quadratic sum) assumes statistical independence from the causes of basic uncertainties. The demonstration of statistical independence of these elementary causes of uncertainty is relatively complex and, in practice, is fairly limited or little used. But if dependencies appear between some causes of elementary uncertainties, we must determine a combined uncertainty in the following manner (the calculation principles are similar to those for indirect uncertainty measurement discussed hereafter):

$$u_{\bar{g}} = \sqrt{\left(\frac{\partial F}{\partial g_1}\right)^2 u_{g_1}{}^2 + \left(\frac{\partial F}{\partial g_2}\right)^2 u_{g_2}{}^2 + \left(\frac{\partial F}{\partial g_i}\right)^2 u_{g_i}{}^2 + .. + \left(\frac{\partial F}{\partial g_n}\right)^2 u_{g_n}{}^2 + 2\sum_{i=1}^{n=1}\left(\frac{\partial F}{\partial g_i}\right)\left(\sum_{j=i}^{n}\left(\frac{\partial F}{\partial g_j}\right)(\pm 1) u_{g_i}{}^2 u_{gj}{}^2\right)}$$

Thus, after determining the sign of the correlation coefficient r (change in the same direction $r > 0$, change in the opposite direction $r < 0$), we calculate the correction terms with $r = 1$ or $r = -1$, as appropriate, and we only retain those that are positive, in order to remain conservative.

G.3.1. Residual errors

Residual errors are the remaining systematic errors that could not be eliminated after correction (see section G.2.2.1).

We can eliminate these residual errors by comparing several measurement methods, with several calibrators. If we suspect that the end result contains errors, we can undertake these comparisons by using the documents from the specific bibliography in section G.6 as guides.

G.3.2. Combined uncertainty

For direct measurements, the measurement method of the desired parameter can implement two previously discussed measurements, type A for experimental evaluations and type B for mathematical assessments. In all cases, combined uncertainty results from the combination of the set of elementary standard uncertainties as defined below:

$$u_C = \sqrt{\sum_{j=1}^{n} u_j^2 + \sum_{k=1}^{p} u_k^2}$$

where u_j = standard uncertainty that occurs due to the influencing quantities and standard uncertainty u_k that occurs due to the reduction of systematic errors.

G.4. Indirect measurement

G.4.1. Principle

Indirect measurement is a measurement method whereby the value of the measured variable is obtained indirectly by measuring other quantities that are functionally related to the quantity being measured (usually through an analytical mathematical relationship). The physical function that gives the value of the quantity to be measured through the elementary quantities $G = F(g_1, g_2, \ldots, g_n)$ is assumed to be exactly known.

G.4.2. Method

G.4.2.1. *Verification of the independence of different quantities (statistical analysis)*

If the same elementary quantity intervenes explicitly or implicitly in different places within function F, it is not unreasonable to group all the terms that include this variable before seeking the independence of these variables or applying a formula to them. This method makes it possible to perform the calculation just once.

Example G.3: We wish to measure a quantity P such that:

$$P = \frac{C_0 . I^2}{(\theta + T)}$$

Where C_0 and θ are two known constants.

A standard resistor R_s is used for measurements of current I and temperature T.

- Current I is determined by using a voltmeter to measure the voltage V_s across the standard resistor R_s according to the relationship:

$$I = \frac{V_S}{R_S}$$

- Temperature T is measured using a resistive bridge with a standard resistor R_s and a calibrated thermal resistor $R_T(T)$ by measuring the ratio $\beta(T) = R_T(T)/R_s$ using the following equation (for 15°C < t < 30°C):

$$T = \alpha . \beta(T)^2 . R_s^{\ 2} - T_0$$

Where α and T_0 are two known constants.

Rather than determine the uncertainty of P from its expression in terms of I and T, it is best to replace the I and T variables in the expression of P by their respective functions V_s and $\beta(T)$, which come from direct measurements. Thus:

$$P = \frac{C_0 . V_S^{\ 2}}{R_S^{\ 2} . \left[\theta + \alpha . \beta(T)^2 . R_S^{\ 2} - T_0\right]}$$

(a) For elementary quantities that result from type A direct measurements and for which a measurement series is therefore available and the measurement sample size of these quantities is the same, it is possible to evaluate the correlation coefficients by calculating the covariances two by two and dividing by the root of the product of variances:

$$C(x, y) = \frac{c\left(\overline{m_x} . \overline{m_y}\right)}{\sqrt{V\left(\overline{m_x}\right) V\left(\overline{m_y}\right)}}$$

With, for large sample sizes (n > 100):

$$V\left(\overline{m_x}\right) = \frac{\sum_{i=1}^{n} x_i^2}{n} - \overline{m_x}^2 \ , \ V\left(\overline{m_y}\right) = \frac{\sum_{i=1}^{n} y_i^2}{n} - \overline{m_y}^2$$

And:

$$c\left(\overline{m_x}, \overline{m_y}\right) = \frac{\sum_{i=1}^{n} \left(x_i - \overline{m_x}\right)\left(y_i - \overline{m_y}\right)}{n}$$

which is generally easier to estimate in the following manner:

$$c\left(\overline{m_x}, \overline{m_y}\right) = \frac{\sum_{i=1}^{n} (x_i . y_i)}{n} - \overline{m_x} . \overline{m_y}$$

And for small sample sizes ($n < 100$):

$$V(\overline{m_x}) = \frac{\sum_{i=1}^{n} x_i^2}{n-1} - \frac{n}{n-1}.\overline{m_x}^2 \quad , \quad V(\overline{m_y}) = \frac{\sum_{i=1}^{n} y_i^2}{n-1} - \frac{n}{n-1}.\overline{m_y}^2$$

$$C(\overline{m_x}, \overline{m_y}) = \frac{\sum_{i=1}^{n} (x_i - \overline{m_x})(y_i - \overline{m_y})}{n-1}$$

which is generally easier to estimate in the following manner:

$$C(\overline{m_x}, \overline{m_y}) = \frac{\sum_{i=1}^{n} (x_i . y_i)}{n-1} - \frac{n}{n-1}.\overline{m_x}.\overline{m_y}$$

NB: If $r(x,y) \approx 0$, the two quantities x and y are considered to be independent. If $(x,y) \approx 1$ or $r(x,y) \approx -1$, the two quantities x and y are considered to be totally dependent. In intermediate cases, the variables are partially dependent.

(b) In other cases (elementary quantities resulting from type B measurements, one variable resulting from a type A measurement and one from a type B measurement, the two quantities resulting from type A measurements but with a number of samples from different measurements, etc.), we cannot use statistics to determine if two variables are correlated, but we may:

- search for physical correlations.

Example G.4: U and I are correlated by Ohm's law, with a correlation of 1; they are completely dependent:

- search for measurement correlations.

Example G.5: ten resistors of 1000 Ohms are measured using the same equipment for measuring the uncertainty that is negligible compared to their value. The values of these resistors are correlated, with a value close to 1. If these resistors are used in an assembly for measuring a given quantity, we will have to take the fact that they are totally dependent into account in order to determine the measurement uncertainty for this magnitude.

In every case, if we suspect (without being sure) that there is a correlation, we assume that the two quantities are totally dependent ($|r(x,y)| = 1$).

After determining the sign of r (change in the same direction $r > 0$, change in the opposite direction $r < 0$), we can calculate the correction terms with $r = 1$ or $r = -1$, as appropriate, and we retain these terms if they are positive (conservative approach).

G.4.2.2. *Measuring elementary quantities*

For all the elementary quantities related to the variable in question, there must be a type A or type B direct measurement estimate, as well as an estimate of uncertainty (see type A direct measurement method and type B direct measurement method); where applicable, for each of these, the systematic errors should be corrected (see method for reducing systematic errors).

G.4.2.3. *Determining combined uncertainty from indirect measurements*

From the function that gives the value of the quantity to be measured through the elementary quantities $G = F(g_1, g_2, ..., g_n)$ and from all the direct measurements of elementary quantities

$\overline{m_{g1}}$, $u_{\overline{m_{g1}}}$; $\overline{m_{g2}}$, $u_{\overline{m_{g2}}}$; ... ; $\overline{m_{gn}}$; $u_{\overline{m_{gn}}}$; one can determine an estimate for the quantity to be measured $\overline{m_G}$, as explained below:

$$\overline{m_G} = F\left(\overline{m_{g1}}, \overline{m_{g2}} \ldots\ldots \overline{m_{gn}}\right)$$

G.4.2.3.1. *In general*

The expression for general uncertainty on $\left(\overline{m_g}, u_{\overline{m_g}}\right)$, without making assumptions about the independence or not of these quantities, is given by:

$$u_{\overline{m_G}} = \sqrt{\sum_{i=1}^{n}\left(\frac{\partial F}{\partial g_i}\right)^2 .u^2\left(\overline{m_{gi}}\right) + 2.\sum_{i=1}^{n-1}\left(\frac{\partial F}{\partial g_i}\right).\left[\sum_{j=i+1}^{n}\frac{\partial F}{\partial g_i}.C\left(\overline{m_{gi}}, \overline{m_{gj}}\right)\right]}$$

The partial derivatives of F relative to g_i are the sensitivity factors of F relative to the uncertainties of g_i. These sensitivity factors are calculated for the estimated values of measurements of g_i.

The correction terms (second sum of the right part of the expression) will be withheld arbitrarily if they represent a significant value (greater than 1/10 of the total variance).

G.4.2.3.2. *Case where some basic variables are independent (r close to 0)*

Whenever two elementary quantities g_i, g_j are independent, $r(g_i, g_j) = 0$, the corresponding covariance $C(\overline{m_{gi}}, \overline{m_{gj}})$ cancels out and the corrector term relating to it disappears. Thus, if there are λ basic linked quantities (r different from 0), the general expression becomes (see section G.2.3.1):

$$u_{\overline{m_G}} = \sqrt{\sum_{i=1}^{n}\left(\frac{\partial F}{\partial g_i}\right)^2 .u_{\overline{m_{gi}}}^2 + 2.\sum_{i=1}^{\lambda-1}\left(\frac{\partial F}{\partial g_i}\right).\left[\sum_{j=i+1}^{\lambda}\left(\frac{\partial F}{\partial g_j}\right).C\left(\overline{m_{gi}}, \overline{m_{gj}}\right)\right]}$$

G.4.2.3.3. *Case of independent variables (r close to 0)*

If all the elementary quantities are independent (see section on checking for the independence of different quantities), the set of covariances $C(\overline{m_{gi}}, \overline{m_{gj}})$ of different sizes is canceled out and the general expression (see section G.2.3.1) is simplified, giving standard uncertainty u_g:

$$u_{\overline{mg}} = \sqrt{\left(\frac{\partial F}{\partial g_1}\right)^2 u_{mg_1}^{\ 2} + \left(\frac{\partial F}{\partial g_2}\right)^2 u_{mg_2}^{\ 2} + \left(\frac{\partial F}{\partial g_i}\right)^2 u_{mg_i}^{\ 2} + \ldots\ldots + \left(\frac{\partial F}{\partial g_n}\right)^2 u_{mg_n}^{\ 2}}$$

Such that:

$$u_{\overline{m_G}} = \sqrt{\sum_{i=1}^{n}\left(\frac{\partial F}{\partial g_i}\right)^2 .u_{\overline{m_{gi}}}^2}$$

Example G.6: Determination of the density ρ_{Object} of an object of mass m and volume V by double weighing in air and in water at T = 25 °C.

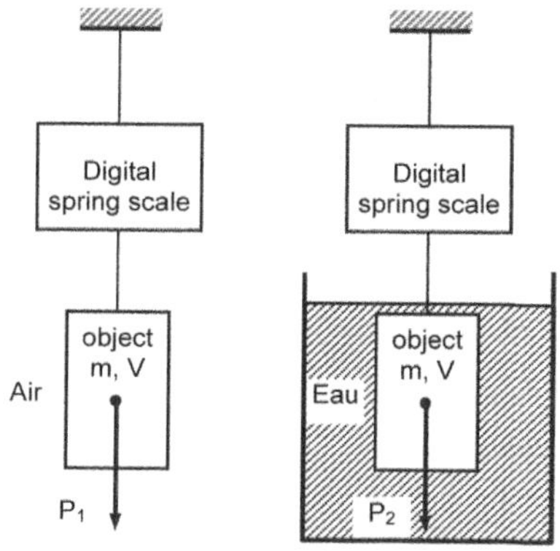

We then get:

$P_1 = m.g = V.\rho_{object}.g$ and $P_2 = m.g - (V.\rho_{water}.g) = (V.\rho_{object}.g) - (V.\rho_{water}.g) = V.g.(\rho_{object} - \rho_{water})$

The unknown variable that we seek is therefore ρ_{object}, so:

$$\frac{P_2}{P_1} = \frac{V.g.(\rho_{object} - \rho_{water})}{V.g.\rho_{object}} = \frac{\rho_{object} - \rho_{water}}{\rho_{object}} \rightarrow \rho_{object} = \frac{P}{P_1 - P_2}.\rho_{water}$$

ρ_{water} is determined by indirect measurement.

The formula shows: ρ_{water} = 998 kg/m^3 at 20 °C and ρ_{water} = 964 kg/m^3 at 90 °C. If a linear evolution is assumed between the two values, we then determine ρ_{water} by measuring temperature (rule of three):

$$\rho_{water} = 998 - \frac{998 - 964}{90 - 20}.(T - 20) \rightarrow \rho_{water} = 998 - \frac{34}{70}.(T - 20) \rightarrow \rho_{water} = a - b.T$$

Where: $\begin{cases} a = 1007,714286 \\ b = 0,485714287 \end{cases}$ and: T in °C ; $20°C \leq T \leq 90°C$

Which ultimately is: $\rho_{object} = \frac{P_1}{P_1 - P_2}.(a - bT)$ $P_{1,2}$ in N

a) Preliminary remarks on this method

(a) P_1 and P_2 are completely dependent because they are commensurate with r = +1 (they change in the same direction). Thus, we get:

$$P_2 = P_1.\frac{\rho_{object} - \rho_{water}}{\rho_{object}}$$

(b) The residual errors are due to:

- imprecision of formula values (are they accurate? Rounded up or truncated to three digits? What about uncertainty?);
- the assumption of linearity of changes in ρ_{water} between 20 °C and 90 °C;
- the fact that we have ignored air buoyancy during weighing in air ($\rho_{air} \approx$ 1.2 kg/m^3);
- the purity of water, which is not necessarily distilled;
- ρ_{object} is given at T, temperature of water at 25 °C here.

b) Calculation of partial derivatives

$$\frac{\partial \rho_{object}(P_1,P_2,T)}{\partial P_1}=\frac{(a-b\overline{T}).(\overline{P_1}-\overline{P_2})-\overline{P_1}.(a-b\overline{T})}{(\overline{P_1}-\overline{P_2})^2}=-\frac{\overline{P_2}.(a-b.\overline{T})}{(\overline{P_1}-\overline{P_2})^2}$$

$$\frac{\partial \rho_{object}(P_1,P_2,T)}{\partial P_2}=\frac{\overline{P_1}.(a-b.\overline{T})}{(\overline{P_1}-\overline{P_2})^2}$$

$$\frac{\partial \rho_{object}(P_1,P_2,T)}{\partial T}=-\frac{b.\overline{P_1}}{(\overline{P_1}-\overline{P_2})}$$

In digital implementation, we measure:

– P_1 = 150 N and P_2 = 105 N

With a spring scale of 0 to 200 N, maximum uncertainty (Type B) ± 0,5% of the full scale or ± 1 N. This gives a standard uncertainty of 0,3333333 ... N;

– T = 25 °C

With a maximum uncertainty thermometer (type B) ±0,3 °C, or an uncertainty of 0,1 °C.

Calculation of the measured value of ρ_{object}:

$$\overline{\rho_{object}}=\frac{\overline{P_1}}{\overline{P_1}-\overline{P_2}}.(a-b\overline{T}) \quad \text{where : } \begin{cases} a=1007{,}714286 \\ b=0{,}485714287 \\ \overline{P_1}=150\text{N} \ ; \ \overline{P_2}=105\text{N} \ ; \ \overline{T}=25°\text{C} \end{cases}$$

$$\rightarrow \overline{\rho_{object}}=3318.571429\text{kg/m}^3$$

c) Calculation of sensitivity factors

- For P_1:

$$\frac{\partial \rho_{object}(P_1,P_2,T)}{\partial P_1}=-\frac{105.(1007{,}714286-0{,}485714287.25)}{(150-105)^2}=-51{,}62222224$$

- For P_2:

$$\frac{\partial \rho_{object}(P_1,P_2,T)}{\partial P_2}=\frac{150.(1007{,}714286-0{,}485714287.25)}{(150-105)^2}=73{,}74603176$$

- For T:

$$\frac{\partial \rho_{object}(P_1,P_2,T)}{\partial T}=-\frac{0{,}485714287.150}{(150-105)}=-1{,}619047623$$

d) Calculating the indirect measurement variance of $\overline{\rho}_{object}$

$$u^2_{\rho_{object}}=\left[\frac{\partial \rho_{object}(P_1,P_2,T)}{\partial P_1}\right]^2.u^2_{\overline{P_1}}+\left[\frac{\partial \rho_{object}(P_1,P_2,T)}{\partial P_2}\right]^2.u^2_{\overline{P_2}}+\left[\frac{\partial \rho_{Object}(P_1,P_2,T)}{\partial T}\right]^2.u^2_{\overline{T}}$$

$$....+2.\frac{\partial \rho_{object}(P_1,P_2,T)}{\partial P_1}.\frac{\partial \rho_{object}(P_1,P_2,T)}{\partial P_2}.C(\overline{P_1},\overline{P_2})$$

As both variables P_1 and P_2 are totally dependent, their covariance is expressed by:

$$C\left(\overline{P_1},\overline{P_2}\right)=r.\sqrt{V\left(\overline{P_1}\right).V\left(\overline{P_2}\right)}=r.\sqrt{u_{\overline{P_1}}^2.u_{\overline{P_2}}^2}=r.u_{\overline{P_1}}.u_{\overline{P_2}}$$

$$\text{where } r=1\rightarrow C\left(\overline{P_1},\overline{P_2}\right)=u_{\overline{P_1}}.u_{\overline{P_2}}$$

Therefore:

$$u_{\rho_{object}}=(-51{,}62222224)^2\,x0{,}11111111+(73{,}74603176)^2.0{,}1111111+(-1{,}619047623)^2.0{,}01$$
$$+2(-51{,}62222224)\,x\,(73{,}74603176).\sqrt{0{,}1111111.0{,}1111111}$$

$$u_{\rho_{object}}=7{,}376371134$$

NB1: Let us note that initially and before performing any rounding up or down, a sufficient number of digits must be kept to complete the calculations.

NB2: When two elementary quantities are totally dependent with a correlation coefficient equal to +1 (r = +1) and all other variables are completely independent (r = 0), the general expression get simplified to:

$$u_{\overline{m_G}}^2=\sum_{i=1}^{n-2}\left(\frac{\partial F}{\partial g_i}\right)^2.u_{\overline{m_{gi}}}^2+\left[\left(\frac{\partial F}{\partial g_{n-1}}\right).u_{\overline{m_{g_{n-1}}}}+\left(\frac{\partial F}{\partial g_n}\right).u_{\overline{m_{g_n}}}\right]^2$$

In the above example, we get:

$$u_{\overline{\rho_{object}}}^2=(-1{,}619047623)^2\,x\,0{,}01+\left[(-51{,}62222224).\sqrt{0{,}11111111}+(73{,}74603176).\sqrt{0{,}11111111}\right]^2$$

Such that: $u_{\rho_{object}}=7{,}376371134$

We get the same result as that obtained through the formula for variance of the indirect measurement of ρ_{object}.

NB3: If two elementary quantities are totally dependent with a correlation coefficient equal to –1 and all other variables are completely independent (r = 0), the general expression also simplifies to:

$$u_{\overline{m_G}}^2=\sum_{i=1}^{n-2}\left(\frac{\partial F}{\partial g_i}\right)^2.u_{\overline{m_{gi}}}^2+\left[\left(\frac{\partial F}{\partial g_{n-1}}\right).u_{\overline{m_{gn-1}}}-\left(\frac{\partial F}{\partial g_n}\right).u_{\overline{m_{gn}}}\right]^2$$

NB4: Sometimes, rather than carrying out a covariance calculation, it is possible to change the measurement method and replace two elementary dependent variables by a set of independent elementary variables. We therefore express the quantity to be measured as a function of these new elementary quantities and we are then reduced to the case in section G.2.3.2. If we manage to replace all elementary dependent quantities by sets of elementary independent variables, we are brought back to the case in section G.2.3.3.

Example G.7: To determine density, ρ_{object} could have been determined through:

$$\rho_{object}=\frac{P_1}{V.g}$$

By directly measuring the quantities P_1 and V (which are independent) and through the use of physical constant tables, g can be determined.

NB5: If the indirect measurement G is exclusively a function of g_i variables that are obtained by direct type A measurements, combined uncertainty calculations can be avoided through the following method:

- carry out direct measurements of the different variables g_i;
- calculate the value $G_j = F(g_{1j}, g_{2j},\ldots, g_{nj})$;
- the process is repeated until we get N estimates of G, in order to apply a statistic;
- ultimately, we determine the average m_G and uncertainty u_{mg} as a direct type A measurement.

We should have reduced systematic errors beforehand by modifying the analytical expression of G so that it takes into account the influencing factors that cause the corrections.

G.4.2.4. *Getting the expanded uncertainty and measurement uncertainty*

Once the standard uncertainty is determined (by direct or indirect method), we determine the "expanded uncertainty", which is more commonly called the "measurement uncertainty".

In all cases, the expanded uncertainty is obtained through the following equation:

$$U = \pm k.u_c$$

where u_c is the combined uncertainty from the composition of standard uncertainties either for a direct measurement or for an indirect measurement.

From the standard combined uncertainty, expanded uncertainty is determined:

- from a Student's distribution with n – 1 DOF, where the number of measurement samples $n \leq 30$;
- from the normal distribution for n > 30.

The confidence interval used to determine the expanded uncertainty will, unless expressly contrarily stated (contractually), be 95% (corresponding to k = 2 in the above formula):

$$U = \pm 2.u_c$$

We consider the measurement uncertainty as the expanded uncertainty. *It is this expanded uncertainty that should be compared to the tolerance of the measurement in question (contractual or regulatory MPE).*

To estimate the confidence interval or the expanded uncertainty from a Student's distribution, we must specify the number of DOF and the confidence level attached to the uncertainty (95% conventionally).

Example G.8: Taking the results from the previous example, the combined uncertainty of the indirect measurement of ρ_{object} was:

$$u_{\rho_{Objevt}} = 7{,}376371134$$

Let us consider that, conventionally, the expanded uncertainty is ± twice the standard uncertainty, so:

$$\pm 2.\ 7{,}376371134 = \pm\ 14{,}75274227$$

The end result, without rounding up nor down, can then be written:

$$\overline{\rho_{object}} = 3318{,}571429 \pm 14{,}5274227\ \text{kg}/\text{m}^3 \text{ at } 25°\text{C}$$

G.4.2.5. *Rule for rounding up or down*

- The final uncertainty only keeps two significant digits and rounding is done by excess: *here* ± 15

- The measurement result is rounded in such a way that the error that occurs due to rounding may be, in absolute value, less than 1/10 of half the extent of the final uncertainty. In addition, the final digit of the measurement and the uncertainty is of the same magnitude, according to the following diagram:

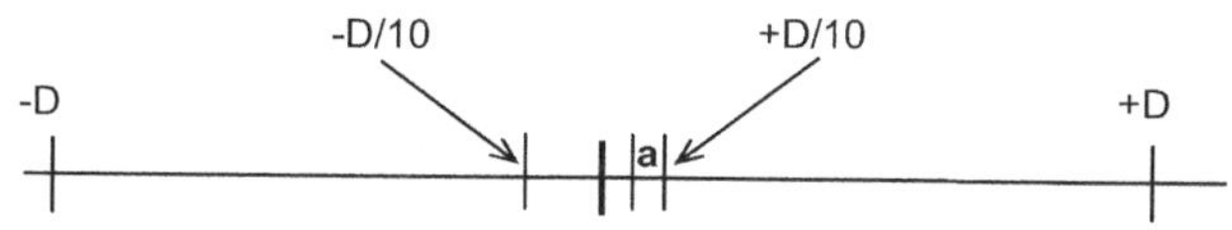

$$|m - a| < \frac{D}{10}$$

Where: m: measurement result without rounding
a: measurement result with rounding
± D: interval of the final uncertainty

So if we round up to 3319, this error will be 0,428571 < 2,3 so the final result will be written:

$$\overline{\rho_{Object}} = 3319 \pm 15\ kg/m^3 \text{ at } 25°C$$

G.5. Example of accelerometer calibration

The reference sensor is calibrated by comparing it with the connected equipment, with the corresponding uncertainty having been documented. The minimum calibration requirement for the reference sensor is calibration in appropriate reference conditions (frequency, amplitude and temperature). Normally, conditions are selected as described in ISO 16063-21.

At a specific frequency and amplitude of acceleration, velocity or displacement, the dynamic range is limited downwards by the noise threshold or the quality of distortion produced by the calibration vibration generator (if no filtering n is used) and upwards by its maximum power.

The measuring set-up is shown in the following diagram:

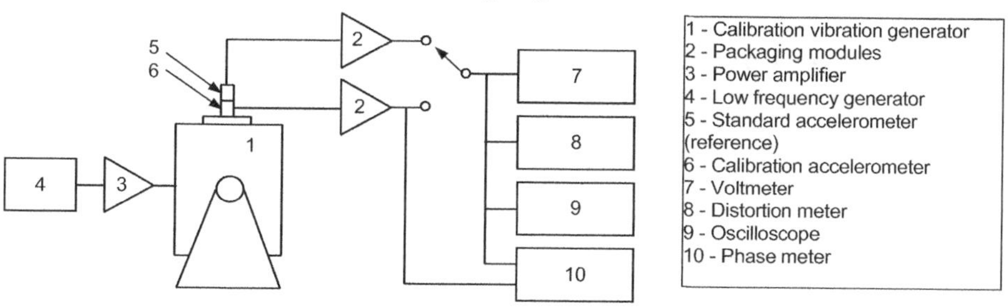

Measurement uncertainty should be expressed by the expanded uncertainty U in accordance with the GUM, based on the approach recommended by the International Committee for Weights and Measures (CIPM). U then provides an interval y – U and y + U in which the Y value, which is the amount subjected to calibration and estimated by y, has a high probability of about 95%.

In fact, the unknown value of the uncertainty is not included in the interval, but the interval contains the true (unknown) value with a probability of 95% (this interval is itself a random magnitude).

To confidently claim that $y - U \leq Y \leq y + U$, the expanded uncertainty U is determined as described previously (type A or B evaluation or indirect measurements). In particular, if there is no specific information on the possible values of a quantity that brings about systematic errors (except that these values are within the limits L– and L+), uniform distribution on the interval [L –; L +] can be used to represent this information. The expected value of [(L–)+(L+)]/2 is to be used for correction.

If an influencing quantity that is considered to be evenly distributed (rectangular or uniform probability density) undergoes preliminary processing, the measurement result must include a specific law (for example: sinusoidal, second or third order polynomial, etc.). This information must be taken into consideration when choosing the corresponding associated distribution model.

Example G.9: The sensitivity of an accelerometer to sinusoidal accelerations in the nominal measurement direction is calculated from the output voltage x of the charge amplifier or the voltage amplifier. The accelerometer is subjected to an amplitude acceleration a, using the relationship:

$$S = x/a$$

Among the various disruptive effects that influence the measurement result during calibration, a transverse amplitude vibration movement component in acceleration a_T can occur, which is transformed into a component of the error x_T upon output, caused by the transverse sensitivity of the accelerometer S_T. It is assumed, in this example, that the acceleration to be measured and the transverse acceleration have the same frequency, which means that there is no difference in the phase angle between the two signals. As the transverse sensitivity is usually a sine function of the angle β between the direction of the maximum transverse sensitivity S_{Tmax} and the direction of the transverse excitation, the error component can be expressed as:

$$e_{x_T} = S_{Tmax}.a_{Tmax}.\cos\beta$$

If the values of the maximum transverse sensitivity S_{Tmax} and the maximum transverse acceleration a_{Tmax} are known, but angle β is not, it is not unreasonable to assume a rectangular distribution of β within the interval $-\pi$, $+\pi$ (which is usually the case when nothing is known about the distribution law, ergo total uncertainty). Thus, the influencing quantity, so the transverse acceleration, for which angle β has a rectangular distribution, leads to a measurement error component e_{x_T} for which the probability density $w\left(e_{\overline{x_T}}\right)$ is described (through the transformation law of probability [ref. I18,T2]) by the relationship:

$$w\left(e_{\overline{x_T}}\right) = \frac{1}{b\pi\sqrt{1-\left(\frac{e_{\overline{x_T}}}{b}\right)^2}} \quad \text{Where: } -b < e_{\overline{x_T}} < +b \text{ and } e_{\overline{x_T}} = \overline{S_{Tmax}.a_{Tmax}}$$

Thus, for an "arc sine" distribution (always the case when the random phase of a signal is involved), we show that the corresponding standard deviation is given by:

$$u\left(e_{\overline{x_T}}\right) = \frac{b}{\sqrt{2}}$$

The expected value is 0 in our case and this is the best estimate of the error $e_{\overline{x_T}}$. While this distribution is symmetrical and centered, it of course has an average of zero. If the purpose of this example was to show that an RV transformation could intervene, we should have chosen an asymmetric distribution.

a) Combined standard uncertainty

The combined standard uncertainty u_c relative to the measurement standard uncertainty of Y can, as we have already seen, be determined by combining the individual standard uncertainties (and covariance, if any) and using the propagation law of uncertainty. This is done according to the equation:

$$u_c(y) = \sqrt{\sum_{i=1}^{N}\left(\frac{\partial f}{\partial x_i}\right)^2 u^2(x_i) + 2\sum_{i=1}^{N-1}\sum_{j=i+1}^{N}\frac{\partial f}{\partial x_i}\frac{\partial f}{\partial x_j}u(x_i, x_j)}$$

As we have already seen, this equation is based on an approximation of a first order Taylor series. Moreover, for independent variables, the second term is zero through $u(x_1, x_2) = 0$.

For the measured variable, we get: $Y = f(X_1, X_2, \ldots, X_N)$

where Y is the measured quantity (the measurand) determined from input variables at X_1, X_2, …, X_N through a functional relationship f. An estimation of the measurand Y, denoted by y, is obtained from the above equation using the estimates x_1, x_2, …, x_n as the values of input variables. Thus, the output estimation that results from the measurement is given by: $y = f(x_1, x_2, \ldots, x_n)$.

The symbols $\partial f / \partial x_i$ are the sensitivity coefficients.

NB: Let us recall that the development of the equation for the propagation law in a first order Taylor series is only applicable if function f is sufficiently linear relative to the change in estimated input x_i within the ranges described by uncertainties $u(x_i)$.

To overcome this obstacle that may exist along with other influencing quantities that act during the calibration of vibration and shock sensors, a suitable model was introduced. To briefly outline this model for the above example, a simplifying factor with

$$\left(1 - \frac{e_{\overline{xT}}}{\overline{x}}\right) \quad \text{where:} \quad \frac{e_{\overline{xT}}}{\overline{x}} << 1$$

is introduced as input quantity X_i into the functional relationship Y that is used to calculate the measurand. The equation of Y specifically defined for this example is then simply reduced to three quantities with: $Y = f(X_1, X_2, X_3)$

where:

- Y is the measurand (of sensitivity S);
- X_1 is the accelerometer output (voltage or load amplifier);
- X_2 is the amplitude of the acceleration;
- $X_3 = \left(1 - e\overline{x_T} / \overline{X}\right)$.

Thus, the relationship of the desired sensitivity can be expressed in the following simplified form:

$$Y = \frac{X_1}{X_2}.X_3$$

This can be established and the Taylor series approximation can be used, leading to a relative combined standard uncertainty. If there are no significant correlations, the latter relationship can be described as:

$$\frac{u_c(y)}{y}=\sqrt{\left(\frac{u(\overline{x_1})}{\overline{x_1}}\right)^2+\left(\frac{u(\overline{x_2})}{\overline{x_2}}\right)^2+\left(\frac{u(\overline{x_3})}{\overline{x_3}}\right)^2}=\sqrt{\left(\frac{u(\overline{x})}{\overline{x}}\right)^2+\left(\frac{u(\overline{a})}{\overline{a}}\right)^2+\left(\frac{u(\overline{e_{x_T}})}{\overline{x}}\right)^2}$$

where:

- $u(\overline{x})/\overline{x}$ is the standard uncertainty relative to the measurement of amplitude of output voltage, $\overline{x}$;

- $u(\overline{a})/\overline{a}$ is the standard uncertainty relative to the measurement of amplitude of acceleration, $\overline{a}$;

- $u(\overline{e_{x_T}}/\overline{x})/\overline{x}$ with, as described in the example: $u(\overline{e_{x_T}})=b/\sqrt{2}$.

Consequently, other factors for which deviations from 1 are similarly expressed by the relative error component of the quantity in question (for example, voltage, overall acceleration or sensitivity) may be introduced as an input quantity (X_4, X_5, ...), thus enabling all the various sources of influence to be considered separately.

b) Expanded or extended uncertainty

Here also, "expanded uncertainty" U is obtained by multiplying u_c by an expansion factor k, with k = 2 as usual, which allows a confidence interval of approximately 95% (1,96 for exactly 95%):

$$U=\pm 2.u_c$$

c) Reporting the result

When reporting the measurement result y of Y, the expanded uncertainty and the value of the expansion factor k (also coverage), if different from k = 2, should be indicated. In this case, the probability associated with the corresponding confidence interval must be indicated.

G.5.1. Calculation of expanded uncertainty under reference conditions

Calculating the expanded uncertainty $U_{rel}(s)$ relative to (S) for amplitude and sensitivity

The relative expanded uncertainty of the measurement of the amplitude of the sensitivity $U_{rel}(S)$ for each of the applied frequencies, for the acceleration and for the amplification parameters (if an amplifier is used) is also calculated from the same relationship, with k = 2:

$$U_{rel}(S)=k.u_{c,rel}(S)$$

$$u_{c,rel}(S)=\frac{1}{S}\sqrt{\sum_i^N\left(\frac{\partial f}{\partial x_i}\right)^2 u_i^2(S)+2\sum_{i=1}^{N-1}\sum_{j=i-1}^{N}\frac{\partial f}{\partial x_i}\frac{\partial f}{\partial x_j}u(x_i,x_j)}$$

$$=\sqrt{\sum_i^N\left(\frac{\partial f}{\partial x_i}\right)^2 u_{rel,i}^2(S)+2\sum_{i=1}^{N-1}\sum_{j=i-1}^{N}\frac{\partial f}{\partial x_i}\frac{\partial f}{\partial x_j}u_{rel}(x_i,x_j)}$$

where S is the sensitivity, $f(x_1, x_2,\ldots, x_n)$ is the estimated sensitivity and x_i are the values of estimated parameters. $u(x_i, x_j)$ is the covariance of (x_i, x_j) (zero if $x_{i,j}$ are uncorrelated). The last part of the expression uses u_{rel} to indicate the relative values.

NB: The following table lists a number of sources of uncertainty. Although it is supposed to contain all significant sources of uncertainty, it cannot be considered as exhaustive as there are a wide range of different calibration systems to which the method can be applied.

i	Standard uncertainty component $u(x_i)$	Source of uncertainty	Relative contribution to the uncertainty $u_{rel,i}(y)$
1	$u(S_1)$	The combined standard uncertainty for the reference sensor and the amplifier associated with specified conditions	$u_{rel,1}(S)$
2	$u(\bar{u}_{R,A})$	Conditioning module gain	$u_{rel,2}(S)$
3	$u(\bar{u}_R)$	Ratio of voltage measurements (often correlated)	$u_{rel,3}(S)$
4	$u(\bar{u}_{R,d})$	Effect of total harmonic distortion on the ratio of voltage measurements*	$u_{rel,4}(S)$
5	$u(\bar{u}_{R,H})$	Effect of noise on*	$u_{rel,5}(S)$
6	$u(\bar{u}_{R,v})$	Effect of transverse vibrations, torsion, flexion on*	$u_{rel,6}(S)$
7	$u(\bar{u}_{R,e})$	Effect of deformations of the accelerometer base on*	$u_{rel,7}(S)$
8	$u(\bar{u}_{R,N})$	Effect of accelerometer assembly conditions (tightening torque, fixing the cable, mass effect, etc.) on*	$u_{rel,8}(S)$
9	$u(\bar{u}_{R,r})$	Effect of accelerator on homogeneity on*	$u_{rel,9}(S)$
10	$u(S_{1,r})$	Stability of the reference sensor over time	$u_{rel,10}(S)$
11	$u(\bar{u}_{R,T})$	Effect of temperature on*	$u_{rel,11}(S)$
12	$u(f)$	Uncertainty on the frequency of vibration	$u_{rel,12}(S)$
13	$u(\bar{u}_{R,L})$	Effect of non-linearity of acceleration on*	$u_{rel,13}(S)$
14	$u(\bar{u}_{R,I})$	Effect of non-linearity of conditioning modules on*	$u_{rel,14}(S)$
15	$u(\bar{u}_{R,G})$	Effect of gravitational acceleration (continuous component) on*	$u_{rel,15}(S)$
16	$u(\bar{u}_{R,B})$	Effect of magnetic field on*	$u_{rel,16}(S)$
17	$u(\bar{u}_{R,E})$	Effect of other environmental agents on*	$u_{rel,17}(S)$
18	$u(S_{RE})$	Residual effects on sensitivity measures (for example, random effect in repeated measurements, experimental standard deviation of the arithmetic mean)	$u_{rel,18}(S)$

*Ratio of voltage measurements.

NB1: If the selected method uses a working sensor, which is calibrated by comparison to the reference sensor, the calculation must be carried out twice and the result from the first calculation must be used as the input for the final calculation.

NB2: The sources of uncertainty can be subdivided and numbered in a different manner than those used in this table, provided that every effect that has a significant influence on the measurement result is taken into account.

G.5.2. Expanded uncertainty on the frequency and amplitude range

Calculation of the expanded uncertainty $U_{rel}(S_t)$

For the transverse, the relative expanded uncertainty for the measurement of the sensitivity amplitude $U_{rel}(S_t)$ is only valid for the calibration frequency, the acceleration and the control parameters are used for conditioning modules (if these modules are part of the calibrated set-up). For the frequency range and the retained amplitude range, the relative expanded uncertainty for the measurement of the complex sensitivity amplitude $U_{rel}(S_t)$ is calculated from the following usual relationship, still with k = 2:

$$u_{c,rel}(S_t) = \frac{u_c(S_t)}{S_t} = \frac{1}{S_t}\sqrt{\sum_i^N \left(\frac{\partial f}{\partial x_i}\right)^2 u_i^2(S_t) + 2\sum_{i=1}^{N-1}\sum_{j=i-1}^{N} \frac{\partial f}{\partial x_i}\frac{\partial f}{\partial x_j} u(x_i, x_j)}$$

where:

$$U_{rel}(S_t) = k.u_{c,rel}(S_t) = 2.u_{c,rel}(S_t)$$

- S_t is the transverse sensitivity determined by the frequency and the amplitude range;
- $f(x_1, x_2,....x_n)$ is the estimated sensitivity;
- x_i are the estimated values for the measurand;
- $u(x_i, x_j)$ is the estimated covariance x, x (zero if uncorrelated).

NB: The following table lists a number of components of uncertainty that are used to determine S_t. Although it is supposed to contain all significant sources of uncertainty, it cannot be considered as exhaustive as there is a wide range of different calibration systems to which the method can be applied.

i	Standard uncertainty component $u_i(x)$	Sources of uncertainty	Contribution to uncertainty $u_i(y)$
1	$u(S)$	Uncertainty of the calculated sensitivity for calibration frequencies, amplitudes and specified parameters	$u_1(S_t)$
2	$u(e_T, A)$	Monitoring of conditioning module (gain adjustments for different amplifications)	$u_2(S_t)$
3	$u(e_L, f, A)$	Frequency response of the conditioner (gain differences for different frequencies)	$u_3(S_t)$
4	$u(e_L, f, T)$	Deviation of the frequency response from the accelerometer with respect to the theoretical curve	$u_4(S_t)$
5	$u(e_L, a, A)$	Effect of amplitude on the gain of the conditioning module (non-linearity)	$u_5(S_t)$
6	$u(e_L, a, T)$	Effect of amplitude on the sensitivity of the accelerometer (non-linearity)	$u_6(S_t)$
7	$u(e_i, A)$	Instability of the gain of the conditioning module and effect of the source impedance	$u_7(S_t)$
8	$u(e_i, T)$	Instability of the sensor sensitivity	$u_8(S_t)$
9	$u(e_E, A)$	Environmental effects on the gain of the conditioning module	$u_9(S_t)$
10	$u(e_E, T)$	Environmental effects on the sensor sensitivity	$u_{10}(S_t)$
11	$u(e_M, T)$	Other effects from set-up parameters (cable fastening, tightening torque, effect of mass loading, etc.) on the sensor sensitivity	$u_{11}(S_t)$

NB: Of course, if for the influencing factors in question and under given calibration conditions, we can demonstrate that a factor or several factors are negligible relative to the main factors, we do not consider them in the final calculation.

G.5.3. Digital application to the example of calculation of uncertainty

To facilitate using the principles mentioned above, let us look at an example based on the configuration in the following diagram (representation of the measurement process).

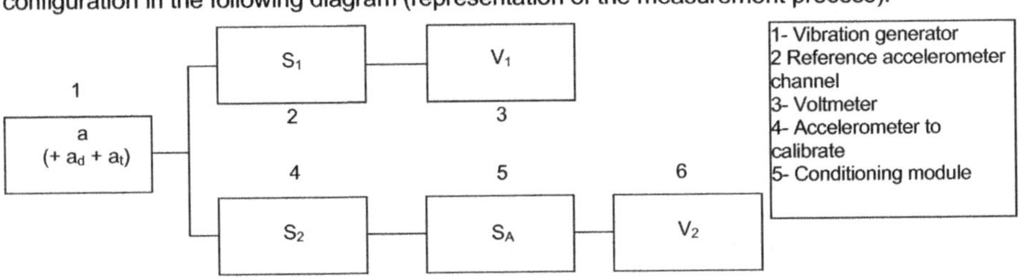

Symbols: the following symbols are used:

a: acceleration in the main direction;
a_v: transverse acceleration to the main direction;
a_d: distortion of the acceleration in the main direction;
S_1: sensitivity of the reference accelerometer channel;

Variable	Description	Relative expanded uncertainty or limits of the error estimation in%	Probability distribution	Factor x_i	Sensitivity coefficient C_i	Relative contribution $u_{rel.i}(y)$ in %
S_1	Calibration of the reference accelerometer channel	0,5	Normal (k = 2)	1/2	1	0,25
$S_{1.S}$	Evolution over 3 years. Manufacturer specification $\leq$ 0.5% / year	0,15	Uniform	$(1/\sqrt{3})$	1	0,087
$S_{A.cal}$	Sensitivity of the conditioning module of the channel to calibrate	0,25	Uniform	$(1/\sqrt{3})$	1	0,14
V_R	Specification of the voltage percentage	0,2	Uniform	$(1/\sqrt{3})$	1	0,12
$I(V_{R.T})$	Influence on V: measured from the temperature change Sensitivity of the reference sensor (23±3) °C < 0.02% per °C Sensor to be calibrated (23±3)°C < 0.1% per °C	0,36	Uniform	$(1/\sqrt{3})$	1	0,21
$I(V_{R.S})$	Maximum permissible difference between the reference level before and after calibration 0.2%	0,2	Uniform	$(1/\sqrt{3})$	1	0,12
$I(V_{R.N})$	Influence of set-up parameters for the sensor to be calibrated (cable, torque, etc.) at its maximum 0.05%	0,05	Uniform	$(1/\sqrt{3})$	1	0,029
$I(V_{R.d})$	Influence of distortion on V_R. Difference in frequency responses between the sensor to be calibrated (PZT) and a reference sensor (quartz) generally –2% per frequency decade The 3rd harmonic at less than 5% assuming a uniform distribution	0,0024 See note 2	Uniform	$(1/\sqrt{3})$	1	0,0014
$I(V_{R.v})$	Influence on V_R of the transverse acceleration Transverse vibration maximum 10% Transverse sensitivity of sensor to be calibrated, max 2% Transverse sensitivity of sensor to be calibrated, max 5%	$\sqrt{S_{v,2}^2 + S_{v,1}^2 + a_T^2}$ = 0.51 See note 3 and [ref. 21 and 22] Explanation of the relationship	Special	$\sqrt{1/18}$ See note 3 for the explanation of factor 1/18	1	0,12
$I(V_{R.e})$	Influence of the deformation of the base on the measurement	0,05	Uniform	$(1/\sqrt{3})$	1	0,029
$I(V_{R.r})$	Influence of the relative movement on the measurement of V_R	0,05	Uniform	$(1/\sqrt{3})$	1	0,029
$I(V_{R.L})$	Influence of non-linearity of the sensors on the measurement of V_R	0,03	Uniform	$(1/\sqrt{3})$	1	0,017
$I(V_{R.I})$	Influence of non-linearity of conditioning modules on the measurement of V_R	0,03	Uniform	$(1/\sqrt{3})$	1	0,017
$I(V_{R.G})$	Influence of gravity on the measurement of V_R	0,00	Uniform	$(1/\sqrt{3})$	1	0,00

[5] RM Aéro 800 01 B (04/86), "Définition et calculs d'erreurs. Incertitude de mesure".

[6] CERESTA (1995), "Aide-mémoire pratique des techniques statistiques".

[7] TAVERNE A., *Connaissance et maîtrise de la statistique*, Editions de l'Organisation, 2 volumes, 1988.

[8] COSTESEQUE M., "Contrôle et analyse statistique. Fiabilité", PLANEC, 1989.

[9] DIXON W.J., MASSEY F.J., *Introduction to statistical analysis*, McGraw-Hill, 1972, 2nd edition.

[10] NF X 07-001 (12/84), "Vocabulaire international des termes fondamentaux et généraux de métrologie".

[11] NF ISO 3534-1, "Statistics – Vocabulary and symbols – Part 1: General statistical terms and terms used in probability".

[12] NF EN ISO 9000, "Quality management systems – Fundamentals and vocabulary".

[13] FD ISO/TR 22971, "Accuracy (trueness and precision) of measurement methods and results".

[14] NF ISO 21748, "Guidance for the use of repeatability, reproducibility and trueness estimates in measurement uncertainty estimation".

[15] EA-4/02 (European Co-operation for Accreditation; see Cofrac), "EA Guidelines for the Evaluation of the Uncertainty of Measurement in Calibration".

[16] FD X07-021/022/015, "Métrologie et applications de la statistique".

[17] ISO/CEI 99, "Guide: International vocabulary of metrology – Basic and general concepts and associated terms (VIM)".

[18] ISO 5348, "Mechanical vibration and shock – Mechanical mounting of accelerometers".

[19] BIPM, IEC, IFCC, ISO, IUPAC, IUPAP, OIML, 19923 – "International vocabulary of basic and general terms in metrology (VIM)".s

[20] MARTENS H.J. VON, ROGAZEWSKI P., "Representation and transfer of the units of vibration quantities in the GDR", *Bulletin OIML*, n° 108, p. 26-37, 1987.

[21] MARTENS H.J. VON, "Evaluation of interferometric vibration measurement", *SPIE*, 4072, p. 82-101, 2000.

[22] JAFFARD P., *Initiation aux méthodes de la statistique et du calcul des probabilités*, Masson, 1976.

Appendix H

Example of MPE Allowances

Let us recall that the maximum permissible error (MPE) is usually either connected to the main variables (to be achieved) that are involved in the operation of the test or connected to influencing variables, which, although unintended, are unavoidable. This influences the measurement of the main variables or the behavior and/or the response of the material under test.

Standardized test methods generally do not use the term MPE. Other terms are used, such as “tolerance” or “accuracy”.

Moreover, the values in standard methods are not always directly used for verification of testing means or measuring devices. This is because they can require a measurand that is deduced by combining several components (different measuring equipment and/or test). This requires each of the constituents to be “allocated” a share of MPE that is judged to be the most appropriate.

Thus, the method for testing “sinusoidal vibrations” from the CEI 60068-2-6 states that the difference between the amplitude of the specified and achieved acceleration should not exceed 15%. This is a “process” tolerance that sets the variation limits for the true value of the variable in question in order to achieve compliance (which can also be interpreted as an overall MPE).

However, this requirement depends on the accuracy of the measurement process:

- MPE1 on the measurement of acceleration including testing means + U(MPE1) influence of measurement conditions;

- MPE2 of the control system + U(MPE2) influence of measurement conditions.

Let us evaluate these two components:

- MPE1 on the acceleration measurement (performed with an acceleration sensor) will depend on:

  - changes in sensitivity with frequency in the useful band of the test (for example 5 to 2,000 Hz), which corresponds to MPE1-1;

  - the calibration uncertainty at the given reference point at 100 Hz, which corresponds to MPE1-2;

  - + U(MPE1) = effect of influencing variables (transverse displacement, distortion, temperature, consistency, torque, etc.), which corresponds to MPE1-3;

- the MPE2 on the control system depends on:

  - the uncertainty of achieving the displayed desired value;

  - + U(MPE2) in the presence of influencing quantities (distortion, etc.).

The limits will therefore have to be allocated to each of the components involved in the overall MPE, such that the combination of uncertainties (the accuracy of the measurement process) has a sufficient capability (matching) with respect to the process tolerance (overall MPE).

a) Determining crucial equipment

Crucial equipment means the "material entity" that has a significant influence on the accuracy of test results. We must check that they conform to their intended use. This verification is carried out by comparing the checked properties to the MPE. For vibration testing, the entities to be distinguished as crucial equipment are usually:

- the single control system, in a closed loop on itself (SC);
- the entire single vibration generator assembly (VGA): clearly, the performance of the VGA will only be associated with the control system used, but the features that are monitored here are independent from the control system;
- the VGA coupled to the control system (VGA & SC);
- the accelerometer control channel (ACC).

Tolerances are generally cited in the standards for the following parameters:

- sinusoidal vibrations:
  - MPE1: the level of transversal displacement;
  - MPE2: the difference between the reference point (control point) and the verification point (closest to the mounting of the equipment under test);
  - MPE3: the difference between the acceleration level achieved at the reference point and the specified acceleration;
  - MPE4: sinusoidal distortion;
  - MPE5: sinusoidal precision on the points of characteristic frequencies (frequency break points in the specified test gauge);
- in random vibrations:
  - MPE1: the noise level of transversal displacement;
  - MPE2: the difference between the noise level at the reference point (control point) to the check point (closest to the mounting of the equipment under test);
  - MPE3: the difference between the noise level at the specified reference point and the achieved noise level;
  - MPE 6: the distribution of the instantaneous noise values;
- for shocks:
  - MPE1: the level of shock of transversal displacement;
  - MPE2: the difference between the level of shock specified at the reference point and the level of impact achieved;
  - MPE8: tolerance on speed variation of the shock.

From the tolerances specified by the above standards, we get the MPEs allocated to different parameters.

We should also include more specific verifications that are drawn from the VGA experiments, to which we have access and which lead to adding two specific points.

Point 1: Periodic verification of measurement devices in the control loops (because such measurements are "judges" for the "measuring" function and "parties" for the "control" function). This involves comparing the measurement in the feedback loop to a measurement obtained through an independent measurement system and ensuring that the gap is less than $MPE7_{sine}$, $MPE7_{noise}$ and $MPE7_{shock}$, which need to be defined. These values are not directly derived from the requirements of standards; they should not be too high so that they can detect unacceptable deviations and they should not be too low in order not to declare non-compliance for an unjustified excessive requirement. The best option is to start carrying out measurements and then set the optimized values. It goes without saying that in sine, the network frequency (50 Hz) in Europe will be part of the characteristic frequencies.

These MPEs will be verified in the following assembly levels:

	$MPE1_{sine}$ $MPE1_{noise}$ $MPE1_{shock}$	$MPE2_{sine}$ $MPE2_{noise}$	$MPE3_{sine}$ $MPE3_{noise}$	$MPE4_{sine}$	$MPE5_{sine}$	$MPE6_{noise}$	$MPE3\text{-}3_{sine}$ $MPE3\text{-}3_{noise}$ $MPE3\text{-}3_{shock}$
SC					X		
VGA	X	X	See below hereafter	X			
VGA&SC						X	X
ACC							

Allowances of MPE induced by $MPE3_{sine}$ and $MPE3_{noise}$

(a) $MPE3_{sine}$

The control accuracy tolerance (the monitored process) is the ±15% tolerated difference between the specified amplitude and performance in sine. The means used to monitor compliance with this tolerance must have a capability of one-third of this tolerance, so ±5% (ISO 5725). However, these means consist of two distinct entities:

- the control channel;

- the control system (acceleration feedback).

So we make an allowance for MPE3 in two parts: the first is the ACC MPE3-1 and the second is the control system MPE3-2:

- MPE3-1 on the measurement for the acceleration control point + U(MPE3-1), which is the value that represents the imprecision resulting from the effect of influencing quantities (such as, essentially, transversal displacement);

- MPE3-2 on the ability of the control system to achieve the required acceleration + U(MPE3-2), which is the value that represents the effect of influencing quantities (such as the sinusoidal vibration signal distortion; the current way to do this in world-class laboratories is to ignore U(MPE3-2)).

Let us evaluate these two components:

- MPE3-1 on the amplitude of the acceleration measurement will depend on:
  - changes in sensitivity with changing frequency in the test frequency range (for example, from 5 to 2,000 Hz), which is denoted MPE3-1-1,
  - the accuracy of calibration at the reference point at 100 Hz, which is denoted MPE3-1-2,
  - + U(MPE3-1-2), which is the effect of the influencing variables (transversal displacement, distortion, temperature, torque sensor, etc.),
- MPE3-2 that characterizes the control system will depend on:
  - the accuracy of the achievement of the specified acceleration (at the reference point) U(MPE3-2-1),
  - + U(MPE3-2-1) effect of influencing variables (bias, etc.), such as distortion of the sinusoidal vibration signal; the current status in world-class laboratories is to ignore U(MPE3-2-1).

Therefore, it will be necessary to assign a value to each elementary MPE mentioned above so that the combination of uncertainty conforms to the overall requirement of $MPE3_{sine}$, which should be limited to 5%. Here, we already see the difficulty of maintaining this objective.

(b) $MPE3_{noise}$

The process outlined above for $MPE3_{sine}$ should be applied in the same way for $MPE3_{noise}$.

$MPE3_{noise}$ is linked to a measuring device that is capable of ensuring a tolerance of 10% difference between the specified effective value and the effective value achieved in noise.

$MPE3_{noise}$ should be limited to 3% (one-third of the 10% that are tolerated for the difference between the achieved and the specified RMS of the vibration level).

We also break $MPE3_{noise}$ down into:

- MPE3-1 on the knowledge of the acceleration given by the ACC + U(MPE3-1), which represents the effect of influencing quantities (such as transversal displacement);

- MPE3-2 the ability of the control system to obtain feedback from the specified vibration + U(MPE3-2), which represents the effect of influencing quantities (the way to do this in world-class laboratories is to ignore U(MPE3-2)).

Let us evaluate these two components:

- MPE3-1 on the knowledge of the acceleration given by the accelerometric channel of the amplitude of the acceleration measurement will depend on:

  - changes in sensitivity with changing frequency in the test frequency range (for example 5 to 2000 Hz), which is denoted MPE3-1-1,
  - the accuracy of calibration at the reference point at 100 Hz, which is denoted MPE3-1-2,
  - + U(MPE3-1-2), which is the effect of influencing quantities (transversal displacement, temperature, sensors, torque, etc.).

NB: As calibrating the accelerometer channel used for noise control is done in sinusoidal mode, the above values are estimated from the results of the calibration carried out in sinusoidal vibration.

- MPE3-2 that characterizes the control system will depend on:

  - the accuracy of achieving the specified acceleration (at the reference point),
  - + U(MPE3-2) effect of influencing quantities, the current way to do this in world-class laboratories is to ignore U(MPE3-2).

It will then be necessary to assign a value to each elementary MPE so that the accuracy combination is consistent with $MPE3_{noise}$, which should be limited to 3%.

NB: SC coupled with the VGA.

Point 2: In the vibration control loop, there is particular concern for the aforementioned fact that the acceleration measurement is "part" of the control process and is simultaneously a "judge" of the acceleration measurement. The consequence, for example, for electrical noise that occurs due to the presence of closed loop circuits is that the electrical noise will not be dissociated from the real-time vibration signal.

It is therefore necessary to periodically (close to annually) compare the acceleration measured by the sensor used to control vibration to a measurement carried out by an independent acceleration measurement channel. The requirement for verification is in a random (one or two control points at the extremes within the normal range, whether this be on the overall RMS value and/or DSP constant) and sinusoidal order (some frequency points with a power supply noise frequency at 50 Hz in Europe and also at the two extremes of the amplitude range that is usually covered); these checks must be carried out for the usual control strategy.

This leads to: MPE3-3-sine + U(MPE3-3-sine)and MPE3-3-noise + U(MPE3-3-noise).

b) Conclusion

MPE allowance therefore aims to determine the final MPE that corresponds to the tolerance of the standard method, taking the measurement uncertainties into account in the compliance declaration, from the composition of the lower levels of MPE that have been previously determined. This is a MPE allowance (MPE can be used to express the accuracy of the equipment).

The accuracy of the measurement results from a combination of the determined MPE and the uncertainties connected to them, with the total being smaller than the tolerance of the process in question. We express the ratio between the two by the capability (in theory, usually 3 or more and up to 6 for certain standards and methods). The capability, as has already been described, is the ratio between tolerance T to be verified and the confidence interval CI of the process or the equipment used C = T/IC. The higher this ratio, the greater the probability (chance) that a result is "compliant".

The approach can be outlined as follows:

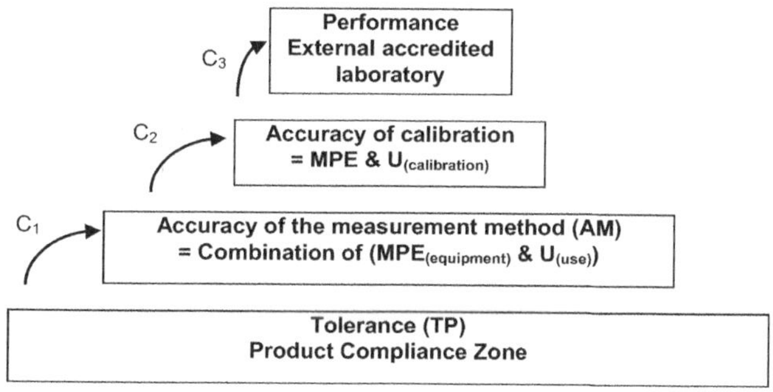

Let us note that the usual approach for estimating uncertainties within a process considers, at every step, the uncertainties of the various equipment and measuring devices that have uncertainties related to associated influencing factors (which is actually the opposite approach). This allows us to also obtain get an overall uncertainty that can then be compared to the tolerance of the process or its MPE. The MPE allowance, however, allows its contribution to the final MPE to be determined at each stage of the process; we can then use this information for further analysis on other measuring and testing methods or we can consider higher performing equipment to meet more demanding needs.

Let us also note that it is always possible to determine the level generated by the system from an external measuring device (this can be the environmental witness, for example), which is of course calibrated by an accredited laboratory, which helps remove doubt about the approach used.

Appendix I

List of Testing Methods in Standard Environments

Tests defined by the DO 160 standard

Designation	Section
Temperature and altitude	4-0
Temperature changes	5-0
Humidity	6-0
Shocks during operation and survivability	7-0
Vibrations	8-0
Explosion-proof tests	9-0
Impermeability to water	10-0
Resistance to fluids	11-0
Sand and dust	12-0
Resistance to fungus	13-0
Salt spray	14-0

Various tests

	Reference texts	
Designation	**Afnor**	**Other**
Resistance to earthquakes, biaxial test by accelerogram		EDF HN 20E53
Resistance to the Class 1E reference accident equipment for nuclear power plants		EDF HM/63-7195/8
Resistance to aging and earthquake		RCC.E
Natural aging	NFT 51165	
Aging test for Class 1E equipment for nuclear power plants		IEEE: STD 323
Resistance to Class 1E equipment for nuclear power plants		IEEE: STD 344
Resistance to earthquakes		UTE C20-420

Tests defined by the GAM EG13 standard

Designation	Method
High temperature	302
Low temperature	303
Thermal shocks	304
Solar radiation	305
Humid heat	306
Immersion	307
Mold	308
Salt spray	309
Rain/sealing	310
Ice	311
Low pressure	312
Sand and dust	313
Fluid contamination	314
Frost	315
Explosive atmosphere	316
Temperature, humidity, altitude	317
Vibrations, temperature, humidity, altitude	318
Acidic atmosphere	319
Vibrations	401
Acoustics	402
Classical shocks	403
Constant acceleration	404
Canon shot	405
Unsecured cargo	406
Stowing of equipment	407
Transport of large sets	408
Hoisting of equipment	409
Stacking of equipment	410
Flexion of equipment	411
Torsion of equipment	412
Acoustics, temperature, vibrations	413
Handling	414
Pyrotechnic shocks	415
Rail impact	416
Shocks in response spectrum	417
Platform displacement	418
Underwater explosion	419
Buffeting vibrations	420
Multi-input vibrations and shocks	421
Ballistic shock	422

60068-1	Environmental testing - Part 1 : General and guidance	10/2013
60068-2-1	Environmental testing - Part 2-1 : Tests - Tests A : Cold	03/2007
60068-2-2	Environmental testing - Part 2-2 : Tests - Tests B : Dry Heat	07/2007
60068-2-5	Environmental testing - Part 2-5 : Tests - Test Sa : Simulated solar radiation at ground level and guidance for solar radiation testing	04/2010Corrigendum 12/2010
60068-2-6	Environmental testing - Part 2-6 : Tests - Test Fc : Vibration (sinusoidal)	12/2007
60068-2-7	Basic environmental testing procedures - Part 2-7 : Tests - Test Ga and guidance : Acceleration, steady state	01/1983 Amendment 1 : 06/1986
60068-2-10	Environmental testing - Part 2-10 : Tests - Test J and guidance : Mould growth	06/2005
60068-2-11	Environmental testing - Part 2 : Tests - Test Ka : Salt mist	01/1981 Corrigendum 12/1999
60068-2-13	Environmental testing - Part 2 : Tests - Test M : Low air pressure	01/1983
60068-2-14	Environmental testing - Part 2-14 : Tests - Test N : Change of temperature	01/2009
60068-2-17	Basic environmental testing procedures - Part 2-17 : Tests - Test Q : Sealing	07/1994
60068-2-18	Environmental testing - Part 2-18 : Tests - Test R and guidance : Water	10/2000
60068-2-20	Environmental testing - Part 2-20 : Tests - Test T : Test methods for solderability and resistance to soldering heat of devices with leads	07/2008
60068-2-21	Environmental testing - Part 2-21 : Tests - Test U : Robustness of terminations and integral mounting devices	06/2006
60068-2-27	Environmental testing - Part 2-27 : Tests - Test Ea and guidance : Shock	02/2008
60068-2-30	Environmental testing - Part 2-30 : Tests - Test Db : Damp Heat, cyclic (12 h + 12 h cycle)	08/2005
60068-2-31	Environmental testing - Part 2-31 : Tests - Test Ec : Rough handling shocks, primarily for equipment - type specimens	05/2008
60068-2-38	Environmental testing - Part 2-38 : Tests - Test Z/AD : Composite temperature/humidity cyclic test	01/2009
60068-2-39	Environmental testing - Part 2 : Tests - Test Z/AMD : Combined sequential cold, low air pressure and damp heat test	01/1976
60068-2-40	Environmental testing - Part 2 : Tests - Test Z/AM : Combined cold/low air pressure tests	01/1976 Amendment 1 : 01/1983
60068-2-41	Environmental testing - Part 2 : Tests - Test Z/BM : Combined dry heat/low air pressure tests	01/1976 Amendment 1 : 01/1983
60068-2-42	Environmental testing - Part 2-42 : Tests - Test Kc : Sulphur dioxide test for contacts and connections	05/2003
60068-2-43	Environmental testing - Part 2-43 : Tests - Test Kd : Hydrogen sulphide test for contacts and connections	05/2003
60068-2-44	Environmental testing - Part 2 : Tests - Guidance on test T : Soldering	01/1995 Corrigendum 08/1995
60068-2-45	Environmental testing - Part 2 : Tests - Test XA and Guidance : Immersion in cleaning solvents	01/1980 Corrigendum 09/1981 Amendment 1 : 02/1993
60068-2-46	Environmental testing - Part 2 : Tests Guidance to Test Kd : Hydrogen sulphide test for contacts and connections	01/1982
60068-2-47	Environmental testing - Part 2-47 : Test - Mounting of specimens for vibration, impact and similar dynamic tests	04/2005

60068-2-49 Environmental testing - Part 2 : Tests Guidance to Test Kc : Sulphur dioxide test for contacts and connections	01/1983
60068-2-52 Environmental testing - Part 2 : Tests - Test Kb : Salt mist, cyclic (sodium chloride solution)	02/1996 Corrigendum 07/1996
60068-2-53 Environmental testing - Part 2-53 : Tests and guidance – Combined climatic (temperature/humidity) and dynamic (vibration/shock) tests	04/2010
60068-2-54 Environmental testing - Part 2-54 : Tests - Test Ta : Solderability testing of electronic components by the wetting balance method	04/2006
60068-2-55 Environmental testing - Part 2-55 : Tests - Test Ee and guidance : loose cargo testing including bounce	02/2013
60068-2-57 Environmental testing - Part 2-57 : Tests - Test Ff : Vibration - Time-history and sine-beat method	04/2013
60068-2-58 Environmental testing - Part 2-58 : Tests - Test Td : Test Methods for solderability, resistance to dissolution of metallization and to soldering heat of surface mounting devices (SMD)	07/2004
60068-2-60 Environmental testing - Part 2 : Tests - Test Ke : Flowing mixed gas corrosion test	12/1995
60068-2-61 Environmental testing - Part 2 : Test Methods Test Z/ABDM : Climatic sequence	07/1991
60068-2-64 Environmental testing - Part 2-64 : Tests - Test Fh : Vibration, broadband random and guidance	04/2008
60068-2-65 Environmental testing - Part 2-65 : Tests - Test Fg : Vibration, Acoustically induced method	02/2013
60068-2-66 Environmental testing - Part 2 : Test Methods - Test Cx : Damp heat, steady state (unsaturated pressurized vapour)	06/1994
60068-2-67 Environmental testing - Part 2 : Tests - Test Cy : Damp heat, steady state, accelerated test primarily intended for components	12/1995
60068-2-68 Environmental testing - Part 2 : Tests - Test L : Dust and sand	08/1994
60068-2-69 Environmental testing - Part 2-69 : Tests - Test Te : Solderability testing of electronic components for surface mounting devices (SMD) by the wetting balance method	05/2007
60068-2-70 Environmental testing - Part 2 : Tests - Test Xb : Abrasion of markings and letterings caused by rubbing of fingers and hands	12/1995
60068-2-74 Environmental testing - Part 2 : Tests - Test Xc : Fluid contamination	06/1999
60068-2-75 Environmental testing - Part 2-75 : Tests - Test Eh : Hammer tests	08/1997
60068-2-77 Environmental testing - Part 2-77 : Tests - Test 77 : Body strength and impact shock	01/1999
60068-2-78 Environmental testing - Part 2-78 : Tests - Test Cab : Damp heat, steady state	10/2012
60068-3-1 Environmental testing - Part 3-1 : Supporting documentation and guidance – Cold and dry heat tests	08/2011
60068-3-2 Environmental testing - Part 3 : Background information - Section Two : Combined temperature/low air pressure tests	01/1976
60068-3-3 Environmental testing - Part 3 : Guidance Seismic test methods for equipment	02/1991
60068-3-4 Environmental testing - Part 3-4 : Supporting documentation and guidance - Damp heat tests	08/2001
60068-3-5 Environmental testing - Part 3-5 : Supporting documentation and guidance - Confirmation of the performance of temperature chambers	08/2001
60068-3-6 Environmental testing - Part 3-6 : Supporting documentation and guidance - Confirmation of the performance of temperature/humidity chambers	08/2001
60068-3-7 Environmental testing - Part 3-7 : Supporting documentation and guidance - Measurements in temperature chambers for tests A and B (with load)	08/2001
60068-5-2 Environmental Testing Part 5: Guide to drafting of test methods - Terms and definitions	12/1990

Extracted from the "Report of expert Group 8 'Environmental engineering', CEN workshop 10, April 2011"

Test	Reference	Version	Aim	Type	Language	Link
Vibration	STANAG 4370 AECTP 401	Edition 3	Vibration test procedures for sine, sine sweep, broad band random and narrow band random plus combinations of these.	NSA	English & French	Download, free of charge, from www.nato.int/cps/en/natolive/stanag.htm
Vibration (incl. combined with temperature and/or humidity)	STANAG 4370 AECTP 401	Edition 3	Vibration test procedures when the vibration is applied in-conjuction with temperature and/or humidity.	NSA	English & French	Download, free of charge, from www.nato.int/cps/en/natolive/stanag.htm
Gunfire	Recommend use of Time History Replication or Shock Response Spectra tests	NA	The simulation of vibration and shock conditions induced by gunfire. This is no longer considered a separate test procedure.	NA	NA	NA
Time History Replication (THR)	Def Stan 00-35 Part 3 Chapter 2-19 (M19)	Issue 4	A test procedure to replicates a specified time history using conventional vibration test equipment. May be used to replicate transients (i.e. gunfire or catapult launch), short duration vibration (i.e. buffet) or continuous vibration (i.e. a measured record).	D Stan (UK)	English	Download, free of charge, from www.dstan.mod.uk/00e.php
Acoustic test Reverberant Chamber	STANAG 4370 AECTP 402	Edition 3	A test procedure to apply acoustic excitation (mostly broad band random) using a Reverberant Chamber. Usually used to simulate unsteady pressures associated with aerodynamic flow regimes.	NSA	English & French	Download, free of charge, from www.nato.int/cps/en/natolive/stanag.htm
Acoustic test Progressive Wave Tube	STANAG 4370 AECTP 402	Edition 3	A test procedure to apply acoustic excitation (mostly broad band random) using a Progressive Wave Tube. Usually used to simulate unsteady pressures associated with aerodynamic flow regimes.	NSA	English & French	Download, free of charge, from www.nato.int/cps/en/natolive/stanag.htm

Acoustic test Combined With Temperature & Vibration	STANAG 4370 AECTP 402	Edition 3	A test procedure to apply acoustic excitation (mostly broad band random) combined with Temperature & Vibration. Usually used to simulate conditions associated experienced by air carried missiles and stores.	NSA	English & French	Download, free of charge, from www.nato.int/cps/en/natolive/stanag.htm
Buffet	Def Stan 00-35 Part 3 Chapter 2-01 (M1)	Issue 4	The simulation of transient vibration conditions arising from aircraft Buffet. This is no longer considered a separate test procedure.	D Stan (UK)	English	Download, free of charge, from www.dstan.mod.uk/00e.php
Multi - Exciter Vibration	Mil Std 810 Procedure 527	Issue G	A test procedure which can be used with vibration tests requiring Multi - Exciters to be used for example to test e. long thin equipment such missiles or large equipment such as large missiles or vehicles.	DoD (US)	English	Download, free of charge, from www.dsp.dla.mil (registration necessary)
Classical waveform shock - basic shock pulses	EN IEC 60068-2-27 Part 2-27 Test Ea and Guidance, Shock	2009	A traditional test procedure to allow Classical waveform shocks to be applied to equipment. Shock include half sine pulses, trailing edge saw tooth and trapezoidal pulses.	IEC, CEN and most National Standards Agencies	English, French and most national languages	Purchase from www.cen.eu or www.iec.ch or national standards agency
Classical waveform shock - Decaying Sinusoids	Def Stan 00-35 Part 3 Chapter 2-03 (M3)	Issue 4	A traditional test procedure to allowing the application of Decaying Sinusoids.	D Stan (UK)	English	Download, free of charge, from www.dstan.mod.uk/00e.php
Drop	EN IEC 60068-2-31 Part 2-31 Test Ec Drop & Topple and EN IEC 60068-2-32 Part 2-32 Test Ed Free Fall	1993 & 2008	A traditional test procedure to allow the simulation of drops and related impacts.	IEC, CEN and most National Standards Agencies	English, French and most national languages	Purchase from www.cen.eu or www.iec.ch or national standards agency
Safety Drop Test for munitions or dangerous goods	STANAG 4375	Issue 2	A special case of the traditional drop test to evaluate the Safety of munitions or dangerous goods.	NSA	English & French	Download, free of charge, from www.nato.int/cps/en/natolive/stanag.htm
Shock Response Spectra (SRS) shock	No procedure recommended. Although versions exist none adequately control the test to ensure repeatability of testing.	NA	A test procedure which allows shocks to be applied and controlled using a Shock Response Spectra (SRS) definition.	NA	NA	NA

Pyrotechnic shock (shock originating from operation of small explosive devices)	STANAG 4370 AECTP 415	Edition 3	The simulation of Pyrotechnic shocks (i.e. shock originating from operation of small explosive devices). This is no longer considered a separate test procedure.	NSA	English & French	Download, free of charge, from www.nato.int/cps/en/natolive/stanag.htm
Rail impact	STANAG 4370 AECTP 416	Edition 3	The simulation of Rail impact conditions occurring when shunting of carriages occurs. This is no longer considered a separate test procedure.	NSA	English & French	Download, free of charge, from www.nato.int/cps/en/natolive/stanag.htm
UNDEX test' ship or submarine shock due to nearby non-contact or underwater explosion	STANAG 4370 AECTP 419	Edition 3	The simulation of nearby non-contact or underwater explosion applied to ships or submarines.	NSA	English & French	Download, free of charge, from www.nato.int/cps/en/natolive/stanag.htm
Ballistic shock	STANAG 4370 AECTP 422	Edition 3	The simulation of high intensity Ballistic shocks	NSA	English & French	Download, free of charge, from www.nato.int/cps/en/natolive/stanag.htm
Catapult Launch / Arrested Landing	Def Stan 00-35 Part 3 Chapter 2-01 (M1)	Issue 4	The simulation of Catapult Launcher Arrested Landing. This is no longer considered a separate test procedure.	D Stan (UK)	English	Download, free of charge, from www.dstan.mod.uk/00e.php
Bump (applicable to components only)	EN IEC 60068-2-31 Part 2-31 Test Eb and Guidance, Bump	1993	A traditional test procedure used to apply a sequence of simple shock pulses to demonstrate ruggedness. Should only be for components and even then better reliability ruggedness improvement methods are available.	IEC, CEN and most National Standards Agencies	English, French and most national languages	Purchase from www.cen.eu or www.iec.ch or national standards agency
Constant acceleration	STANAG 4370 AECTP 404	Edition 3	A traditional test procedure used to apply quasi-static acceleration loadings on equipment usually using a centrifuge. Today mostly undertaken by analysis.	NSA	English & French	Download, free of charge, from www.nato.int/cps/en/natolive/stanag.htm

Bounce/loose cargo	STANAG 4370 AECTP 406	Edition 3	A traditional test procedure to replicate the effects of loose cargo bouncing on a cargo bed. The test is limited by the test equipment which has little ability to control the severities. The use of vibration testing has largely replaced this test and modern vehicles have mostly eliminated the occurrence of bounce.	NSA	English & French	Download, free of charge, from www.nato.int/cps/en/natolive/stanag.htm
Materiel tie-down	No procedure can be currently recommended	NA	A test to apply tie-down loading to equipment. Today mostly undertaken by analysis.	NA	NA	NA
Motion Platform	No procedure can be currently recommended	NA	A test to investigate the effects on equipment of platform motion. Today commonly undertaken by analysis.	NA	NA	NA
Large assembly transport (use of test tracks instead of laboratory tests)	STANAG 4370 AECTP 408	Edition 3	A procedure to allow large assemblies to be transported across test tracks instead of laboratory tests. Test tracks allow increased severities (surface and vehicle speed) to be adopted.	NSA	English & French	Download, free of charge, from www.nato.int/cps/en/natolive/stanag.htm
Materiel, lifting, stacking, bending and racking	STANAG 4370 AECTP 409, 410, 411 & 412	Edition 3	A traditional tests to demonstrate packaged equipment is able to be lifted, stacked, withstanding bending loads and racking.	NSA	English & French	Download, free of charge, from www.nato.int/cps/en/natolive/stanag.htm
High Temperature – Constant (for COTS equipment)	EN IEC 60068-2-2 Part 2-2 Test B Dry Heat	2007	A test procedure to subject equipment to constant High Temperature useful for proving survival or function particularly for COTS equipment. The constant high temperature test can rapidly degrade energetic materials.	IEC, CEN and most National Standards Agencies	English, French and most national languages	Purchase from www.cen.eu or www.iec.ch or national standards agency

High Temperature – diurnal or cyclic	STANAG 4370 AECTP 302	Edition 3	A test procedure to subject equipment to diurnal or cyclic High Temperature conditions. These are more closely aligned with actual conditions. This is more applicable to defence equipment that constant temperature.	NSA	English & French	Download, free of charge, from www.nato.int/cps/en/natolive/stanag.htm
Low Temperature – Constant (for COTS equipment)	EN IEC 60068-2-1 Part 2-1 Test A Cold	2007	A test procedure to subject equipment to constant Low Temperature useful for proving survival or function particularly for COTS equipment.	IEC, CEN and most National Standards Agencies	English, French and most national languages	Purchase from www.cen.eu or www.iec.ch or national standards agency
Low Temperature - diurnal or cyclic	STANAG 4370 AECTP 303	Edition 3	A test procedure to subject equipment to diurnal or cyclic Low Temperature conditions. These are more closely aligned with actual conditions. This is more applicable to defence equipment that constant temperature.	NSA	English & French	Download, free of charge, from www.nato.int/cps/en/natolive/stanag.htm
Solar Radiation *(encompassing thermal and actinic degradation)*	STANAG 4370 AECTP 305	Edition 3	A test procedure to apply Solar Radiation to equipment. This may be used to evaluate thermal or actinic degradation.	NSA	English & French	Download, free of charge, from www.nato.int/cps/en/natolive/stanag.htm
Thermal Shock	Def Stan 00-35 Part 3 Chapter 3-03 (CL3)	Issue 4	A test procedure to apply a Thermal Shock to equipment. Mostly used for airborne equipment but can be used to simulate moving land equipment from warn to cold area (indoors / outdoors or vice versa), although with smaller temperature change.	D Stan (UK)	English	Download, free of charge, from www.dstan.mod.uk/00e.php
Humidity – constant cyclic and aggravated	STANAG 4370 AECTP 306	Edition 3	A test procedure to evaluate the effects of high Humidity using either a constant, cyclic or aggravated cycle.	NSA	English & French	Download, free of charge, from www.nato.int/cps/en/natolive/stanag.htm

Pressure (at altitude as well as normal, extreme and abnormal pressure change during air transport/ carriage)	STANAG 4370 AECTP 312	Edition 3	A test procedure to evaluate the effects of pressure, mostly low pressure arising from operation at high altitude as well as normal, extreme and abnormal pressure change during air transport and carriage.	NSA	English & French	Download, free of charge, from www.nato.int/cps/en/natolive/stanag.htm
Combined Temperature, Humidity Altitude (usually associated air carriage) - General Procedure	STANAG 4370 AECTP 317	Edition 3	A procedure combining the application of Temperature, Humidity and Pressure. Usually associated with air carriage. This is a general procedure useful when failure modes are not known.	NSA	English & French	Download, free of charge, from www.nato.int/cps/en/natolive/stanag.htm
Combined Temperature, Humidity Altitude (usually associated air carriage) - Specific Applications	EN IEC 60068-2-39 Part 2-39 Test Z/AMD Combined sequential cold, low air pressure and damp heat test	1999	A procedure combining the application of Temperature, Humidity and Pressure. Usually associated with air carriage. This procedure is more useful when specific aspects of the effects need to be investigated.	IEC, CEN and most National Standards Agencies	English, French and most national languages	Purchase from www.cen.eu or www.iec.ch or national standards agency
Icing (not including impact of ice on aircraft and missiles in flight)	STANAG 4370 AECTP 311	Edition 3	This test procedure can be used to investigate survival and operation of equipment when subject to ice build up. However, it does not include impact of ice on aircraft and missiles in flight.	NSA	English & French	Download, free of charge, from www.nato.int/cps/en/natolive/stanag.htm
Freeze / Thaw	STANAG 4370 AECTP 315	Edition 3	A test to investigate to damage induced by water freezing and thawing in a cyclic manner.	NSA	English & French	Download, free of charge, from www.nato.int/cps/en/natolive/stanag.htm
Immersion (including accidental immersion and vehicle fording) - Fording	STANAG 4370 AECTP 307	Edition 3	A test procedure to investigate the effects of immersion on equipment. This includes procedure is suitable for reproducing immersion occurring as a result of a vehicle fording. A river, water course or sea.	NSA	English & French	Download, free of charge, from www.nato.int/cps/en/natolive/stanag.htm

Immersion (including accidental immersion and vehicle fording) - Immersion	Def Stan 00-35 Part 3 Chapter 3-29 (CL29)	Issue 4	A test procedure to investigate the effects of Immersion on equipment. This includes procedure is suitable for reproducing the effects of accidental Immersion.	D Stan (UK)	English	Download, free of charge, from www.dstan.mod.uk/00e.php
Mould growth	STANAG 4370 AECTP 308	Edition 3	A test procedure to investigate whether a material is not degraded by mould growing on and within it. Usually undertaken on material samples only and only when resistance cannot be established from existing databases.	NSA	English & French	Download, free of charge, from www.nato.int/cps/en/natolive/stanag.htm
Salt spray	Def Stan 00-35 Part 3 Chapter 4-02 (CN2)	Issue 4	A test procedure to investigate the effects of salt fog on equipment. The test may be used to demonstrate functionality or resistance to corrosion. However, slightly different tests may be required for these two different uses. This is an accelerated test difficult to relate to real life durations.	D Stan (UK)	English	Download, free of charge, from www.dstan.mod.uk/00e.php
Rain and water tightness	STANAG 4370 AECTP 310	Edition 3	A test procedure to investigate the effects of Rain and to establish rain proofness and water tightness	NSA	English & French	Download, free of charge, from www.nato.int/cps/en/natolive/stanag.htm
Condensation and Dripproofness.	Def Stan 00-35 Part 3 Chapter 3-28 (CL28)	Issue 4	A test procedure to investigate the effects of Condensation and Dripproofness. Usually used for equipment located in enclosed and partly enclosed locations.	D Stan (UK)	English	Download, free of charge, from www.dstan.mod.uk/00e.php
Sand and dust	STANAG 4370 AECTP 313	Edition 3	A test procedure to investigate the erosive, penetrative and clogging effects of sand and dust. The standard test procedure uses chemically inert sand and dust so does not allow the investigation of chemical interactions.	NSA	English & French	Download, free of charge, from www.nato.int/cps/en/natolive/stanag.htm

Contamination by fluids	STANAG 4370 AECTP 314	Edition 3	A test procedure to investigate the chemical interactions between equipment and contaminates.	NSA	English & French	Download, free of charge, from www.nato.int/cps/en/natolive/stanag.htm
Explosive atmosphere (Safety regulations may require a specific (IP) rating rather than this test)	STANAG 4370 AECTP 316	Edition 3	A test procedure to investigate the Explosive atmosphere. However, this test may not be adequate to meet all national regulations relating to explosive atmospheres.	NSA	English & French	Download, free of charge, from www.nato.int/cps/en/natolive/stanag.htm
Acidic atmosphere	STANAG 4370 AECTP 319	Edition 3	A test procedure to investigate the effects of Acidic (usually man made) atmosphere on equipment.	NSA	English & French	Download, free of charge, from www.nato.int/cps/en/natolive/stanag.htm

Appendix J

Control Strategies – Simplified Summary

J.1. Storing the last control spectrum

One final consequence of the length of time of the pretest loop and startup phases is the need to store the last control spectrum before controlled or untimely shutdown.

This memory storage allows us to resume testing at a nominal level, bypassing the pre-equalization and incremental recovery cycles.

Storing and visualizing spectra

In addition to automatically storing the control spectrum, digital systems can also do this for the pilot spectrum and for a number of measurement points by simply calling this function during the test. The visualization may be done at the end of the test or, optionally, during it, but this considerably lengthens the loop time and is therefore not advisable permanently for each loop; however, it is required in order to do a visual check.

The same goes for visualizing the pilot spectrum, which, on some systems, is at the expense of the loop time. The only available information that is compatible with the maximum speed of equalization is, in this case, the RMS value of the signal in the range of vibrations.

"Fully digital" systems have a rather limited accuracy that is very difficult to know before having practiced with them. Moreover, they impose a "blind" vibration, so without viewing the spectrum in real time, in order to keep the equalization rate properties.

In order to select a console, the following points should be taken into consideration by the supplier:

- the loop time;
- the equalization time from "n" dB to "k" dB "KDB";
- the signal to real noise ratio;
- the effective bandwidth;
- the accuracy of drawn diagrams and the nature and quality of the information;
- the climb mode at the nominal level;
- the real feedback dynamic.

J.2. Feedback properties in sinusoidal state

(a) Frequency range and resolution

The frequency range of these systems is perfectly compatible with vibration tests since it covers the range from 0.1 to 10 kHz with a resolution of at least one millihertz.

The resolution depends on the number of periods that are taken into account when calculating the fundamental (and incidentally, for each of the harmonic components). The number of harmonic components included will depend on the value of the cut-off frequency, which generally switches during scanning in order to occupy some predefined frequency values.

(b) Scanning speed and resolution

The scanning speed is to be chosen between linear or logarithmic modes with values of up to 1,000 Hz/s or 50 octaves/min, with the resolution being from 10 mHz to 10 Hz for 2 octaves/min. Such values cover most cases.

(c) Output dynamic

This is the maximum dynamic that the machine is able to compensate for in one resonance. It is important to know because it truly reflects the ability of the machine to correct a variation from "n" dB to "p" dB.

(d) Correction speed

It is important to know the correction speed (or compression speed for analog systems) that is linked to the test conditions, a value in dB/s that only means anything if one specifies to how many dB the correction is achieved. The system initiates a corrective action when a level in the test sets off alarms and the correction speed is then the loop time.

(e) Accuracy of the output amplitude

The accuracy of the output amplitude has no significance in itself and a value of 0.1 dB will only be at the expense of the correction speed because, for example, for 50 corrections per second, it would be 50 dB/s, meaning that it is unable to control a somewhat difficult test. Moreover, the need is still there for manual centering of the correction dynamics in order for both peaks and troughs of the loop's transfer function to benefit from it.

(f) Control strategy

The possibilities for choosing the detection mode are the following values, which can be taken as a reference (reminder):

- the overall RMS value;
- the value of the filtered fundamental in $\Delta F/F$ or ΔF constant;
- the average value;
- the peak value.

In addition, the detection mode may depend on the outcome of processing several measurement channels. In this case, the maximum value, the minimum value (that must always be linked to an external safety device that controls the thrust requested from the driver in case of antiresonance) or a generally quadratic average (that minimizes the low level action) of channels, as well as "Notching" (possibility of limiting the level to a value in advance for a particular point of the equipment) can be selected.

(g) Acquisition strategy

Control on the true RMS value is only possible if the sample lets the signal harmonic through for the highest frequency to be considered, otherwise the control is no longer on the RMS value but on the filtered value.

(h) Restitution during testing

The system must be able to provide a certain amount of information on the course of the test, during it and at the end. Thus, the variation of the pilot signal can be visualized with the control signal. This information is also stored in order to make them reappear at the end of the test, especially when the latter is stopped due to exceeding the tolerance limits. Alternatively, one can sometimes visualize the transfer function amplitude and phase between two filtered channels.

(i) Test profile

This profile can currently be composed of at least 30–40 segments that represent a displacement, a speed or an acceleration, and feedback can be received on any one of these parameters:

- the loop time (frequency richness reflected in the return loop);
- the safety aspect.

NB: The correction speed expresses the catch-up that the feedback is able to achieve within a set time. The larger this speed, the more the system is able to catch up with rapid changes.

For pilot and monitoring means, we refer the reader to Appendix 2 of RE.Aero 601 11.

However, it is important to consider that:

- in analog, the reaction time is extremely short and the reactions are extremely brutal (high risk of accidental shock)

- in digital, the response time is increased but switchover to redundant sensors is easy. The reactions of the system are soft and the risk of accidental shock is very low.

J.3. Feedback properties in random state

(a) Frequency range

The first parameter that characterizes the generation of vibrations is the achievable frequency range: this is how one can feed the setup in a range from the first channel to 5,000 Hz, with a certain number of lower values also being feasible such as 10, 50, 100, 250, 500, 1 000 or 2 000 Hz.

However, if the target band is smaller than the band being used, it is necessary to ensure that the system takes the spectral content in the desired frequency range into account, by automatically setting everything that is the outside of this range to zero.

(b) Resolution

This can be defined as the quotient of the frequency range by the number of bands (channels) (typically 641,024). This setting, which is always present in records, must be distinguished from the actual bandwidth that allows for weighting and which can also be different during generation and analysis.

(c) Generated signal

The generated signal may be randomly or pseudo-randomly selected. The first signal is one that is generally implied when one speaks of a "noise test" and is the most used; nevertheless, and especially during "pretest" phases, when the system acquires the transfer function of the setup, it is not uncommon to use the second signal, which involves sending only one line per frequency band, with predetermined levels.

This signal has the advantage of being deterministic and the estimation of the transfer function is done with greater accuracy for a given time. Thus it allows us, by imposing accuracy, to minimize the number of loops necessary for obtaining the characteristics of the feedback channel.

(d) Feedback dynamic

Four different dynamics can be found in a machine:

- a D1 dynamic, which concerns the analysis of the accelerometer signal within a range of voltages that are selected manually or by an autoranging device;

- a D2 dynamic, which is inherent to the feedback channel that includes computing operations and a Fourier transform;

- a D3 dynamic, for digital-analog output conversion, which is also indicated by the manufacturer with a usual value of 60 dB;

- a D4 dynamic, eligible for the reference spectrum.

Let us note that we always have: D4 < D1 < D2 < D3.

Of these four values, the one with the most potential is the D2 dynamic because it truly characterizes the ability of the machine to compensate for irregularities in the spectrum of "n" dB ("n" being equal to D2 at most in the most favorable cases).

(e) Background noise

A natural limitation of the dynamics is usually background noise, but the declared values are difficult to verify since in the absence of any signal, the output is often physically shortcircuited.

J.4. Alarms

(a) Number of segments

The spectrum is defined from a number of steps that involve hinge frequencies (also called characteristic frequencies in some standards) and the corresponding levels in physical units. It also includes barriers which the achieved spectrum must remain within.

(b) Alarm and stoppage per segment

These consist of two sorts: those that trigger an alarm (alarm) and those that stop the system (stoppage).

Alarms can have a different value by excess or by default and be independent from one segment to another.

For highly nonlinear specimens, it is often only possible to apply these safety devices when a significant level is reached (approximately equal to the nominal level). This occurs in order to overcome the challenges of feedback in the phases of ascent.

(c) Alarm and stoppage on the overall RMS value

In random state, a second system quality control mode consists of measuring the overall RMS value. An alarm level and a stoppage level can be, depending on the systems, different by default or by excess.

If the RMS value is calculated within the loop by summing the spectrum or through autocorrelation, this requires waiting at least one loop time before a safety mechanism is triggered, which may be dangerous in some cases.

Let us note that these safety mechanisms are almost always at the expense of the loop time and that regardless, the computer is not able to ensure its own safety.

(d) Other safety features

There are other independent safety features of the loop and their effectiveness is often instantaneous; they concern the loss of the return signal (cable break, accelerometer lift-off), as well as orders chosen by the user that may lead to too large a displacement, an abnormal temperature or a parameter of the equipment under test that falls outside of tolerance levels.

These safety features are completely ineffective in case of failure of the system itself. Therefore, they do not use devices that act directly on the moving coil, such as electromagnetic brakes.

J.5. Control during shocks

(a) Waveform

References can be a sine wave, a saw-tooth wave with initial or final slope, a triangle, a rectangle or a waveform programmed by the user, which presents a great advantage for simulations of earthquakes, explosions, transport, etc.

(b) Amplitude and width

The amplitude and the width may also be selected from a very wide range: from 1/100 g_n to 1,000 g_n and 0.1 ms to 1 s.

(c) Shaking

It is possible to transform shocks into bumps, for which the rate can go up to 10 or so per second. To the extent that the required severity is within the possibilities of the shaker, this excitation mode saves time and is considerably simpler than eccentric or guillotine machines. In order not to excessively tire out the tested material through repeated shocks, the time between each shock depends on the lowest mode of the material and the estimated overvoltage coefficient (maximum of 50 for a mechanical system) so that the next shock occurs with enough attenuation from the preceding one, or with a particular normative requirement. For example, for a low resonance frequency of 30 Hz and a Q value of 50, the duration may be (1/30) × 50 = 1.7 s.

The limitations imposed by low performance of shakers in this field very quickly reduce the extension field of this technique.

(d) Operation synoptic

Initially, the system acquires the transfer function of the setup. To do this, three techniques are possible from different signals: pseudo-random, pulse, real low level shock form.

(e) Amplitude of side lobes

From the moment when the shapes are defined and mainly for that which follows, which is that of acceleration, the difficulty lies in staying within established tolerances.

On all systems, the amplitude of the side lobes is adjustable from 10 to 100% of the shock amplitude; it can even be automatically optimized based on the performance of the shaker. In all cases, before the application of shocks, the system must provide an indication of the displacement and the speed imposed on the movable elements in order to enable the user to check that they comply with the capacities of the setup.

Beware of low-frequency SRS distortions that may overtest or undertest the VLF modes of tested components, see [ref. I18].

(f) Equalizing accuracy

When all the settings necessary for shocks have been defined, they can be directly applied at the nominal level. Indeed, the acquisition of the transfer function is carried out at low level or with a different spectral distribution from that of the shock, which requires updating during the test.

(g) Change in speed

Some standards require measuring the speed change during the impact. If the acceleration is used and is subjected to a double integration, the laboratory should have evaluated the impact of offsets (static and/or dynamic) on the compliance with the specified tolerance.

(h) Restitution

The acquisition of shocks is generally only performed on a single channel; it is therefore only possible to view the shape of the acceleration of the shock that was sent. As the application of the shock is in "open loop", the response from certain points of the equipment under test can only be measured at a rate of one measurement by shock.

Achieving shocks on shakers shows great progress in the field of environmental testing because it significantly simplifies the problems of handling, interfacing and implementation.

However, this technique quickly reaches its limits, not because of generation systems (which carry out their role perfectly), but because of the poor performance of electrodynamic vibration generators from the point of view of maximum acceleration, thrust, speed and especially, maximum permissible displacement.

In some cases, electrohydraulic vibration generators can take over or even freefall impact shock machines. These are the only machines that minimize distortion in VLF of SRS of the theoretical shock, due to the value of –1 g of the pre- and postshock (if the shock has rebound) relative to the magnitude of the main shock. Of course, displacement and fall time is important but it does not penalize the construction of testing means.

Appendix K

Mathematical Elements Involved in the Estimation of Uncertainties (Supplementary)

In this appendix, we propose supplementary simplified formulas that are, in theory, used in the estimation of uncertainties. The books in the bibliography detail these approaches accurately; however, we thought it would be useful to give an overview of the relationships in this appendix by discussing some examples or retrieving information from the literature.

K.1. Calculation of variances, simplified formulas

From a sample of n results x_1, x_2, ..., x_n, we estimate the population variance in the following manner.

(a) For a large sample size, n > 100:

$$V\left(\overline{m_x}\right)=\frac{1}{n}.\sum_{i=1}^{n}\left(x_i-\overline{m_x}^2\right)$$

$$\rightarrow V\left(\overline{m_x}\right)=\frac{1}{n}.\left[\left(x_1^2-2x_1\overline{m_x}+\overline{m_x}^2\right)+\left(x_2^2-2x_2\overline{m_x}+\overline{m_x}^2\right)+....+\left(x_n^2-2x_n\overline{m_x}+\overline{m_x}^2\right)\right]$$

$$\rightarrow V\left(\overline{m_x}\right)=\frac{1}{n}.\left[\left(\sum_{i=1}^{n}x_i^2\right)-\left(2.\overline{m_x}.\sum_{i=1}^{n}x_i\right)+n.\overline{m_x}^2\right]\rightarrow V\left(\overline{m_x}\right)=\frac{\sum_{i=1}^{n}x_i^2}{n}-\frac{2.\overline{m_x}.n.\overline{m_x}}{n}+\overline{m_x}^2$$

$$\boxed{V\left(\overline{m_x}\right)=\frac{\sum_{i=1}^{n}x_i^2}{n}-\overline{m_x}^2}$$

(b) For a small sample size, n < 100:

$$V\left(\overline{m_x}\right)=\frac{1}{n-1}.\sum_{i=1}^{n}\left(x_i-\overline{m_x}^2\right)$$

$$\rightarrow V\left(\overline{m_x}\right)=\frac{1}{n-1}.\left[\left(x_1^2-2x_1\overline{m_x}+\overline{m_x}^2\right)+\left(x_2^2-2x_2\overline{m_x}+\overline{m_x}^2\right)+....+\left(x_n^2-2x_n\overline{m_x}+\overline{m_x}^2\right)\right]$$

$$\rightarrow V\left(\overline{m_x}\right)=\frac{1}{n-1}.\left[\left(\sum_{i=1}^{n}x_i^2\right)-\left(2.\overline{m_x}.\sum_{i=1}^{n}x_i\right)+n.\overline{m_x}^2\right]\rightarrow V\left(\overline{m_x}\right)=\frac{\sum_{i=1}^{n}x_i^2}{n-1}-\frac{2.\overline{m_x}.n.\overline{m_x}}{n-1}+\frac{n.\overline{m_x}^2}{n-1}$$

$$V(\overline{m_x}) = \frac{\sum_{i=1}^{n} x_i^2}{n-1} - \frac{n.\overline{m_x}^2}{n-1}$$

K.2. Problem of average of averages

What is the average m_x for a set of N samples of n_j measurements x_{1j}, x_{2j}, ..., x_{nj} when one only has the averages m_{xj} and sizes n_j of each sample?

NB: The averages m_{xj} of N samples of n_j measurements x_{1j}, x_{2j}, ..., x_{nj} are calculated as follows:

$$\overline{m_{x1}} = \frac{1}{n_1}.\sum_{j=1}^{n_1} x_{1j} \text{ , } \overline{m_{x2}} = \frac{1}{n_2}.\sum_{j=1}^{n_2} x_{2j} \ldots\ldots\ldots\ldots, \overline{m_{xN}} = \frac{1}{n_N}.\sum_{j=1}^{n_N} x_{Nj}$$

and we wish calculate m_x such that: $\overline{m_x} = \dfrac{\sum_{j=1}^{n_1} x_{1j} + \sum_{j=1}^{n_2} x_{2j} + \ldots\ldots + \sum_{j=1}^{n_N} x_{Nj}}{n_1 + n_2 + \ldots + n_N}$

from N equations above, we get:

$$\sum_{j=1}^{n_1} x_{1j} = n_1.\overline{m_{x1}} \text{ , } \sum_{j=1}^{n_2} x_{2j} = n_2.\overline{m_{x2}} \ldots\ldots\ldots\ldots\ldots\ldots \sum_{j=1}^{n_N} x_{Nj} = n_N.\overline{m_{xN}}$$

that is substituted into $\overline{m_x} \to \overline{m_x} = \dfrac{n_1.\overline{m_{x1}} + n_2.\overline{m_{x2}} + \ldots + n_N.\overline{m_{xN}}}{n_1 + n_2 + \ldots + n_N}$

$$\overline{m_x} = \frac{1}{\sum_{j=1}^{N} n_j}.\sum_{j=1}^{N} \left(n_j.\overline{m_{xj}}\right)$$

K.3. Problem of average of variances

What is the variance $V(\overline{m_x})$ of a set of N samples of n_j measurements x_{1j}, x_{2j}, ..., x_{nj} when we only have the averages $\overline{m_{xj}}$ and sizes n_j of each of the samples?

- if N is a large sample size, $n_j > 100$, their variances $V(\overline{m_{xi}})$ are calculated as follows:

$$V(\overline{m_{x_1}}) = \left(\frac{1}{n_1}.\sum_{j=1}^{n_1} x_{1j}^2\right) - \overline{m_{x_1}}^2, V(\overline{m_{x_2}}) = \left(\frac{1}{n_2}.\sum_{j=1}^{n_2} x_{2j}^2\right) - \overline{m_{x_2}}^2 \ldots\ldots\ldots V(\overline{m_{x_N}}) = \left(\frac{1}{n_N}.\sum_{j=1}^{n_N} x_{Nj}^2\right) - \overline{m_{x_N}}^2$$

we seek to calculate $V(\overline{m_x})$ such that:

$$V(\overline{mx}) = \left[\frac{1}{n_1 + n_2 + \ldots + n_N}.\left(\sum_{j=1}^{n_1} x_{1j}^2 + \sum_{j=1}^{n_2} x_{2j}^2 + \ldots\ldots + \sum_{j=1}^{n_N} x_{Nj}^2\right)\right] - \overline{mx}^2$$

from N equations above, we get:

$$\sum_{j=1}^{n_1} x_{1j}^2 = n_1.\left(V\left(\overline{m_{x1}}\right) + \overline{m_{x1}}^2\right), \sum_{j=1}^{n_2} x_{2j}^2 = n_2.\left(V\left(\overline{m_{x2}}\right) + \overline{m_{x2}}^2\right) \ldots\ldots\ldots\ldots \sum_{j=1}^{n_N} x_{Nj}^2 = n_N.\left(V\left(\overline{m_{xN}}\right) + \overline{m_{xN}}^2\right)$$

that is substituted into $V(\overline{m_x})$:

$$\rightarrow V(\overline{m_X}) = \frac{1}{n_1 + n_2 + \ldots + n_N} \cdot \left[n_1\left(V(\overline{m_{X_1}}) + \overline{m_{X_1}}^2 \right) + \ldots + n_N\left(V(\overline{m_{X_N}}) + \overline{m_{X_N}}^2 \right) \right] - \overline{m_X}^2$$

$$V(\overline{m_X}) = \frac{\sum_{j=1}^{N} \left[n_j \cdot \left(V(\overline{m_{Xj}}) + \overline{m_{Xj}}^2 \right) \right]}{\sum_{j=1}^{N} n_j} - \overline{m_X}^2$$

- if N is a small sample size, $n_j < 100$, but is significant, $n_j > 10$, and if the population is large, $Sn_j > 100$, their variances $V(\overline{m_{xj}})$ are calculated as follows:

$$V(\overline{m_{X_1}}) = \left(\frac{1}{n_1 - 1} \cdot \sum_{j=1}^{n_1} x_{1j}^2 \right) - n_1 \cdot \overline{m_{X_1}}^2 ; \ldots\ldots\ldots\ldots ; V(\overline{m_{X_N}}) = \left(\frac{1}{n_N - 1} \cdot \sum_{j=1}^{n_1} x_{Nj}^2 \right) - n_N \cdot \overline{m_{X_N}}^2$$

we seek to calculate $V(\overline{m_x})$ such that:

$$V(\overline{m_X}) = \left[\frac{1}{n_1 + n_2 + \ldots + n_N} \left(\sum_{j=1}^{n1} x_{1j}^2 + \ldots\ldots\ldots + \sum_{j=1}^{nN} x_{Nj}^2 \right) \right] - \overline{m_X}^2$$

from N equations above, we get:

$$\sum_{j=1}^{n1} x_{1j}^2 = (n_1 - 1) \cdot V(\overline{m_{X_1}}) + n_1 \cdot \overline{m_{X_1}}^2 ; \ldots\ldots\ldots ; \sum_{j=1}^{nN} x_{Nj}^2 = (n_N - 1) \cdot V(\overline{m_{X_N}}) + n_N \cdot \overline{m_{X_N}}^2$$

that is substituted into $V(\overline{m_x})$:

$$V(\overline{m_X}) = \frac{1}{n_1 + n_2 + \ldots + n_N} \left\{ \left[(n_1 - 1) V(\overline{m_{X_1}}) + n1 \cdot \overline{m_{X_1}}^2 \right] + \ldots + \left[(n_N - 1) V(\overline{m_{X_N}}) + nN \cdot \overline{m_{X_N}}^2 \right] \right\} - \overline{m_x}^2$$

$$V(\overline{m_X}) = \frac{\sum_{j=1}^{N} \left[(n_j - 1) V(\overline{m_{Xj}}) + n_j \cdot \overline{m_{Xj}}^2 \right]}{\sum_{j=1}^{N} n_j} - \overline{m_X}^2$$

- if N is a small sample size, $n_j < 100$, but is significant, $n_j > 10$, and if the population is large, $Sn_j > 100$, their variances $V(m_{xj})$ are calculated as follows:

$$V(\overline{m_{X_1}}) = \left(\frac{1}{n_1 - 1} \cdot \sum_{j=1}^{n_1} x_{1j}^2 \right) - n_1 \cdot \overline{m_{X_1}}^2 ; \ldots ; V(\overline{m_{X_N}}) = \left(\frac{1}{n_N - 1} \cdot \sum_{j=1}^{n_1} x_{Nj}^2 \right) - n_N \cdot \overline{m_{X_N}}^2$$

we seek to calculate $V(\overline{m_x})$ such that:

$$V(\overline{m_X}) = \frac{1}{(n_1 + n_2 + \ldots + n_N) - 1} \cdot \left[\left(\sum_{j=1}^{n1} x1j^2 + \ldots + \sum_{j=1}^{nN} xNj^2 \right) - \left(n_1 + n_2 + \ldots + nN \right) \cdot \overline{m_X}^2 \right]$$

from N equations above, we get:

$$\sum_{j=1}^{n1} x_{1j}^2 = (n_1 - 1).V\left(\overline{mx_1}\right) + n_1.\overline{mx_1}^2 \;;\ldots\ldots;\; \sum_{j=1}^{nN} x_{Nj}^2 = (n_N - 1).V\left(\overline{mx_N}\right) + n_N.\overline{mx_N}^2$$

that is substituted into $V(m_x)$:

$$\rightarrow V\left(\overline{mx}\right) = \frac{1}{(n_1 + n_2 + \ldots + n_N) - 1}\cdot\left\{\left[(n_1 - 1)V\left(\overline{mx_1}\right) + n_1.\overline{mx_1}^2\right] + \ldots + \left[(n_N - 1)V\left(\overline{mx_N}\right) + n_N.\overline{mx_n}^2\right]\right\}$$
$$- (n_1 + n_2 + \ldots + n_N).\overline{mx}^2$$

$$\boxed{V\left(\overline{mx}\right) = \frac{\sum_{j=1}^{N}\left[(n_j - 1)V\left(\overline{mxj}\right) + n_j.\overline{mxj}^2\right]}{\left(\sum_{j=1}^{N} n_j\right) - 1} - \frac{\sum_{j=1}^{N} n_j}{\left(\sum_{j=1}^{N} n_j\right) - 1}\,\overline{mx}^2}$$

K.4. Calculating the average "on the fly"

From a sample of n results $x_1, x_2, \ldots, x_n$, we only have the average m_{xn} and the sample size n. We want calculate the new average $m_{xn+1}$from a new result x_{n+1}.

The average m_{xn} is calculated as follows: $\overline{m_{x_n}} = \frac{\sum_{i=1}^{n} x_j}{n} \rightarrow \sum_{i=1}^{n} x_j = n.\overline{m_{x_n}}$

and we wish to calculate m_{xn+1} such that: $\overline{m_{x_{n+1}}} = \frac{\sum_{i=1}^{n+1} x_j}{n+1} \rightarrow \overline{m_{x_{n+1}}} = \frac{(x_{n+1}) \sum_{i=1}^{n+1} x_j}{n+1}$

$$\boxed{\overline{m_{x_{n+1}}} = \frac{(x_{n+1}) + n.\overline{m_{x_n}}}{n+1}}$$

This prevents the need to accumulate all measurement values in order to calculate the average over continuous measurements.

K.5. Calculating the variance "on the fly"

From a sample of n results $x_1, x_2, \ldots, x_n$, we only have the average $\overline{m_{xn}}$, variance $V(\overline{m_{xn}})$ and sample size n. We want calculate the new variance $V(\overline{m_{xn+1}})$ from a new result x_{n+1}.

For a large sample size, the variance $V(\overline{m_{xn}})$ is calculated in the following manner:

$$V\left(\overline{mx_n}\right) = \frac{\sum_{i=1}^{n} x_j^2}{n} - \overline{m_{x_n}}^2 \rightarrow \sum_{i=1}^{n} x_j^2 = n.\left[V\left(\overline{mx_n}\right) + \overline{m_{x_n}}^2\right]$$

and we wish to calculate $V(\overline{m_{x_{n+1}}})$ such that:

$$V\left(\overline{m_{x_{n+1}}}\right)=\frac{\sum_{i=1}^{n+1} x_j^2}{n+1}-\overline{m_{x_{n+1}}}=\frac{x_{n+1}^2+\sum_{i=1}^{n+1} x_j^2}{n+1}-\overline{m_{x_{n+1}}}^2$$

$$=\frac{x_{n+1}^2+n.\left[V\left(\overline{m_{x_n}}\right)+\overline{m_{x_n}}^2\right]}{n+1}-\overline{m_{x_{n+1}}}^2$$

On the other hand: $\overline{m_{x_{n+1}}}=\dfrac{\left(x_{n+1}\right)+n.\overline{m_{x_n}}}{n+1}$

$$\rightarrow V\left(\overline{m_{x_{n+1}}}\right)=\frac{\left(x_{n+1}^2\right)+n.\left[V\left(\overline{m_{x_n}}\right)+\overline{m_{x_n}}^2\right]}{n+1}-\left[\frac{\left(x_{n+1}\right)+n.\overline{m_{x_n}}}{n+1}\right]^2$$

$$\rightarrow V\left(\overline{m_{x_{n+1}}}\right)=\frac{1}{n+1}.\left(x_{n+1}^2\right)+\frac{n}{n+1}.V\left(\overline{m_{x_n}}\right)+\frac{n}{n+1}.\overline{m_{x_n}}^2-\frac{1}{(n+1)^2}.\left(x_{n+1}^2\right)$$
$$-\frac{2n}{(n+1)^2}.x_{n+1}.\overline{m_{x_n}}-\frac{n^2}{(n+1)^2}.\overline{m_{x_n}}^2$$

$$\rightarrow V\left(\overline{m_{x_{n+1}}}\right)=\frac{(n+1)}{(n+1)^2}.\left(x_{n+1}^2\right)+\frac{n(n+1)}{(n+1)^2}.V\left(\overline{m_{x_n}}\right)+\frac{n(n+1)}{(n+1)^2}.\overline{m_{x_n}}^2-\frac{1}{(n+1)^2}.\left(x_{n+1}^2\right)$$
$$-\frac{2n}{(n+1)^2}.x_{n+1}.\overline{m_{x_n}}-\frac{n^2}{(n+1)^2}.\overline{m_{x_n}}^2$$

$$\rightarrow V\left(\overline{m_{x_{n+1}}}\right)=\frac{n}{(n+1)^2}.\left(x_{n+1}^2\right)+\frac{n(n+1)}{(n+1)^2}.V\left(\overline{m_{x_n}}\right)+\frac{n}{(n+1)^2}.\overline{m_{x_n}}^2-\frac{2n}{(n+1)^2}.x_{n+1}.\overline{m_{x_n}}$$

$$\rightarrow V\left(\overline{m_{x_{n+1}}}\right)=\frac{n}{(n+1)^2}.\left[(n+1).V\left(\overline{m_{x_n}}\right)+\left(\overline{m_{x_n}}^2-2.\overline{m_{x_n}}.x_{n+1}+x_{n+1}^2\right)\right]$$

$$\boxed{V\left(\overline{m_{x_{n+1}}}\right)=\frac{n}{(n+1)}.\left[V\left(\overline{m_{x_n}}\right)+\frac{\left(\overline{m_{x_n}}-x_{n+1}\right)^2}{n+1}\right]}$$

This prevents the need to accumulate all the measurement values in order to calculate the variance over continuous measurements.

In this demonstration, it is assumed that the sample size was greater than 100 (which is generally the case for “on the fly” calculations).

K.6. Testing normality of distribution, elimination of outliers

Normality Test, extract from CERESTA [ref. 6].

Significance test	III – Non-parametric tests	2. Fitness test

The *fitness test* involves defining a *decision rule* on the validity of a hypothesis concerning the *overall agreement of an empirical distribution with a theoretical distribution* that is in theory specified or adjusted for the observations (from the decision theory).

1. Terms of use

- A random sample is taken from an unknown population (A) where n independent observations are divided into k classes.
- A type of theoretical reference law of distribution (T) is chosen for which the parameters can in theory be fixed or estimated through an effective method.

2. Criterion

$$\hat{\chi}^2 = \sum_i^k \frac{(O_i - T_i)^2}{T_i}$$

where O_i and T_i, respectively, refer to each of the k classes, the observed and expected numbers for the reference distribution.

NB: This method assumes that none of the T_i numbers are less than 5. If necessary, we will regroup the classes and ultimately retain the number of classes actually constituted for the value of k in order to validly apply the definition of the criterion.

3. Test interpretation

"α" is the risk of the first test case and $\hat{\chi}^2(\nu)$ is the Pearson variable, where:

$$\nu = k - 1 - r$$

for the setting and where *r* is the number of the reference law parameters that were estimated using the observations. We then test:

$$H_0\ (A = T) \text{ versus } H_1\ (A \neq T)$$

And:

$$\hat{\chi}^2 = \sum_i^k \frac{(O_i - T_i)^2}{T_i} < \hat{\chi}_{1-\alpha}(\nu)$$

We then accept or reject H_0.

Example K.1: A random sample of n = 40 independent samples is taken from an unknown population. The values are:

x_i	1	2	3	4	5	6
n_i	1	4	14	12	7	2

We perform the fitness test of the distribution for a normal distribution by adopting the risk of first case α = 0,05.

Elements for calculations

(a) Determining the parameters of the normal distribution that is adjusted to the distribution

$$\bar{x} = \frac{\sum_i n_i x_i}{n} = \frac{146}{40} = 3,65$$

$$s = \sigma = \sqrt{\frac{1}{n-1}\sum_i n_i\left(x_i - \bar{x}\right)^2} = \sqrt{\frac{1}{n-1}\left[\sum_i n_i x_i^2 - \frac{\left(\sum_i n_i x_i\right)^2}{n}\right]} = \sqrt{\frac{582 - 532,9}{39}} = 1,122$$

(b) Determining the theoretical values.

Class limits	$u_i = \frac{(a_i - \bar{x})}{\sigma}$	$F(u_i)$	$D_i = F(u_i) - F(u_i - 1)$	$T = nD_i$	O_i	Classes
$-\infty$	–	0				
1,5	–1,916	0,0277	0,0277	6,11 (})	5	1
2,5	–1,025	0,1527	0,1250			
3,5	–0,134	0,4467	0,2940	11,76	14	2
4,5	0,758	0,7758	0,3791	13,76	12	3
5,5	1,649	0,9504	0,1745	8,97 (})	9	4
$+\infty$	–	1	0,0497			
Total			1,000	40	40	

$\alpha = 0,05$; $\nu = 4 - 1 - 2 = 1$

Test: $$\hat{\chi}^2 = \sum_i \frac{(O_i - T_i)^2}{T_i} = 0,731 < \chi^2_{0,95}(1) = 3,84$$

So we accept the assumption of normality of the population variable.

K.7. Henry line

This is a convenient and fast procedure for verifying the assumption of normality of a population from which a sample is observed.

This graphical procedure is not, strictly speaking, a *statistical test* such as those discussed here. Nevertheless, drawing a *Henry graph* may be sufficient in many cases where we must ensure the normality of a sampling distribution.

The principle of the method is based on the following linear link:

$$u = \frac{x - \bar{x}}{\sigma}$$

between a normal variable X (average $m = \bar{x}$; standard deviation σ) and a reduced variable U.

To facilitate drawing of the points, we use *Gaussian arithmetic* paper for which one of the scales is a *transformed* scale, directly graduated in F(x), the distribution of the normal distribution function.

a) Individual observations in ascending order

$x_1 \leq x_2 \leq ...... \leq x_i \leq .... \leq x_n \quad (n > 10)$

$F_i = \dfrac{i}{n+1}$ is formed and it brings the series of points (x_i, F_i) on to the functional paper.

If, with possible abstraction of extremes, it seems feasible to manually adjust a line around which the points (x_i, F_i) are poorly dispersed, then the *presumption* of normality is confirmed and we can directly determine:

- average m: value of x for F = 0,5;

- standard deviation σ: half the difference between the values of x corresponding to F = 0,84 and F = 0,16.

b) Observations grouped into classes

The values of F used here correspond to *cumulative empirical frequencies* of i associated classes in order to plot the points with the upper limits a_{i+1}. The rest of the procedure is carried out as in the previous case.

Example K.2: Individual series

i	1	2	3	4	5	6	7	8	9	10
F_i %	9	18	27	36	45	55	64	73	82	91
x_i	20	25	27	33	35	40	41	48	52	55

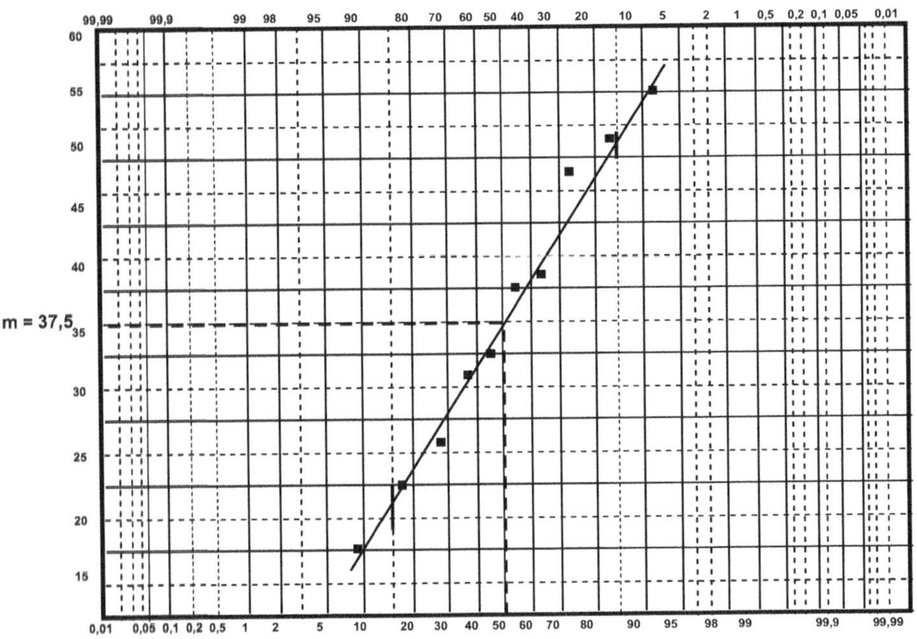

We read: m = 37,5; $\sigma = \dfrac{51-24}{2} = 13,5$

Under the assumption of normality, direct calculation would give:

$$m = \frac{20+25+27+33+35+40+41+48+52+55}{10} = \frac{376}{10} = 37,6$$

$$\sigma^2 = \frac{(20)^2 + ... + (55)^2}{9} - \frac{10.(37,6)^2}{9} = \frac{15382}{9} - \frac{14137}{9} = 138 \rightarrow \sigma = 11,8$$

This would a good match for the results, with the biggest mistake of course being in the estimation of the slope of the line and thus the standard deviation σ.

Example of normal paper for drawing Henry lines.

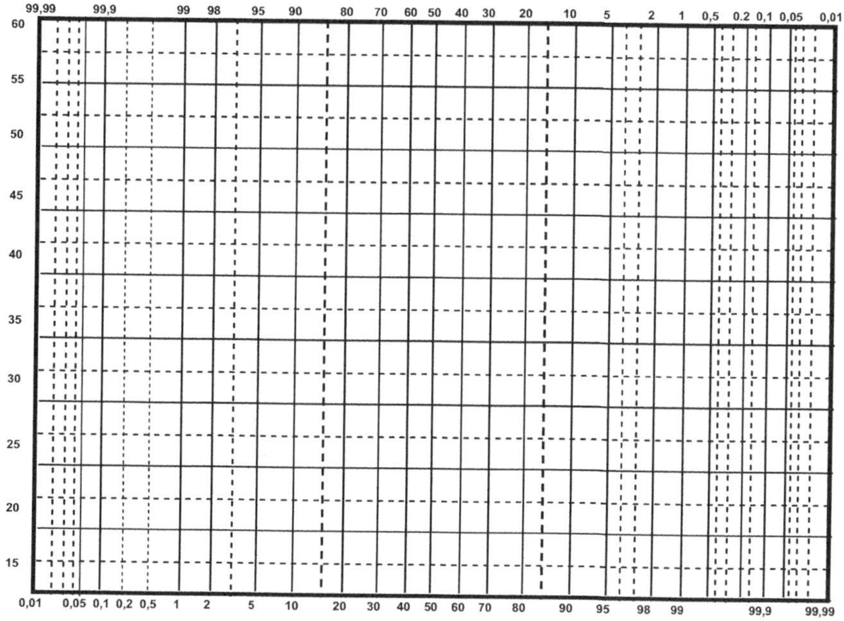

K.8. Dixon test, detection of outliers

For samples of size $N \leq 25$, we detect the presence of outliers when:

$$\text{Max}(Q_1, Q_2) \geq a + b.N^c$$

Sample size	Q_1	Q_2	a	b	c
$3 \leq N \leq 7$	$\frac{X_2\text{-}X_1}{X_N\text{-}X_1}$	$\frac{X_N\text{-}X_{N\text{-}1}}{X_N\text{-}X_1}$	0,18975704	2,7367561	–1,2452274
$8 \leq N \leq 10$	$\frac{X_2\text{-}X_1}{X_{N\text{-}1}\text{-}X_1}$	$\frac{X_N\text{-}X_{N\text{-}1}}{X_N\text{-}X_2}$	–0,18411559	1,8755055	–0,5
$11 \leq N \leq 13$	$\frac{X_3\text{-}X_1}{X_{N\text{-}1}\text{-}X_1}$	$\frac{X_N\text{-}X_{N\text{-}2}}{X_N\text{-}X_2}$	0,04530918	8,0981086	–1,5
$14 \leq N \leq 25$	$\frac{X_3\text{-}X_1}{X_{N\text{-}2}\text{-}X_1}$	$\frac{X_N\text{-}X_{N\text{-}2}}{X_N\text{-}X_3}$	0,21734606	5,4540159	–1,1320184

where N = sample size and X_1, X_2, ..., X_{N-1}, X_N are N measured values in ascending order.

This table is taken from [ref. 9] and carried out a regression on the critical data values at 90% cumulative probabilities. Thus, this test can detect the presence of outliers at 90% confidence at either end of small sample size.

K.9. Calculation of covariance and correlation factor

a) Calculating covariances, simplified formulas

- For two large sample sizes, n > 100:

$$C(\overline{m_x},\overline{m_y})=\frac{\sum_{i=1}^{n}(x_i-\overline{m_x})(y_i-\overline{m_y})}{n}$$

$$C(\overline{m_x},\overline{m_y})=\frac{1}{n}.\left[(x_1-\overline{m_x})(y_1-\overline{m_y})+.......+(x_n-\overline{m_x})(y_n-\overline{m_y})\right]$$

$$C(\overline{m_x},\overline{m_y})=\frac{1}{n}.\left[(x_1.y_1-\overline{m_x}.y_1-\overline{m_y}.x_1+\overline{m_x}.\overline{m_y})+...+(x_n.y_n-\overline{m_x}.y_n-\overline{m_y}.x_n+\overline{m_x}.\overline{m_y})\right]$$

$$C(\overline{m_x},\overline{m_y})=\frac{1}{n}.\begin{bmatrix}(x_1.y_1+x_2.y_2+....+x_ny_n)-\overline{m_x}(y_1+...+y_n)-...-\overline{m_y}(x_1+...+x_n)\\ +n.\overline{m_x}.\overline{m_y}\end{bmatrix}$$

$$C(\overline{m_x},\overline{m_y})=\frac{\sum_{i=1}^{n}(x_i.y_i)}{n}-\overline{m_x}.\frac{\sum_{i=1}^{n}y_i}{n}-\overline{m_y}.\frac{\sum_{i=1}^{n}x_i}{n}+\overline{m_x}.\overline{m_y}$$

$$\boxed{C(\overline{m_x},\overline{m_y})=\frac{\sum_{i=1}^{n}(x_i.y_i)}{n}-\overline{m_x}.\overline{m_y}}$$

- For two small sample sizes, n < 100:

$$C(\overline{m_x},\overline{m_y})=\frac{\sum_{i=1}^{n}(x_i-\overline{m_x})(y_i-\overline{m_y})}{n-1}$$

$$C(\overline{m_x},\overline{m_y})=\frac{1}{n-1}.\left[(x_1-\overline{m_x})(y_1-\overline{m_y})+.....+(x_n-\overline{m_x})(y_n-\overline{m_y})\right]$$

$$C(\overline{m_x},\overline{m_y})=\frac{1}{n-1}.\left[(x_1y_1-\overline{m_x}.y_1-\overline{m_y}.x_1+\overline{m_x}.\overline{m_y})+...+(x_ny_n-\overline{m_x}.y_n-\overline{m_y}.x_n+\overline{m_x}.\overline{m_y})\right]$$

$$C(\overline{m_x},\overline{m_y})=\frac{1}{n-1}.\left[(x_1y_1+....+x_ny_n)-\overline{m_x}(y_1+...+y_n)-.....-\overline{m_y}(x_1+...+x_n)+n.\overline{m_x}.\overline{m_y}\right]$$

$$C(\overline{m_x},\overline{m_y})=\frac{\sum_{i=1}^{n}(x_i.y_i)}{n-1}-\overline{m_x}.\frac{\sum_{i=1}^{n}y_i}{n-1}-\overline{m_y}.\frac{\sum_{i=1}^{n}x_i}{n-1}+\frac{n.\overline{m_x}.\overline{m_y}}{n-1}$$

$$\boxed{C(\overline{m_x},\overline{m_y})=\frac{\sum_{i=1}^{n}(x_i.y_i)}{n-1}-\frac{n.\overline{m_x}.\overline{m_y}}{n-1}}$$

b) Correlation factor, extract from CERESTA [ref. 6]

(a) Simple regression

Studies using *simple regression* (see above) have helped us define the *linear correlation coefficient* r between variable x (independent) and variable y (dependent) as part of a normal model.

On this occasion, we established that $\frac{(1-r^2)SCE_t}{n-2}$ was providing an unbiased estimation of the experimental (or residual) error σ^2, which is the dispersion of the points around the correction line (D). The cloud of points that represent the observed sample therefore gets nearer to (D) the closer r^2 is to 1. At the limit where $r^2 = 1$, all points are on (D) and the connection is said to be *functional*; we can predict the value y_0 that corresponds to a given value x_0 with some certainty.

(b) The intensity of the connection or dependence

Often, however, one is interested in the *intensity of the connection or dependence* between two random variables X and Y, which play symmetrical roles, unlike in regression where only one variable is held as random.

Such is the purpose of *correlation*, which aims to study the *correlation coefficient* between two variables based on observations from a random sample from a given population.

NB: Correlation and linear regression should not be confused. Nevertheless, the correlation coefficient does not necessarily measure the strict independence between two random variables (except in particular cases), but instead it measures a degree of similarity, also called dependence (hence the confusion), which must therefore be handled with care. An Excel spreadsheet has some functions that allow correlation calculations to be carried out more easily.

(c) The model

- We assume that the X and Y values are observed *without error*, which in practice is equivalent to inevitable corresponding measurement errors being of a magnitude that is significantly lower than the variability of variables.

- The *probabilistic model* is that of the *normal distribution in two dimensions*, which depends on five parameters (see probabilistic models 7.2 of [ref. 6]).

(d) Using the results

The following table provides estimations for the model parameters that are based on a sample of n random observations. Below, we will only consider individual pairs (x, y). For observations grouped into classes in a correlation table, one may refer to the corresponding formulas that are presented for simple regression (section IV of [ref.6]).

$$\bar{x} = \frac{\sum_i x_i}{n} \qquad \text{for} \quad \mu_x$$

$$\bar{y} = \frac{\sum_i y_i}{n} \qquad \text{for} \quad \mu_y$$

$$s_x^2 = \frac{\sum_i (x_i - \bar{x})_i^2}{n-1} \qquad \text{for} \quad \sigma_x$$

$$s_y^2 = \frac{\sum_i (y_i - \bar{y})_i^2}{n-1} \qquad \text{for} \quad \sigma_y$$

$$r = \frac{\sum_i (x_i - \bar{x})(y_i - \bar{y})}{\sqrt{\left(\sum_i (x_i - \bar{x})_i^2\right)\left(\sum_i (y_i - \bar{y})_i^2\right)}} \qquad \text{for} \quad \rho$$

NB: In fact, the unbiased estimation of ρ^2 is $\frac{(n-1)r^2 - 1}{n-2}$ but in practice, correlation studies are based on samples such that we can retain the estimate r of ρ (for large n).

c) Significance of correlation

We have shown, through the study of simple regression, that under the assumption $H_0(\rho = 0)$:

$\frac{r\sqrt{n-2}}{\sqrt{1-r^2}}$ follows a Student distribution at $\nu = n - 2$ degree of freedom (DOF) and table No. X indicates the limit values of $|r|$ (depending on $\nu = n - 2$ and risk α, to ultimately accept or reject the null hypothesis $H_0(\rho = 0)$.

d) Fisher transform

The random variable Fisher transform: $z = th^{-1}(r) = \frac{1}{2}\text{Log}\frac{1+r}{1-r}$ follows a law that is similar to a *normal distribution* of *parameters*:

$$m_z = 0 \qquad \text{and} \qquad \sigma_z = \frac{1}{\sqrt{n-3}}$$

Such a transformation can determine a *confidence interval* for the value of ρ, at $(1 - \alpha)$:

$r_1 \leq \rho \leq r_2$ where:

$$r_1 = th\left[z - t_{1-\alpha/2}(n-2).\frac{1}{\sqrt{n-3}}\right]$$

$$r_2 = th\left[z + t_{1-\alpha/2}(n-2).\frac{1}{\sqrt{n-3}}\right]$$

$$z = th^{-1}(r) = \frac{1}{2}\text{Log}\frac{1+r}{1-r}$$

Similarly, the problem found when comparing a correlation coefficient ρ to a given value ρ_0, as well as that for the comparison of two correlation coefficients can be solved by performing corresponding tests from the Fisher z variable treated as a normal variable (see hypothesis tests).

For example, for the assumption $H_0(\rho_1 = \rho_2)$, we use a reduced normal variable:

$$z = \frac{z_1 - z_2}{\sqrt{\frac{1}{n_1 - 3} + \frac{1}{n_2 - 3}}}$$

where z_1 and z_2 are the Fisher transforms for each of the two studied samples.

Example K.3: By taking digital data from the multiple regression example (p. 215 in the reference), we can easily calculate:

$r_{yx_1}=0,6854$ $\qquad$ $r_{yx_2}=0,7597$ $\qquad$ $r_{x_1x_2}=0,2759$

Table No. X enables us to conclude that only the second linear correlation coefficient is significant at a 5% risk level, from which we can deduce that the two studied explanatory variables are independent, which are the favorable conditions for specifying a model with two explanatory variables (lack of *colinearity*).

For $r_{yx_2} = 0,7597$, we get $z = th^{-1}(r) = 0,995$ and with $t_{0,95}(6) = 1,943$

$$\begin{cases} r_1 = th\left(0,9955 - 1,943/\sqrt{5}\right) = 0,126 \\ r_2 = th\left(0,9955 + 1,943/\sqrt{5}\right) = 0,953 \end{cases} \quad \text{Hence} \rightarrow \quad 0,13 \pounds \rho_{yx_2} \pounds 0,95$$

K.10. Independence test, extract from CERESTA [ref. 6]

Non-parametric independence tests for two characters.

The *independence test* involves defining a *decision rule* on the validity of the hypothesis relative to the *independence of two characters* A and B (quantitative or qualitative).

a) Terms of use

A random sample of n units taken independently from a population defined by two classification criteria A and B, which are described by k_A and k_B classes or modalities.

b) Criterion

$$\hat{\chi}^2 = \sum_{i,j} n \frac{\left(n_{i,j} - \frac{n_i.n_j}{n}\right)^2}{n_i.n_j}$$

Where $n_{i,j}$ is the number of observations with both characteristics of classes (or modalities) A_i and B_j (i from 1 to k_A; from 1 to k_B):

$$n_{i.} = \sum_j n_{i,j} \qquad \text{and} \qquad n_{.j} = \sum_i n_{i,j}$$

c) Interpretation of the test

α is the risk of the first test case and $\chi^2(\nu)$ is the Pearson variable with:

$\nu = (k_A - 1)(k_B - 1)$ as a parameter.

We are testing: H_0(A independent from B) versus H_1(A dependent on B):

$$\text{If} \qquad \hat{\chi}^2 \le \chi^2_{1-\alpha}(\nu) \qquad \begin{cases} \text{We accept } H_0 \\ \text{Otherwise we reject } H_0 \end{cases}$$

Example K.4: A random sample of n = 500 units collected independently from a population defined by two classification criteria A and B, respectively, defined as $k_A = 3$ and $k_B = 4$ modalities give the following breakdown:

	B_1	B_2	B_3	B_4	Total
A_1	12	48	59	31	150
A_2	20	65	71	44	200
A_3	18	37	70	25	150
Total	50	150	200	100	500

We must carry out the independence test for characters A and B by taking the risk of the first test case $\alpha = 0{,}05$.

Calculation elements

The following table shows theoretical numbers such that $\dfrac{n_i.n_j}{n}$ (numbers in brackets), which are compared to the "observed" numbers $n_{i,j}$ (summarized in the table).

	B1	B2	B3	B4	Total
A1	12 (15)	48 (45)	59 (60)	3(30)	150
A2	20 (20)	65 (60)	71 (80)	44 (40)	200
A3	18 (15)	37 (45)	70 (60)	25 (30)	150
Total	50	150	200	100	500

$\alpha = 0{,}05$ $\qquad \nu = (k_A - 1).(k_B - 1) = (3 - 1).(4 - 1) = 6$

Test

$$\hat{\chi}^2 = \frac{(12\text{-}15)^2}{15} + \frac{(48\text{-}45)^2}{45} + \ldots + \frac{(70\text{-}60)^2}{60} + \frac{(25\text{-}30)^2}{30} = 7{,}20$$

Such as: $\qquad \hat{\chi}^2 = 7{,}20 \; < \; \chi^2_{0,95} \; = \; 12{,}6$

So we accept the hypothesis of independence of characters A and B in the population.

K.11. Table of cumulative normal distribution

z	X	Area	z	X	Area
-3,25	μ - 3,25σ	0,0006	-1,00	μ - 1,00σ	0,1587
-3,20	μ - 3,20σ	0,0007	-0,95	μ - 0,95σ	0,1711
-3,15	μ - 3,15σ	0,0008	-0,90	μ - 0,90σ	0,1841
-3,10	μ - 3,10σ	0,0010	-0,85	μ - 0,85σ	0,1977
-3,05	μ - 3,05σ	0,0011	-0,80	μ - 0,80σ	0,2119
-3,00	μ - 3,00σ	0,0013	-0,75	μ - 0,75σ	0,2266
-2,95	μ - 2,95σ	0,0016	-0,70	μ - 0,70σ	0,2420
-2,90	μ - 2,90σ	0,0019	-0,65	μ - 0,65σ	0,2578
-2,85	μ - 2,85σ	0,0022	-0,60	μ - 0,60σ	0,2743
-2,80	μ - 2,80σ	0,0026	-0,55	μ - 0,55σ	0,2912
-2,75	μ - 2,75σ	0,0030	-0,50	μ - 0,50σ	0,3083
-2,70	μ - 2,70σ	0,0035	-0,45	μ - 0,45σ	0,3264
-2,65	μ - 2,65σ	0,0040	-0,40	μ - 0,40σ	0,3446
-2,60	μ - 2,60σ	0,0047	-0,35	μ - 0,35σ	0,3632
-2,55	μ - 2,55σ	0,0054	-0,30	μ - 0,30σ	0,3821
-2,50	μ - 2,50σ	0,0062	-0,25	μ - 0,25σ	0,4013
-2,45	μ - 2,45σ	0,0071	-0,20	μ - 0,20σ	0,4207
-2,40	μ - 2,40σ	0,0082	-0,15	μ - 0,15σ	0,4404
-2,35	μ - 2,35σ	0,0094	-.0,10	μ - 0,10σ	0,4602
-2,30	μ - 2,30σ	0,0107	-0,05	μ - 0,05σ	0,4801
-2,25	μ - 2,25σ	0,0122			
-2,20	μ - 2,20σ	0,0139			
-2,15	μ - 2,15σ	0,0158	0,00	μ	0,5000
-2,10	μ - 2,10σ	0,0179			
-2,05	μ - 2,05σ	0,0202			
-2,00	μ - 2,00σ	0,0228	0,05	μ + 0,05σ	0,5199
-1,95	μ - 1,95σ	0,0256	0,10	μ + 0,05σ	0,5398
-1,90	μ - 1,90σ	0,0287	0,15	μ + 0,05σ	0,5596
-1,85	μ - 1,85σ	0,0322	0,20	μ + 0,05σ	0,5793
-1,80	μ - 1,80σ	0,0359	0,25	μ + 0,05σ	0,5987
-1,75	μ - 1,75σ	0,0401	0,30	μ + 0,05σ	0,6179
-1,70	μ - 1,70σ	0,0446	0,35	μ + 0,05σ	0,6368
-1,65	μ - 1,65σ	0,0495	0,40	μ + 0,05σ	0,6554
-1,60	μ - 1,60σ	0,0548	0,45	μ + 0,05σ	0,6736
-1,55	μ - 1,55σ	0,0606	0,50	μ + 0,05σ	0,6915
-1,50	μ - 1,50σ	0,0668	0,55	μ + 0,05σ	0,7088
-1,45	μ - 1,45σ	0,0735	0,60	μ + 0,05σ	0,7257
-1,40	μ - 1,40σ	0,0808	0,65	μ + 0,05σ	0,7422
-1,35	μ - 1,35σ	0,0885	0,70	μ + 0,05σ	0,7580
-1,30	μ - 1,30σ	0,0968	0,75	μ + 0,05σ	0,7734
-1,25	μ - 1,25σ	0,1056	0,80	μ + 0,05σ	0,7881
-1,20	μ - 1,20σ	0,1151	0,85	μ + 0,05σ	0,8023
-1,15	μ - 1,15σ	0,1251	0,90	μ + 0,05σ	0,8159
-1,10	μ - 1,10σ	0,1357	0,95	μ + 0,05σ	0,8289
-1,05	μ - 1,05σ	0,1469	1,000	μ + 0,05σ	0,8413

z	X	Area	z	X	Area
1,05	μ + 1,05σ	0,8531	-4,265	μ - 4,265σ	0,00001
1,10	μ + 1,10σ	0,8643	-3,719	μ - 3,719σ	0,00010
1,15	μ + 1,15σ	0,8749	-3,090	μ - 3,090σ	0,00100
1,20	μ + 1,20σ	0,8849	-2,576	μ - 2,576σ	0,00500
1,25	μ + 1,25σ	0,8944	-2,326	μ - 2,326σ	0,01000
1,30	μ + 1,30σ	0,9032	-2,054	μ - 2,054σ	0,0200
1,35	μ + 1,35σ	0,9115	-1,960	μ - 1,960σ	0,0250
1,40	μ + 1,40σ	0,9192	-1,881	μ - 1,881σ	0,0300
1,45	μ + 1,45σ	0,9265	-1,751	μ - 1,751σ	0,0400
1,50	μ + 1,50σ	0,9332	-1,645	μ - 1,645σ	0,0500
1,55	μ + 1,55σ	0,9394	-1,555	μ - 1,555σ	0,0600
1,60	μ + 1,60σ	0,9452	-1,476	μ - 1,476σ	0,0700
1,65	μ + 1,65σ	0,9505	-1,405	μ - 1,405σ	0,0800
1,70	μ + 1,70σ	0,9554	-1,341	μ - 1,341σ	0,0900
1,75	μ + 1,75σ	0,9599	-1,282	μ - 1,282σ	0,1000
1,80	μ + 1,80σ	0,9641	-1,036	μ - 1,036σ	0,1500
1,85	μ + 1,85σ	0,9678	-0,842	μ - 0,842σ	0,2000
1,90	μ + 1,90σ	0,9713	-0,674	μ - 0,674σ	0,2500
1,95	μ + 1,95σ	0,9744	-0,524	μ - 0,524σ	0,3000
2,00	μ + 2,00σ	0,9772	-0,385	μ - 0,385σ	0,3500
2,05	μ + 2,05σ	0,9798	-0,253	μ - 0,253σ	0,4000
2,10	μ + 2,10σ	0,9821	-0,126	μ - 0,126σ	0,4500
2,15	μ + 2,15σ	0,9842	0,00	μ	0,5000
2,20	μ + 2,20σ	0,9861	0,126	μ + 0,126σ	0,5500
2,25	μ + 2,25σ	0,9878	0,253	μ + 0,253σ	0,6000
2,30	μ + 2,30σ	0,9893	0,385	μ + 0,385σ	0,6500
2,35	μ + 2,35σ	0,9906	0,524	μ + 0,524σ	0,7000
2,40	μ + 2,40σ	0,9818	0,674	μ + 0,674σ	0,7500
2,45	μ + 2,45σ	0,9929	0,842	μ + 0,842σ	0,8000
2,50	μ + 2,50σ	0,9938	1,036	μ + 1,036σ	0,8500
2,55	μ + 2.55σ	0,9946	1,282	μ + 1,282σ	0,9000
2,60	μ + 2.60σ	0,9953	1,341	μ + 1,341σ	0,9100
2,65	μ + 2.65σ	0,9960	1,405	μ + 1,405σ	0,9200
2,70	μ + 2.70σ	0,9965	1,476	μ + 1,476σ	0,9300
2,75	μ + 2.75σ	0,9970	1,555	μ + 1,555σ	0,9400
2,80	μ + 2.80σ	0,9974	1,645	μ + 1,645σ	0,9500
2,85	μ + 2.85σ	0,9978	1,751	μ + 1,751σ	0,9600
2,90	μ + 2.90σ	0,9981	1,881	μ + 1,881σ	0,9700
2,95	μ + 2.95σ	0,9984	1,960	μ + 1,960σ	0,9750
3,00	μ + 3.00σ	0,9987	2,054	μ + 2,054σ	0,9800
3,05	μ + 3,05σ	0,9989	2,326	μ + 2,326σ	0,9900
3,10	μ + 3,10σ	0,9990	2,576	μ + 2,576σ	0,9950
3,15	μ + 3,15σ	0,9992	3,090	μ + 3,090σ	0,9990
3,20	μ + 3,20σ	0,9993	3,719	μ + 3,719σ	0,9999
3,25	μ + 3,25σ	0,9994	4,265	μ + 4,265σ	0,99999

K.12. χ^2 table: this table gives χ such that P(K > x) = p

p / ddl	0,999	0,995	0,99	0,98	0,95	0,9	0,8	0,2	0,1	0,05	0,02	0,01	0,005	0,001
1	0,0000	0,0000	0,0002	0,0006	0,0039	0,0158	0,0642	1,6424	2,7055	3,8415	5,4119	6,6349	7,8794	10,8276
2	0,0020	0,0100	0,0201	0,0404	0,1026	0,2107	0,4463	3,2189	4,6052	5,9915	7,8240	9,2103	10,5966	13,8155
3	0,0243	0,0717	0,1148	0,1848	0,3518	0,5844	1,0052	4,6416	6,2514	7,8147	9,8374	11,3449	12,8382	16,2662
4	0,0908	0,2070	0,2971	0,4294	0,7107	1,0636	1,6488	5,9886	7,7794	9,4877	11,6678	13,2767	14,8603	18,4668
5	0,2102	0,4117	0,5543	0,7519	1,1455	1,6103	2,3425	7,2893	9,2364	11,0705	13,3882	15,0863	16,7496	20,5150
6	0,3811	0,6757	0,8721	1,1344	1,6354	2,2041	3,0701	8,5581	10,6446	12,5916	15,0332	16,8119	18,5476	22,4577
7	0,5985	0,9893	1,2390	1,5643	2,1673	2,8331	3,8223	9,8032	12,0170	14,0671	16,6224	18,4753	20,2777	24,3219
8	0,8571	1,3444	1,6465	2,0325	2,7326	3,4895	4,5936	11,0301	13,3616	15,5073	18,1682	20,0902	21,9550	26,1245
9	1,1519	1,7349	2,0879	2,5324	3,3251	4,1682	5,3801	12,2421	14,6837	16,9190	19,6790	21,6660	23,5894	27,8772
10	1,4787	2,1559	2,5582	3,0591	3,9403	4,8652	6,1791	13,4420	15,9872	18,3070	21,1608	23,2093	25,1882	29,5883
11	1,8339	2,6032	3,0535	3,6087	4,5748	5,5778	6,9887	14,6314	17,2750	19,6751	22,6179	24,7250	26,7568	31,2641
12	2,2142	3,0738	3,5706	4,1783	5,2260	6,3038	7,8073	15,8120	18,5493	21,0261	24,0540	26,2170	28,2995	32,9095
13	2,6172	3,5650	4,1069	4,7654	5,8919	7,0415	8,6339	16,9848	19,8119	22,3620	25,4715	27,6882	29,8195	34,5282
14	3,0407	4,0747	4,6604	5,3682	6,5706	7,7895	9,4673	18,1508	21,0641	23,6848	26,8728	29,1412	31,3193	36,1233
15	3,4827	4,6009	5,2293	5,9849	7,2609	8,5468	10,3070	19,3107	22,3071	24,9958	28,2595	30,5779	32,8013	37,6973
16	3,9416	5,1422	5,8122	6,6142	7,9616	9,3122	11,1521	20,4651	23,5418	26,2962	29,6332	31,9999	34,2672	39,2524
17	4,4161	5,6972	6,4078	7,2550	8,6718	10,0852	12,0023	21,6146	24,7690	27,5871	30,9950	33,4087	35,7185	40,7902
18	4,9048	6,2648	7,0149	7,9062	9,3905	10,8649	12,8570	22,7595	25,9894	28,8693	32,3462	34,8053	37,1565	42,3124
19	5,4068	6,8440	7,6327	8,5670	10,1170	11,6509	13,7158	23,9004	27,2036	30,1435	33,6874	36,1909	38,5823	43,8202
20	5,9210	7,4338	8,2604	9,2367	10,8508	12,4426	14,5784	25,0375	28,4120	31,4104	35,0196	37,5662	39,9968	45,3147
21	6,4467	8,0337	8,8972	9,9146	11,5913	13,2396	15,4446	26,1711	29,6151	32,6706	36,3434	38,9322	41,4011	46,7970
22	6,9830	8,6427	9,5425	10,6000	12,3380	14,0415	16,3140	27,3015	30,8133	33,9244	37,6595	40,2894	42,7957	48,2679
23	7,5292	9,2604	10,1957	11,2926	13,0905	14,8480	17,1865	28,4288	32,0069	35,1725	38,9683	41,6384	44,1813	49,7282
24	8,0849	9,8862	10,8564	11,9918	13,8484	15,6587	18,0618	29,5533	33,1962	36,4150	40,2704	42,9798	45,5585	51,1786
25	8,6493	10,5197	11,5240	12,6973	14,6114	16,4734	18,9398	30,6752	34,3816	37,6525	41,5661	44,3141	46,9279	52,6197
26	9,2221	11,1602	12,1981	13,4086	15,3792	17,2919	19,8202	31,7946	35,5632	38,8851	42,8558	45,6417	48,2899	54,0520
27	9,8028	11,8076	12,8785	14,1254	16,1514	18,1139	20,7030	32,9117	36,7412	40,1133	44,1400	46,9629	49,6449	55,4760
28	10,3909	12,4613	13,5647	14,8475	16,9279	18,9392	21,5880	34,0266	37,9159	41,3371	45,4188	48,2782	50,9934	56,8923
29	10,9861	13,1211	14,2565	15,5745	17,7084	19,7677	22,4751	35,1394	39,0875	42,5570	46,6927	49,5879	52,3356	58,3012
30	11,5880	13,7867	14,9535	16,3062	18,4927	20,5992	23,3641	36,2502	40,2560	43,7730	47,9618	50,8922	53,6720	59,7031
40	17,9164	20,7065	22,1643	23,8376	26,5093	29,0505	32,3450	47,2685	51,8051	55,7585	60,4361	63,6907	66,7660	73,4020
50	24,6739	27,9907	29,7067	31,6639	34,7643	37,6886	41,4492	58,1638	63,1671	67,5048	72,6133	76,1539	79,4900	86,6608
60	31,7383	35,5345	37,4849	39,6994	43,1880	46,4589	50,6406	68,9721	74,3970	79,0819	84,5799	88,3794	91,9517	99,6072
70	39,0364	43,2752	45,4417	47,8934	51,7393	55,3289	59,8978	79,7146	85,5270	90,5312	96,3875	100,4252	104,2149	112,3169
80	46,5199	51,1719	53,5401	56,2128	60,3915	64,2778	69,2069	90,4053	96,5782	101,8795	108,0693	112,3288	116,3211	124,8392
90	54,1552	59,1963	61,7541	64,6347	69,1260	73,2911	78,5584	101,0537	107,5650	113,1453	119,6485	124,1163	128,2989	137,2084
100	61,9179	67,3276	70,0649	73,1422	77,9295	82,3581	87,9453	111,6667	118,4980	124,3421	131,1417	135,8067	140,1695	149,4493
120	77,7551	83,8516	86,9233	90,3667	95,7046	100,6236	106,8056	132,8063	140,2326	146,5674	153,9182	158,9502	163,6482	173,6174
140	93,9256	100,6548	104,0344	107,8149	113,6593	119,0293	125,7581	153,8537	161,8270	168,6130	176,4709	181,8403	186,8468	197,4508
160	110,3603	117,6793	121,3456	125,4400	131,7561	137,5457	144,7834	174,8283	183,3106	190,5165	198,8464	204,5301	209,8239	221,0190
180	127,0111	134,8844	138,8204	143,2096	149,9688	156,1526	163,8682	195,7434	204,7037	212,3039	221,0772	227,0561	232,6198	244,3705
200	143,8428	152,2410	156,4320	161,1003	168,2786	174,8353	183,0028	216,6088	226,0210	233,9943	243,1869	249,4451	255,2642	267,5405
250	186,5541	196,1606	200,9386	206,2490	214,3916	221,8059	231,0128	268,5986	279,0504	287,8815	298,0388	304,9396	311,3462	324,8324
300	229,9634	240,6634	245,9725	251,8637	260,8781	269,0679	279,2143	320,3971	331,7885	341,3951	352,4246	359,9064	366,8444	381,4252
400	318,2596	330,9028	337,1553	344,0781	354,6410	364,2074	376,0218	423,5895	436,6490	447,6325	460,2108	468,7245	476,6064	493,1318
500	407,9470	422,3034	429,3875	437,2194	449,1468	459,9261	473,2099	526,4014	540,9303	553,1268	567,0698	576,4928	585,2066	603,4460
600	498,6229	514,5289	522,3651	531,0191	544,1801	556,0560	570,6680	628,9433	644,8004	658,0936	673,2703	683,5156	692,9816	712,7712
700	590,0480	607,3795	615,9075	625,3175	639,6130	652,4973	668,3308	731,2805	748,3591	762,6607	778,9721	789,9735	800,1314	821,3468
800	682,0665	700,7250	709,8969	720,0107	735,3623	749,1852	766,1555	833,4557	851,6712	866,9114	884,2789	895,9843	906,7862	929,3289
900	774,5698	794,4750	804,2517	815,0267	831,3702	846,0746	864,1125	935,4987	954,7819	970,9036	989,2631	1001,6296	1013,0364	1036,8260

NB: For large values of DOF, one can also use the following formula:

$$\chi^2_\beta = n\left(1 - \frac{2}{9n} + z_\alpha\sqrt{\frac{2}{9n}}\right)^3$$

where z_α is the standard deviation for 99% probability (see following table) and n is the number of DOF. For example, for 99% and 60 DOF:

$$\chi^2_{0,99} = 60[1 - 0,00370 + 2,326(0,06086)]^3 = 60(1,1379)^3 = 88,4$$

for 88,3794 in the previous table (column 0,01 for 99%).

K.13. Student distribution table

Number of DOF (number of measurements –1)	±k.s At 95,5% confidence	±k.s At 99,7% confidence
2	4,53	19,21
3	3,31	9,22
4	2,87	6,62
5	2,65	5,51
6	2,52	4,90
7	2,43	4,53
8	2,37	4,28
9	2,32	4,09
10	2,28	3,96
11	2,25	3,85
12	2,23	3,76
13	2,21	3,69
14	2,20	3,64
15	2,18	3,59
16	2,17	3,54
17	2,16	3,51
18	2,15	3,48
19	2,14	3,45
20	2,13	3,42
25	2,11	3,33
30	2,09	3,27
35	2,07	3,23
40	2,06	3,20
45	2,06	3,18
50	2,05	3,16
100	2,025	3,077
$+\infty$	2	3

These numbers may be estimated from n = 2 to 100 DOF using the following formulas:

- at 95,5%:

$k = \exp\{\ln(2) + a/n + b/n^c\}$; with: a = 1,2012469 ; b = 0,74238821 ; c = 1,7875574

- at 99,7%:

$k = \exp\{\ln(3) + a/n + b/n^c\}$; with: a = 2,36073; b = 2,2600822; c = 1,7405413

Example K.5: With n = 50 and 99,7%: k = exp(1,098612289 + 0,0472146 + 0,002494588) = 3,154 for 3,16 given from the table.

Bibliography

Books

I1 - *Mechanical Vibration and Shock Analysis, volumes 1–5*
C. Lalanne
ISTE Ltd., London and John Wiley & Sons, New York 2009

I2 - *Shock and Vibration Handbook, volumes 1–3*
C. Harris and C. Crede
McGraw-Hill, New York 1961

I3 - *Shock and Vibration Engineering, volume 1*
C.T. Morrow
John Wiley & Sons, New York 1963

I4 - *Structural Dynamics in Industry*
A. Girard and N. Roy
ISTE Ltd., London and John Wiley & Sons, New York 2007

I5 - *Elements of Vibration Analysis* (2nd edition)
L. Meirovitch
McGraw-Hill, New York 1986

I6 - *Shock Spectrum Analysis and Structural Design*
A.C. Keller, B.G. Pollard and E.A. Andren
Spectral Dynamics Corporation, San Diego 1986

I7 - *Essais et moyens d'essais au laboratoire et dans l'industrie*
M.R. Moreau
Technique et documentation, Lavoisier, Paris 1982

I8 - *Méthodes et techniques de traitement du signal et applications aux mesures physiques, volumes 1 and 2*
J. Max
Masson, Paris 1981

I9 - *Sup'Aero Analyse des structures par éléments finis* (2nd edition)
J.F. Imbert
Cépaduès éditions, Toulouse 1979

I10 - *Mesures dynamiques, volumes 1–3*
ENDEVCO 1984

I11 - *Analogies et modèles électriques*, no. 2154
G. Ney
Ecole supérieure d'électricité SUPELEC 1969

I12 - *Calcul analogique et hybride*, no. 2074
N. Steinberg
Ecole supérieure d'électricité SUPELEC 1967

I13 - *Dynamique de la commande linéaire*
J.C. Gilles, P. Decaulne and M. Pellegrin
Dunod, Paris 1967

I14 - *Traitement numérique du signal*
M. Bellanger
CENT ENST, Masson, Paris 1981

I15 - *Etude et détermination des systèmes hydrauliques*
M. Guillon
Dunod, Paris 1961

I16 - *Formulaire technique de mécanique générale*
J. Muller
Lavoisier, Paris 1964

I17 - *Vibration Analysis For Electronic Equipment* (3rd edition)
D.S. Steinberg
John Wiley & Sons, New York 2000

I18 - *Chocs mécaniques et applications aux matériels*
Y. Mori
Hermes Science–Lavoisier, Paris 2006

I19 - Electronique pour le traitement du signal, 8 volumes
Y. Mori
Hermes Science–Lavoisier, Paris 2006

I20 - *Formulaire vibrations (A410)*
J. Plusquellec
Techniques de l'Ingénieur 2011

I21 - *Simulation des vibrations mécaniques par Matlab, Simulink et Ansys*
M. Thomas and F. Laville
Presses de l'Université du Québec 2007

I22 - *Mechanical Vibration and Shock Measurements*
J.T. Broch
Brüel et Kjaer, Naerum 1972

I23 - *Théorie des asservissements linéaires*
J.C. Gilles, P. Decaulne and M. Pellegrin
Dunod, Paris 1967

Courses

f1 - Cours SOPEMEA, 50/B/M/P100
Notions sur les mouvements vibratoires 1968

f2 - Séminaire de vibrations SOPEMEA
2nd session, Mme Gaudry

f3 - Stage ADERA ASTE
Théorie des chocs et vibrations
C. Lalanne 1973

f4 - Centre de formation de l'ADERA avec le concours de l'ASTE
Simulation des vibrations et des chocs, 2nd part

f5 - Formation VI INTESPACE
Environnements vibratoires réels. Analyse et spécifications 1997
B. Colin, A. Girard, M. Jousset and C. Lalanne

f6 - Formation CEA (méthode d'élaboration des spécifications d'environnement mécanique, vibrations aléatoires)
C. Lalanne 1994

f7 - Stages de perfectionnement – Société des amis de l'ENSAE et de l'ENSTA
MO5 Analyse des structures par éléments finis 1984
MO6 Analyse dynamique des structures 1985
A. Girard, J.F. Imbert

f8 - Les analogies électriques
Stage de perfectionnement en environnement
M. Gaultier de la SACTRA

f9 - Introductory and advanced course on vibrations, Bruël & Kjaër 1989

Journals, articles, conferences and bulletins

r1 - Application notes, primers, handbooks and technical reviews, Bruël & Kjaër
- "Mechanical vibration and shock measurements" 1980
- "Frequency analysis" 1977
- "The discrete Fourier transform and FFT analysers" 1979
- "Digital filters and FFT technique in real-time analysis" 1978

r2 - ASTE (Association pour les Sciences et les Techniques de l'Environnement) conference records Journées 1980 and 1982 (4 volumes)

r3 - Articles by C. Lalanne for CEA CESTA ASTE documentation

r4 - IMEX information D.T.E IMEX France
- Bulletins no. 4 and 5
- "Symposium chocs et mesure des chocs" 1967
- "Méthode d'essais de chocs au moyen de vibrateurs électrodynamiques" G. Hug

r5 - Design concepts documentation X MOUNT, in mechanical vibrations and shocks

r6 - Notes techniques Diverses ENDEVCO

r7 - GERB technical documentation relating to isolation

r8 - Les équipements électroniques en environnement vibratoires, thesis
J. Canteleux 1982

r9 - "Efficient FPGA Implementation of Gaussian noise Generator for Communication channel emulation", *The 7th international Conference on Electronics Circuits and Systems*, ICES 2000, vol. 1, pp. 336–369
J.L. Danger, A. Ghazel, E. Boutillon and H. Laamari 2000

r10 - Conception et mise en œuvre de module de contrôle de puissance et de génération de bruit pour conditionner les signaux RF d'un simulateur de constellation GPS et Galiléo, master's thesis
R. Loukil 2010

Standards

n1 - Norm NF EN ISO/CEI 17025 2005

n2 - CEI NF EN 60068-2-64 2008

n3 - ISO norms: in particular, 5344 – 1980F
Moyens d'essais électrodynamiques utilisés pour la génération de vibrations. Méthode de description des caractéristiques

n4 - International and national military standards such as:
- GAM EG 13 in France;
- the general mechanical index in GAMEG13 which provides invaluable information;
- MIL STD in the United States.

n5 - Guide technique d'accréditation – Essais en environnement mécanique et climatique
Document COFRAC - LAB GTA 38 Rév.00

n6 - ISO Norm CEI 99

Index

Other titles from

iSTE

in

Mechanical Engineering and Solid Mechanics

2016

BOREL Michel, VÉNIZÉLOS Georges
Movement Equations 1: Location, Kinematics and Kinetics
(Non-deformable Solid Mechanics Set – Volume 1)

Movement Equations 2: Mathematical and Methodological Supplements
(Non-deformable Solid Mechanics Set – Volume 2)

BOYARD Nicolas
Heat Transfer in Polymer Composite Materials

CARDON Alain, ITMI Mhamed
New Autonomous Systems
(Reliability of Multiphysical Systems Set – Volume 1)

DAHOO Pierre Richard, POUGNET Philippe, EL HAMI Abdelkhalak
Nanometer-scale Defect Detection Using Polarized Light
(Reliability of Multiphysical Systems Set – Volume 2)

DE SAXCÉ Géry, VALLÉE Claude
Galilean Mechanics and Thermodynamics of Continua

DORMIEUX Luc, KONDO Djimédo
Micromechanics of Fracture and Damage (Micromechanics Set – Volume 1)

EL HAMI Abdelkhalak, RADI Bouchaib
Stochastic Dynamics of Structures (Mathematical and Mechanical Engineering Set – Volume 2)

GOURIVEAU Rafael, MEDJAHER Kamal, ZERHOUNI Noureddine
From Prognostics and Health Systems Management to Predictive Maintenance 1: Monitoring and Prognostics (Reliability of Multiphysical Systems Set – Volume 4)

KHARMANDA Ghias, EL HAMI Abdelkhalak
Reliability in Biomechanics (Reliability of Multiphysical Systems Set – Volume 3)

Biomechanics: Optimization, Uncertainties and Reliability (Reliability of Multiphysical Systems Set – Volume 5)

MOLIMARD Jérôme
Experimental Mechanics of Solids and Structures

RADI Bouchaib, EL HAMI Abdelkhalak
Material Forming Processes: Simulation, Drawing, Hydroforming and Additive Manufacturing (Mathematical and Mechanical Engineering Set – Volume 1)

2015

KARLIČIĆ Danilo, MURMU Tony, ADHIKARI Sondipon, MCCARTHY Michael
Non-local Structural Mechanics

SAB Karam, LEBÉE Arthur
Homogenization of Heterogeneous Thin and Thick Plates

2014

ATANACKOVIC M. Teodor, PILIPOVIC Stevan, STANKOVIC Bogoljub, ZORICA Dusan
Fractional Calculus with Applications in Mechanics: Vibrations and Diffusion Processes

ATANACKOVIC M. Teodor, PILIPOVIC Stevan, STANKOVIC Bogoljub, ZORICA Dusan
Fractional Calculus with Applications in Mechanics: Wave Propagation, Impact and Variational Principles

CIBLAC Thierry, MOREL Jean-Claude
Sustainable Masonry: Stability and Behavior of Structures

ILANKO Sinniah, MONTERRUBIO Luis E., MOCHIDA Yusuke
The Rayleigh–Ritz Method for Structural Analysis

LALANNE Christian
Mechanical Vibration and Shock Analysis – 5-volume series – 3rd edition
Sinusoidal Vibration – volume 1
Mechanical Shock – volume 2
Random Vibration – volume 3
Fatigue Damage – volume 4
Specification Development – volume 5

LEMAIRE Maurice
Uncertainty and Mechanics

2013

ADHIKARI Sondipon
Structural Dynamic Analysis with Generalized Damping Models: Analysis

ADHIKARI Sondipon
Structural Dynamic Analysis with Generalized Damping Models: Identification

BAILLY Patrice
Materials and Structures under Shock and Impact

BASTIEN Jérôme, BERNARDIN Frédéric, LAMARQUE Claude-Henri
Non-smooth Deterministic or Stochastic Discrete Dynamical Systems: Applications to Models with Friction or Impact

EL HAMI Abdelkhalak, BOUCHAIB Radi
Uncertainty and Optimization in Structural Mechanics

KIRILLOV Oleg N., PELINOVSKY Dmitry E.
Nonlinear Physical Systems: Spectral Analysis, Stability and Bifurcations

LUONGO Angelo, ZULLI Daniele
Mathematical Models of Beams and Cables

SALENÇON Jean
Yield Design

2012

DAVIM J. Paulo
Mechanical Engineering Education

DUPEUX Michel, BRACCINI Muriel
Mechanics of Solid Interfaces

ELISHAKOFF Isaac *et al.*
Carbon Nanotubes and Nanosensors: Vibration, Buckling and Ballistic Impact

GRÉDIAC Michel, HILD François
Full-Field Measurements and Identification in Solid Mechanics

GROUS Ammar
Fracture Mechanics – 3-volume series
Analysis of Reliability and Quality Control – volume 1
Applied Reliability – volume 2
Applied Quality Control – volume 3

RECHO Naman
Fracture Mechanics and Crack Growth

2011

KRYSINSKI Tomasz, MALBURET François
Mechanical Instability

SOUSTELLE Michel
An Introduction to Chemical Kinetics

2010

BREITKOPF Piotr, FILOMENO COELHO Rajan
Multidisciplinary Design Optimization in Computational Mechanics

DAVIM J. Paulo
Biotribolgy

PAULTRE Patrick
Dynamics of Structures

SOUSTELLE Michel
Handbook of Heterogenous Kinetics

2009

BERLIOZ Alain, TROMPETTE Philippe
Solid Mechanics using the Finite Element Method

LEMAIRE Maurice
Structural Reliability

2007

GIRARD Alain, ROY Nicolas
Structural Dynamics in Industry

GUINEBRETIÈRE René
X-ray Diffraction by Polycrystalline Materials

KRYSINSKI Tomasz, MALBURET François
Mechanical Vibrations

KUNDU Tribikram
Advanced Ultrasonic Methods for Material and Structure Inspection

SIH George C. *et al.*
Particle and Continuum Aspects of Mesomechanics

Printed and bound by CPI Group (UK) Ltd, Croydon, CR0 4YY
16/09/2026
14963079-0002